AF615922

MOTION ANALYSIS AND IMAGE SEQUENCE PROCESSING

THE KLUWER INTERNATIONAL SERIES IN ENGINEERING AND COMPUTER SCIENCE

VLSI, COMPUTER ARCHITECTURE AND DIGITAL SIGNAL PROCESSING

Consulting Editor

Jonathan Allen

Latest Titles

Introduction to the Design of Transconductor-Capacitor Filters, J. E. Kardontchik
ISBN: 0-7923-9195-0

The Synthesis Approach to Digital System Design, P. Michel, U. Lauther, P. Duzy
ISBN: 0-7923-9199-3

Fault Covering Problems in Reconfigurable VLSI Systems, R.Libeskind-Hadas, N. Hassan, J. Cong, P. McKinley, C. L. Liu
ISBN: 0-7923-9231-0

High Level Synthesis of ASICs Under Timing and Synchronization Constraints D.C. Ku, G. De Micheli
ISBN: 0-7923-9244-2

The SECD Microprocessor, A Verification Case Study, B.T. Graham
ISBN: 0-7923-9245-0

Field-Programmable Gate Arrays, S.D. Brown, R. J. Francis, J. Rose, Z.G. Vranesic
ISBN: 0-7923-9248-5

Anatomy of A Silicon Compiler, R.W. Brodersen
ISBN: 0-7923-9249-3

Electronic CAD Frameworks, T.J. Barnes, D. Harrison, A.R. Newton, R.L. Spickelmier
ISBN: 0-7923-9252-3

VHDL for Simulation, Synthesis and Formal Proofs of Hardware, J. Mermet
ISBN: 0-7923-9253-1

Wavelet Theory and its Applications, R. K. Young
ISBN: 0-7923-9271-X

Digital BiCMOS Integrated Circuit Design, S.H.K. Embabi, A. Bellaouar, M.I Elmasry
ISBN: 0-7923-9276-0

Design Automation for Timing-Driven Layout Synthesis, S. S. Sapatnekar, S. Kang
ISBN: 0-7923-9281-7

Acoustical and Environmental Robustness in Automatic Speech Recognition, A. Acero
ISBN: 0-7923-9284-1

Logic Synthesis and Optimization, T. Sasao
ISBN: 0-7923-9308-2

Sigma Delta Modulators: Nonlinear Decoding Algorithms and Stability Analysis, S. Hein, A. Zakhor
ISBN: 0-7923-9309-0

High-Level Synthesis for Real-Time Digital Signal Processing: The Cathedral-II Silicon Compiler
J. Vanhoof, K. Van Rompaey, I. Bolsens, G. Goosens, H. De Man
ISBN; 0-7923-9313-9

MOTION ANALYSIS AND IMAGE SEQUENCE PROCESSING

edited by

M. Ibrahim Sezan
Eastman Kodak Company

Reginald L. Lagendijk
Delft University of Technology

KLUWER ACADEMIC PUBLISHERS

Boston/Dordrecht/London

Distributors for North America:
Kluwer Academic Publishers
101 Philip Drive
Assinippi Park
Norwell, Massachusetts 02061 USA

Distributors for all other countries:
Kluwer Academic Publishers Group
Distribution Centre
Post Office Box 322
3300 AH Dordrecht, THE NETHERLANDS

Library of Congress Cataloging-in-Publication Data
Motion analysis and image sequence processing / edited by M. Ibrahim Sezan, Reginald L. Lagendijk.
p. cm. -- (The Kluwer international series in engineering and computer science ; SECS 0220. VLSI, computer architecture and digital signal processing)
Includes bibliographical references and index.
ISBN 0-7923-9329-5
1. Image processing. 2. Motion perception (Vision) I. Sezan, M. Ibrahim. II. Lagendijk, Reginald L. III. Series: Kluwer international series in engineering and computer science ; SECS 0220. IV. Series: Kluwer international series in engineering and computer science. VLSI, computer architecture and digital signal processing.
TA1632 . M68 1993
621 . 39 ' 9--dc20 92-46330
CIP

Printed on acid-free paper.

Printed in the United States of America

Contents

List of Contributors

K. Aizawa
Electrical Engineering Department
University of Tokyo
Tokyo, 113 Japan

P. Anandan
David Sarnoff Research Center
Princeton NJ 08543-5300, USA

J. G. Apostolopoulos
Advanced Television Signal Proc. Group
MIT
Cambridge MA 02139, USA

A. Basso
Signal Processing Laboratory
EPFL-Ecublens
CH-1015 Lausanne, Switzerland

R. A. F. Belfor
Department of Electrical Engineering
Delft University of Technology
2600 GA Delft, The Netherlands

J. R. Bergen
David Sarnoff Research Center
Princeton NJ 08543-5300, USA

J. Biemond
Department of Electrical Engineering
Delft University of Technology
2600 GA Delft, The Netherlands

M. Buck
Daimler-Benz AG
D-7900 Ulm, Germany

T. M. Chin
Rosenstiel School of Marine Science
University of Miami
Miami FL 33149, USA

C-S. Choi
Dept. of Information Communication
Myong Ji University
Kyunggido, 499-728, Korea

N. Diehl
Daimler-Benz AG
D-7900 Ulm, Germany

E. Dubois
INRS-Telecommunications
Verdun, Canada H3E 1H6

A. T. Erdem
Eastman Kodak Company
Rochester NY 14650-1816, USA

B. Girod
Academy of Media Arts Cologne
W-5000 Koln 1, Germany

K. Hanna
David Sarnoff Research Center
Princeton NJ 08543-5300, USA

H. Harashima
Electrical Engineering Department
University of Tokyo
Tokyo, 113 Japan

R. Hingorani
AT&T Bell Laboratories
Murray Hill NJ 08544, USA

T. S. Huang
Coordinated Science Laboratory
University of Illinois
Urbana IL 61801, USA

W. C. Karl
Laboratory for Information and Decision Systems
MIT
Cambridge MA 02139, USA

C. S. Kim
Department of Electrical Engineering
University of Washington
Seattle WA 98195, USA

J. Kim
ECSE Department
Rensselaer Polytechnic Institute
Troy NY 12180-3590, USA

J. Konrad
INRS-Telecommunications
Verdun, Canada H3E 1H6

F. Kossentini
School of Electrical Engineering
Georgia Institute of Technology
Atlanta, GA 30332-0250, USA

M. Kunt
Signal Processing Laboratory
EPFL-Ecublens
CH-1015 Lausanne, Switzerland

R. L. Lagendijk
Department of Electrical Engineering
Delft University of Technology
2600 GA Delft, The Netherlands

F. Lari
Electrical Engineering Department
University of California
Berkeley CA 94720, USA

W. Li
Signal Processing Laboratory
EPFL-Ecublens
CH-1015 Lausanne, Switzerland

J. Lim
Advanced Television Signal Proc. Group
MIT
Cambridge MA 02139, USA

M. Luettgen
Laboratory for Information and Decision Systems
MIT
Cambridge MA 02139, USA

M. Mattavelli
Signal Processing Laboratory
EPFL-Ecublens
CH-1015 Lausanne, Switzerland

R. M. Mersereau
School of Electrical Engineering
Georgia Institute of Technology
Atlanta, GA 30332-0250, USA

Y. Neuvo
Signal Processing Laboratory
Tampere University of Technology
SF-33101 Tampere, Finland

A. Nicoulin
Signal Processing Laboratory
EPFL-Ecublens
CH-1015 Lausanne, Switzerland

M. K. Ozkan
Thomson Consumer Electronics
Indianapolis IN 46201-2598, USA

A. Popat
The Media Laboratory
MIT
Cambridge MA 02139, USA

F. Rocca
Dipartimento di Elettronica
e Informazione
Politecnico di Milano
20133 Milano, Italy

M. I. Sezan
Eastman Kodak Company
Rochester NY 14650-1816, USA

M. J. T. Smith
School of Electrical Engineering
Georgia Institute of Technology
Atlanta, GA 30332-0250, USA

A. M. Tekalp
Department of Electrical Engineering
University of Rochester
Rochester NY 14627, USA

K. K. Truong
Atlanta Signal Processors
Atlanta GA 30332, USA

S. Tubaro
Dipartimento di Elettronica
e Informazione
Politecnico di Milano
20133 Milano, Italy

T. Viero
Signal Processing Laboratory
Tampere University of Technology
SF-33101 Tampere, Finland

A. Willsky
Laboratory for Information and Decision
Systems
MIT
Cambridge MA 02139, USA

J. W. Woods
ECSE Department
Rensselaer Polytechnic Institute
Troy NY 12180-3590, USA

A. Zakhor
Electrical Engineering Department
University of California
Berkeley CA 94720, USA

Preface

An image or video sequence is a series of two-dimensional (2-D) images sequentially ordered in time. Image sequences can be acquired, for instance, by video, motion picture, X-ray, or acoustic cameras, or they can be synthetically generated by sequentially ordering 2-D still images as in computer graphics and animation. The use of image sequences in areas such as entertainment, visual communications, multimedia, education, medicine, surveillance, remote control, and scientific research is constantly growing as the use of television and video systems are becoming more and more common. The boosted interest in digital video for both consumer and professional products, along with the availability of fast processors and memory at reasonable costs, has been a major driving force behind this growth.

Before we elaborate on the two major terms that appear in the title of this book, namely *motion analysis* and *image sequence processing*, we like to place them in their proper contexts within the range of possible operations that involve image sequences. In this book, we choose to classify these operations into three major categories, namely (i) image sequence processing, (ii) image sequence analysis, and (iii) visualization. The interrelationship among these three categories is pictorially described in Figure 1 below in the form of an "image sequence triangle". Image sequence processing, which is essentially a sequence-in and sequence-out operation, refers to the operations of filtering, spatiotemporal interpolation and subsampling, and compression of image sequences, aimed at improving the (visual) image quality, conversion between different video formats, and bandwidth-efficient representation of image sequences, respectively.

The second category is image sequence analysis, referring to those operations that generate some type of data from image sequences for the purpose of information retrieval or interpretation. Motion analysis is an important branch of image sequence analysis. It refers to estimation of the image motion (the projection of the 3-D motion onto the 2-D image plane, or optical flow), inference of the 3-D object motion, depth information, as well as the surface characteristics of objects present in the actual 3-D scene. Motion information is utilized not only in various analysis tasks such as segmentation, pattern recognition and tracking, and scene interpretation, but is also instrumental in image sequence processing, as pointed out in Figure 1. The third category of operations that involves image sequences is visualization. It is concerned with generating image sequences on the basis of data which is not readily in the form of an image sequence. Computer graphics, virtual reality, and several emerging methods for scientific data visualization belong to this category. This book focuses on motion analysis and image sequence processing, and the use of motion analysis in developing efficient and powerful image sequence processing algorithms.

At this point, one may raise the following questions: Since image sequences

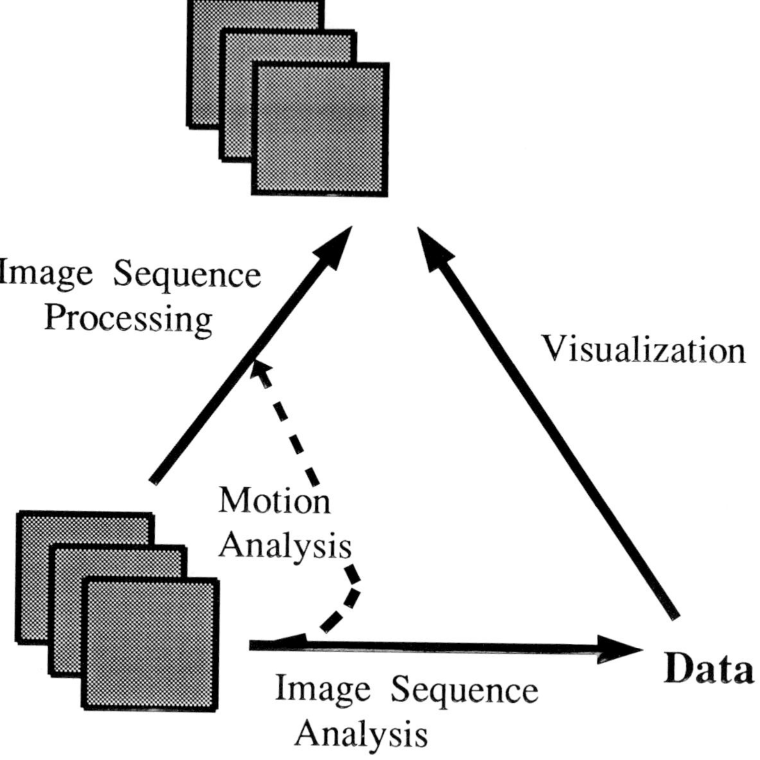

Fig. 1. The "image sequence triangle" depicting possible operations that involve image sequences.

can be viewed as being a series of still images, why are we concerned with special algorithms for image sequences? Can't we apply the vast variety of well-known processing and analysis algorithms that have been developed for still images to image sequences on a frame by frame basis? The need for algorithms especially designed for image *sequences* can be rationalized by the following two arguments: (i) time-varying phenomena and motion cannot be inferred from individual still pictures, and (ii) more efficient and powerful processing algorithms that utilize temporal information, such as the interframe motion vector field and temporal correlations, in addition to the spatial information can be developed for image sequences. As we shall see, the chapters in this book do indeed support these arguments.

The boosted interest in fundamental and applied research in the area of motion analysis and image sequence processing has led to a significant growth in the number of conferences, workshops, and specialized journals in this field. In contrast to this growth in meetings and periodicals, there are currently only three existing monographs on motion analysis [1, 2, 3], and two edited books on image sequence analysis (including motion analysis) and processing [4, 5]. Both books on motion analysis and image sequence processing are edited by Prof. T.S. Huang, and the most recent one dates back to 1983. Considering the amount of new developments that took place in the field of motion analysis and image sequence processing during the last decade, we have felt the need for an up-to-date book. This edited book contains coherent and rigorous discussions of recent fundamental developments in the field of motion analysis and image sequence processing. As editors, we have strived to bring together prominent and active researchers from leading international research institutes and universities. Further, we have tried to obtain contributions such that the entire breadth of the field is covered. When we consider the contents of the 15 chapters contained in this book, we hope that, in all modesty, we have succeeded in reaching our goals; this is indeed for the reader to decide.

The structure of the book and the areas that it covers are depicted in Figure 2. Chapters 1 to 4 discuss new developments in the area of motion analysis. Several of these chapters also discuss possible applications of motion analysis in image sequence processing. The rest of the chapters are on image sequence processing and can be classified into three main groups. Chapters 5 to 7 discuss various aspects of the spatiotemporal representation of image sequences. The principles and theories discussed in these chapters can be regarded as being of fundamental importance for de-interlacing, frame rate conversion, compression, and filtering techniques. Next, in Chapters 8 to 11, the emphasis is placed on data compression of image sequences using a variety of recent techniques with the common goal of representing image sequences as bandwidth efficient as possible. Chapter 15 also falls in this category, but takes a more system oriented approach in which not only compression techniques, but several other aspects of digital TV transmission and compression systems are discussed as well. Chapters 12 to 14 focus on filtering of image sequences with emphasis on noise suppression and deblurring. In the following, we briefly introduce the chapters.

Chapters 1, 2, 3 and 4 discuss algorithms for estimating the 2-D image motion, i.e., the optical flow. Chapters 3 and 4 also discuss the use of this motion information in developing motion-compensated algorithms for image sequence processing. As it is pointed out in these first four chapters, estimation of image motion is a mathematically ill-posed problem where a solution may not exist (e.g., in the case of occlusions), and even when a solution can be found, it may not be unique, for instance due to the finite aperture problem or in the presence of identical objects moving independently. To alleviate the effects of ill-posedness, motion estimation algorithms utilize a priori information and constraints about the motion field. Motion estimation algorithms differ with respect to the type of a priori information and constraints they use as well as the computational framework within which they perform the estimation.

Motion estimation algorithms discussed in the first four chapters are all model-based in the sense that a priori information and constraints about the

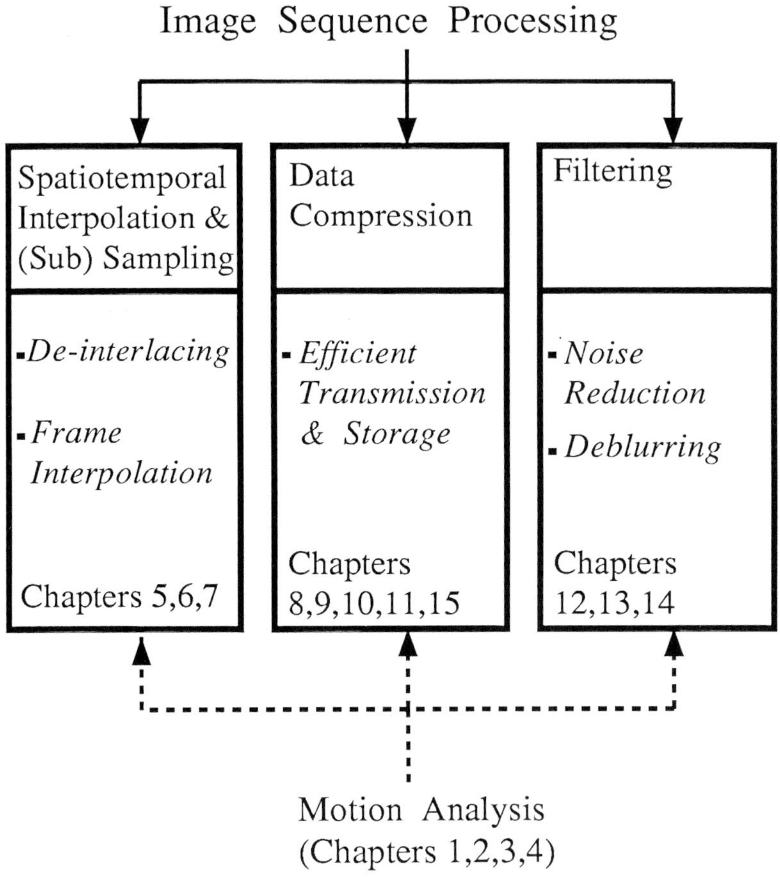

Fig. 2. Illustration of the structure and the contents of the book.

motion field are expressed in terms of underlying models for the motion vector field. In Chapter 1, Anandan, Bergen, Hanna and Hingorani discuss fully parametric, quasi-parametric and non-parametric models of the motion vector field. They propose a common hierarchical computational framework for these three types of deterministic models where the estimation criterion is minimization of an error measure that implies the conservation of image brightness constraint, i.e., the difference between the brightness of image areas that correspond to each other in the sense of motion, in consecutive frames, is minimized. The affine and planar surface models (both parametric), the rigidly moving object model (quasi-parametric), and the spatial smoothness model (non-parametric) are utilized in a hierarchical computational framework where the hierarchy is formed using a Laplacian pyramid structure. The authors furnish examples illustrating the performance of these models in estimating motion in real-life image sequences.

Another model-based approach to motion estimation is proposed by Chin, Luettgen, Karl and Willsky in Chapter 2. The authors start with the classical Horn-Schunck formulation of the motion estimation problem where the conservation of brightness and the smoothness constraints are utilized in posing the motion estimation problem as a constrained optimization problem. They show that this formulation is equivalent to an estimation-theoretic formulation where the spatial smoothness constraint can be interpreted as a prior probabilistic spatial model for the motion field, namely a Brownian motion model, and the estimation takes the form of a maximum likelihood problem. Then, the authors propose a novel extension to this formulation by imposing a temporal smoothness constraint on the motion field. This new constraint imposes a temporal coherence constraint on the motion field in addition to the spatial coherence implied by the spatial smoothness constraint. The extended estimation problem is solved using a near-optimum Kalman filtering approach. Finally, the authors propose the use of multiscale probabilistic models for the motion field which in turn makes it possible to use the computational framework of efficient multigrid algorithms.

Dubois and Konrad, in Chapter 3, propose modeling the motion field as a vector Markov random field. They utilize the equivalence of a Markov random field and a random field governed by a Gibbs distribution and perform the estimation in a maximum a posteriori probability (MAP) framework using simulated annealing. The Gibbs distribution model makes it possible to establish a straightforward relationship between the a priori quantitative characteristics (e.g., smoothness or discontinuity) of the Markov random field and its parameters via the neighbor system, the cliques, and the potential functions. The proposed motion field model includes an occlusion process and a motion discontinuity process which are in turn modeled by scalar Markov random fields that are described by appropriate Gibbs distributions. One of the important features of the formulation proposed by the authors is the fact that it takes into account the multispectral nature of the image data by assuming that the frames at each time instant are samples of a vector random field, e.g., a 3-D field in the case of color image sequences. In the second half of their chapter, Dubois and Konrad discuss the use of the estimated motion field in various image sequence processing tasks, such as motion-compensated prediction for data compression, motion-compensated spatiotemporal interpolation, and motion-compensated noise suppression. It is worthwhile to note that in the particular case of motion-compensated frame interpolation, the authors' modeling of motion trajectories allows for accelerated interframe motion. This is in contrast to the widely used assumption of linear motion trajectories in solving this interpolation problem.

In sequences that are acquired by a camera, image motion can be attributed to (i) camera motion (global motion) and (ii) independent object motion (local motion). Estimation of the global motion and its use in global-motion compensated prediction in video compression applications are the subject matters of Chapter 4 by Zakhor and Lari. Global motion includes zoom, rotation (around an axis parallel to the image plane, i.e., pan, or around the camera axis), and translation along the camera axis or within a plane normal to the camera axis. Camera motion can be modeled using parametric models that describe the change in the image-plane coordinates of objects (assumed to be stationary), from one frame to another, as the camera moves. Hence, image coordinates at frame $t+\delta$ can be predicted from frame t in forming a global-motion compensated prediction. The authors discuss models for zoom and rotation, and translation and rotation, and propose algorithms to estimate the model parameters for these two cases. The proposed algorithms are two-step algorithms where the first step is the estimation of local motion using edge matching, and the second step is utilization of the local motion information in estimating the global motion parameters. It is important to note that in the case of translation, the depth map should also be estimated prior to the estimation of the global motion parameters. Handling the translational component of the camera motion as well as the use of edge matching are novel features of this chapter. The authors also discuss the computational aspects and performance of edge matching and block matching algorithms in the context of global motion estimation.

In Chapters 5, 6 and 7, the spatiotemporal representation of image sequences and the role of motion information therein is discussed. In these chapters, the emphasis is not immediately on applications but on the fundamental issues that are encountered in deciding what minimal spatial and temporal bandwidth should be considered as appropriate in representing a given image sequence. A common theme of these chapters is the basic issue of spatial and temporal subsampling and spatiotemporal interpolation of spatially or temporally subsampled sequences by making use of the motion information. Properties of estimated motion fields play an important role, but not the motion estimation process itself. Applications encountered in these chapters are de-interlacing, frame rate conversion, temporal prediction, and data compression.

In Chapter 5, Girod looks at various ways the estimated motion vectors are used in different applications. The question of what role the human perception plays in motion-compensated processing of image sequences is considered. It is shown that motion compensation is extremely important because of smooth pursuit eye movements. Next, the issue of motion estimation accuracy and the related issue of fundamental limits on the effects of motion compensation are addressed in the case of motion-compensated de-interlacing and motion-compensated prediction. Again, the concepts are given more emphasis than particular implementation details. The material in this chapter can be regarded as forming a bridge between the "pure" motion estimation chapters and those discussing processing of image sequences using the estimated motion information.

The problem of recovering missing image frames is discussed in Chapter 6 by Tubaro and Rocca. Image frames may be unavailable either because they have been skipped at an earlier stage to achieve data compression, or because the given sequence follows a standard that utilizes a relatively low frame rate, as in the case of motion picture film sequences. Straightforward approaches to temporal interpolation like frame repetition introduce undesirable artifacts that can be greatly avoided by motion-compensated interpolation. However, the use of motion information introduces some problems itself. The problem of designing a suitably smooth motion estimator for interpolation, as well as the related problem of segmenting a sequence in meaningful regions is discussed in this chapter. The results of the techniques developed are illustrated in case of video telephony sequences.

Chapter 7 by Belfor, Lagendijk and Biemond, where spatiotemporal subsampling of image sequences and the use of motion information in subsampling and interpolation are discussed, links the two preceding chapters to subsequent chapters on data compression of image sequences. In this chapter, the problem of subsampling a given image sequence in order to achieve a reduction in the required number of samples used in representing that sequence is discussed in detail. Particular attention is paid to the interpolation process that recovers the full-resolution image sequence from a subsampled version. In this context, the accuracy limits on estimated motion vectors is investigated. The first part of the chapter reviews the basics of subsampling and interpolation, leading to the concept of motion-compensated filtering and interpolation. The fundamental concepts of the MUSE and the HD-MAC television systems are introduced along these lines. In the second part, a novel approach to motion adaptive sub-Nyquist subsampling is discussed. This approach alleviates problems introduced by so-called critical velocities. A practical implementation is discussed and applied to test sequences.

Several recent developments in data compression of image sequences are discussed in Chapters 8, 9, 10 and 11. Data compression (sometimes simply referred to as "coding") of image sequences has been a topic of great interest in the last decade. Although the field of image sequence coding is far too wide to cover in its entire breadth within the scope of this book, several interesting recent developments in this area are discussed. In particular, Chapters 8 and 9 discuss the application of subband coding and vector quantization in motion-compensated compression schemes operating at moderate to low bit rates. Model-based coding approaches that aim at extremely low bit rates are the subject of Chapters 10 and 11.

It is well-known how transform coding (usually the discrete cosine transform (DCT)) can be applied in conjunction with motion-compensated prediction. In fact, this combination has been the basis for most of the video codecs realized in hardware in the recent years, as well as the standard algorithms such as the H.261 standard and the forthcoming Moving Pictures Experts Group (MPEG) standard. The question of how to apply motion compensation in conjunction with subband coding is still an issue that has not been answered quite satisfactorily. In Chapter 8, Nicoulin, Mattavelli, Li, Basso, Popat and Kunt discuss the basic ingredients of a motion-compensated subband coder for medium bit rates. After introducing the block diagram, attention is shifted to all essential details. Several new ideas are launched in the areas of motion compensation in conjunction with a temporal DCT transform, the design of an adaptive arithmetic coder, quantizer design, and the overall rate control. Experimental results obtained at a bit rate of 9 Mbits/sec are shown and discussed .

Chapter 9 by Mersereau, Smith, Kim, Kossentini and Truong concentrates on how recent developments in the field of vector quantization (VQ) can be applied to compression of video sequences. First a comprehensive overview of the major classes of VQ approaches is given, such as finite state VQ, residual VQ, cache VQ, and subband residual VQ. A comparison of these methods is presented in the case of intraframe coding of image frames. In the second part of the chapter, attention is shifted to ways in which VQ can be combined with motion compensation. This leads to several conceptual solutions for VQ-based compression schemes where several new ideas, such as the use of cache VQ in hierarchical video codecs, are discussed. Applications are taken out of the area of low bit-rate coding for video telephony.

The previous two chapters have discussed compression methods that are essentially based on stochastic signal models. In contrast, the next two chapters focus on semantic approaches in which the analysis of the actual contents (or structure) of a time-varying scene and the synthesis of an original scene on the basis of extracted scene parameters (visualization) play a central role. Model-based methods aim at achieving extremely high data compression. The tools that are made use of in model-based compression are quite similar to those utilized in the areas of "Image Sequence Analysis" and "Visualization" (see Figure 1). Still, since the overall purpose of image analysis and visualization in model-based coding is data compression of image sequences, these chapters do fit within the scope of "Image Sequence Processing".

Chapter 10 by Buck and Diehl presents a broad overview of the different view points in model-based coding and discusses recent developments in this field. The emphasis in this chapter is on image modeling, and several considerations relevant to this are discussed in detail. For instance, the issues of what kind of model (explicit vs. parametric, or surface vs. volume, etc.) should be chosen, how much a priori information is available about the scene semantics, and the importance of real-time or in-line operation are addressed. A variety of modeling approaches is described in this chapter, including shape from disparity, wire frame adaptation, facial models, implicit models, scene segmentation, and finally the use of motion information.

The subject of facial motion analysis and synthesis is discussed in greater detail in Chapter 11 by Aizawa, Choi, Harashima, and Huang. In developing model-based codecs for facial image sequences, two subproblems need to be addressed, namely (i) the facial modeling and synthesis, and (ii) the analysis of face movements. The first part of the chapter details a method for facial modeling using the so-called "generic face model" based on a wire frame. Much attention is paid to synthesizing accurate facial expressions using this generic model assuming that the parameters describing an expression are known. The second part of the chapter concentrates on the analysis of face movements, which in general is much more difficult than the synthesis problem. A two-stage analysis technique for facial motion is discussed, where the head motion parameters are estimated first and then the facial expression is determined on the basis of a locally estimated motion field. A number of experimental results are given in order to verify the proposed analysis-synthesis techniques.

Filtering of image sequences for the purpose of noise suppression and/or deblurring is discussed in Chapters 12, 13 and 14. The problem of noise suppression is addressed in Chapters 12 and 14. Chapter 13 deals with deblurring image sequences that suffer from blur as well as noise contamination. The filters developed in Chapters 12 and 13 are linear filters whereas nonlinear filters, namely median filters, are considered in Chapter 14. The use of median filters in forming predictions in predictive data compression schemes is also discussed in Chapter 14.

In Chapter 12, Woods and Kim propose a 3-D spatiotemporal extension of the 2-D reduced update Kalman filter (RUKF). They use a 3-D scalar image model and filter the image sequence line by line and then frame by frame. One of the fundamental assumptions of 3-D RUKF is temporal stationarity, which is indeed invalid in the presence of motion. The motion-compensated RUKF proposed by the authors alleviates this problem by operating on the motion-compensated frames where temporal stationarity is a realistic assumption. Several important aspects of using RUKF in practice are also discussed in Chapter 12.

An extension of single-frame Wiener filtering for deblurring in the presence of noise (i.e., restoration) is discussed by Ozkan, Sezan Erdem and Tekalp in Chapter 13. The authors derive the general expression for a restoration filter in the frequency domain that operates on multiple frames simultaneously. This filter admits an efficient implementation that requires the inversion of $N \times N$ matrices, where N is the number of frames that are simultaneously restored. The multiframe Wiener filter becomes extremely efficient in certain special cases. An important special case occurs when the interframe motion is a global relative shift. In this case, by incorporating the motion information into the multiframe formulation, the authors derive the motion-compensated Wiener filter where analytic solutions can be found without the need for matrix inversion.

Three-dimensional, spatiotemporal median filter structures are discussed by Viero and Neuvo in Chapter 14 for noise suppression and predictive data compression. A distinct character of this chapter is the fact that none of the filter structures explicitly utilize motion information. Following a review of median operators and their extensions, derivation of 3-D weighted median filters for noise suppression are discussed. Next, the authors present several median-based predictors for predictive data compression. Median predictors combine linear predictors and median operations to maintain robustness in the presence of transmission errors. The median structures proposed in this chapter lend themselves to simple implementations and good performance.

In the last chapter of the book (Chapter 15), the subject of data compression is revisited from the perspective of designing a digital advanced television (ATV) system. Apostolopoulos and Lim present a comprehensive overview of the fundamental principles of video compression in the first part of this chapter. Next, the impact of several important system issues on the choice of compression algorithms, such as the requirements for system extensibility, interoperability and scalability, and the usefulness of VCR functionality, are discussed. The video compression subsystem of the Channel Compatible Digi-Cipher (CCDC) digital high definition television (HDTV) system, recently proposed by MIT and GI for possible adoption as the US standard, is then discussed as an example for a practical system that employs the fundamental principles of video compression.

M. I. Sezan, Rochester NY, USA
R. L. Lagendijk, Delft, The Netherlands
November 1992.

References

[1] A. Singh, *Optic Flow Computation.* Los Alamitos, CA: IEEE Computer Society Press, 1991.

[2] D. W. Murray and B. F. Buxton, *Experiments in the Machine Interpretation of Visual Motion.* Cambridge, MA: MIT Press, 1990.

[3] D. J. Fleet, *Measurement of Image Velocity.* Norwell, MA: Kluwer Academic Publishers, 1992.

[4] T. S. Huang, ed., *Image Sequence Analysis.* Berlin: Springer Verlag, 1981.

[5] T. S. Huang, ed., *Image Sequence Processing and Dynamic Scene Analysis.* Berlin: Springer Verlag, 1983.

Acknowledgments

We thank all the authors for their valuable contributions that made this book possible. We like to acknowledge, in particular, their synergy and close attention to our tight deadlines, which enabled us to complete this project in eighteen months after its initiation. In our opinion, the great enthusiasm we have shared with the authors during the course of this project is one of the clear indications of the timeliness of this book. We also like to acknowledge the work of all researchers who have made significant contributions to the field of motion analysis and image sequence processing but are not direct contributors of this book. We are thankful to Terry Lund and Mike Kriss of Imaging Research Laboratories at Eastman Kodak Company for their continuing encouragement during the course of this project. R.L. Lagendijk was a visiting research scientist at the Imaging Research Laboratories during the initiation of this project. Thanks are extended to Bob Holland and the staff at Kluwer for their support. Finally, we are indebted to our wives and children, Sugako, Meliz, and Marleen, Annick and Dave for their constant encouragement and patience.

1

Hierarchical Model-Based Motion Estimation

P. Anandan, J. R. Bergen, K. J. Hanna

David Sarnoff Research Center, Princeton, NJ

Rajesh Hingorani

AT&T Bell Laboratories, Murray Hill, NJ

1.1 Introduction

A large body of work in image processing and computer vision over the last 10 or 15 years has been concerned with the estimation of motion of pixels between pairs of images. As can be seen from the contents of this book, the motivation of this work is actually quite diverse, with intended applications ranging from data compression to registration of remotely sensed data to robotics and vehicle navigation. In tandem with this diversity of motivation is a diversity of representation of motion information: from optical flow, to affine or other parametric transformations, to 3-D egomotion plus range or other structure. The purpose of this chapter is to describe a common framework within which all of these computations can be represented.

This unification is possible because all of these problems can be viewed from the perspective of image registration. That is, given an image sequence, compute a representation of the motion field that best aligns pixels in one frame of the sequence with those in the next. The various approaches mentioned above differ in terms of the assumptions they make about the spatial structure (or *model*) of the motion field. Often, the problem is formulated as that of minimizing some type of an error norm computed from measurements made from the two images. For the different models, the minimization is with respect to different sets of parameters.

The framework presented in this chapter unifies these different computations by taking the following approach: A particular error norm is chosen (in our case this represents intensity constancy within Laplacian pyramid images) which is written as a function of the motion field. The motion field is then represented in terms of the particular set of parameters that are appropriate for a given model. The error norm is minimized with respect to these parameters within a hierarchical coarse-fine refinement framework.

The various motion models used in the different algorithms can be divided into three categories: (i) fully parametric, (ii) quasi-parametric, and (iii) non-parametric. In general, a parametric model will be *global* in the sense of describing (or constraining) the motion field over an extended image region. In many cases, there may also be additional variation within the field which may be described (approximately) by a *local model*[1].

1.1.1 Motion Models

Because optical flow computation is an underconstrained problem, *all* motion estimation algorithms involve additional assumptions about the structure of the motion computed. In many cases, however, this assumption is not expressed explicitly as such, rather it is presented as a regularization term in an objective function [15, 17] or described primarily as an assumption (e.g., uniform translation within a small region) that enables the computation of local motion vectors [19, 4, 2, 21]. Each of these implicitly represents a type of model of the underlying motion field. These assumptions tend to generic in that they apply to a wide class of situations. But for the same reason, the accuracy of the resulting motion fields tends to be limited.

It has been recognized that an explicit representation of the motion model may lead to more accurate computation of motion fields. A number of researchers have investigated the "direct" estimation of rigid body motion parameters without the prior estimation of optical flow [18, 22, 14]. Others have developed techniques for using parametric models within local regions [7, 9]. These tend to apply under somewhat restricted conditions. The description "direct methods" actually applies equally to both types of techniques.

We divide the motion models into three categories: (i) fully parametric models, (ii) quasi-parametric models, and (iii) non-parametric models. Fully parametric models are applicable when the the motion of individual pixels within a region in terms of a parametric form consisting of a small number of parameters. The problem of motion estimation then reduces to that of estimating the values of these parameters. Since all the pixels within a region can contribute to this estimation, highly accurate results may be obtained. Although these models may be exactly applicable under limited circumstances, they are often good approximations for a wide range of situations. For instance, the motion of the image of a planar surface under orthographic projection can be described as an affine transformation. But the same model forms a good approximation to motion of images of distant shallow surfaces under perspective projection, a situation not uncommon in many aerial image sequences. Similarly, the motion of the image of a planar surface under perspective projection can be described by a eight-parameter quadratic transformation.

Quasi-parametric models involve representing the motion of a pixel as a

[1] Because this model will be used in a multiresolution data structure, it is "local" in a slightly unconventional sense that will be discussed below.

combination of a parametric component that is valid for the entire region and a local component which varies from pixel to pixel. For instance, the rigid motion model belongs to this class: the six egomotion parameters constrain the local flow vector to lie along a specific line, while the local depth value determines the exact value of the flow vector at each pixel. Since the egomotion constraint is globally applicable, it may be possible to estimate those parameters accurately by combining all of the available information, whereas the accuracy of the local depth estimates depends on the degree to which the local image structures can be uniquely matched between frames.

By non-parametric models, we mean those such as are commonly used in optical flow computation, i.e. those involving the use of some type of a smoothness or uniformity constraint. In these cases, some type of a constraint is imposed on the local variation within the flow field, but the resulting field cannot be described in a simple parametric form. As a result a different flow vector must be estimated for each pixel. As might be expected, the resulting flow field is usually limited in accuracy, but the model is applicable under a much wider range of circumstances.

A parallel taxonomy of motion models can be constructed by considering local models that constrain the motion in the neighborhood of a pixel and global models that describe the motion over the entire visual field. This distinction becomes especially useful in analyzing hierarchical approaches where the meaning of "local" changes as the computation moves through the multiresolution hierarchy. In this scheme fully parametric models are global models, non-parametric models such as smoothness or uniformity of displacement are local models, and quasi-parametric models involve both a global and a local model. The reason for describing motion models in this way is that it clarifies the relationship between different approaches and allows consideration of the range of possibilities in choosing a model appropriate to a given situation. Purely global (or fully parametric) models in essence trivially imply a local model so no choice is possible. However, in the case of quasi- or non-parametric models, the local model can be more or less complex. Also, it makes clear that by varying the size of local neighborhoods, it is possible to move continuously from a partially or purely local model to a purely global one.

The reasons for choosing one model or another are generally quite intuitive, though the exact choice of model is not always easy to make in a rigorous way. In general, parametric models constrain the local motion more strongly than the less parametric ones. A small number of parameters (e.g., six in the case of affine flow) are sufficient to completely specify the flow vector at every point within their region of applicability. However, they tend to be applicable only within local regions, and in many cases, are approximations to the actual flow field within those regions (although they may be very good approximations). From the point of view of motion estimation, such models allow the precise estimation of motion at locations containing no image structure, provided the region contains at least a few locations with significant image structure.

Quasi-parametric models constrain the flow field less, but nevertheless constrain it to some degree. For instance, for rigidly moving objects under perspective projection, the rigid motion parameters (same as the egomotion parameters in the case of observer motion), constrain the flow vector at each point to lie along a line in the velocity space. One dimensional image structure (e.g., an edge) is generally sufficient to precisely estimate the motion of that point. These models tend to be applicable over a wide region in the image, perhaps even the entire image. If the local structure of the scene can be further parametrized (e.g., planar surfaces under rigid motion), the model becomes fully parametric within the region.

Non-parametric models require local image structure that is two-dimensional (e.g., corner points, textured areas). However, with the use of a smoothness constraint it is usually possible to "fill-in" where there is inadequate local information. The estimation process is typically more computationally expensive than the other two cases. These models are more generally applicable (not requiring parametrizable scene structure or motion) than the other two classes.

1.1.2 Hierarchical estimation

Hierarchical approaches have been used by various researchers e.g., see [2, 10, 11, 24, 20]). More recently, a theoretical analysis of hierarchical motion estimation was described in [8] and the advantages of using parametric models within such a framework have also been discussed in [5].

Arguments for use of hierarchical (i.e. pyramid based) estimation techniques for motion estimation have usually focused on issues of computational efficiency. A matching process that must accommodate large displacements can be very expensive to compute. Simple intuition suggests that if large displacements can be computed using low resolution image information great savings in computation will be achieved. Higher resolution information can then be used to improve the accuracy of displacement estimation by incrementally estimating small displacements (see, for example, [2]). However, it can also be argued that it is not only *efficient* to ignore high resolution image information when computing large displacements, in a sense it is *necessary* to do so. This is because of aliasing of high spatial frequency components undergoing large motion. Aliasing is the source of false matches in correspondence solutions or (equivalently) local minima in the objective function used for minimization. Minimization or matching in a multiresolution framework helps to eliminate problems of this type. Another way of expressing this is to say that many sources of non-convexity that complicate the matching process are not stable with respect to scale.

Thus, the motivation for using hierarchical processing is twofold: to eliminate false matches by using "large scale" structures, and to achieve a computationally efficient estimation. With only a few exceptions ([5, 9]), much of previous work on hierarchical motion estimation has concentrated on the estimation smooth optical flow fields, or according to our taxonomy, non-parametric models. The use of other types motion models (e.g., parametric and quasi-parametric models) within a hierarchical computational framework has generally been ignored.

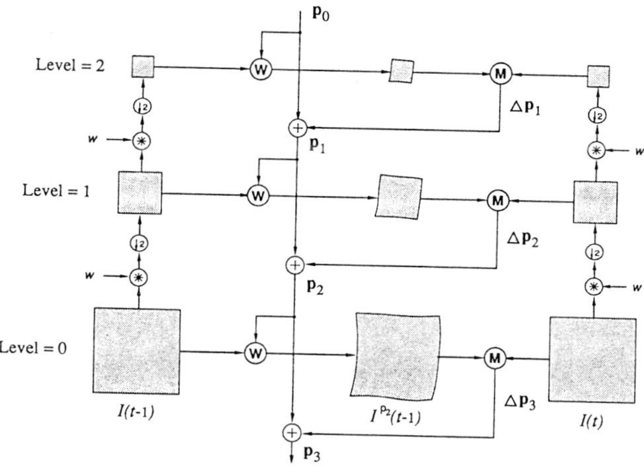

Figure 1.1: The hierarchical motion estimation framework.

However, the arguments given above for use of a multiresolution, hierarchical approach apply equally to more structured models of image motion.

The remainder of this chapter consists of an overview of the hierarchical motion estimation framework, a description of each of the four models and their application to specific examples, and a discussion of the overall approach and its applications.

1.2 Hierarchical Motion Estimation

Figure 1.1 describes the hierarchical motion estimation framework. The basic components of this framework are: (i) pyramid construction, (ii) motion estimation, (iii) image warping, and (iv) coarse-to-fine refinement.

Since one of the motivations for using hierarchical estimation is the use of large scale structures in order to eliminate false matches, and smaller scale structures to refine the estimates, a multi-scale representation of the images that allows this is needed. In particular, the representation should be similar across the scales, and be continuously related. In other words, the structures at neighboring scales at the same spatial location should be correlated.

Our implementation of the framework uses Laplacian pyramid [6] for representing the image structures at different scales used for matching. Besides satisfying the requirements of similarity and continuity across scales, these representations also provide a certain degree of invariance with respect to photometric variations of the input image.

The estimation process used in the current implementation is the minimization of Sum-of-squared-differences (SSD) of the Laplacian filtered intensity images. This is done separately at each level, starting at a user specified coarse-level and refining the estimates down to the finest level (the resolution of the input images). Within each level a Gauss-Newton minimization algorithm is employed to iteratively refine the estimates. The choice of this strategy for numerical estimation is motivated by considerations of uniformity of computation across different models, and computational efficiency. However, in order to assure convergence to the correct solution, it is necessary to guarantee good initial values. Often such values may be obtained from prior data concerning sensor calibration and motion.

The basic assumption behind SSD minimization is *intensity constancy* as applied to the Laplacian pyramid images. Thus,

$$I(\mathbf{x},t) = I(\mathbf{x}-\mathbf{u}(\mathbf{x}),t-1)$$

where $\mathbf{x} = (x, y)$ denotes the spatial image position of a point, I the (Laplacian pyramid) image intensity and $\mathbf{u}(\mathbf{x}) = (u(x,y), v(x,y))$ denotes the image velocity at that point. The SSD error measure for estimating the flow field within a region is:

$$E(\{\mathbf{u}\}) = \sum_{\mathbf{x}} (I(\mathbf{x},t) - I(\mathbf{x}-\mathbf{u}(\mathbf{x}),t-1))^2 \qquad (1.1)$$

where the sum is computed over all the points within the region and $\{\mathbf{u}\}$ is used to denote the entire flow field within that region. In general this error (which is actually the sum of individual errors) is not quadratic in terms of the unknown quantities $\{\mathbf{u}\}$, because of the complex pattern of intensity variations. Hence, we typically have a non-linear minimization problem at hand.

Note that the basic structure of the problem is independent of the choice of a motion model. The model is in essence a statement about the function $\mathbf{u}(\mathbf{x})$. To make this explicit, we can write,

$$\mathbf{u}(\mathbf{x}) = \mathbf{u}(\mathbf{x}; \mathbf{p}_m), \qquad (1.2)$$

where $\mathbf{p}_m$ is a vector representing the model parameters.

If $\{\mathbf{u}\}_i$ is the current estimate of the flow field during the ith iteration, the incremental estimate $\{\delta\mathbf{u}\}$ can be obtained by minimizing the quadratic error measure

$$E(\{\delta\mathbf{u}\}) = \sum_{\mathbf{x}} (\Delta I + \nabla I \cdot \delta\mathbf{u}(\mathbf{x}))^2 , \qquad (1.3)$$

where

$$\Delta I(\mathbf{x}) = I(\mathbf{x},t) - I(\mathbf{x}-\mathbf{u}_i(\mathbf{x}),t-1),$$

that is the difference between the two images at corresponding pixels, after taking the current estimate into account.

As such, the minimization problem described in Equation 1.3 is underconstrained. The different motion models constrain the flow field in different ways. When these are used to describe the flow field, the estimation problem can be reformulated in terms of the unknown (incremental) model parameters. The details of these reformulations are described in the various sections corresponding to the individual motion models.

The third component, image warping, is achieved by using the current values of the model parameters to compute a flow field, and then using this flow field to warp $I(t-1)$ towards $I(t)$, which is used as the reference image. Our current warping algorithm uses bilinear interpolation. The warped image (*as against the original second image*) is then used for the computation of the error ΔI for further estimation[2].

The final component, coarse-to-fine refinement, propagates the current motion estimates from one level to the next level where they are then used as initial estimates. For the parametric component of the model, this is easy; the values of the parameters are simply transmitted to the next level. However, when a local model is also used, that information is typically in the form of a dense image (or images)—e.g., a flow field or a depth map. This image (or images) must be propagated via a pyramid expansion operation as described in [6]. The global parameters in combination with the local information can then be used to generate the flow field necessary to perform the initial warping at this next level.

1.3 Motion Models

This section describes the use of four different models within our hierarchical estimation farmework. Two of these, namely affine and planar surface flow under perspective projection, are parametric models. The third one, the instantaneous flow field induced by a rigidly moving object, is quasi-parametric. The fourth is the non-parametric model of general smoothly varying flow.

In each case, first the model is described in a mathematical form and the parameters to be estimated are identified. The estimation procedure for the model is then described. While there are strong similarities between the various estimation algorithms, there are also important differences in the strategy as a result of the particular differences between the models. These differences arise because of the variation between the types of models, as well as because of computational considerations associated with particular models. Finally, illustrative experimental results on real images are shown.

1.3.1 Affine Flow

The Model:

Under orthographic projection, a planar surface undergoing 3D translation and linear deformation gives rise to a 2D affine transformation in the image plane. The affine transformation is also a good first-order approximation of a distant object undergoing 3D translation and linear deformation and subtending a limited field of view.

$$\begin{aligned} u(x,y) &= a_1 + a_2x + a_3y \\ v(x,y) &= a_4 + a_5x + a_6y \end{aligned} \qquad (1.4)$$

[2]We have avoided using the standard notation I_t in order to avoid any confusion about this point.

Using vector notation this can be rewritten as follows:

$$\mathbf{u}(\mathbf{x}) = \mathbf{X}(\mathbf{x})\mathbf{a} \tag{1.5}$$

where $\mathbf{a}$ denotes the vector $(a_1, a_2, a_3, a_4, a_5, a_6)^T$, and

$$\mathbf{X}(\mathbf{x}) = \begin{bmatrix} 1 & x & y & 0 & 0 & 0 \\ 0 & 0 & 0 & 1 & x & y \end{bmatrix}$$

Thus, the motion of the entire region is completely specified by the parameter vector $\mathbf{a}$, which is the unknown quantity that needs to be estimated.

The Estimation Algorithm:

Let $\mathbf{a}_i$ denote the current estimate of the affine parameters. After using the flow field represented by these parameters in the warping step, an incremental estimate $\delta\mathbf{a}$ can be determined. To achieve this, we insert the parametric form of $\delta\mathbf{u}$ into Equation 1.3, and obtain an error measure that is a function of $\delta\mathbf{a}$.

$$E(\delta\mathbf{a}) = \sum_{\mathbf{x}} \left(\Delta I + (\nabla I)^T \mathbf{X}\delta\mathbf{a}\right)^2 \tag{1.6}$$

Minimizing this error with respect to $\delta\mathbf{a}$ leads to the equation:

$$\left[\sum \mathbf{X}^T(\nabla I)(\nabla I)^T\mathbf{X}\right]\delta\mathbf{a} = -\sum \mathbf{X}^T(\nabla I)(\Delta I). \tag{1.7}$$

The summation index in the above equations ranges over an area of interest, within which the affine transformation is assumed to be valid. The conditioning of the matrix that is inverted on the right-hand side of equation 1.7 gives an indication of the uniqueness of the estimated parameters. If the matrix is singular (or poorly conditioned), that is usually an indication that the image structure within the region of interest does not have sufficient variation in order to allow unique estimates. This is a generalization of the "aperture problem" for translational motion estimation.

Experiments with the affine motion model:

To demonstrate use of the affine flow model, we show its performance on an aerial image sequence. Two frames of the original sequence are shown in Figure 1.2a and Figure 1.2b. The unprocessed difference between two frames of this sequence is shown in Figure 1.2c. Figure 1.2d shows the result of estimating an affine transformation using the hierarchical warp motion approach, and then using this to compensate for camera motion induced flow. Although the terrain is not perfectly flat, we still obtain encouraging compensation results. In this example the simple difference between the compensated and original image is sufficient to detect and locate a helicopter in the image. We use extensions of the approach, like integration of compensated difference images over time, to detect smaller objects moving more slowly with respect to the background.

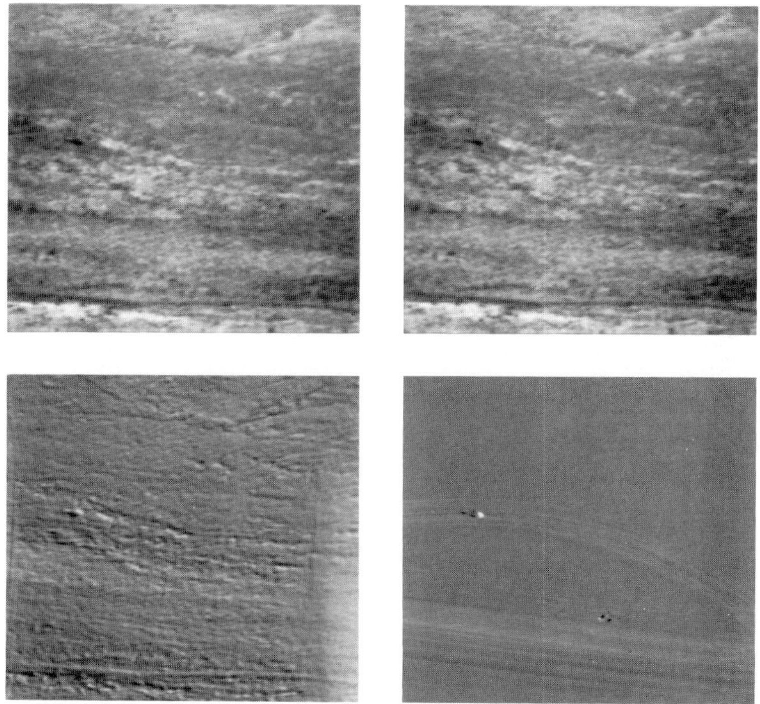

Figure 1.2: Affine motion estimation:
Top left: Frame 1.
Top right: Frame 2.
Bottom left: Raw difference.
Bottom right: Compensated difference.

1.3.2 Planar Surface Motion Under Perspective Projection

The Model:

It is generally known that the instantaneous motion induced by a rigidly moving planar surface under perspective projection can be described as a second order function of image coordinates involving eight independent parameters (e.g., see [16]). In this section we provide a brief derivation of this description and make some observations concerning its estimation.

We begin by observing that the image motion induced by a rigidly moving object (in this case a plane), can be written as:

$$\mathbf{u}(\mathbf{x}) = \frac{1}{Z(\mathbf{x})}\mathbf{A}(\mathbf{x})\mathbf{t} + \mathbf{B}(\mathbf{x})\omega \tag{1.8}$$

where $Z(\mathbf{x})$ is the distance from the camera of the point (i.e., depth) whose image position is $\mathbf{x} = (x, y)$, and

$$\mathbf{A}(\mathbf{x}) = \begin{bmatrix} -f & 0 & x \\ 0 & -f & y \end{bmatrix}$$

$$\mathbf{B}(\mathbf{x}) = \begin{bmatrix} (xy)/f & -(f^2+x^2)/f & y \\ (f^2+y^2)/f & -(xy)/f & -x \end{bmatrix}.$$

The $\mathbf{A}$ and the $\mathbf{B}$ matrices depend only on the image positions and the focal length f and not on the unknowns: $\mathbf{t}$, the translation vector, ω the angular velocity vector, and Z.

A planar surface can be described by the equation

$$k_1X + k_2Y + k_3Z = 1 \tag{1.9}$$

where (k_1, k_2, k_3) relate to the surface slant, tilt, and the distance of the plane from the origin of the chosen coordinate system (in this case, the camera origin). Dividing throughout by Z, and using the relationships $(x/f = X/Z, y/f = Y/Z)$ we get

$$\frac{1}{Z} = k_1\frac{x}{f} + k_2\frac{y}{f} + k_3.$$

Using $\mathbf{k}$ to denote the vector (k_1, k_2, k_3) and $\mathbf{r}$ to denote the vector $(x/f, y/f, 1)$ we obtain

$$\frac{1}{Z(\mathbf{x})} = \mathbf{r}(\mathbf{x})^T\mathbf{k}.$$

Substituting this into Equation 1.8 gives

$$\mathbf{u}(\mathbf{x}) = (\mathbf{A}(\mathbf{x})\mathbf{t})\left(\mathbf{r}(\mathbf{x})^T\mathbf{k}\right) + \mathbf{B}(\mathbf{x})\omega \tag{1.10}$$

This flow field is quadratic in $(\mathbf{x})$ and can be written also as

$$\begin{aligned} u(\mathbf{x}) &= a_1 + a_2x + a_3y + a_7x^2 + a_8xy \\ v(\mathbf{x}) &= a_4 + a_5x + a_6y + a_7xy + a_8y^2 \end{aligned} \tag{1.11}$$

where the 8 coefficients $(a_1, \ldots, a_8)$ are functions of the motion parameters $\mathbf{t}, \omega$ and the surface parameters $\mathbf{k}$. Since this 8-parameter form is rather well-known (e.g., see [16]) we omit its details.

If the egomotion parameters are known, then the three parameter vector $\mathbf{k}$ can be used to represent the motion of the planar surface. Otherwise the 8-parameter representation can be used. In either case, the flow field is linear in the unknown parameters.

The problem of estimating planar surface motion has been extensively studied before [22, 1, 25]. In particular, Negahdaripour and Horn [22] suggest iterative methods for estimating the motion and the surface parameters, as well as a method of estimating the 8 parameters and then decomposing them into the five rigid motion parameters the three surface parameters in closed form. Besides the embedding of these computations within the hierarchical estimation framework, we also take a slightly different approach to the problem.

We assume that the rigid motion parameters are already known or can be estimated (e.g., see Section 1.3.3 below). Then, the problem reduces to that of estimating the three surface parameters $\mathbf{k}$. There are several practical reasons to prefer this approach: First, in many situations the rigid motion model may be more globally applicable than the planar surface model, and can be estimated using information from all the surfaces undergoing the same rigid motion. Second, unless the region of interest subtends a significant field of view, the second order components of the flow field will be small, and hence the estimation of the eight parameters will be inaccurate and the process may be unstable. On the other hand, the information concerning the three parameters $\mathbf{k}$ is contained in the first order components of the flow field, and (if the rigid motion parameters are known) their estimation will be more accurate and stable.

The Estimation Algorithm:

Let $\mathbf{k}_i$ denote the current estimate of the surface parameters, and let $\mathbf{t}$ and ω denote the motion parameters. These parameters are used to construct an initial flow field that is used in the warping step. The residual information is then used to determine an incremental estimate $\delta\mathbf{k}$.

By substituting the parametric form of $\delta\mathbf{u}$

$$\begin{aligned}
\delta\mathbf{u} &= \mathbf{u} - \mathbf{u_0} \\
&= (\mathbf{A}(\mathbf{x})\mathbf{t})\left(\mathbf{r}(\mathbf{x})^T(\mathbf{k_0} + \delta\mathbf{k})\right) + \mathbf{B}(\mathbf{x})\omega - (\mathbf{A}(\mathbf{x})\mathbf{t})\left(\mathbf{r}(\mathbf{x})^T\mathbf{k_0}\right) - \mathbf{B}(\mathbf{x})\omega \\
&= (\mathbf{A}(\mathbf{x})\mathbf{t})\,\mathbf{r}(\mathbf{x})^T\delta\mathbf{k} \qquad (1.12)
\end{aligned}$$

in Equation 1.3, we can obtain the incremental estimate $\delta\mathbf{k}$ as the vector that minimizes:

$$E(\delta\mathbf{k}) = \sum_{\mathbf{x}}(\left(\Delta I + (\nabla I)^T(\mathbf{A}\mathbf{t})\mathbf{r}^T\delta\mathbf{k}\right)^2 \qquad (1.13)$$

Minimizing this error leads to the equation:

$$\left[\sum \mathbf{r}(\mathbf{t}^T\mathbf{A}^T)(\nabla I)(\nabla I)^T(\mathbf{A}\mathbf{t})\mathbf{r}^T)\right]\delta\mathbf{k} = -\sum \mathbf{r}(\mathbf{t}^T\mathbf{A}^T)(\nabla I)\Delta I \qquad (1.14)$$

This equation can be solved to obtain the incremental estimate $\delta\mathbf{k}$.

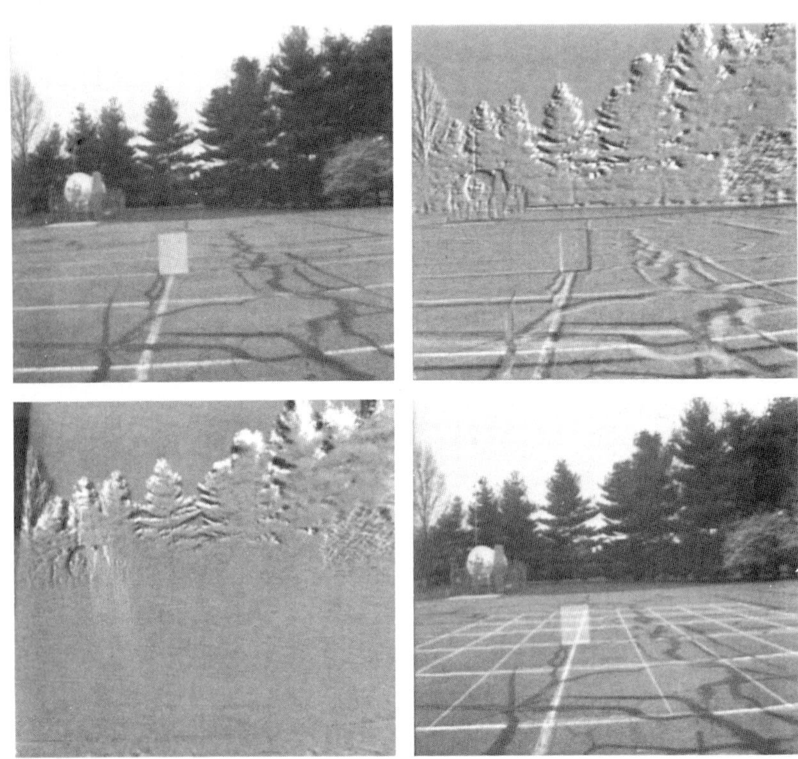

Figure 1.3: Planar surface motion estimation.
Top left: Original image.
Top right: Raw difference.
Bottom left: Difference after planar compensation.
Bottom right: Planar grid superimposed on the original image.

Experiments with the planar surface motion model:

We demonstrate the application of the planar surface model using images from an outdoor sequence. One of the input images is shown in Figure 1.3a, and the difference between both input images is shown in Figure 1.3b. After estimating the camera motion between the images using the algorithm described in Section 1.3.3, we applied the planar surface estimation algorithm to a manually selected image window placed roughly over a region on the ground plane. These parameters were then used to warp the second frame towards the first (this process should align the ground plane alone). The difference between this warped image and the original image is shown in Figure 1.3c. The figure shows compensation of the ground plane motion, leaving residual parallax motion of the trees and other objects in the background. Finally, in order to demonstrate the plane-fit, we graphically projected a rectangular grid onto that plane. This is shown superimposed on the input image in Figure 1.3d.

1.3.3 Rigid Body Motion Model

The Model:

The motion of arbitrary surfaces undergoing rigid motion cannot usually be described by a single global model. We can however make use of the global rigid body model if we combine it with a local model of the surface. In this section, we provide a brief derivation of the global and the local models. Hanna [12] provides further details and results, and also describes how the local and global models interact at corner-like and edge-like image structures.

As described in Section 1.3.2, the image motion induced by a rigidly moving object can be written as:

$$\mathbf{u}(\mathbf{x}) = \frac{1}{Z(\mathbf{x})}\mathbf{A}(\mathbf{x})\mathbf{t} + \mathbf{B}(\mathbf{x})\omega \tag{1.15}$$

where $Z(\mathbf{x})$ is the distance from the camera of the point (i.e., its depth), whose image position is $(\mathbf{x})$, and

$$\mathbf{A}(\mathbf{x}) = \begin{bmatrix} -f & 0 & x \\ 0 & -f & y \end{bmatrix}$$

$$\mathbf{B}(\mathbf{x}) = \begin{bmatrix} (xy)/f & -(f^2+x^2)/f & y \\ (f^2+y^2)/f & -(xy)/f & -x \end{bmatrix}.$$

The $\mathbf{A}$ and the $\mathbf{B}$ matrices depend only on the image positions and the focal length f and not on the unknowns: $\mathbf{t}$, the translation vector, ω the angular velocity vector, and Z. Equation 1.15 relates the parameters of the global model, ω and $\mathbf{t}$, with parameters of the local scene structure, $Z(\mathbf{x})$.

A local model we use is the frontal-planar model, which means that over a local image patch, we assume that $Z(\mathbf{x})$ is constant. An alternative model uses the assumption that $\delta Z(\mathbf{x})$—the difference between a previous estimate and a refined estimate—is constant over each local image patch.

The Estimation Algorithm:

The estimation process for this (and other) quasi-parametric model breaks down into two interleaved parts: estimation of the global parameters, and the estimation of the local parameters. Given current estimates of the global and local parameters, refinements for these parameters can be computed during each iteration step of the minimization process.

The iteration process proceeds as before in a coarse-to-fine fashion. The final estimates at a coarse level are simply used as initial values for the estimation at the next finer level.

Let the current global parameters be denoted as $\mathbf{t}_i$ and ω_i and the current local parameters as $Z_i(\mathbf{x})$. As in the other models, we can use the model parameters to construct an initial flow field, $\mathbf{u}_i(\mathbf{x})$, which is used to warp one of the image frames towards the next. The residual error between the warped image and the original image to which it is warped is used to refine the parameters of the local and global models. We now show how these models are refined.

We begin by writing equation 1.15 in an incremental form so that

$$\delta\mathbf{u}(\mathbf{x}) = \frac{1}{Z(\mathbf{x})}\mathbf{A}(\mathbf{x})\mathbf{t} + \mathbf{B}(\mathbf{x})\omega - \frac{1}{Z_i(\mathbf{x})}\mathbf{A}(\mathbf{x})\mathbf{t_i} - \mathbf{B}(\mathbf{x})\omega_\mathbf{i} \tag{1.16}$$

Inserting the parametric form of $\delta\mathbf{u}$ into Equation 1.3 we obtain the pixel-wise error as

$$E(\mathbf{t},\omega,1/Z) = \left(\Delta I + \frac{(\nabla I)^T\mathbf{A}\mathbf{t}}{Z} + (\nabla I)^T\mathbf{B}\omega - \frac{(\nabla I)^T\mathbf{A}\mathbf{t}_i}{Z_i} - (\nabla I)^T\mathbf{B}\omega_i\right)^2. \tag{1.17}$$

This estimation problem involves determining one unknown range value for each pixel and six unknown global parameters. Since the global parameter $\mathbf{t}$ and the local parameters $1/Z(\mathbf{x})$ appear in a linear relationship throughout the equations, these parameters can be estimated only upto an unknown scale factor. Without loss of generality, we can assume that the magnitude of translation is unity. Also, if we consider each pixel as contributing a constraint to the overall estimation problem, it is easy to see that there are fewer constraints than the number of unknowns. Hence, the estimation problem is underconstrained.

This situation is typical of quasi-parametric models and can be handled by introducing a model for the variation of the local parameters. We assume that $1/Z$ is constant over 5×5 image patches centered on each image pixel at the current pyramid level. Note that this is *not* the same as the model that the surfaces are piecewise planar. Because of the overlap between neighboring patches overlap and the use of this assumption represents the model that the inverse range-map varies "smoothly" across the image.

There are several ways to solve this bilinear estimation problem. We have chosen the approach of algebraically eliminating the local parameters in favor of the global ones. Consider the local component of the error measure,

$$E_{local} = \sum_{5\times 5} E(\mathbf{t},\omega,1/Z). \tag{1.18}$$

Differentiating equation 1.17 with respect to $1/Z(\mathbf{x})$ and setting the result to zero, we get

$$1/Z = \frac{-\sum_{5\times5}(\nabla I)^T\mathbf{At}\left(\Delta I - (\nabla I)^T\mathbf{At}_i/Z_i) + (\nabla I)^T\mathbf{B}\omega - (\nabla I)^T\mathbf{B}\omega_i\right)}{\sum_{5\times5}\left((\nabla I)^T\mathbf{At}\right)^2} \tag{1.19}$$

To refine the global model, we minimize the error in Equation 1.17 summed over the entire image:

$$E_{global} = \sum_{Image} E(\mathbf{t}, \omega, 1/Z). \tag{1.20}$$

We insert the expression for $1/Z(\mathbf{x})$ given in Equation 1.19—*not the current numerical value of the local parameter*—into Equation 1.20. The result is an expression for E_{global} that is non-quadratic in $\mathbf{t}$ but quadratic in ω . We recover refined estimates of $\mathbf{t}$ and ω by performing one Gauss-Newton minimization step using the previous estimates of the global parameters, $\mathbf{t_i}$ and $\omega_\mathbf{i}$, as starting values. Expressions are evaluated numerically at $t = t_i$ and $\omega = \omega_i$.

We then repeat the estimation algorithm several times at each image resolution.

Experiments with the rigid body motion model:

We have chosen an outdoor scene to demonstrate the rigid body motion model. Figure 1.4a shows one of the input images, and Figure 1.4b shows the difference between the two input images. The algorithm was performed beginning at level 3 (subsampled by a factor of 8) of a Laplacian pyramid. The local surface parameters $1/Z$ were all initialized to zero, and the rigid-body motion parameters were initialized to $\mathbf{t_0} = (\mathbf{0}, \mathbf{0}, \mathbf{1})^\mathrm{T}$ and $\omega = (\mathbf{0}, \mathbf{0}, \mathbf{0})^\mathrm{T}$. The model parameters were refined 10 times at each image resolution. Figure 1.4c shows an image of the recovered local surface parameters $1/Z(\mathbf{x})$ such that bright points are nearer the camera than dark points. The recovered inverse ranges are plausible almost everywhere, except at the image border and near the recovered focus of expansion. The bright dot at the bottom right hand side of the inverse range map corresponds to a leaf in the original image that is blowing across the ground towards the camera. Figure 1.4d shows the difference image between the second image and the first image after being warped using the final estimates of the rigid-body motion parameters and the local surface parameters. Figure 1.4e shows a table of rigid-body motion parameters that were recovered at the end of each resolution of analysis.

More experimental results and a detailed discussion of the algorithm's performance on various types of scenes can be found in [12].

Resolution	Ω	T
.	(.0000,.0000,.0000)	(.0000,.0000,1.0000)
32 × 30	(.0027,.0039,-.0001)	(-.3379,-.1352,.9314)
64 × 60	(.0038,.0041,.0019)	(-.3319,-.0561,.9416)
128 × 120	(.0037,.0012,.0008)	(-.0660,-.0383,.9971)
256 × 240	(.0029,.0006,.0013)	(-.0255,-.0899,.9956)

Figure 1.4: Egomotion based flow model.
Top left: Original image from an outdoor sequence.
Top right: Raw difference.
Bottom left: Inverse range map.
Bottom right: Difference after ego-motion compensation.
Table: Rigid body parameters recovered at each resolution.

1.3.4 General Flow Fields

The Model:

Unconstrained general flow fields are typically not described by any global parametric model. The observations made concerning the local parameters in the case of quasi-parametric models apply to this problem as well, with the difference that there is no global model that describes the motion of the scene elements.

Different local models have been used to facilitate the estimation process, including constant flow within a local neighborhood (at multiple resolutions) and locally smooth or continuous flow. The former facilitates direct local estimation [19, 21], whereas the latter model requires iterative relaxation techniques [17] It is also not uncommon to use the combination of these two types of local models (e.g., [3, 10]).

We have chosen the same local model as in the case of range values for the egomotion model, namely constant flow within 5×5 pixel windows at each level of the pyramid. This model was used by Lucas and Kanade [19] but here it is embedded as a local model at multiple resolutions.

The Estimation Algorithm:

Assume that we have an approximate flow field from previous levels (or previous iterations at the same level). Assuming that the incremental flow vector $\delta\mathbf{u}$ is constant within the 5×5 window, Equation 1.3 can be written as

$$E(\delta u) = \sum_{\mathbf{x}} (\Delta I + \nabla I^T \delta\mathbf{u})^2 \qquad (1.21)$$

where the sum is taken within the 5×5 window. Minimizing this error with respect to $\delta\mathbf{u}$ leads to the equation,

$$\left[\sum (\nabla I)(\nabla I)^T\right] \delta\mathbf{u} = -\sum \nabla I \Delta I. \qquad (1.22)$$

We make some observations concerning the singularities of this relationship. If the summing window consists of a single element, the 2×2 matrix on the left-hand-side is an outer product of a 2×1 vector and hence has a rank of atmost unity. In our case, when the summing window consists of 25 points, the rank of the matrix on the left-hand-side will be two unless the directions of the gradient vectors ∇I everywhere within the window coincide. This situation is the general case of the *aperture effect*.

In our implementation of this technique, the flow estimate at each point is obtained by using a 5×5 window centered around that point. This amounts to assuming implicitly that the flow field varies smoothly over the image.

Experiments with the general flow model:

We demonstrate the general flow algorithm on an image sequence containing several independently moving objects, a case for which the other motion models described here are not applicable. Figure 1.5a shows one image of the original

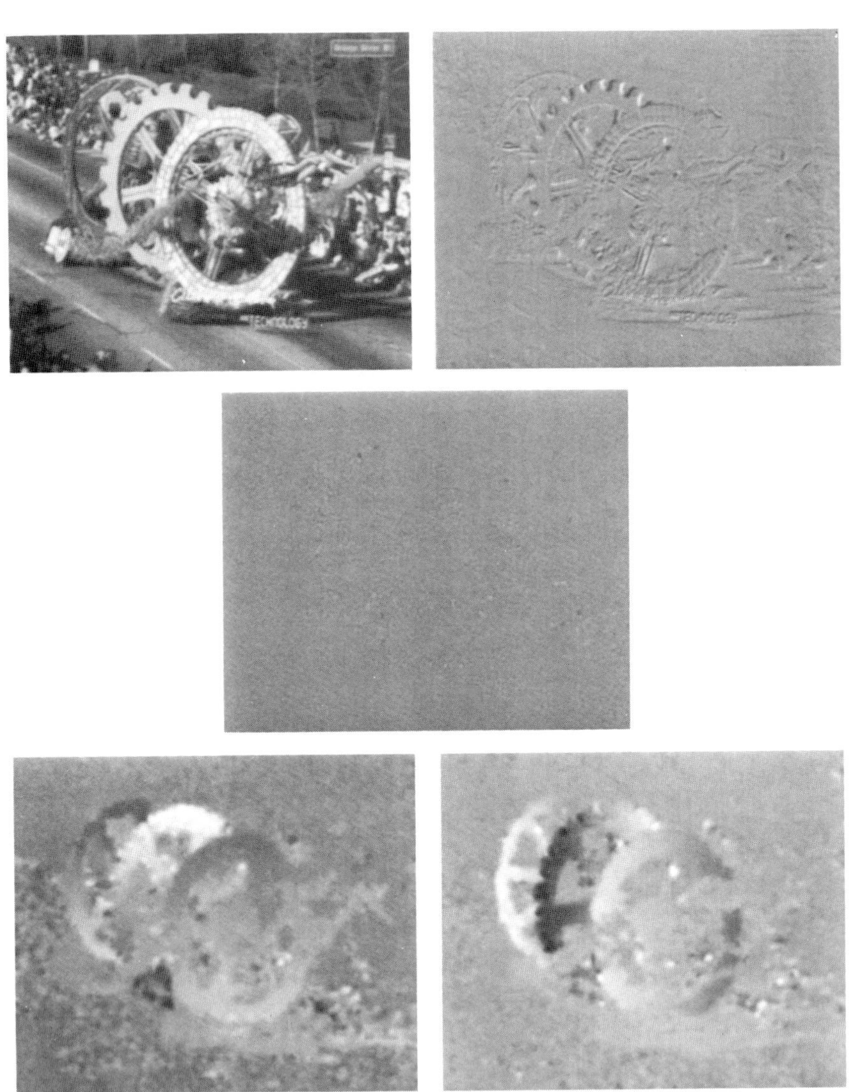

Figure 1.5: Optical flow estimation.
Top left: Original image.
Top right: Raw difference.
Middle: Difference after motion compensation.
Bottom left: Horizontal component of the recovered flow field.
Bottom right: Vertical component of the recovered flow field.

sequence. Figure 1.5b shows the difference between the two frames that were used to compute image flow. Figure 1.5c shows little difference between the compensated image and the other original image. Figure 1.5d shows the horizontal component of the computed flow field, and figure 1.5e shows the vertical component. In local image regions where image structure is well-defined, and where the local image motion is simple, the recovered motion estimates appear plausible. Errors predictably occur however at motion boundaries. Errors also occur in image regions where the local image structure is not well-defined (like some parts of the road), but for the same reason, such errors do not appear as intensity errors in the compensated difference image.

1.4 Discussion

Thus far, we have described a hierarchical framework for the estimation of image motion between two images using various models. Our motivation was to generalize the notion of direct estimation to model-based estimation and unify a diverse set of model-based estimation algorithms into a single framework. The framework also supports the combined use of parametric global models and local models which typically represent some type of a smoothness or local uniformity assumption.

One of the unifying aspects of the framework is that the same objective function (SSD) is used for all models, but the minimization is performed with respect to different parameters. As noted in the introduction, this is enabled by viewing all these problems from the perspective of image registration.

It is interesting to contrast this perspective (of model-based image registration) with some of the more traditional approaches to motion analysis. One such approach is to compute image flow fields, which involves combining the local brightness constraint with some sort of a global smoothness assumption, and then interpret them using appropriate motion models. In contrast, the approach taken here is to use the motion models to constrain the flow field computation. The obvious benefit of this is that the resulting flow fields may generally be expected to be more consistent with models than general smooth flow fields. Note, however, that the framework also includes general smooth flow field techniques, which can be used if the motion model is unknown.

In the case of models that are not fully parametric, local image information is used to determine local image/scene properties (e.g., the local range value). However, the accuracy of these can only be as good as the available local image information. For example, in homogeneous areas of the scene, it may be possible to achieve perfect registration even if the surface range estimates (and the corresponding local flow vectors) are incorrect. However, in the presence of significant image structures, these local estimates may be expected to be accurate. On the other hand, the accuracy of the global parameters (e.g., the rigid motion parameters) depends only on having sufficient and sufficiently diverse local information across the entire region. Hence, it may be possible to obtain reliable estimates of these global parameters, even though estimated local information may not be reliable everywhere within the region. For fully parametric models, this problem does not exist.

1.5 Future Work

There are three obvious ways in which the current work is incomplete; these offer immediate opportunities for future work:

1. A systematic (perhaps even formal) analysis of the behaviour of the estimation process over multiple scales. Although there is considerable experiential and anecdotal evidence suggesting that the coarse-fine estimation process offers some unique advantages in dealing with complex error surfaces, there has been little by way of systematic study.

2. The current work does not address the issue of evaluating the suitability of a particular model for a given problem. An important area for further study is the development of metrics for evaluating the performance of a model when applied to a given problem, and an approach to gradually apply more complex models starting with simple ones.

3. The extension of the current work to deal with multiple motions (or more simply even some way to handle model outliers during the estimation process). One approach would be apply robust estimation strategies for this purpose. An alternative formulation involves computing partial estimates using local information and checking for consistency while using the global model (e.g., some sort of a generalization of Schunck's constraint line clustering [23]).

Besides these, an important and interesting area for further study is how to extend this registration framework to handle multiple frame sequences. In particular, for models involving object shape (e.g., planar surface, egomotion), we may be able to compute more accurate registration and shape parameters via multiple frame analysis. A preliminary effort undertaken in this direction can be found in [13].

1.6 Conclusion

The image registration problem addressed in this paper occurs in a wide range of image processing applications, far beyond the usual ones considered in computer vision (e.g., navigation and image understanding). These include image compression via motion compensated encoding, spatiotemporal analysis of remote sensing type of images, image database indexing and retrieval, and possibly object recognition. One way to state this general problem is as that of recovering the coordinate system that relate two images of a scene taken from two different viewpoints. In this sense, the framework proposed here unifies motion analysis across these different applications as well.

Acknowledgements:

Many individuals have contributed to the ideas and results presented here. These include Peter Burt, Jim Muller, and Leonid Oliker from the David Sarnoff Research Center, and Shmuel Peleg from Hebrew University.

References

[1] G. Adiv. Determining three-dimensional motion and structure from optical flow generated by several moving objects. *IEEE Trans. on Pattern Analysis and Machine Intelligence*, 7(4):384–401, July 1985.

[2] P. Anandan. A unified perspective on computational techniques for the measurement of visual motion. In *International Conference on Computer Vision*, pages 219–230, London, May 1987.

[3] P. Anandan. A computational framework and an algorithm for the measurement of visual motion. *International Journal of Computer Vision*, 2:283–310, 1989.

[4] J. R. Bergen and E. H. Adelson. Hierarchical, computationally efficient motion estimation algorithm. *J. Opt. Soc. Am. A.*, 4:35, 1987.

[5] J. R. Bergen, P. J. Burt, R. Hingorani, and S. Peleg. Computing two motions from three frames. In *International Conference on Computer Vision*, Osaka, Japan, December 1990.

[6] P. J. Burt and E. H. Adelson. The laplacian pyramid as a compact image code. *IEEE Transactions on Communication*, 31:532–540, 1983.

[7] P.J. Burt, J.R. Bergen, R. Hingorani, R. Kolczinski, W.A. Lee, A. Leung, J. Lubin, and H. Shvaytser. Object tracking with a moving camera, an application of dynamic motion analysis. In *IEEE Workshop on Visual Motion*, pages 2–12, Irvine, CA, March 1989.

[8] P.J. Burt, R. Hingorani, and R. J. Kolczynski. Mechanisms for isolating component patterns in the sequential analysis of multiple motion. In *IEEE Workshop on Visual Motion*, pages 187–193, Princeton, NJ, October 1991.

[9] Stefan Carlsson. Object detection using model based prediction and motion parallax. In *Stockholm workshop on computational vision*, Stockholm, Sweden, August 1989.

[10] J. Dengler. Local motion estimation with the dynamic pyramid. In *Pyramidal systems for computer vision*, pages 289–298, Maratea, Italy, May 1986.

[11] W. Enkelmann. Investigations of multigrid algorithms for estimation of optical flow fields in image sequences. *Computer Vision, Graphics, and Image Processing*, 4339:150–177, 1988.

[12] K. J. Hanna. Direct multi-resolution estimation of ego-motion and structure from motion. In *Workshop on Visual Motion*, pages 156–162, Princeton, NJ, October 1991.

[13] K. J. Hanna. Combining stereo and motion vision data for autonomous navigation. In *AUVS-92 Proceedings*, Huntsville, Alabama, June 1992.

[14] J. Heel. Direct estimation of structure and motion from multiple frames. Technical Report 1190, MIT AI LAB, Cambridge, MA, 1990.

[15] E. C. Hildreth. *The Measurement of Visual Motion.* The MIT Press, 1983.

[16] B. K. P. Horn. *Robot Vision.* MIT Press, Cambridge, MA, 1986.

[17] B. K. P. Horn and B. G. Schunck. Determining optical flow. *Artificial Intelligence*, 17:185–203, 1981.

[18] B. K. P. Horn and E. J. Weldon. Direct methods for recovering motion. *International Journal of Computer Vision*, 2(1):51–76, June 1988.

[19] B.D. Lucas and T. Kanade. An iterative image registration technique with an application to stereo vision. In *Image Understanding Workshop*, pages 121–130, 1981.

[20] L. Matthies, R. Szeliski, and T. Kanade. Kalman filter-based algorithms for estimating depth from image-sequences. In *International Conference on Computer Vision*, pages 199–213, Tampa, FL, 1988.

[21] H. H. Nagel. Displacement vectors derived from second order intensity variations in intensity sequences. *Computer Vision, Pattern recognition and Image Processing*, 21:85–117, 1983.

[22] S. Negahdaripour and B.K.P. Horn. Direct passive navigation. *IEEE Trans. on Pattern Analysis and Machine Intelligence*, 9(1):168–176, January 1987.

[23] B. G. Schunck. Image flow segmentation and estimation by constraint line clustering. *PAMI*, 11(10):1010–1027, 1989.

[24] A. Singh. An estimation theoretic framework for image-flow computation. In *International Conference on Computer Vision*, Osaka, Japan, November 1990.

[25] A.M. Waxman and K. Wohn. Contour evolution, neighborhood deformation and global image flow: Planar surfaces in motion. *International Journal of Robotics Research*, 4(3):95–108, Fall 1985.

2

An Estimation Theoretic Perspective on Image Processing and the Calculation of Optical Flow

T. M. Chin

RSMAS, U. of Miami, Miami FL, USA

M. R. Luettgen, W. C. Karl, A. S. Willsky

M.I.T., Cambridge MA, USA

2.1 Introduction

Many problems of image processing and image sequence analysis involve both great computational complexity and the accommodation of noise and uncertainty through the indirect observation of quantities of interest. In this chapter we describe several aspects of an estimation theoretic approach to such problems. The vehicle for our development is the estimation of the apparent velocity field of a sequence of images. This apparent velocity field, known as the optical flow, appears as an important quantity in both the qualitative and quantitative analysis of image sequences. For example, knowledge of the optical flow is used in the detection of object boundaries and the segmentation of visual scenes [1, 2], the derivation of 3-D motion and structure [3, 4], and the compression of image sequences for efficient transmission [5, 6].

We take an estimation-theoretic perspective to the computation of optical flow, using and extending the formulation of Rougee et al. [7, 8] in both the temporal and spatial directions. In particular, we use model-based interpretations of the various components arising in the estimation theoretic setting to allow us to develop novel extensions to existing approaches. First, we consider the imposition of a *temporal coherence* to the flow obtained by modeling the evolution of the vector optical flow process with a linear state equation and then applying a recursive Kalman filter to the observations obtained from the image sequence. The classical (Horn and Schunck [9]) formulation of the optical flow estimation problem contains no such formal requirement of temporal coherence. The inclusion of such a constraint allows the reliable and robust estimation of optical flow under conditions difficult for the classical approach. For example, in situations where a single image pair contains insufficient information to recover the flow field due to the"aperture problem," the integration of observations over a longer time frame can yield reasonable results.

Applications of Kalman filtering to various formulations of optical flow estimation [10, 11] as well as to other low-level reconstruction problems in computational vision [12] have been proposed. In these previous approaches, however, the apparently computationally daunting task of implementing the Kalman filtering equations, and in particular the error covariance equations, on even moderately-sized images resulted in the use of drastically simplified and suboptimal filter specifications. Specifically, the uncertainty in the dynamic model for the time-varying unknown field, and hence the uncertainty in the estimate itself, is not formally represented or properly propagated in these approaches. In an exact implementation of a Kalman filter, such uncertainty, as captured in the estimation error covariance matrix, is propagated along with the estimate itself [13, 14, 15] and allows for the optimal fusing of the current estimate with new observations. The filtering algorithm presented in this paper employs a more systematic and rational approximation of the Kalman filter than those previously reported. This approximation is based on the propagation of approximate local *models* of the estimation error covariance. These results provide, to our knowledge, the first implementation of the *complete* Kalman filtering equations for space-time problems of this scale, and the only example of successful, near optimal, propagation of covariance matrices of this size. The mathematical details of our approximation techniques can be found in [16] in the more general context of low-level visual reconstruction.

Second, we use the observation that both the single and multi-frame problems can be formulated as spatial estimation problems, wherein sets of observations are fused with prior spatial field models, to motivate the use of a recently developed class of *multiscale* statistical models in their solution. What makes these multiscale field models especially interesting is 1) that there exist extremely efficient, multigrid-type estimation algorithms based on them and 2) that a large number of degrees of freedom exist in their specification, allowing them to approximate a wide range of different flows, including, as least conceptually, any Markov Random Field based flow. Together, these qualities imply that the utilization of such multiscale spatial models for spatial estimation problems, and in particular for the optical flow problem, provides a flexible yet *extremely* efficient estimation framework. Preliminary examples of our results are provided showing factors of 10-100 computational improvement over conventional methods. Finally, such models provide multiscale representations of the flow field and, though we have not used it here, also provide the possibility of optimal integration of *multiscale measurements*.

In this chapter we focus on a particular image processing problem, namely the computation of optical flow. However, the model-based approaches used here are more generally applicable to the wide range of space-time estimation problems arising in image sequence processing.

2.2 Optical Flow Estimation

2.2.1 Single-frame formulation

The 2-D vector field of the apparent motion of brightness patterns in an image is referred to as the optical flow [9]. One commonly used way to obtain information about the optical flow field $x(z_1, z_2, t) \equiv [dz_1/dt,\ dz_2/dt]^T$ at a given point in space (z_1, z_2) and time t was presented by Horn and Schunck in [9]. This approach is based on the assumption that changes in scene brightness in the image sequence are due only to motion. This assumption leads to the so called *brightness constraint equation* [9]:

$$0 = \frac{d}{dt}E(z_1, z_2, t) = \frac{\partial}{\partial t}E(z_1, z_2, t) + \nabla E(z_1, z_2, t) \cdot x(z_1, z_2, t) \tag{2.1}$$

where $E(z_1, z_2, t)$ is the image intensity as a function of time and space and $\nabla E = [\partial E/\partial z_1,\ \partial E/\partial z_2]$, is the gradient of the image intensity.

The brightness constraint equation (2.1) does not completely specify the flow field since it provides only one linear constraint for the two unknown components of $x(z_1, z_2, t)$ at each point. This is usually referred to as the aperture problem [9]. One way to obtain a unique solution is to *regularize* the problem by imposing an additional *smoothness constraint.* Specifically, one formulates the following optimization problem [9]:

$$\underset{x(z_1, z_2, t)}{\operatorname{argmin}} \iint \nu\ \|\frac{\partial}{\partial t}E(z_1, z_2, t) + \nabla E(z_1, z_2, t) \cdot x\|^2 + \|\nabla x\|^2\ dz_1 dz_2 \tag{2.2}$$

The smoothness constraint is captured by the second term which penalizes large gradients in the optical flow and is necessary to make the formulation mathematically well-posed [17]. This term also represents our prior belief about the flow field, implying that the computed flow should vary smoothly over space. Such *spatial coherence* of the flow vectors reflects the smoothness and stiffness of the object surface in the scene [18]. The constant $\nu(z_1, z_2, t)$ allows one to tradeoff between the relative importance in the cost function of the brightness and smoothness constraint terms.

Before proceeding let us analyze the smoothness constraint in more detail. Note that the penalty associated with the smoothness constraint term in (2.2) is equal to the integral of the squared norm of the field gradient over the image plane. In a one-dimensional context, such a constraint would penalize each of the (one-dimensional) fields in Figure 2.1 equally. Intuitively, the smoothness constraint has a fractal nature, and in fact this can be demonstrated in a much more precise sense, as we show in Section 2.3.1.

2.2.2 Multi-frame formulation

The formulation (2.2) processes the data (i.e. the gradients of the image intensity) a frame at a time, yielding flow estimates independently over time. The

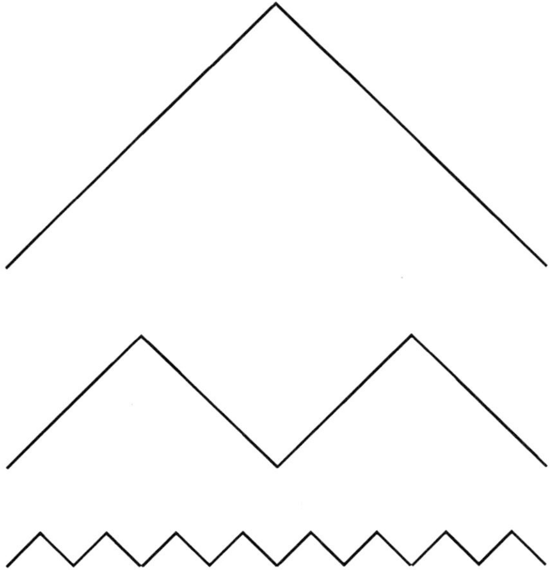

Figure 2.1: Depiction of three fields which are equally favored by the smoothness constraint, illustrating how this penalty provides a fractal prior model for the optical flow.

imposition of *temporal coherence* [19] to the flow field can be considered in addition to the spatial coherence enforced by (2.2) in order to utilize more data for each flow vector estimate. Temporal coherence imposes an inertia condition on the flow field, favoring gradual changes in the optical flow vectors over time. Temporal coherence models of optical flow are applicable to a wide range of motions in natural scenes, as most movements display inertia of some type. To obtain such a *multi-frame* formulation of the optical flow computation problem we use a simple temporal extension of (2.2) [20, 12]. In particular, we find the flow field $\widehat{x}(z_1, z_2, \tau)$ which is the solution to the following problem at time $t = \tau$:

$$\underset{x(z_1,z_2,t)}{\operatorname{argmin}} \int_0^{\tau}\iint \nu \left\| \frac{\partial E}{\partial t} + \nabla E(z_1, z_2, t) \cdot x \right\|^2 + \|\nabla x\|^2 + \rho \left\| \frac{\partial}{\partial t} x \right\|^2 \, dz_1 dz_2 \, dt \quad (2.3)$$

The multi-frame formulation (2.3) is obtained from the single-frame formulation (2.2) by the addition of a quadratic term involving the first order temporal derivative. Note that the full solution to the optimization problem (2.3) leads to a reconstructed space-time field $\widehat{x}(z_1, z_2, t)$ in which the reconstruction at any time takes advantage of all available constraints over the entire time interval $0 \leq t \leq \tau$. We are only interested in the value of the solution to (2.3) at the current time $t = \tau$, corresponding to the best causal or filtered estimate of the flow field. Filtered estimates are desirable in applications where the optical flow needs to be calculated as soon as each frame in the image sequence becomes available; however, obtaining such estimates corresponds to solving a *different*

3-D optimization problem for each τ as τ increases. Such a solution clearly results in a greatly increased computational burden over what is required for the single-frame solution of (2.2), making direct solution of the optimization problem (2.3) prohibitive.

2.3 Discretization and Probabilistic Interpretation

In the first part of this section we present a discrete formulation of the single-frame optical flow problem. In this context we develop two *model-based* interpretations of the single-frame problem which will be central to the results of Sections 2.4 and 2.5. In particular, we illustrate 1) how the smoothness constraint can be interpreted as a prior probabilistic spatial model for the flow field and 2) how the inverse of the covariance of the field estimate may be naturally interpreted as another spatial model, this time for the posterior estimation error of the field. The second part of the section shows how we may model the temporal coherence constraint of (2.3) by a discrete dynamic equation, which may then be coupled with a set of observation equations obtained from the single-frame case to yield an equivalent state estimation problem.

2.3.1 Single-frame case

In practice, brightness measurements are only available over a discrete set of points in space and time. Thus, the temporal and spatial derivative terms must be approximated with finite differences, and the optical flow is only estimated on a discrete space-time grid. There are a number of important issues which arise due to the discretization which we do not discuss here; we refer the reader to [21] for a detailed treatment. We will assume here that we have normalized the space-time coordinates so that the optical flow is to be estimated on the set of integers $(z_1, z_2) \in \{(i,j)|i,j \in \{1,\cdots,2^M\}\}$, where M is also an integer, so the total number of image points is $N \equiv 4^M$. The assumption that the grid is square and that the number of rows is equal to a power of two makes the development of Section 2.5 easier, but is not essential. Now at grid point (i,j) at time t let us define $y(i,j) \equiv (-\partial E(i,j,t)/\partial t)$ to be the measured temporal brightness derivative, $x(i,j)$ to be the desired optical flow vector, and $C(i,j) = \nabla E(i,j,t)$ to be the spatial gradient of the image brightness (also measured). Then we may write the brightness constraint (2.1) at the point (i,j) at time t as:

$$y(i,j) \quad = \quad C(i,j)\,x(i,j) \tag{2.4}$$

The brightness constraints (2.4) at all grid points can now be grouped into one large set of linear equations to concisely capture the optical flow information contained in the image sequence. Let $\mathbf{x}(t)$ be a vector containing the optical flow values $x(i,j)$ at all the grid points at time t (using, say, a lexicographic ordering) and $\mathbf{y}(t)$ be the associated vector of the samples $y(i,j)$. Similarly, let $\mathbf{C}(t)$ and $\mathbf{W}(t)$ be block diagonal matrices whose diagonal elements are the samples $C(i,j)$ and $\nu(i,j)$, respectively, taken in the same order at time t. A discrete version of the single-frame formulation (2.2) is then given by:

$$\widehat{\mathbf{x}}_{SC}(t) = \underset{\mathbf{x}(t)}{\text{argmin}} \left\{ \|\mathbf{y}(t) - \mathbf{C}(t)\mathbf{x}(t)\|^2_{\mathbf{W}(t)} + \|\mathbf{S}\mathbf{x}(t)\|^2_I \right\} \tag{2.5}$$

where $\|\mathbf{x}\|^2_{\mathbf{W}}$ denotes the weighted norm $\mathbf{x}^T\mathbf{W}\mathbf{x}$, I represents the identity matrix, and $\mathbf{S}$ is the matrix first-order spatial difference operator. Note that the discrete nature of the problem alluded to above implies that we must actually approximate the samples $-\partial E(i,j,t)/\partial t$ and $\nabla E(i,j,t)$, and thus $\mathbf{C}$ and $\mathbf{y}$ by finite differences. The spatially varying entries of $\mathbf{W}$ can actually be used to reflect our confidence in these approximations. We refer the reader to [21] for further details.

Spatial models

An estimation-theoretic formulation of the optimization problem in (2.5) can now be developed, and we will use it to show that the *statistically optimal* estimate of the optical flow, given a particular set of measurements, is identical to the smoothness constraint solution given in (2.5). Specifically, solving the quadratic minimization problem (2.5) is equivalent to solving a maximum likelihood (ML) estimation problem [15] for $\mathbf{x}(t)$ with the following *observation equations*:

$$\mathbf{y}(t) = \mathbf{C}(t)\mathbf{x}(t) + \mathbf{r}_1(t) \qquad \mathbf{r}_1(t) \sim \left(\mathbf{0}, \mathbf{W}^{-1}(t)\right) \tag{2.6}$$

$$0 = \mathbf{S}\,\mathbf{x}(t) + \mathbf{r}_2(t) \qquad \mathbf{r}_2(t) \sim (\mathbf{0}, I) \tag{2.7}$$

where we have used the notation $\mathbf{x} \sim (\mathbf{m}, \mathbf{H})$ to denote a Gaussian random vector $\mathbf{x}$ whose mean and covariance are $\mathbf{m}$ and $\mathbf{H}$, respectively, so $\mathbf{r}(t) \equiv [\mathbf{r}_1(t), \mathbf{r}_2(t)]^T$ is a zero-mean Gaussian random noise process. Thus, the maximum likelihood problem formulation results in the same solution as the smoothness constraint formulation when $\mathbf{S}$ is used to define an additional set of noisy measurements.

By formulating the problem in this estimation-theoretic framework, we can use (2.7) to interpret the smoothness constraint as a prior probabilistic spatial *model* for the flow field. Specifically, we can rewrite (2.7) as:

$$\mathbf{S}\mathbf{x}(t) = -\mathbf{r}_2(t) \tag{2.8}$$

Recalling that $\mathbf{S}$ is an approximation to the gradient operator, we see that (2.8) is nothing more than a spatial difference equation model for $\mathbf{x}(t)$ driven by the spatial white noise field $\mathbf{r}_2(t)$. In particular, this prior model represents the optical flow field as composed of independent, two-dimensional Brownian motions[1] [7, 8]. Then, the statistically optimal estimate of the flow field, given the measurements (2.6) and the Brownian motion prior model, is the same as the optical flow estimate given by (2.5). The estimation-theoretic interpretation

[1]More precisely, to avoid biasing the optical flow estimates towards zero, we only assume that the *gradients* of the optical flow field components are equal to the *gradients* of the Brownian motion processes. This avoids placing a constraint on the DC (i.e. average) value of the optical flow and focuses only on imposing a preference for smoothness in the flow.

simply allows us to interpret the smoothness constraint as a Brownian motion model. In one-dimension, Brownian motion is a statistically self-similar, fractal process with a $1/f^2$ generalized spectrum [22], and for this reason the smoothness constraint is often referred to as a "fractal prior" [12]. We will return to this interpretation in Section 2.5 where we discuss a multiscale modeling approach to the single frame problem. In particular, we will replace the prior model (2.8) by a similar but multiscale prior model, which leads to dramatic computational savings.

Next, let us consider another model based interpretation of the single-frame problem (2.5) that will be useful in treating the multi-frame problem. The ML estimate for the optical flow, $\hat{\mathbf{x}}(t)$, based on the measurements (2.6),(2.7) is obtained as the solution of the following inverse problem:

$$\left(\mathbf{C}^T(t)\mathbf{W}(t)\mathbf{C}(t) + \mathbf{S}^T\mathbf{S}\right)\hat{\mathbf{x}}(t) = \mathbf{C}^T(t)\mathbf{W}(t)\mathbf{y}(t) \tag{2.9}$$

The equations in (2.9) represent a discrete version of the coupled Poisson equations of the Horn and Schunck formulation. The matrix operator

$$\mathbf{L}(t) = (\mathbf{C}^T(t)\mathbf{W}(t)\mathbf{C}(t) + \mathbf{S}^T\mathbf{S}) \tag{2.10}$$

on the left hand side of (2.9) has a sparse, *nearest neighbor* (a nested block tri-diagonal) structure [23], whose sparseness enables us to use efficient iterative procedures, such as multigrid methods [24], in the solution of (2.9). Also, this sparse matrix corresponds to the *information matrix* (the inverse of the covariance matrix) associated with the posterior estimation error $\mathbf{d}(t) \equiv \mathbf{x}(t) - \hat{\mathbf{x}}(t)$. In particular $\mathbf{L}(t)$ can naturally be considered to specify an implicit Markov Random Field model for the estimation error process $\mathbf{d}(t)$ of the following form:

$$\mathbf{L}(t)\,\mathbf{d}(t) = \zeta(t), \qquad \zeta(t) \sim (\,\mathbf{0}, \mathbf{L}(t)\,) \tag{2.11}$$

The nearest neighbor structure of $\mathbf{L}(t)$ in (2.11) or (2.9) reflects a corresponding local structure to the statistical model for the estimated field error covariance. We will use this observation in Section 2.4 to develop tractable yet near optimal filtering algorithms.

2.3.2 Multi-frame case

Now we consider the multi-frame extension of the single-frame formulation given in (2.6),(2.7). The continuous optimization problem (2.3) can be considered to be an *optimal smoothing* problem based on the following temporal, linear Gauss-Markov dynamic system for $\mathbf{x}(t)$ [16]:

$$\frac{\partial}{\partial t}x(z_1, z_2, t) = q(t) \tag{2.12}$$

where $q(t)$ is a Gaussian white noise process of zero mean and intensity ρ^{-1}. For such an optimal smoothing problem, *two-filter* methods (i.e. obtained by running a Kalman filter in each of the causal and anti-causal directions) are applicable [7]. In general we wish to compute only the most recent estimate $\hat{\mathbf{x}}(z_1, z_2, \tau)$

from (2.3) for each $\tau \geq 0$. Such an estimate can be obtained by a single causal Kalman filter. Specifically, a *discrete* version of this multi-frame problem can be formulated as a state estimation problem for the dynamic system whose dynamic equation is

$$\mathbf{x}(t) = \mathbf{x}(t-1) + \mathbf{q}(t), \quad \mathbf{q}(t) \sim \left(\mathbf{0}, \rho^{-1} I \right) \tag{2.13}$$

coupled with the observations given by (2.6),(2.7). The process noise $\mathbf{q}(t)$ is uncorrelated over time and captures the uncertainty in the dynamic model (2.13). This Gauss-Markov dynamic model, a discrete version of (2.12), indicates that the optical flow evolves in time as the accumulation of a random perturbation at each time frame. While we will be concerned with temporal dynamics of the form (2.13), naturally more complicated dynamic models, corresponding to different temporal coherence terms in (2.3), could be used.

2.4 Sequential Multi-Frame Estimation

In this section we consider state estimation for the dynamic system represented by (2.13),(2.6),(2.7). Conceptually, we may use well-developed optimal sequential estimation algorithms, such as the Kalman filter and its variants, for solution of this multi-frame optical flow estimation problem. One such algorithm, that will prove convenient for us, is the following implementation of the *information form* [13, 15] of the Kalman filter [16]:

- prediction stage

$$\overline{\mathbf{L}}(t) = \rho I - \rho^2 \left(\widehat{\mathbf{L}}(t-1) + \rho I \right)^{-1} \tag{2.14}$$

$$\overline{\mathbf{x}}(t) = \widehat{\mathbf{x}}(t-1) \tag{2.15}$$

$$\overline{\mathbf{z}}(t) = \overline{\mathbf{L}}(t)\overline{\mathbf{x}}(t) \tag{2.16}$$

- update stage

$$\widehat{\mathbf{L}}(t) = \overline{\mathbf{L}}(t) + \mathbf{C}^T(t)\mathbf{W}(t)\mathbf{C}(t) + \mathbf{S}^T\mathbf{S} \tag{2.17}$$

$$\widehat{\mathbf{z}}(t) = \overline{\mathbf{z}}(t) + \mathbf{C}^T(t)\mathbf{W}(t)\mathbf{y}(t) \tag{2.18}$$

$$\widehat{\mathbf{L}}(t)\widehat{\mathbf{x}}(t) = \widehat{\mathbf{z}}(t) \tag{2.19}$$

where $\overline{\mathbf{x}}(t)$ is the one-step predicted estimate and $\widehat{\mathbf{x}}(t)$ is the updated estimate using the new data available at time t. Also, $\overline{\mathbf{L}}(t)$ and $\widehat{\mathbf{L}}(t)$ denote the predicted and updated information matrices, respectively. Note that the updated estimate $\widehat{\mathbf{x}}(t)$ in (2.19) is specified implicitly, as for the single-frame case (2.9).

2.4.1 Suboptimal Kalman filtering

The number of pixels, N, in a frame of a typical image sequence is on the order of 10^4 to 10^6. Such a large number of points makes direct implementation of the optimal information Kalman filter (2.14)–(2.19) impractical as the associated information matrices $\overline{\mathbf{L}}(t)$ and $\widehat{\mathbf{L}}(t)$ of the optimal filter will have on the order of 10^8 to 10^{12} elements. The storage and manipulation of such large matrices is clearly prohibitive, necessitating the use of a suboptimal method. The sub-optimal filtering algorithm presented below employs a systematic and rational approximation of Kalman filter, which is based on the propagation of approximate local *models* of the estimation error covariance, as discussed in connection with (2.11).

To develop our sub-optimal filter, consider the set of equations (2.14)-(2.19). First consider the update stage of the Kalman filter. If $\overline{\mathbf{L}}(t)$ possesses a sparse and banded nearest neighbor structure, as was true for the single-frame problem, then (2.17) will preserve this structure in $\widehat{\mathbf{L}}(t)$ since, as we pointed out in connection with (2.9), $\mathbf{C}^T(t)\mathbf{W}(t)\mathbf{C}(t) + \mathbf{S}^T\mathbf{S}$ also possesses this structure. In particular, if this is the case, then (2.19) may still be solved efficiently for the updated estimate $\widehat{\mathbf{x}}(t)$, and in fact this step would have *exactly* the same computational complexity as in the single-frame case. Thus, we desire to preserve such a sparse and banded structure in $\overline{\mathbf{L}}(t)$.

Now consider the prediction stage. Unfortunately, even if $\widehat{\mathbf{L}}(t-1)$ in (2.14) is initially sparse and banded, the predicted information matrix $\overline{\mathbf{L}}(t)$ will not be due to the matrix inverse on the right hand side of this equation. In addition, finding the inverse of this matrix is a prohibitively complex procedure. What we desire in the present framework, then, is a sparse and banded *approximation* to $\overline{\mathbf{L}}(t)$ that may be efficiently computed.

As detailed in [16, 21], such an approximation may indeed be obtained by expanding the matrix inverse on the right hand of (2.14) in a series as follows:

$$\overline{\mathbf{L}}(t) \quad = \quad \rho I - \rho^2(\Omega^{-1} - \Omega^{-1}\Delta\Omega^{-1} + \Omega^{-1}\Delta\Omega^{-1}\Delta\Omega^{-1} - \cdots) \qquad (2.20)$$

where Ω is a block diagonal matrix whose 2×2 diagonal blocks are identical to the corresponding diagonal blocks of the matrix $\widehat{\mathbf{L}}(t-1) + \rho I$ while $\Delta = \widehat{\mathbf{L}}(t-1) + \rho I - \Omega$ is given by the remaining off-diagonal part of $\widehat{\mathbf{L}}(t-1) + \rho I$. Note that Ω^{-1} is block diagonal. The series (2.20) may now be truncated to any desired number of terms to obtain an approximation to the exact expression of the desired level of accuracy. The more terms are kept, the less sparse and banded the approximation will become. Thus, there is a tradeoff between accuracy and computational efficiency. Our experience has shown that retaining only the first two terms yields excellent results. In particular, we obtain our near-optimal filter by replacing the optimal prediction step (2.14) by the following two-term approximation:

$$\overline{\mathbf{L}}(t) \quad = \quad \rho I - \rho^2(\Omega^{-1} - \Omega^{-1}\Delta\Omega^{-1}) \qquad (2.21)$$

Unlike (2.14), the suboptimal prediction step (2.21) does indeed preserve the desired nearest neighbor structure in the (approximated) information matrix $\overline{\mathbf{L}}(t)$.

It can be verified straightforwardly that propagating the information matrix in the approximate filter as in (2.17) and (2.21) costs only $O(N)$ flops per frame and has a local, modular computational structure suitable for parallel implementation. Throughout the filtering procedure, the approximated information matrices maintain the nearest neighbor structure and have only $O(N)$ non-zero elements. Thus, the approximate filter has significant computational and storage advantages over the optimal Kalman filter, which normally requires $O(N^2)$ storage elements and $O(N^3)$ flops per frame of data.

A useful way to understand our approximation is provided by an examination of the update stage of the Kalman filter. In this part of the filter we are fusing the information from the previous prediction stage, as captured by $\overline{\mathbf{L}}(t)$ and $\overline{\mathbf{z}}(t)$ (or equivalently $\overline{\mathbf{x}}(t)$), with the new observation. In particular, $\overline{\mathbf{L}}(t)$ can naturally be thought of as specifying a prior model for the error $\mathbf{e}(t) \equiv \mathbf{x}(t) - \overline{\mathbf{x}}(t)$ in the current estimate of the following form:

$$\overline{\mathbf{L}}(t)\,\mathbf{e}(t) = \zeta(t), \qquad \zeta(t) \sim \left(\mathbf{0}, \overline{\mathbf{L}}(t)\right) \tag{2.22}$$

which is just the counterpart of (2.11) for the dynamic problem. This model is then combined with the new observation to produce the best estimate $\widehat{\mathbf{e}}(t)$ of this error. The updated estimate $\widehat{\mathbf{x}}(t)$ in (2.19) is then equal to $\overline{\mathbf{x}}(t) + \widehat{\mathbf{e}}(t)$. The update stage is thus just a static spatial estimation problem, where (2.22) represents a prior model just before the inclusion of new data. That is, by writing the observation equations (2.6),(2.7) concisely as $\mathbf{g}(t) = \mathbf{H}(t)\mathbf{x}(t) + \mathbf{r}(t)$, where $\mathbf{g}(t) \equiv [\mathbf{y}(t)^T,\ \mathbf{0}^T]^T$, $\mathbf{H}(t) \equiv [\mathbf{C}(t)^T,\ \mathbf{S}(t)^T]^T$, and $\mathbf{r} = [\mathbf{r}_1^T,\ \mathbf{r}_2^T]^T$, the estimate $\widehat{\mathbf{e}}(t)$ can be obtained by solving the following static spatial estimation problem:

$$\begin{bmatrix} \mathbf{0} \\ \mathbf{g}(t) - \mathbf{H}(t)\overline{\mathbf{x}}(t) \end{bmatrix} = \begin{bmatrix} \overline{\mathbf{L}}(t) \\ \mathbf{H}(t) \end{bmatrix} \mathbf{e}(t) + \begin{bmatrix} -\zeta(t) \\ \mathbf{r}(t) \end{bmatrix}, \tag{2.23}$$

which is statistically equivalent to obtaining the updated estimate $\widehat{\mathbf{x}}(t)$ of the unknown $\mathbf{x}(t)$ given the prediction $\overline{\mathbf{x}}(t)$ and observation $\mathbf{g}(t)$. Since the implicit model is specified by $\overline{\mathbf{L}}(t)$, our approximation of this matrix by a sparse matrix of the given nearest neighbor structure in (2.21) corresponds naturally to the specification of an approximate, *reduced-order* model for the spatial error process. In particular, this approximation may be viewed as the imposition of a Markov Random Field structure of fixed spatial extent on the flow field estimation-error [16]. Our approximation thus has a rational basis in estimation-theoretic considerations.

2.4.2 Numerical experiments

We demonstrate the beneficial effects of the temporal coherence constraint, formulated as the dynamic model (2.13), and the efficacy of our near-optimal filter for optical flow estimation by numerical example in this section. Recall that, for images of realistic dimension, such as we consider here, exact implementation of the optimal Kalman filtering equations is impossible and thus we apply the temporal coherence constraint (2.13) via our suboptimal filter of Section 2.4.1. A detailed comparison of the suboptimal and true optimal filters demonstrating the near-optimality of our approximation can be found in [16]. Here, a synthetic image sequence of a moving brightness pattern is processed by various multi-frame and single-frame optical flow estimation methods, and the improvements gained by using the particular temporal coherence constraint (2.13), as implemented by the filter we presented in Section 2.4.1, are compared to the conventional methods. Specifically, the following two methods are considered:

- **SF** (Single Frame)
  This method is a discrete version of the single-frame computational approach proposed by Horn and Schunck [9]. Each frame of optical flow is computed independently, i.e., without any provision for temporal integration of data, by solving the inversion problem (2.9) for $\hat{\mathbf{x}}(t)$.

- **TCS** (Temporal Coherence, Suboptimally computed)
  This method is the suboptimal but computationally efficient version of the optimal Kalman filter; as described in Section 2.4.1, the prediction step (2.14) of the Kalman filter is approximated as (2.21).

Variants of these methods arise in different computational environments. Specifically, the inversion steps (2.9) for **SF** and (2.19) for **TCS** can be implemented by one of the following computational procedures, leading to variations in the algorithms above:

- **ic** (iterative inversion, iterations to convergence)
  In practice, the inversion problems are solved iteratively. We use Gauss-Seidel iterations in the experiments here. Needless to say, this iterative solution will converge to the true solution in the limit.

- **is** (iterative inversion, single iteration)
  In time sequential processing, it is natural to initialize the iterative inversion at time t with the estimate obtained at time $t-1$, providing a reasonably good estimate for time t even before the first iteration. By slightly "updating" this initial guess with a single (or a small number of) Gauss-Seidel iteration(s) at the present time, a fairly accurate estimate of the flow field can emerge after continuing the process over several time frames [9], although such estimates are suboptimal in the statistical sense.

In this section, each computational method is made explicit by the name of its main algorithm suffixed by the name of the variation, e.g., **TCS-ic**, **SF-is**, etc. Also, in each experiment, the initial frame of optical flow estimate is computed identically by the **SF-ic** method for every participating computational method in order to highlight the differences in the temporal effects of each method.

The method **SF-is** deserves special attention. This method is the approach to multi-frame optical flow estimation suggested by Horn and Schunck in [9]. It performs only one Gauss-Seidel iteration for the inverse problem (2.9) at each t but uses the estimate from the *previous* frame, $\hat{\mathbf{x}}(t-1)$, to initialize the current iteration. Unlike the **SF-ic** method, therefore, this method *does* have some provision for propagating the estimates temporally. Note that if, instead of only a single Gauss-Seidel step, the iterations are allowed to converge for each frame of data, the resulting flow estimates would have lost all information from the previous frame and become exactly the same as the **SF-ic** estimates. Although the **SF-is** method is *ad hoc* in terms of its temporal integration of data, its ease in implementation is appealing from a practical point of view.

One of the advantages of using a temporal coherence constraint in optical flow estimation is the improvement in the estimates due to the reduced effect of measurement noise through the averaging of the noisy data over time. Another, less obvious advantage, is the temporal accumulation of complementary information regarding the flow vectors. Reconstruction of optical flow using only spatial data integration (i.e., the **SF-is** method) cannot be performed correctly when a complete set of the information necessary to estimate the flow vectors is not contained in each data frame. Specifically, since diversity in the orientations of the measured spatial gradients is necessary to resolve the aperture problem, optical flow computation methods employing only a spatial coherence constraint will have difficulties dealing with cases where all the spatial gradients happen to be oriented in nearly the same direction (including the cases where most of the spatial gradient vectors have small magnitudes). Addition of a temporal coherence constraint can often relieve such difficulties by allowing the use of information from adjacent image frames. The example we give below demonstrates both the fact that the temporal constraint is instrumental in correctly estimating the flow in such cases and that it aids in noise suppression.

Stagnation flow experiment

In this experiment we consider estimation of the motion of a non-rigid body using the **SF-ic** and **TCS-ic** methods as well as the **SF-is** and **TCS-is** methods.

1. *The image sequence.*
   Figure 2.2 shows a flow pattern whose velocity vector at point (z_1, z_2) is given by $(\alpha s_1, -\alpha s_2)$ for $\alpha = 0.1$, where the coordinate origin is at the midpoint of the bottom edge of the figure. This type of flow (for an arbitrary constant α) is useful for a local characterization of *stagnation flow* [25], i.e., the flow of fluid obstructed perpendicularly by a solid object. A sequence of 64×48 images are synthesized based on such a velocity field. Figure 2.3 presents four images from the sequence. Note that the direction of the predominant contrasts in each image changes from mostly vertical in the early frames to mostly horizontal in later frames, implying that some type of temporal coherence constraint is necessary for correct estimation of the flow from this image sequence. We have corrupted the images by adding an independent Gaussian noise with a variance of 9 to each pixel and then requantizing the resulting pixel values to 256 grey levels.

2. *The flow estimates and estimation errors.*
   A 9×9 unit uniform stencil is used to spatially smooth the images before brightness gradients are computed. The computational parameters $\rho = 400$ and $\nu = 40$ have been used. Figure 2.4 shows frame 18 of the estimated flow vectors computed by the **SF-ic** and **TCS-ic** methods. The **SF-ic** method, without any provision for temporal data integration, has completely failed to reconstruct the flow field, while the **TCS-ic** method has performed a reasonable reproduction of the flow in Figure 2.2. The

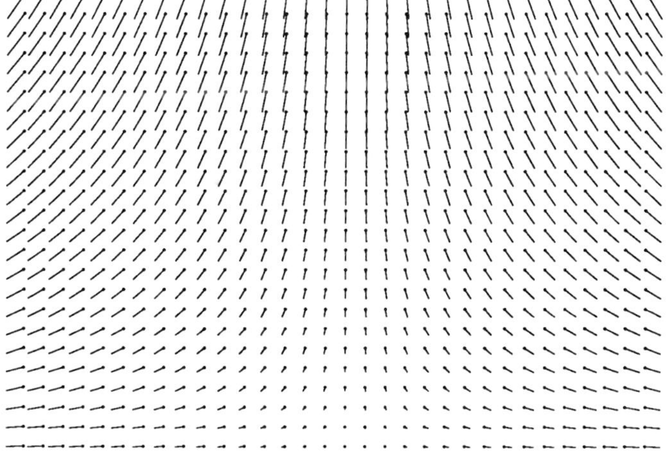

Figure 2.2: The true flow in the Stagnation Flow experiment. Every other flow vector along each axes is shown with a magnification factor of 4 for clarity.

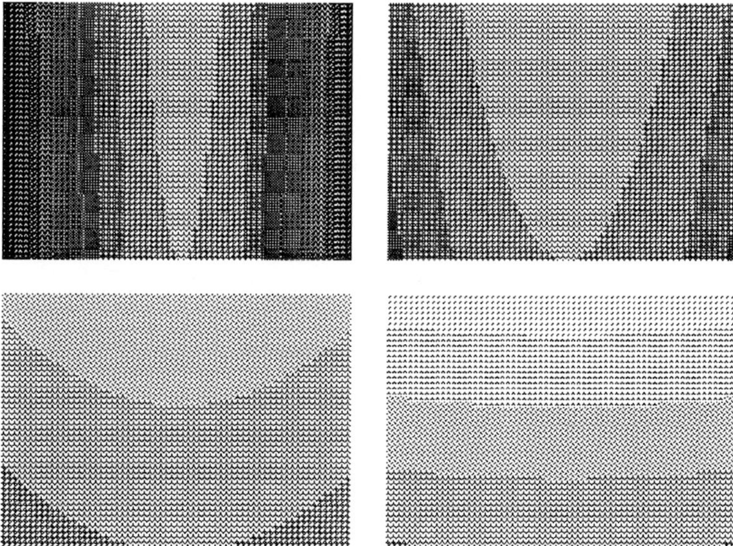

Figure 2.3: The Stagnation Flow image sequence. Frames 0 and 7 (top row) as well as 14 and 21 (bottom row) are shown.

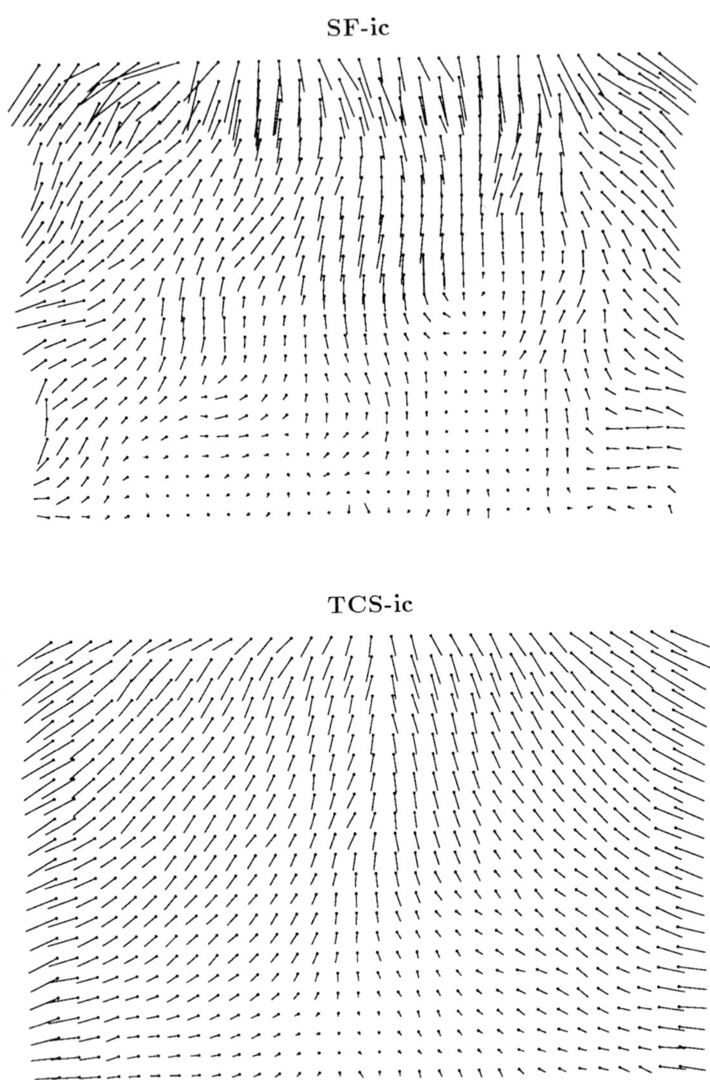

Figure 2.4: The optical flow estimates for frame 18 of the Stagnation Flow sequence by the **SF-ic** and **TCS-ic** methods.

flows computed by the **SF-is** and **TCS-is** are shown on Figure 2.5, which also displays the importance of temporal coherence in reconstruction. Figure 2.6 displays the percent average estimation error for each t,

$$\frac{\|\hat{\mathbf{x}}(t) - \mathbf{x}(t)\|^2}{\|\mathbf{x}(t)\|^2} \times 100, \tag{2.24}$$

where $\mathbf{x}(t)$ is the true flow and $\hat{\mathbf{x}}(t)$ is the estimated flow, for the four methods. These errors are consistent with our previous observations. Again, superior performance of the **TCS**-type methods over the **SF**-type methods is displayed rather dramatically by the error curves.

2.5 Multiscale Model-Based Estimation

One of the major computational bottlenecks of the Kalman filtering algorithm of Section 2.4 is the spatial estimation problem represented by (2.19). In Section 2.4.1 we were able to transform this step of the multi-frame problem back to the level of complexity of the single-frame Horn and Schunck formulation as a result of our reduced-order approximation to the field model (2.21), which resulted in a sparse and banded structure of the matrix $\hat{\mathbf{L}}(t)$. While such an approximation succeeds in making the multi-frame problem tractable while preserving near optimality of the resulting estimates and indeed represents, to our knowledge, the first implementation of the *complete* Kalman filtering equations to problems of this scale, the resulting inverse problem still leads to computationally intensive algorithms. Specifically, the associated set of equations (2.19), while sparse, is extremely large, corresponding to discrete versions of elliptic partial differential equations [9]. The standard approaches to solving such large sets of linear equations, such as the Gauss-Seidel [9, 26], multigrid [24], and successive over-relaxation (SOR) [27] algorithms, are iterative, requiring increasing numbers of iterations (and thus increasing per pixel computational load) as the image size grows.

In this section we examine a novel approach to such large, computationally intensive spatial estimation problems wherein we combine a *multiscale prior model* of the field with a set of field observations. Recall from Section 2.4.1 that we may view the update step of the optimal Kalman filter as such a static spatial estimation problem wherein a prior spatial field model (2.22) is combined with a set of observations $\mathbf{g}(t)$. The use of a multiscale modeling paradigm leads to extremely efficient estimation algorithms which hold the promise of *dramatically* reducing the computation required to solve such problems. For example, the resulting multiscale algorithm is not iterative and in fact requires a fixed number of floating point operations per pixel *independent of image size.* Thus, the computational savings associated with the new approach actually increases as the image size grows. For simplicity and to illustrate the issues involved we will focus only on the single-frame case here. For clarity we will drop the time index t from the notation for the rest of this section, with the understanding that all quantities are taken at this point in time.

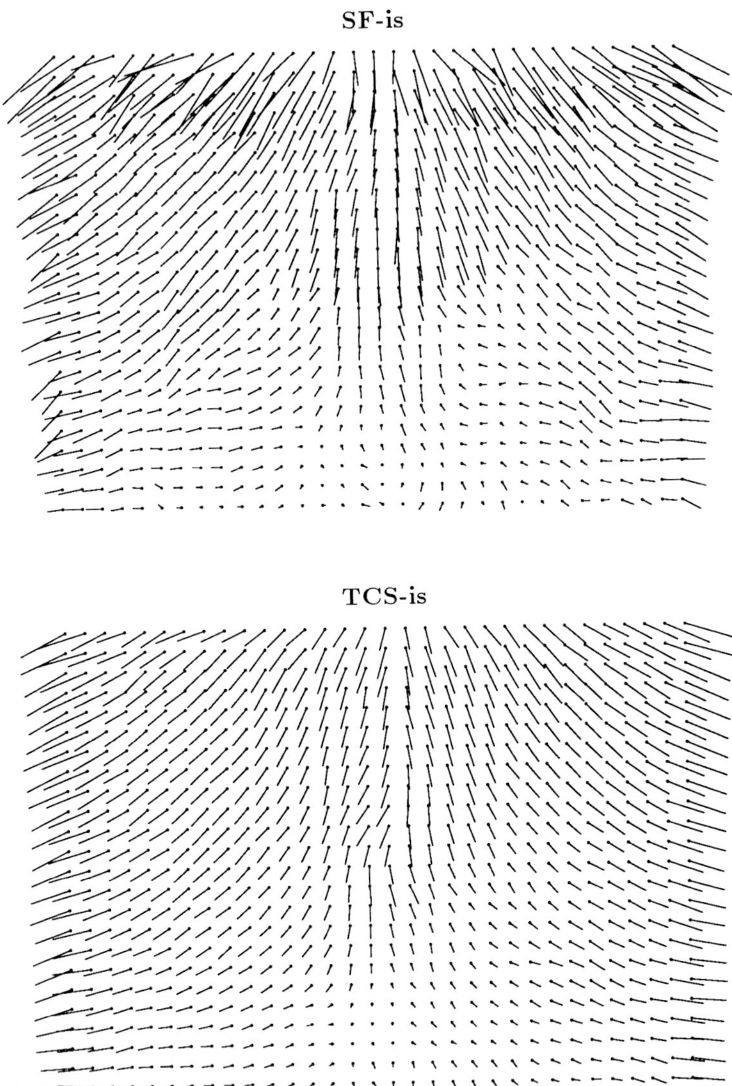

Figure 2.5: The optical flow estimates for the frame 18 of the Stagnation Flow sequence by the **SF-is** method and **TCS-is** method.

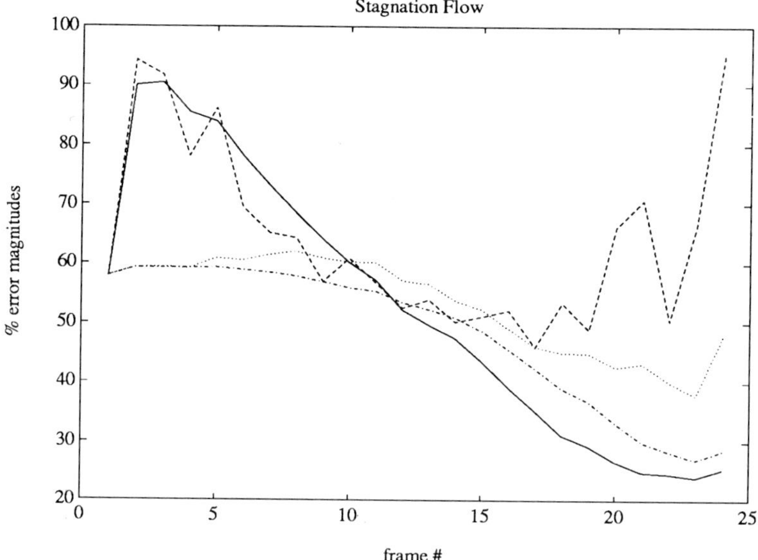

Figure 2.6: The estimation errors by the **TCS-ic** (solid-line), **SF-ic** (dashed-line), **SF-is** (dotted-line), and **TCS-is** (dash-dot line) methods for the Stagnation Flow experiment.

Recall, we argued in Section 2.3.1 that the single-frame optical flow estimates corresponding to (2.5) could be viewed as arising from the combination of the prior statistical spatial model (2.8), corresponding to the smoothness constraint, and the observations (2.6). Now, there is nothing special about the prior model (2.8) associated with the smoothness constraint. Thus we are lead to the idea of using a different class of prior models, capable of capturing a wide range of phenomenon, and in particular of yielding behavior which is similar in nature to that corresponding to the smoothness constraint, but which lead to computationally more attractive problem formulations. That is, we want to change the smoothness constraint term $\mathbf{x}^T\mathbf{S}^T\mathbf{S}\mathbf{x}$ in (2.5) to something similar, say, $\mathbf{x}^T\mathbf{\Lambda}\mathbf{x} \approx \mathbf{x}^T\mathbf{S}^T\mathbf{S}\mathbf{x}$ (where $\mathbf{\Lambda}$ is a symmetric positive semi-definite matrix) such that the resulting optimization problem is easy to solve. If we factor $\mathbf{\Lambda}$ as $\mathbf{\Lambda} = \bar{\mathbf{S}}^T\bar{\mathbf{S}}$ then we can interpret the new constraint as a prior probabilistic model just as we did with the smoothness constraint. In addition, there is a precise interpretation of what we have done as a Bayesian estimation problem. Specifically, if $\mathbf{\Lambda}$ is invertible, then the use of this new constraint in place of the smoothness constraint is equivalent to modeling the flow field probabilistically as $\mathbf{x} \sim (\mathbf{0}, \mathbf{\Lambda}^{-1})$, since in this case the Bayes' least squares estimate of the flow field $\mathbf{x}$, given this prior model and the measurements in (2.6) is provided by:

$$\hat{\mathbf{x}}_{BLSE} = \underset{\mathbf{x}}{\operatorname{argmin}} \left\{ (\mathbf{y} - \mathbf{C}\mathbf{x})^T \mathbf{W} (\mathbf{y} - \mathbf{C}\mathbf{x}) + \mathbf{x}^T \mathbf{\Lambda} \mathbf{x} \right\} \tag{2.25}$$

The normal equations corresponding to (2.25) are given by:

$$(\mathbf{C}^T\mathbf{W}\mathbf{C} + \mathbf{\Lambda})\hat{\mathbf{x}}_{BLSE} = \mathbf{C}^T\mathbf{W}\mathbf{y} \tag{2.26}$$

Comparison of the problem formulations (2.5) and (2.25), or of the normal equations (2.9) and (2.26), makes it apparent how the two problem formulations are related. The choice of the new prior model corresponding to $\mathbf{\Lambda}$ is now clearly at the heart of the problem. We introduce our class of new models next.

2.5.1 A class of multiscale models

The models we utilize to replace the smoothness constraint prior model were recently introduced in [28, 29, 30, 31]. These models represent the flow field at multiple scales, i.e. for a set of scales $m = 0, \ldots, M$, with $m = 0$ being the coarsest scale and $m = M$ the finest scale, we define a set of optical flow fields indexed by scale and space, namely $x_m(i,j)$. At the m-th scale, the field consists of 4^m flow vectors, as illustrated in Figure 2.7, capturing features of the optical flow field discernible at that scale (i.e. finer-resolution features of the field appear only in finer-scale representations). Thus, the coarsest version of the flow field consists of just a single vector, corresponding to the average value of the optical flow over the entire spatial domain of interest, and successively finer versions consist of a geometrically increasing number of vectors. At the finest level, the flow field is represented on a grid with the same resolution as the image brightness data. In particular, $x_M(i,j)$ corresponds to the optical flow vector $x(i,j)$ in (2.4).

Abstractly, we are representing the flow field on the *quadtree structure* illustrated in Figure 2.8. Pyramidal data structures such as the quadtree naturally arise in image processing algorithms which have a multiscale component. For instance, successive filtering and decimation operations lead to images defined on such a hierarchy of grids in the Laplacian pyramid coding algorithm of Burt and Adelson [32] and in the closely related wavelet transform decomposition of images [33]. Also, the multigrid approaches to low level vision problems discussed by Terzopoulos [24] involve relaxation on a similar sequence of grids.

The model we introduce in this section describes in a probabilistic manner how the optical flow field $x(i,j) = x_M(i,j)$ is constructed by adding detail from one scale to the next. Just as the smoothness constraint prior model (2.8) describes probabilistic constraints among values of the optical flow at different spatial locations, our multiscale model describes such constraints among values at different *scales*. That is, our model describes the probabilistic evolution of $x_m(i,j)$ as the scale m evolves from coarse to fine. For notational convenience in describing such models, we denote nodes on the quadtree with a single abstract index s which is associated with the 3-tuple (m,i,j) where, again, m is the scale and (i,j) is a spatial location in the grid at the m-th scale. It is also useful to define an *upward shift operator* $\bar{\gamma}$. In particular, the *parent* of node s is denoted $s\bar{\gamma}$ (see Figure 2.8). We note that the operator $\bar{\gamma}$ is not one-to-one; it is in fact four-to-one since each node will have four "offspring" at the next scale. For instance, if s corresponds to any of the nodes in the upper left quadrant of the

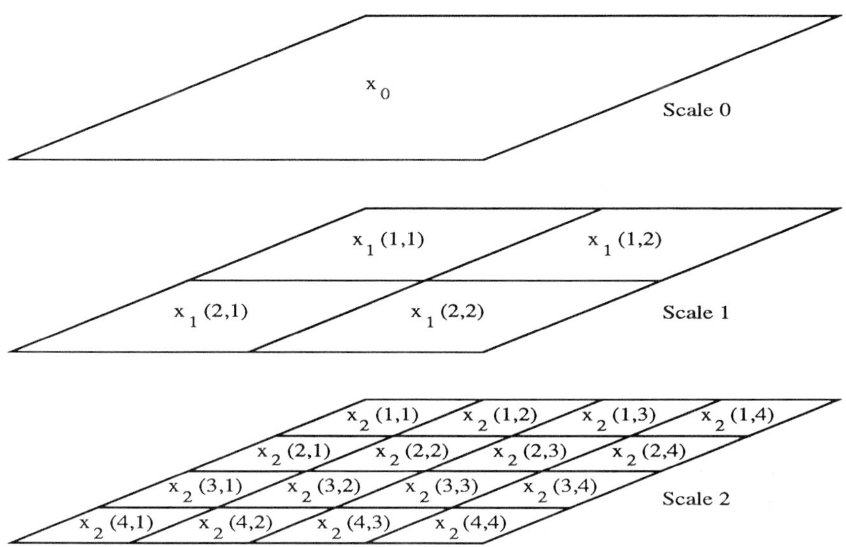

Figure 2.7: The structure of a multiscale optical flow field is depicted. The components of the field are denoted $x_m(i,j)$ where m refers to the scale and the pair (i,j) denotes a particular grid location at a given scale. At the coarsest scale, there is a single flow vector and, more generally, at the m-th scale there are 4^m vectors.

second level grid (see Figure 2.7), i.e. nodes $(2,1,1),(2,2,1),(2,1,2)$ or $(2,2,2)$, then $s\bar{\gamma}$ corresponds to their parent on the first level, namely node $(1,1,1)$.

We are now in a position to define the class of multiscale models which describe the evolution of a multiscale stochastic processes indexed by nodes on the quadtree. Specifically, a stochastic quadtree process $x(s)$ is described recursively by:

$$x(s) \quad = \quad A(s)x(s\bar{\gamma}) + B(s)w(s) \tag{2.27}$$

under the following assumptions:

$$x_0 \quad \sim \quad (\mathbf{0}, P_0) \tag{2.28}$$

$$w(s) \quad \sim \quad (\mathbf{0}, I) \tag{2.29}$$

The vectors $x(s)$ and $w(s)$ are referred to as the state and driving noise terms. The state variable x_0 at the root node of the tree provides an initial condition for the recursion. The driving noise is white in both space and scale, and is uncorrelated with the initial condition. Interpreting each level as a representation of a two-dimensional field, we see that (2.27) describes the evolution of the process from coarse to fine scales. The term $A(s)x(s\bar{\gamma})$ represents interpolation down to the next level, and $B(s)w(s)$ represents higher resolution detail added as the process evolves from one scale to the next. In the application of interest here, $x(s) = x_m(i,j)$, where $s = (m,i,j)$, and thus $A, B \in \Re^{2\times 2}$. Such a model corresponds in essence to a first-order recursion in scale for optical flow.

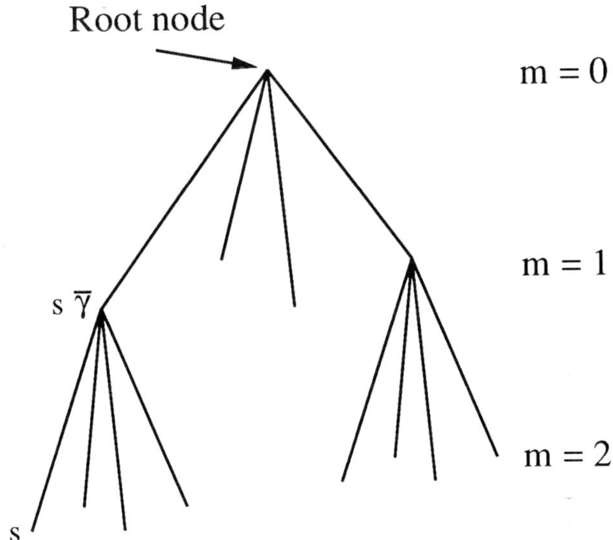

Figure 2.8: Quadtree structure on which the multiscale processes are defined. The abstract index s refers to a node in the quadtree; $s\bar{\gamma}$ refers to the parent of node s.

Measurements of the finest level optical flow field are available from the brightness constraint. In particular, at the grid point (i,j) at the finest level M, we have a measurement equation corresponding to that in (2.4):

$$y(i,j) = C(i,j)\, x_M(i,j) + v(i,j) \quad v(i,j) \sim \Big(0, \nu^{-1}(i,j)\Big) \tag{2.30}$$

where $C(i,j) \in \Re^{1\times 2}$ is the spatial brightness gradient at location (i,j) and the white Gaussian observation noise is assumed to be independent of the initial condition x_0 and the driving noise $w(s)$ in (2.27)-(2.29). Of course, we can group the state variables $x(s)$ at the finest level into a vector $\mathbf{x}_M$ as well as the corresponding measurements $y(s)$ and spatial gradient terms $C(s)$ in the same way as we did to get (2.6):

$$\mathbf{y} = \mathbf{C}\,\mathbf{x}_M + \mathbf{v} \quad \mathbf{v} \sim \Big(\mathbf{0}, \mathbf{W}^{-1}\Big) \tag{2.31}$$

We now have exactly the framework which led to the statement of (2.25) as an alternative to the smoothness constraint formulation (2.5). In particular, the modeling equations (2.27)-(2.29) indicate that at the finest level of the quadtree, the flow field vectors will be a set of jointly Gaussian random variables $\mathbf{x}_M \sim (\mathbf{0}, \mathbf{\Lambda}^{-1})$, where $\mathbf{\Lambda}^{-1}$ is implicitly given by the parameters in (2.27)-(2.29). The Bayes' least squares estimate of $\mathbf{x}_M$ given the measurements in (2.31) and the prior model (2.27)-(2.29) is then given by:

$$\hat{\mathbf{x}}_M = \underset{\mathbf{x}_M}{\operatorname{argmin}}\, (\mathbf{y} - \mathbf{C}\mathbf{x}_M)^T \mathbf{W} (\mathbf{y} - \mathbf{C}\mathbf{x}_M) + \mathbf{x}_M^T \mathbf{\Lambda} \mathbf{x}_M \tag{2.32}$$

The multiscale modeling framework thus provides an alternative to the smoothness constraint formulation of (2.5).

What remains to be done is to specify a model within this class that has characteristics similar to those of the smoothness constraint prior model. In particular, for our multiscale model based on (2.27)–(2.29) to approximate the smoothness constraint prior we would like to choose our model parameters so that we have $\mathbf{S}^T\mathbf{S} \approx \mathbf{\Lambda}$. The observation in Section 2.3.1 that the prior model (2.8) implied by the operator $\mathbf{S}$ in (2.5) corresponds to a Brownian motion "fractal prior" suggests one approach to choosing the model parameters. In particular, the one-dimensional Brownian motion has a $1/f^2$ generalized spectrum [22]. It has been demonstrated that such processes are well approximated by multiscale models such as ours in one dimension if geometrically decreasing powers of noise are added at each level m of the process [30, 34]. In particular, this motivates the choice of $B(s) = b4^{-\mu m(s)}I$ in (2.27), where b and μ are scalar constants. The constant b directly controls the overall noise power in the process. Also, as discussed in [34], the choice of μ controls the power law dependence of the generalized spectrum of the process at the finest resolution as well as the fractal dimension of its sample paths. Specifically, this spectrum has a $1/f^{2\mu}$ dependence. Thus, the choice of $\mu = 1$ would correspond to a Brownian-like fractal process. To achieve greater flexibility in both the modeling and estimation, we allow μ to be a parameter that can be varied. In addition, recall that in the smoothness constraint formulation, $\mathbf{S}^T\mathbf{S}$ was not invertible because of the implicit assumption of infinite prior variance on the DC value of the optical flow field. In our multiscale regularization context, this would correspond to setting the initial covariance P_0 equal to infinity in (2.28). This can be done without difficulty in the estimation algorithms described next, but we have found that it is generally sufficient to simply choose P_0 to be a large multiple of the identity.

We have now specified a class of models which will allow us to approximate the smoothness constraint prior model. The simple multiscale structure of these models leads to very efficient algorithms for computing the optimal estimate of the state given a set of measurements. One of these algorithms, which we refer to as the Multiscale Regularization (MR) algorithm, was developed in [28, 29, 30, 35] for one-dimensional signals, and its extension to images is described in [36].

The MR algorithm computes the Bayes least squares estimate of the state vectors (2.27) given the measurements (2.30) in two steps. The first step is an *upward* or *fine-to-coarse* sweep on the quadtree, which propagates the measurement information in parallel, level by level, from the fine scale nodes up to the root node. This step produces the best estimate at each node given all the data in the subtree under that node. At the top of the tree, one obtains the smoothed estimate of the root node, that is, the estimate of this node based on *all* of the data. The smoothed estimate and associated error covariance at the root node then provide initialization for the next step. This second step is a *downward* or *coarse-to-fine* sweep which propagates the global measurement information now at the root node back down, and throughout the tree. The result at each node is the least squares estimate $\hat{x}^s(s)$ of the state $x(s)$ based on all of the data. The resulting estimates at the finest level of the quadtree provide the solution to (2.32). The resulting algorithm is just a generalization of the Rauch-Tung-Striebel smoothing algorithm [37] *in scale.* The details of the upward and downward sweeps are discussed in greater detail in [30, 35, 36].

2.5.2 Numerical experiments

Here we demonstrate the substantial computational benefit that can be achieved through the use of our multiscale modeling paradigm. To specify the MR algorithm completely we need to choose the parameters of the model. We utilize the following parameterization:

$$x(s) = x(s\bar{\gamma}) + (b4^{-\mu m(s)})w(s) \tag{2.33}$$

$$y(s) = C(s)x(s) + v(s) \tag{2.34}$$

$$w(s) \sim (\mathbf{0}, I) \tag{2.35}$$

$$v(s) \sim (0, \nu^{-1}(s)) \tag{2.36}$$

$$x_0 \sim (\mathbf{0}, pI) \tag{2.37}$$

From (2.33) and (2.35) we see that the two components of the optical flow field are modeled as independent sets of random variables, and that each will have a fractal-like characteristic due to the choice of the driving noise gain $B(s)$ (as discussed in the previous section). We view μ and b as free model parameters which can be varied to control the degree and type of regularization in much the same way that the parameter ν in the smoothness constraint formulation (2.2) is used to tradeoff between the data dependent and regularization terms in the optimization functional.

As discussed previously, the measurements $y(s)$ and measurement matrix $C(s)$ come directly from the image temporal and spatial gradients, which are available at the finest level of the quadtree. In the experiments described below, we use a simple two-image difference to approximate the temporal gradient. The spatial gradient is computed by smoothing the image with a 3×3 Gaussian kernel followed by a central difference approximation. The additive noise variance is given by $\nu^{-1}(s)$. We have found empirically that the choice $\nu^{-1}(s) = max(||C(s)||^2, 10)$ works well. This choice effectively penalizes large spatial gradients, which are likely points of occlusion where the brightness constraint equation will not hold [38]. The parameter p in the prior covariance of the root node was set to $p = 100$. The distribution (2.37) on the root node effectively says that we are modeling the optical flow field components as zero mean random processes. The prior covariance reflects our confidence in this assumption. Since we do not believe that *any* prior assumption on the mean of optical flow field components can be justified, we set the parameter p such that the implied standard deviation is much larger than the sizes of the flow fields we expect to see.

We compare our approach computationally and visually to the the Gauss-Seidel (GS) and successive over-relaxation (SOR) algorithms, which can be used to compute the solution of the smoothness constraint formulation given by (2.5). Straightforward analysis shows that the GS and SOR algorithms require 14 and

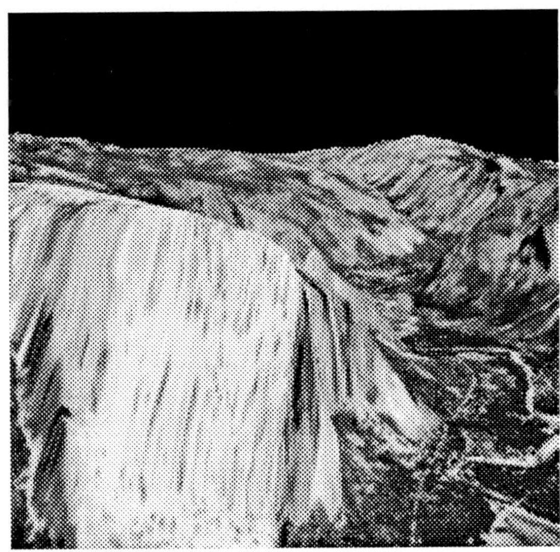

Figure 2.9: First frame of "Yosemite" sequence.

18 floating point operations (flops) per pixel per iteration respectively. The number of iterations required for convergence of the iterative algorithms grows with image size [27]. For reasonable size images (say, 512 × 512), SOR may require on the order of hundreds of iterations to converge, so that the total computation per pixel can be on the order of $10^3 - 10^4$ flops. On the other hand, the MR algorithm requires 76 flops per pixel. Note that the MR algorithm is *not* iterative. Thus, the computational gain associated with the MR algorithm can be on the order of one to two orders of magnitude. Details may be found in [36].

Yosemite sequence experiment

This example is a synthetic 256 × 256 image sequence which simulates the view obtained by flying through the Yosemite Valley[2]. The first image in the sequence is shown in Figure 2.9 along with the actual flow field in Figure 2.10. The flow computed via the MR algorithm is shown in Figure 2.11 and the smoothness constraint solution is shown in Figure 2.12. The smoothness constraint flow estimates required 250 SOR iterations in this example, representing a factor of 60 more computation than the MR estimates. Note the substantial increase over the previous example in the number of iterations required for the SOR algorithm to converge. The number of iterations required for convergence depends on several things, including the parameter ν, the image gradient characteristics and the image size. Theoretical analysis in [27] shows that the SOR algorithm requires

[2]This sequence was synthesized by Lyn Quam of SRI International.

Figure 2.10: Yosemite sequence true optical flow.

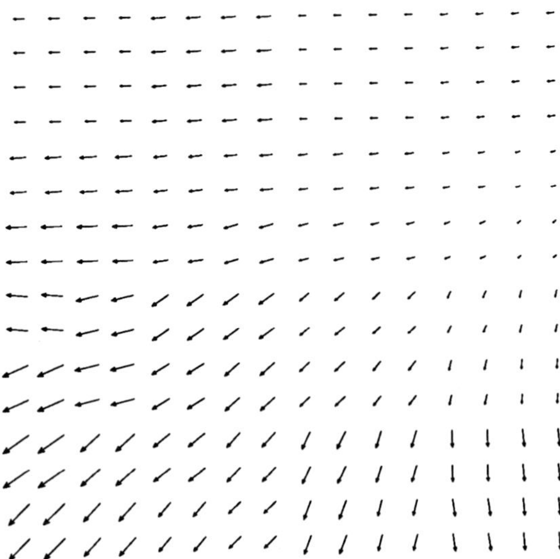

Figure 2.11: MR algorithm flow estimates: $b = 10, \mu = 2.5$.

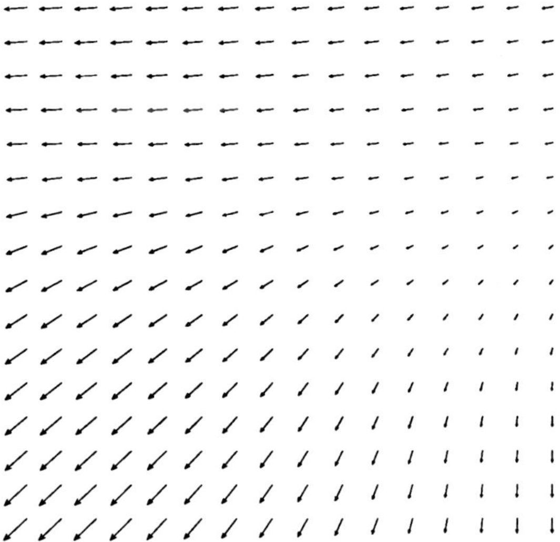

Figure 2.12: SOR algorithm flow estimates: $R = 500^2$, 250 iterations. The SOR algorithm required a factor of 60 more computation in this example.

on the order of n iterations for an $n \times n$ image. Thus, we expect substantially more computational savings as the image size increases.

A root mean square (rms) error comparison of the algorithms is shown in Figure 2.13. As expected, the SOR algorithm is significantly faster than the GS algorithm (they will converge to the same result since they are solving the same partial differential equation). The rms error in the MR flow estimates is depicted as a straight line, since the algorithm is not iterative. Neither of the estimates coincides with the actual optical flow, but they do have comparable rms error as in the previous example. In addition, the figure illustrates the computational advantage of the MR algorithm. In particular, the SOR algorithm is still reducing the rms error in its flow estimates after 300 iterations, at which point the MR algorithm requires a factor of $300/4.2 = 71.4$ less computation.

This image sequence contains a problem often encountered in real images: regions of constant intensity. The problem is that the lack of gradient information in that region implies that the optical flow is not well defined. The smoothness constraint and multiscale prior models provide a means of interpolating out into these regions. The result of this is apparent in the top portion of Figures 2.11 and 2.12. An advantage of the MR formulation is that it accomplishes this extrapolation at an appropriately coarser, and hence computationally simpler, scale.

Note the the MR and SC flow estimates are not identical due to differences in the prior models. If there is particular interest in obtaining the SC solution, the question arises of using the MR solution as an initial guess for the iterative

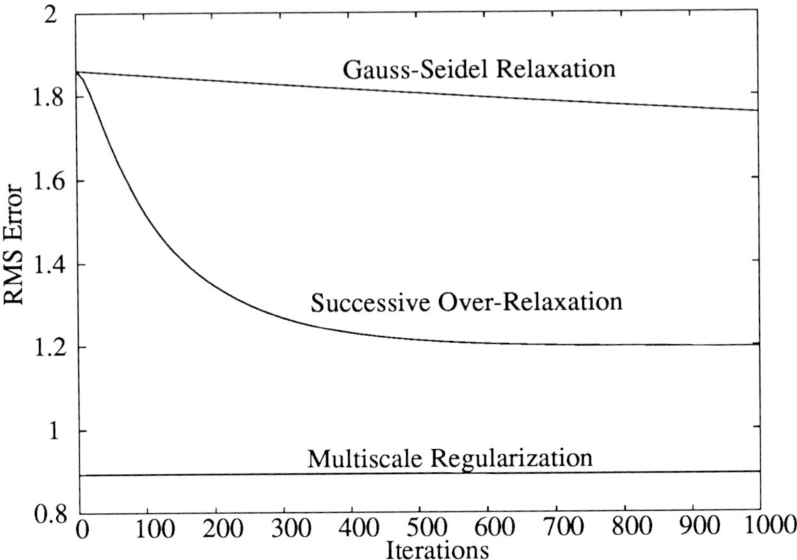

Figure 2.13: Rms Error Comparison of MR, SOR and Gauss-Seidel (GS) algorithm flow estimates for the Yosemite sequence.

algorithms which compute the SC solution. Note that the difference between the SC and MR flow estimates is associated with the non-smooth, high frequency aspects of the MR flow at block edges. It is precisely these high frequency components that are quickly removed by SOR or GS algorithms computing the the smoothness constraint solution and suggests that the MR algorithm would provide an excellent *pre-conditioner* for the iterative algorithms.

2.6 Conclusions

We have taken an estimation-theoretic perspective to image sequence processing using as our vehicle the computation of optical flow. Our results made use of the interpretation of various components of standard formulations of this problem as *statistical models*. First we presented a near-optimal Kalman filter for the estimation of optical flow under a temporal coherence constraint and based on the propagation of approximate local *models* of the estimation error covariance. This filter provides, to our knowledge, the first implementation of the *complete* Kalman filtering equations for space-time problems of this scale, and the only example of successful, near optimal, propagation of covariance matrices of this size.

Next we used the observation that both the single and multi-frame problems can be formulated as spatial estimation problems, wherein sets of observations are fused with prior spatial field models, to motivate the use of a recently developed class of *multiscale* statistical models in their solution. The algorithms arising from this class of models allows *dramatic* speedups in computational speed. For simplicity we concentrated on the single frame case, though we could also use for the static spatial estimation problem occurring in the multi-frame situation. A particularly interesting and open question is how to directly propagate one of the multiscale models used in Section 2.5 in time.

Acknowledgements

This work was partially supported by the Charles Stark Draper Laboratory IR&D Program under Grant DL-H-418524, the Office of Naval Research under Grant N00014-91-J-1004, the National Science Foundation under Grant MIP-9015281, and the Army Research Office under Grant DAAL03-86-K-0171.

References

[1] A. Meygret and M. Thonnat, "Segmentation of optical flow and 3D data for the interpretation of mobile objects," in *Third International Conference on Computer Vision*, pp. 238–245, IEEE Computer Society Press, 1990. Osaka, Japan.

[2] F. Heitz, P. Perez, E. Memin, and P. Bouthemy, "Parallel visual motion analysis using multiscale Markov random fields," in *Proceedings of Workshop on Visual Motion*, IEEE Computer Society Press, 1991. Princeton, NJ.

[3] H. C. Longuet-Higgins and K. Prazdny, "The interpretation of a moving retinal image," *Proceedings of the Royal Society of London B*, vol. 208, pp. 385–397, 1980.

[4] S. Negahdaripour and B. K. Horn, "Direct passive navigation," *IEEE Transactions on Pattern Analysis and Machine Intelligence*, vol. PAMI-9, 1987.

[5] N. Baaziz and C. Labit, "Multigrid motion estimation on pyramidal representations for image sequence coding," Tech. Rep. 572, IRISA, Feb. 1991.

[6] D. Walker and K. Rao, "Improved pel-recursive motion compensation," *IEEE Transactions on Communications*, vol. 32, pp. 1128–1134, 1984.

[7] A. Rougee, B. C. Levy, and A. S. Willsky, "An estimation-based approach to the reconstruction of optical flow," Tech. Rep. LIDS-P-1663, Laboratory for Information and Decision Systems, Massachusetts Institute of Technology, 1987.

[8] A. Rougee, B. Levy, and A. S. Willsky, "Reconstruction of two dimensional velocity fields as a linear estimation problem," in *Proceedings 1st International Conference on Computer Vision*, (London, England), pp. 646–650, 1987.

[9] B. K. P. Horn and B. G. Schunck, "Determining optical flow," *Artificial Intelligence*, vol. 17, pp. 185–203, 1981.

[10] M. J. Black and P. Anandan, "A model for the detection of motion over time," in *Third International Conference on Computer Vision*, pp. 33–37, IEEE Computer Society Press, 1990. Osaka, Japan.

[11] A. Singh, "Incremental estimation of image-flow using a Kalman filter," in *Proceedings of Workshop on Visual Motion*, pp. 36–43, IEEE Computer Society Press, 1991. Princeton, NJ.

[12] R. Szeliski, *Baysian Modeling of Uncertainty in Low-level Vision.* Norwell, Massachuesetts: Kluwer Academic Publishers, 1989.

[13] B. D. O. Anderson and J. B. Moore, *Optimal Filtering.* Englewood Cliffs, N.J.: Prentice-Hall, 1979.

[14] A. Gelb, ed., *Applied Optimal Estimation.* Cambridge, MA: MIT Press, 1974.

[15] F. L. Lewis, *Optimal Estimation.* New York: John Wiley & Sons, 1986.

[16] T. M. Chin, W. C. Karl, and A. S. Willsky, "Sequential filtering for multi-frame visual reconstruction." to appear in *Signal Processing*, Aug. 1992.

[17] M. Bertero, T. Poggio, and V. Torre, "Ill-posed problems in early vision," *Proceedings of the IEEE*, vol. 76, pp. 869–889, 1988.

[18] E. C. Hildreth, "Computations underlying the measurement of visual motion," *Artificial Intelligence*, vol. 23, pp. 309–354, 1984.

[19] N. M. Grzywacz, J. A. Smith, and A. L. Yuille, "A common theoretical framework for visual motion's spatial and temporal coherence," in *Proceedings of Workshop on Visual Motion*, pp. 148–155, IEEE Computer Society Press, 1989. Irvine, CA.

[20] L. H. Matthies, R. Szeliski, and T. Kanade, "Kalman filter-based algorithms for estimating depth from image sequences," *International Journal of Computer Vision*, vol. 3, 1989.

[21] T. M. Chin, *Dynamic Estimation in Computational Vision.* PhD thesis, Massachusetts Institute of Technology, 1991.

[22] B. Mandelbrot and H. V. Ness, "Fractional Brownian motions, fractional noises and applications," *SIAM Review*, vol. 10, pp. 422–436, 1968.

[23] B. C. Levy, M. B. Adams, and A. S. Willsky, "Solution and linear estimation of 2-D nearest-neighbor models," *Proceedings of the IEEE*, vol. 78, pp. 627–641, 1990.

[24] D. Terzopoulos, "Image analysis using multigrid relaxation models," *IEEE Transactions on Pattern Analysis and Machine Intelligence*, vol. PAMI-8, pp. 129–139, 1986.

[25] M. C. Potter and J. F. Foss, *Fluid Mechanics.* Okemos, Michigan: Great Lakes Press, 1982.

[26] G. Strang, *Introduction to Applied Mathematics.* Wellesley, MA: Wellesley-Cambridge Press, 1986.

[27] C.-C. J. Kuo, B. C. Levy, and B. R. Musicus, "A local relaxation method for solving elliptic PDE's on mesh connected arrays," *SIAM J. Sci. Stat. Comput.*, vol. 8, pp. 550–573, 1987.

[28] K. C. Chou, A. S. Willsky, A. Benveniste, and M. Basseville, "Recursive and iterative estimation algorithms for multiresolution stochastic processes," in *Proc. of the IEEE Conference on Decision and Control*, Dec. 1989.

[29] K. C. Chou, *A stochastic modeling approach to multiscale signal processing.* PhD thesis, Massachusetts Institute of Technology, 1991.

[30] K. C. Chou, A. S. Willsky, and A. Benveniste, "Multiscale recursive estimation, data fusion and regularization." submitted to IEEE Transactions on Automatic Control, 1992.

[31] S. C. Clippingdale and R. G. Wilson, "Least squares image estimation on a multiresolution pyramid," in *Proc. of the 1989 International Conference on Acoustics, Speech, and Signal Processing*, 1989.

[32] P. Burt and E. Adelson, "The Laplacian Pyramid as a compact image code," *IEEE Transactions on Communications*, vol. 31, pp. 482–540, 1983.

[33] S. Mallat, "Multi-frequency channel decomposition of images and wavelet models," *IEEE Transactions on Acoustics, Speech, and Signal Processing*, vol. 37, pp. 2091–2110, 1989.

[34] G. Wornell, "A Karhunen-Loeve like expansion for 1/f processes," *IEEE Transactions on Information Theorey*, vol. 36, pp. 859–861, 1990.

[35] K. C. Chou, A. S. Willsky, and R. Nikoukhah, "Multiscale systems, Kalman filters and Riccati equations." submitted to IEEE Transactions on Automatic Control, 1992.

[36] M. R. Luettgen, W. C. Karl, and A. S. Willsky, "Optical flow computation via multiscale regularization." submitted to IEEE Transactions on Image Processing, 1992.

[37] H. E. Rauch, F. Tung, and C. T. Striebel, "Maximum likelihood estimates of linear dynamic systems," *AIAA Journal*, vol. 3, pp. 1445–1450, 1965.

[38] E. Simoncelli, E. Adelson, and D. Heeger, "Probability distributions of optical flow," in *Proceedings of the IEEE Conference on Computer Vision and Pattern Recognition*, (Maui, Hawaii), June 1991.

3

Estimation of 2-D Motion Fields from Image Sequences with Application to Motion-Compensated Processing

E. Dubois and J. Konrad

INRS-Télécommunications, Verdun, Québec, Canada H3E 1H6

3.1 Introduction

In this chapter we are concerned with the estimation of 2-D motion from time-varying images and with the application of the computed motion to image sequence processing. Our goal for motion estimation is to propose a general formulation that incorporates object acceleration, nonlinear motion trajectories, occlusion effects and multichannel (vector) observations. To achieve this objective we use Gibbs-Markov models linked together by the Maximum *A Posteriori* Probability criterion which results in minimization of a multiple-term cost function. The specific applications of motion-compensated processing of image sequences are prediction, noise reduction and spatiotemporal interpolation.

Estimation of motion from dynamic images is a very difficult task due to its *ill-posedness* [4]. Despite this difficulty, however, many approaches to the problem have been proposed in the last dozen years [27],[24],[40]. This activity can certainly be attributed in large measure to the importance of motion in the processing and coding of image sequences. Below we explain why motion is important in these tasks.

Efficient encoding of time-varying images is essential to provide economical use of network or storage facilities in the provision of video services. Image sequences can be compressed by independent coding of each frame (intraframe coding) or by straightforward extension of spatial coding techniques to three dimensions (e.g., 3-D transform coding). However, such approaches ignore the fact that the majority of new information (innovations) in a time-varying image is carried by the motion. The correlation of image intensity or color is very high along the direction of motion. Thus, the knowledge of motion helps in removing significant interimage redundancy, as is the case in predictive or hybrid (predictive/transform or predictive/subband) coding compensated for motion. In fact, these techniques include most algorithms currently used in videoconferencing

[41],[42] and proposed for High Definition Television (HDTV) [21],[60].

Sampling structure conversion and noise reduction are two examples of image sequence processing that can benefit greatly from the knowledge of motion. In the case of conversion, two scenarios are possible. In one situation a missing image must be recovered. Again, due to the high correlation along motion trajectories, motion-compensated interpolation is the most effective tool [53], [15]. In the other situation, only part of an image must be reconstructed. Since some spatial information is available and since motion estimates are occasionally unreliable, methods can be devised which combine "smart" spatial interpolation [50],[18] with motion-compensated interpolation [49]. Noise reduction also benefits from high correlation along motion trajectories. One-dimensional filtering applied along the direction of motion does not alter image features since they are highly correlated, but it attenuates noise since its samples are highly uncorrelated [16],[54].

One remark is in order at this point. Optimal motion fields, in the sense of final image quality, are not necessarily identical for motion-compensated predictive coding and for motion-compensated interpolation. This is due to the fact that the sole task of a predictor is to find the best match of a pixel value given previous images. To find this match any motion trajectory providing minimum prediction error is appropriate. This trajectory does not have to correspond to the real 2-D motion of objects in the image. On the contrary, the task of a temporal interpolator is to calculate a pixel value which is consistent with other pixel values along a motion trajectory so that continuity and smoothness of object motion are preserved. Thus, an estimate of the real 2-D motion is needed. It is a different situation, however, when motion fields need to be transmitted along with the prediction error. Optimal motion fields for prediction require relatively high bit rates since they lack correlation. To reduce the bit rate in this case, joint optimization of the prediction error and of motion correlation can be carried out [56]. Interestingly, since real 2-D motion is usually quite smooth (except for motion discontinuities), it is often estimated by combining a prediction-like error with a smoothness constraint imposed on the estimated motion field. Thus, methods used to estimate 2-D motion for the purpose of temporal interpolation are also applicable to predictive coding with simultaneous transmission of motion data.

The remainder of this chapter is divided into two parts. The first part is devoted to the problem of estimating 2-D motion from image sequences, while the second one treats the motion-compensated processing of dynamic images. The main focus of this contribution is the general formulation of the motion estimation and motion-compensated filtering problems. No experimental results are presented; many can be found in our previous publications cited in the text.

3.2 Estimation of 2-D Motion

In this section we define motion trajectories and other related terms, we formulate the estimation problem and we describe the models used. Then, we

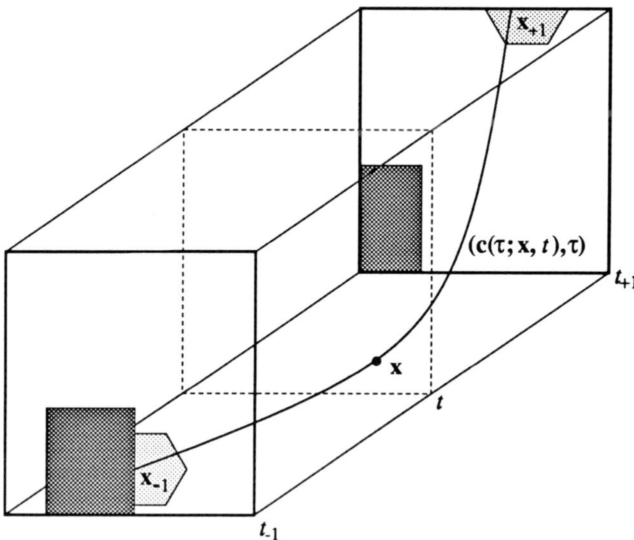

Figure 3.1: Trajectory of a projected scene point in 3-D xyt space; a point is uncovered at t_{-1} and leaves the image at t_{+1}. Note that $(\mathbf{c}(\tau;\mathbf{x},t),\tau) = (x(\tau), y(\tau), \tau)$, where τ is a time instant from the interval (t_{-1}, t_{+1}).

incorporate the proposed models into the formulation to obtain a cost function to be minimized. Finally, we discuss suitable solution methods, including the multiresolution approach.

3.2.1 Definitions

We assume that time-varying images from which motion is to be calculated are natural, i.e., obtained by a camera that projects a 3-D scene onto a rectangular portion $\mathcal{W}$ of the 2-D image plane. The special case of synthetic images, such as those generated by computer graphics, is not considered here. We make a further restricting assumption that every point in the image corresponds to a single point in the 3-D scene. Obviously, this assumption is violated for transparent or reflective surfaces. The relative motion of the scene and camera results in 2-D motion on the image plane of projections of scene points and a consequent time variation of the image. Let $\mathbf{x} = (x, y)$ denote the spatial coordinate of an image point [1]. Since coordinates x and y depend on time t, it is useful to consider the trajectory of an image point in a conceptual 3-D xyt space. An example of a 3-D trajectory $(x(t), y(t), t)$ of an image point drawn in such a space is shown in Figure 3.1.

Let the function $\mathbf{c}(\tau;\mathbf{x},t)$ mathematically describe a trajectory in the image plane, i.e., let $\mathbf{c}(\tau;\mathbf{x},t)$ be the spatial position at time τ of an image point which

[1] In this chapter, boldface characters indicate vector quantities, including 2-D spatial coordinates and parameter vectors.

at time t was located at $\mathbf{x}$ [15]. $\mathbf{c}(\tau;\mathbf{x},t)$ describes a 2-D trajectory in the image plane, while $(\mathbf{c}(\tau;\mathbf{x},t),\tau) = (x(\tau), y(\tau), \tau)$ describes a 3-D trajectory in the xyt space. Clearly, there is a unique mapping between the two trajectories. For each $\mathbf{x}$ at t, the corresponding trajectory starts at time $t_i(\mathbf{x},t)$ and ends at time $t_f(\mathbf{x},t)$. The trajectory shown in Figure 3.1 corresponds to a point on the hexagon uncovered at time $t_i(\mathbf{x},t) = t_{-1}$ that leaves the image at $t_f(\mathbf{x},t) = t_{+1}$. For $\tau \neq t$, we can define a subset $\mathcal{V}(\tau;t)$ of the image at t consisting of pixels that are visible over the entire time interval between t and τ:

$$\mathcal{V}(\tau;t) = \{\mathbf{x} : t_i(\mathbf{x},t) \leq \tau \leq t_f(\mathbf{x},t)\}. \tag{3.1}$$

For $\tau > t$, $\mathcal{W} - \mathcal{V}(\tau;t)$ is the set of pixels occluded or leaving the image between t and τ, while for $\tau < t$, $\mathcal{W} - \mathcal{V}(\tau;t)$ is the set of pixels newly exposed or introduced into the image between τ and t. A more detailed treatment of motion trajectories in the xyt space can be found in [15].

The shape of the trajectories $\mathbf{c}(\tau;\mathbf{x},t)$ depends on the nature of object motion. Define the instantaneous velocity $\mathbf{v}$ of a pixel at $(\mathbf{x},t)$ as follows:

$$\mathbf{v}(\mathbf{x},t) = \left.\frac{d\mathbf{c}(\tau;\mathbf{x},t)}{d\tau}\right|_{\tau=t}. \tag{3.2}$$

If the velocity $\mathbf{v}$ is constant along the motion trajectory passing through $(\mathbf{x},t)$, then 2-D and 3-D trajectories are linear. In general, however, image points undergo acceleration. If an image point accelerates along a straight line, the 2-D trajectory in the image plane is linear. However, the same point traces out a nonlinear trajectory in the xyt space. In the most complex case an image point may accelerate along a nonlinear 2-D trajectory, thus tracing a nonlinear 3-D trajectory in the xyt space.

Trajectory $\mathbf{c}(\tau;\mathbf{x},t)$ mathematically describes motion of a point in the image plane. This motion may be very complex, thus needing a complex underlying model. Often, however, a simple model, such as the assumption of linear trajectories, is sufficient. For linear motion we use the concept of a *displacement.* Given an image at time t, the displacement field $\mathbf{d}$ is a collection of 2-D vectors describing pixel movements between times t and τ. We define $\mathbf{d}$ only for pixels visible between t and τ

$$\mathbf{d}(\tau;\mathbf{x},t) = \begin{cases} \mathbf{x} - \mathbf{c}(\tau;\mathbf{x},t), & \text{if } \tau < t; \\ \mathbf{c}(\tau;\mathbf{x},t) - \mathbf{x}, & \text{if } \tau > t; \end{cases} \qquad \mathbf{x} \in \mathcal{V}(\tau;t). \tag{3.3}$$

Note that for $\tau > t$, $\mathbf{d}(\tau;\mathbf{x},t)$ is a forward displacement field, while for $\tau < t$ it is a backward displacement field. The displacement field $\mathbf{d}(\tau;\mathbf{x},t)$ can be calculated from the velocity field by integration

$$\mathbf{d}(\tau;\mathbf{x},t) = \int_t^\tau \mathbf{v}(\mathbf{c}(s;\mathbf{x},t),s)ds, \qquad \mathbf{x} \in \mathcal{V}(\tau;t). \tag{3.4}$$

For motion with constant velocity $\mathbf{v}(\mathbf{c}(\tau;\mathbf{x},t),\tau) = \mathbf{v}(\mathbf{x},t)$, the displacement is simply $\mathbf{d}(\tau;\mathbf{x},t) = \mathbf{v}(\mathbf{x},t)\cdot(\tau - t)$. Thus, it follows from (3.3) that

$$\mathbf{c}(\tau;\mathbf{x},t) = \mathbf{x} + \mathbf{v}(\mathbf{x},t)\cdot(\tau - t), \qquad \mathbf{x} \in \mathcal{V}(\tau;t). \tag{3.5}$$

Consequently, for linear motion the task is to find, for each pixel $(\mathbf{x}, t)$, the two components v_x and v_y of the velocity $\mathbf{v}(\mathbf{x}, t)$.

We are interested here in the estimation of segments of motion trajectories $\mathbf{c}(\tau; \mathbf{x}, t)$ for τ over some time interval containing t, where $(\mathbf{x}, t)$ is defined on a sampling lattice $\Lambda_c \subset R^3$ [14]. These trajectories, which we refer to as *2-D motion*, correspond to the term *optical flow* often used in computer vision [27]. Usually, they are estimated from intensity or luminance images. However, there is no particular reason to use only luminance for motion estimation, especially if the resulting motion fields are applied to full-color images at the coding or processing stage. Thus, we consider a more general case where motion is estimated from color images. We hope to improve the estimated motion quality in this way, and also to reduce the residual error to be transmitted (for example, in motion-compensated DPCM). Consequently, we develop an approach to motion estimation that is based on vector data. Let $\mathbf{u}$ be the *true* underlying K-component image that is continuous in amplitude and in coordinates, and let $\mathbf{g}$ be the observed K-component image, i.e., $\mathbf{g} = [g_1, g_2, ..., g_K]$. Let each g_k be sampled on lattice Λ_{g_k} $(k = 1, .., K)$. This representation is very general, since individual g_k's $(k = 1, ..., K)$ can be components of a color image such as RGB, $YC1C2$, YIQ, can be derived from a spectral decomposition of an image, e.g., in the form of sub-bands, or even can come from a completely new set of measurements, such as infrared data. For simplicity, we consider only orthorhombic lattices Λ_c and Λ_{g_k} with sampling periods (T_c^h, T_c^v, T_c) and $(T_{g_k}^h, T_{g_k}^v, T_g)$, respectively [14]. Note that we assume identical temporal sampling for all components of $\mathbf{g}$, as is the case in the color representations mentioned above.

To make our formulation complete, we take into account occlusion effects present in dynamic images. We do so by defining an *occlusion field* $o(\mathbf{x}, t)$ with samples on the lattice Λ_c. Every occlusion tag o can take one of several possible occlusion states, e.g., moving/stationary (visible), occluded, newly exposed. The number of such states is finite and depends on the number of images used in the estimation. To estimate the trajectories $\mathbf{c}$ in practice, we model them by parametric functions over the time interval of interest. Since parameters of these functions may change rapidly at object boundaries, we permit such a variation by using the concept of motion discontinuity. We define a motion discontinuity field $l(\mathbf{x}, t)$ over a union of shifted lattices $\Psi_l = \psi_h \cup \psi_v$, where $\psi_h = \Lambda_c + [0, T_c^v/2, 0]^T$ and $\psi_v = \Lambda_c + [T_c^h/2, 0, 0]^T$ are orthorhombic cosets [14] specifying positions of horizontal and vertical discontinuities, respectively. l is often called a *line field* or a *line process* [19], while a single sample is called a *line element.*

3.2.2 Formulation

Motion present in images is not directly observable; we can only see the effect of motion and not the motion itself. Consequently, it cannot be measured but must be estimated instead. The process of estimating motion is difficult because the problem is *ill-posed* [4], i.e., the solution to the problem may not exist, may be non-unique or may be discontinuous with respect to the data. The most common and thus cumbersome problem is non-uniqueness of the solution. Even in the case of a perfect acquisition system, i.e., noise-free, continuous image obtained with an ideal pinhole camera, many different motion fields may be consistent with the observed data. In practice, however, the imaging system is not ideal, and the obtained data are subject to filtering, nonlinearity, sampling error and noise of a real image acquisition system. This introduces uncertainty and further complicates the estimation process. Finally, in some cases the motion estimation problem may have no solution or it may be discontinuous with respect to the data [33] as in the presence of transparent components and reflective surfaces. Thus, we cannot simply calculate motion. The most we can do is to estimate a motion field that is "best" in a certain sense or "most likely". This usually involves some prior assumptions about motion properties. We address this issue later by specifying appropriate *motion models.*

The uncertainty involved in image acquisition suggests a possible statistical approach to motion estimation. For example, image acquisition and motion attributes can be modeled by stochastic processes and used in a statistical criterion [40]. For certain types of criteria, such as the Minimum Expected Cost (MEC) [36], only a statistical formulation can be considered. For other criteria, such as the Maximum *A Posteriori* Probability (MAP) [40], a cost function results that can also be proposed directly without any recourse to statistical formulation [27],[26].

In the following sections we employ *Markov random field* (MRF) models to both observations $\mathbf{g}$ along motion trajectories and motion descriptors $(\mathbf{c}, o, l)$. Although the usual characterization of a MRF through initial and transitional probabilities is complex and cumbersome, thanks to the *Hammersley-Clifford theorem* [5] it is known that a random field has Markovian properties *if and only if* it is governed by a *Gibbs distribution.* This distribution is uniquely specified by a *neighborhood system*, *cliques* and a *potential function*, as well as by two constants: *partition function* and *natural temperature*[2]. In the case of MAP estimation [35], Gibbs distributions lead directly to a cost function to be optimized that is a weighted sum of certain "energies". In this chapter, we use this probabilistic formulation to arrive at our cost functions; however, similar cost functions could be determined without the application of the statistical framework by the direct use of these energies.

Let subscript t denote the restriction of a field to time instant t, e.g., $\mathbf{g}_t$. Also, let $(\Lambda.)_t$ be a restriction of the lattice $\Lambda.$ to time t and to the window $\mathcal{W}$ (image window in which estimates are sought) simultaneously. Let $\mathcal{I}_t$ denote a finite set of time instants of images $\mathbf{g}$ used to estimate trajectory $\mathbf{c}_t$; $\mathcal{I}_t = \{\tau : \mathbf{g}_\tau$ is used in estimation of $\mathbf{c}_t\}$. Two examples of the set $\mathcal{I}_t$ are shown in Figure 3.2. In order to carry out estimation of trajectories $\mathbf{c}_t$ in practice, they need to belong to a finite-dimensional space. This can be achieved by expressing each trajectory over a given interval of time as a function of a finite number of

[2]In order to facilitate understanding of the models, the Appendix at the end of this chapter contains a brief review of Markov random fields, Gibbs distributions and the relationship between them.

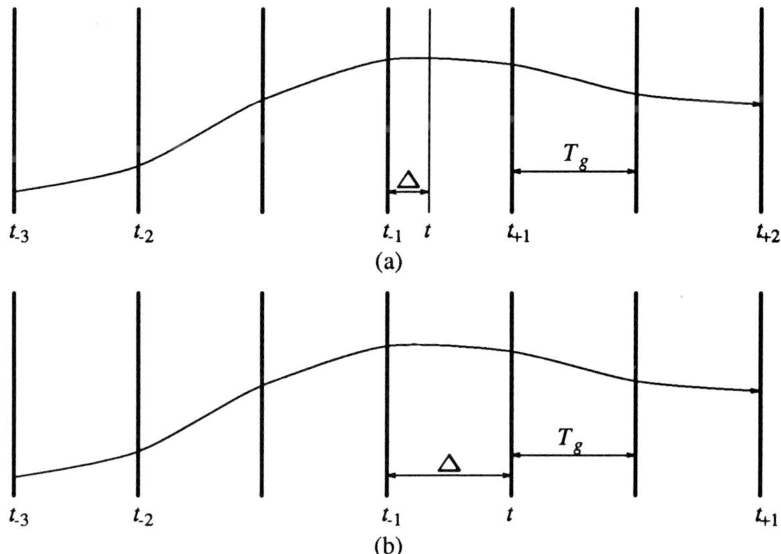

Figure 3.2: Relative temporal position t of motion trajectories with respect to positions of images: (a) $0 < \Delta < T_g$, $\mathcal{I}_t = \{t_{-3}, t_{-2}, t_{-1}, t_{+1}, t_{+2}\}$; (b) $\Delta = T_g$, $\mathcal{I}_t = \{t_{-3}, t_{-2}, t_{-1}, t, t_{+1}\}$. Note that not all images need to be used in the process of motion estimation.

parameters[3]. Then, we can assume that trajectories $\mathbf{c}_t$ are samples from a vector random field $\mathbf{C}_t$. Similarly, we assume that images $\mathbf{g}_t$ are samples from vector random field $\mathbf{G}_t$, and that occlusion fields o_t and motion discontinuity fields l_t are samples from scalar random fields O_t and L_t, respectively.

Our goal is to determine the *most likely* triplet $(\mathbf{c}_t, o_t, l_t)$ corresponding to the true underlying image $\mathbf{u}$ based on observations $\mathcal{G}_t = \{\mathbf{g}_\tau : \tau \in \mathcal{I}_t\}$. Thus, according to the MAP criterion, we seek the triplet $(\mathbf{c}_t, o_t, l_t)$ that maximizes the conditional probability distribution $p(\mathbf{c}_t, O_t = o_t, L_t = l_t|\mathcal{G}_t)$. In general, this function is a mixed probability density/probability mass function, corresponding to a continuous-valued $\mathbf{c}_t$ and discrete-valued o_t and l_t. We cannot obtain an exact expression for $p(\mathbf{c}_t, O_t = o_t, L_t = l_t|\mathcal{G}_t)$, but based on certain models we can get a good approximation whose maximization will yield excellent motion field estimates.

Our approach is based on the following factorization of the conditional probability distribution using Bayes' rule:

$$\begin{aligned} p(\mathbf{c}_t, O_t = o_t, L_t = l_t|\mathcal{G}_t) \;=\;& p(\mathcal{G}_t^n|\mathbf{c}_t, o_t, l_t, \mathbf{g}_{t_n}) \cdot \\ & p(\mathbf{c}_t|o_t, l_t, \mathbf{g}_{t_n}) \cdot p(O_t = o_t|l_t, \mathbf{g}_{t_n}) \cdot \\ & p(L_t = l_t|\mathbf{g}_{t_n})/p(\mathcal{G}_t^n|\mathbf{g}_{t_n}), \end{aligned} \tag{3.6}$$

[3] Details of this parametrization are provided in Section 3.2.5; until then, explicit dependence of $\mathbf{c}_t$ and $\mathbf{C}_t$ on parameters will not be used.

where $t_n \in \mathcal{I}_t$ is an arbitrarily chosen time instant and $\mathcal{G}_t^n = \{\mathbf{g}_\tau : \tau \in \mathcal{I}_t - \{t_n\}\}$. The MAP estimate is then given by

$$(\widehat{\mathbf{c}}_t, \widehat{o}_t, \widehat{l}_t) = \arg \max_{(\mathbf{c}_t, o_t, l_t)} \quad p(\mathcal{G}_t^n | \mathbf{c}_t, o_t, l_t, \mathbf{g}_{t_n}) \cdot p(\mathbf{c}_t | o_t, l_t, \mathbf{g}_{t_n}) \cdot \\ p(O_t = o_t | l_t, \mathbf{g}_{t_n}) \cdot p(L_t = l_t | \mathbf{g}_{t_n}). \tag{3.7}$$

Note that since the denominator in (3.6) is not a function of $(\mathbf{c}_t, o_t, l_t)$, it can be ignored in the maximization (3.7).

We now discuss the models that allow us to specify the constituent probabilities in (3.7). The first probability $p(\mathcal{G}_t^n | \mathbf{c}_t, o_t, l_t, \mathbf{g}_{t_n})$ is determined by the *structural model* relating motion to the observed image, and the remaining probabilities are determined by the *motion model.*

3.2.3 Observation process

The data from which motion is estimated is usually obtained by an image acquisition system. Thus, the observed image $\mathbf{g}$ is related to the true underlying image $\mathbf{u}$ by a fairly complex random operator called the *observation process.* A model of the observation process is needed to specify the *structural model* with respect to the underlying image $\mathbf{u}$. Typically, the observation process is considered to consist of a nonlinear shift-variant filtering, a random perturbation, and a spatiotemporal sampling. It can be expressed as follows:

$$g_k(\mathbf{x}, t) = \mathcal{H}_k[u_k(\mathbf{x}, t)] \circ n_k(\mathbf{x}, t), \qquad \mathbf{x} \in (\Lambda_{g_k})_t, \ k = 1, ..., K \tag{3.8}$$

where $\mathcal{H}_k$ denotes a nonlinear, shift-variant filter, n_k is a random noise term and $\circ$ is an operator combining the filtered image with noise. Note that in equation (3.8) absence of channel cross-talk is assumed. Usually, the above model is simplified to a linear, shift-invariant filter and additive noise:

$$g_k(\mathbf{x}, t) = h_k(\mathbf{x}, t) * u_k(\mathbf{x}, t) + n_k(\mathbf{x}, t), \qquad \mathbf{x} \in (\Lambda_{g_k})_t, \ k = 1, ..., K \tag{3.9}$$

which is much easier to analyze. In the above equation, h_k's form a bank of K filters used to model the point-spread function (PSF) of the imaging system and $*$ denotes convolution. Despite the above simplification, impulse responses h_k $(k = 1, ..., K)$ are still needed. Unless precise parameters of the imaging system are known or can be estimated, the use of mismatched filters in the model can do more harm than good. To simplify subsequent developments we constrain ourselves, without loss of generality, to the case of no filtering, i.e., $h_k(\mathbf{x}, t) = \delta(\mathbf{x}, t)$ for $k = 1, ..., K$.

The degree of sophistication used in modeling the observation process undoubtedly has an impact on the estimated motion fields. It is not clear, however, whether it is more beneficial to increase the complexity of the observation process above or of the *motion model* discussed below, especially in the view of the usual lack of knowledge of imaging system parameters.

3.2.4 Structural model and matching error

Since motion is not observable and cannot be measured directly, it is imperative that a relationship between motion and image sequence be established. This relationship, called the *structural model*, expresses assumptions about the properties of objects undergoing motion. On one hand, such complex descriptors as size, shape or structure of an object can be used. Alternatively, simple characteristics such as brightness, color or their derivatives can be employed. While the former are usually more stable (constant) under motion than the latter, their evaluation is a problem in itself. They are mostly used in long-range correspondence tasks where stability of features over extended periods of time is essential. The second type of characteristics, insufficiently stable over long intervals, is usually appropriate over short time.

We are interested in resolving the correspondence problem over relatively short time intervals. Under this assumption, the most frequent hypothesis made is that image brightness along motion trajectories is constant [48], [27], which can be expressed as follows:

$$u(\mathbf{x},t) = u(\mathbf{c}(\tau;\mathbf{x},t),\tau), \qquad \mathbf{x} \in \mathcal{V}(\tau;t). \tag{3.10}$$

Since chromatic properties of moving objects seem to be at least as stable as their brightness, it has been suggested to use additional color constraints [59],[44]. Recently, a structural model for multiple cues, where representation using luminance and chrominances is a special case, has been proposed in the case of linear motion [38],[34]. This model can be generalized to arbitrary trajectories as follows:

$$\mathbf{u}(\mathbf{x},t) = \mathbf{u}(\mathbf{c}(\tau;\mathbf{x},t),\tau), \qquad \mathbf{x} \in \mathcal{V}(\tau;t), \tag{3.11}$$

where $\mathbf{u}$ is the true underlying image with K components.

To deal with the departure from brightness or color constancy due to effects such as change of illumination, two approaches have been proposed. In the first approach [22],[45] the difference between both sides of equation (3.10) is modeled by a smooth function representing a slowly varying illumination error. In the second approach [58], [4] it is the spatial variation of intensity rather than its value that is assumed constant along motion trajectories.

Equations (3.10) and (3.11) express the structural model for the underlying image $\mathbf{u}$ over a continuum of spatiotemporal locations. The observed images $\mathbf{g}$, however, are corrupted and sampled versions of $\mathbf{u}$. To take these effects into account assume that we first find an estimate $\tilde{\mathbf{g}}$ of $\mathbf{u}$ by a suitable operation on the observed data $\mathbf{g}$, i.e., $\tilde{\mathbf{g}} = \mathcal{P}(\mathbf{g})$. This operation usually involves spatial interpolation, such as proposed in [31],[50], and possibly a noise reduction (linear or non-linear filtering). The effect of the operator $\mathcal{P}$ can be expressed as $\tilde{\mathbf{g}} = \mathbf{u} + \mathbf{e}$, where $\mathbf{e}$ is a K-component estimation error. Exploiting equation (3.11) we can write:

$$\begin{aligned} \tilde{\mathbf{g}}(\mathbf{x},t) - \tilde{\mathbf{g}}(\mathbf{c}(\tau;\mathbf{x},t),\tau) &= \\ \mathbf{e}(\mathbf{x},t) - \mathbf{e}(\mathbf{c}(\tau;\mathbf{x},t),\tau) &= \chi(\mathbf{x},t), \qquad \mathbf{x} \in \mathcal{L}_\tau, \end{aligned} \tag{3.12}$$

where $\mathcal{L}_\tau = (\Lambda_c)_t \cap \mathcal{V}(\tau;t)$ is the set of all pixels on lattice Λ_c at time t visible in the window $\mathcal{W}$ between t and τ. When no occlusion effects are present, then $\mathcal{L}_\tau = (\Lambda_c)_t \cap \mathcal{W} = (\Lambda_c)_t$. $\chi(\mathbf{x},t)$ is a noiselike K-component term governed by a probability distribution depending on the statistics of the estimation error $\mathbf{e}$. We assume that $\chi(\mathbf{x},t)$ is independent of $\tilde{\mathbf{g}}$.

The fact that pairwise differences of the interpolated image along motion trajectories have the properties of random noise is used to model the conditional distribution $p(\mathcal{G}_t^n|\mathbf{c}_t, o_t, l_t, \mathbf{g}_{t_n})$ from equation (3.7). Specifically, we assume that this conditional probability depends only on the variability of $\tilde{\mathbf{g}}$ along motion trajectories, and that this variability is independent for each distinct trajectory, i.e., for each $\mathbf{x} \in (\Lambda_c)_t$. Thus, we assume that conditional distribution along each trajectory is Gibbsian:

$$p_{\mathbf{x}}(\mathcal{G}_t^n|\mathbf{c}_t, o_t, l_t, \mathbf{g}_{t_n}) = \frac{1}{Z_g} e^{-U_g(\tilde{\mathbf{g}}_{\mathbf{x}}^{\mathbf{c}_t}, o_t)/\beta_g}, \tag{3.13}$$

where $\tilde{\mathbf{g}}_{\mathbf{x}}^{\mathbf{c}_t} = \{\tilde{\mathbf{g}}(\mathbf{c}(\tau;\mathbf{x},t),\tau) : \tau \in \mathcal{I}_t\}$ is the set of interpolated observations along the trajectory through $(\mathbf{x},t)$, $U_g(\tilde{\mathbf{g}}_{\mathbf{x}}^{\mathbf{c}_t}, o_t)$ is an energy function that measures departure of these interpolated observations from the structural model, and Z_g, β_g are constants called partition function and natural temperature, respectively (see the Appendix for a discussion of Gibbs distributions).

The total conditional distribution is hence a product of distributions[4] (3.13)

$$p(\mathcal{G}_t^n|\mathbf{c}_t, o_t, l_t, \mathbf{g}_{t_n}) = \prod_{\mathbf{x}\in(\Lambda_c)_t} p_{\mathbf{x}}(\mathcal{G}_t^n|\mathbf{c}_t, o_t, l_t, \mathbf{g}_{t_n}) = \frac{1}{Z_g'} e^{-U_g'(\mathcal{G}_t, \mathbf{c}_t, o_t)/\beta_g}, \tag{3.14}$$

where Z_g' is a product of Z_g's for all $\mathbf{x} \in (\Lambda_c)_t$, and the total energy function is

$$U_g'(\mathcal{G}_t, \mathbf{c}_t, o_t) = \sum_{\mathbf{x}\in(\Lambda_c)_t} U_g(\tilde{\mathbf{g}}_{\mathbf{x}}^{\mathbf{c}_t}, o_t). \tag{3.15}$$

In the above formulation, we assume that the trajectory $\mathbf{c}(\tau;\mathbf{x},t)$ extends through the whole set $\mathcal{I}_t$, but that the occlusion function $o(\mathbf{x},t)$ identifies which pixels $(\mathbf{c}(\tau;\mathbf{x},t),\tau)$ are visible at $(\mathbf{x},t)$; only these will contribute to the energy $U_g(\tilde{\mathbf{g}}_{\mathbf{x}}^{\mathbf{c}_t}, o_t)$.

The occlusion function $o(\mathbf{x},t)$ is a discrete function with a finite number of states depending on the cardinality of the set $\mathcal{I}_t$. For each spatiotemporal position $(\mathbf{x},t)$ the following set

$$\mathcal{I}_t^{\mathbf{x}} = \{\tau \in \mathcal{I}_t : \mathbf{x} \in \mathcal{L}_\tau\} \tag{3.16}$$

can be defined. This set, called a visibility set, contains time instants from $\mathcal{I}_t$ at which pixel $(\mathbf{x},t)$ is visible. Two examples of occlusion states and visibility sets for 3- and 5-image estimation are given in Table 3.1. In the case of estimation from 5 images, not all possible visibility/non-visibility combinations are defined, but rather only the most likely ones.

[4] Although the representation (3.14) is intuitively plausible, we have given here only qualitative arguments justifying it. A more formal approach can be followed using a spatiotemporal neighborhood system, with potentials that are non-zero only on spatiotemporal cliques oriented along the trajectories, and where the potentials involve spatial interpolation to get $\tilde{\mathbf{g}}$ values.

State	Description	Visibility set $\mathcal{I}_t^{\mathbf{x}}$ for pixel at $(\mathbf{x}, t)$
$\mathcal{M}$	moving/stationary	$\{t_{-1}, t, t_{+1}\}$
$\mathcal{E}$	exposed	$\{t, t_{+1}\}$
$\mathcal{C}$	covered	$\{t_{-1}, t\}$

(a)

State	Description	Visibility set $\mathcal{I}_t^{\mathbf{x}}$ for pixel at $(\mathbf{x}, t)$
$\mathcal{M}$	moving/stationary	$\{t_{-2}, t_{-1}, t, t_{+1}, t_{+2}\}$
$\mathcal{E}$	exposed in (t_{-1}, t)	$\{t, t_{+1}, t_{+2}\}$
$\mathcal{E}_{-1}$	exposed in (t_{-2}, t_{-1})	$\{t_{-1}, t, t_{+1}, t_{+2}\}$
$\mathcal{C}$	covered in (t, t_{+1})	$\{t_{-2}, t_{-1}, t\}$
$\mathcal{C}_{+1}$	covered in (t_{+1}, t_{+2})	$\{t_{-2}, t_{-1}, t, t_{+1}\}$

(b)

Table 3.1: Tables of occlusion states and visibility sets for the case of $\Delta = T_g$: (a) estimation from 3 images, $\mathcal{I}_t = \{t_{-1}, t, t_{+1}\}$, (b) estimation from 5 images, $\mathcal{I}_t = \{t_{-2}, t_{-1}, t, t_{+1}, t_{+2}\}$. The single-element set $\{t\}$ is not included in (a) because it means that a pixel is visible only at t. Similarly, 11 combinations are excluded from (b) because of trajectory discontinuities.

Note that in $U_g(\tilde{\mathbf{g}}_{\mathbf{x}}^{\mathbf{c}_t}, o_t)$, dependence on motion discontinuities l_t has been omitted since the information about such discontinuities is already conveyed through motion field $\mathbf{c}_t$. The line field l_t is not a descriptor of motion, but rather an artificial concept that permits introduction of abrupt changes in trajectory parameters.

There is considerable flexibility in the choice of the form of $U_g(\tilde{\mathbf{g}}_{\mathbf{x}}^{\mathbf{c}_t}, o_t)$, based on the structural model. We do this by defining a one-dimensional neighborhood system on $\mathcal{I}_t$, and choosing appropriate cliques and clique potentials. The energy takes the form

$$U_g(\tilde{\mathbf{g}}_{\mathbf{x}}^{\mathbf{c}_t}, o_t) = \sum_{\theta_g \in \Theta_g} V_g(\theta_g, \tilde{\mathbf{g}}_{\mathbf{x}}^{\mathbf{c}_t}, o_t) \tag{3.17}$$

where θ_g is a clique and Θ_g is the set of all cliques determined by the neighborhood system, and V_g is the potential function. Since the energy function expresses variability along a trajectory, we do not use single-element cliques in any of our models. For two-element cliques of the form $\theta_g = \{\tau_1, \tau_2\}$, a suitable potential function is

$$V_g^{(2)}(\{\tau_1, \tau_2\}, \tilde{\mathbf{g}}_{\mathbf{x}}^{\mathbf{c}_t}, o_t) = \|\tilde{\mathbf{g}}(\mathbf{c}(\tau_1; \mathbf{x}, t), \tau_1) - \tilde{\mathbf{g}}(\mathbf{c}(\tau_2; \mathbf{x}, t), \tau_2)\|^2, \tag{3.18}$$

for some norm $\|\cdot\|$ on the K-dimensional observation space. This potential penalizes deviation from constancy along the trajectory. For the norm, we can use any quadratic form $\mathbf{g}^T\mathbf{M}\mathbf{g}$ where $\mathbf{M}$ is a positive-definite matrix, although a diagonal matrix $\mathbf{M} = diag(m_1, ..., m_K)$, allowing a different relative weighting for each component is probably sufficient for most applications. There are many possibilities for potentials on cliques of three or more elements; some examples are given below.

The flexibility of this model can be illustrated with the following examples that generalize or are equivalent to existing approaches.

A. $\Lambda_c \subset \Lambda_g$, $\mathcal{I}_t^{\mathbf{x}} = \mathcal{I}_t = \{t - T_g, t\}$. This corresponds to the case of motion-compensated prediction of the image field at time t based on the image at $t - T_g$, with no treatment of newly exposed areas. In this case, there is only one two-element clique, $\{t, t - T_g\}$, and by choosing the potential function from (3.18), the energy function is given by

$$U_g^A(\tilde{\mathbf{g}}_{\mathbf{x}}^{\mathbf{c}_t}, o_t) = V_g^{(2)}(\{t - T_g, t\}, \tilde{\mathbf{g}}_{\mathbf{x}}^{\mathbf{c}_t}, o_t), \quad \mathbf{x} \in (\Lambda_c)_t. \tag{3.19}$$

Since the above model uses two-element cliques, only linear trajectories need to be used. Then, for velocity $\mathbf{v}(\mathbf{x}, t)$ with trajectory given by (3.5) and for scalar data g, the above energy simplifies to the very well-known squared displaced pixel difference

$$U_g = (g(\mathbf{x}, t) - g(\mathbf{x} - \mathbf{v}(\mathbf{x}, t)T_g, t - T_g))^2. \tag{3.20}$$

This energy has been used extensively in various motion estimation algorithms, of which those presented in [10],[48],[30] are just few examples.

B. $\Lambda_c \cap \Lambda_g = \emptyset$, $t_{-1} < t < t_{+1}$, $\mathcal{I}_t^{\mathbf{x}} = \mathcal{I}_t = \{t_{-1}, t_{+1}\}$. This corresponds to the estimation of motion at a time between two existing observed image fields based on these two fields only, and is applicable to frame rate conversion (temporal interpolation). Occlusions are not considered. Again, there is only one possible two-element clique, $\{t_{-1}, t_{+1}\}$, and a suitable energy function is

$$U_g^B(\tilde{\mathbf{g}}_{\mathbf{x}}^{\mathbf{c}_t}, o_t) = V_g^{(2)}(\{t_{-1}, t_{+1}\}, \tilde{\mathbf{g}}_{\mathbf{x}}^{\mathbf{c}_t}, o_t), \quad \mathbf{x} \in (\Lambda_c)_t. \tag{3.21}$$

For linear trajectories this energy has been used in [39] for scalar data and in [38],[34] for vector data.

C. $\Lambda_c \subset \Lambda_g$, $\mathcal{I}_t^{\mathbf{x}} = \mathcal{I}_t = \{t_{-1}, t, t_{+1}\}$. Here, we want to estimate a motion field at the position of an existing image field using three image fields, without consideration of occlusions. This could be applied in interpolative coding, as in MPEG [41]. If we consider only a first-order neighborhood system (i.e., with neighborhoods $\eta_g(t_{-1}) = \{t\}$, $\eta_g(t) = \{t_{-1}, t_{+1}\}$, $\eta_g(t_{+1}) = \{t\}$), then there are two two-element cliques, $\{t_{-1}, t\}$ and $\{t, t_{+1}\}$. A suitable energy function for these cliques is given by

$$U_g^{C1}(\tilde{\mathbf{g}}_{\mathbf{x}}^{\mathbf{c}_t}, o_t) = \alpha_1 V_g^{(2)}(\{t_{-1}, t\}, \tilde{\mathbf{g}}_{\mathbf{x}}^{\mathbf{c}_t}, o_t) \; + \; \alpha_2 V_g^{(2)}(\{t, t_{+1}\}, \tilde{\mathbf{g}}_{\mathbf{x}}^{\mathbf{c}_t}, o_t), \quad \mathbf{x} \in (\Lambda_c)_t. \tag{3.22}$$

where α_i's are weights.

If we expand the neighborhood system to second order, then every time position is a neighbor of every other, and we have the additional two-element clique $\{t_{-1}, t_{+1}\}$. We can thus augment the energy (3.22) with the additional potential to give

$$U_g^{C2}(\tilde{\mathbf{g}}_{\mathbf{x}}^{\mathbf{c}_t}, o_t) = U_g^{C1}(\tilde{\mathbf{g}}_{\mathbf{x}}^{\mathbf{c}_t}, o_t) + \alpha_3 V_g^{(2)}(\{t_{-1}, t_{+1}\}, \tilde{\mathbf{g}}_{\mathbf{x}}^{\mathbf{c}_t}, o_t), \quad \mathbf{x} \in (\Lambda_c)_t. \tag{3.23}$$

The resulting energy is proportional to the sample variance of the three intensities along the trajectory if the weights α_i are equal.

The second-order neighborhood system also includes a three-element clique $\{t_{-1}, t, t_{+1}\}$, for which the potential can be chosen rather arbitrarily. If our structural model permits linear intensity variation along a trajectory, for example, we can consider a potential of the form

$$\begin{aligned} V_g^{(3)}(\{t_{-1}, t, t_{+1}\}, \tilde{\mathbf{g}}_{\mathbf{x}}^{\mathbf{c}_t}, o_t) \quad = \quad & \|\tilde{\mathbf{g}}(\mathbf{x}, t) - ((1-\Delta_t)\tilde{\mathbf{g}}(\mathbf{c}(t_{-1}; \mathbf{x}, t), t_{-1}) + \\ & \Delta_t \tilde{\mathbf{g}}(\mathbf{c}(t_{+1}; \mathbf{x}, t), t_{+1}))\|^2, \end{aligned} \tag{3.24}$$

where $\Delta_t = (t - t_{-1})/(t_{+1} - t_{-1})$. This can be interpreted as the norm of the difference between the observation at time t and the linearly interpolated value based on the observations at times t_{-1} and t_{+1}. The overall energy function can be a linear combination of (3.23) and (3.24):

$$U_g^{C3}(\tilde{\mathbf{g}}_{\mathbf{x}}^{\mathbf{c}_t}, o_t) = U_g^{C2}(\tilde{\mathbf{g}}_{\mathbf{x}}^{\mathbf{c}_t}, o_t) + \gamma V_g^{(3)}(\{t_{-1}, t, t_{+1}\}, \tilde{\mathbf{g}}_{\mathbf{x}}^{\mathbf{c}_t}, o_t), \ \mathbf{x} \in (\Lambda_c)_t, \tag{3.25}$$

where γ must be determined according to the structural model. Previous work has used $\gamma = 1$ and $\alpha_i = 0, i = 1, ..., K$ [13].

D. $\Lambda_c \subset \Lambda_g$, $\mathcal{I}_t^{\mathbf{x}} \subset \mathcal{I}_t = \{t_{-1}, t, t_{+1}\}$. This example follows Example C, but with consideration of the occlusion states, as given in Table 3.1(a). In this case, if $t_i \notin \mathcal{I}_t^{\mathbf{x}}$, the potential for any two-element clique involving t_i is set to zero. Similarly, if $\mathcal{I}_t^{\mathbf{x}} \neq \mathcal{I}_t$, the potential $V_g^{(3)}$ can be set to zero, since the two-element clique remaining is sufficient. Thus, a general expression for the energy function is [13]

$$U_g^D(\tilde{\mathbf{g}}_{\mathbf{x}}^{\mathbf{c}_t}, o_t) = \begin{cases} U_g^{C3}(\tilde{\mathbf{g}}_{\mathbf{x}}^{\mathbf{c}_t}, o_t), & \text{if } o(\mathbf{x}, t) = \mathcal{M}; \\ V_g^{(2)}(\{t, t_{+1}\}, \tilde{\mathbf{g}}_{\mathbf{x}}^{\mathbf{c}_t}, o_t), & \text{if } o(\mathbf{x}, t) = \mathcal{E}; \\ V_g^{(2)}(\{t_{-1}, t\}, \tilde{\mathbf{g}}_{\mathbf{x}}^{\mathbf{c}_t}, o_t), & \text{if } o(\mathbf{x}, t) = \mathcal{C}; \end{cases} \quad \mathbf{x} \in (\Lambda_c)_t. \tag{3.26}$$

3.2.5 Motion models

So far trajectories $\mathbf{c}_t$ have been written as general functions of $\mathbf{x}$ and t, which could potentially belong to an infinite-dimensional space. To make the estimation problem tractable, we assume that each motion trajectory $\mathbf{c}(\tau; \mathbf{x}, t)$ over a certain time interval containing t can be described by a parametric function

$\mathbf{c}^p(\tau; \mathbf{x}, t)$ uniquely identified by the parameter vector $\mathbf{p}$. With this assumption $\mathbf{c}_t$'s belong to a finite-dimensional space, as assumed in previous sections.

In typical images, motion fields are usually smooth functions of spatial position $\mathbf{x}$, except at motion boundaries. Thus, we model trajectories $\mathbf{c}_t^p$ by continuous-valued vector MRFs $\mathbf{C}_t^p$ [46],[35], and also we model occlusions o_t and motion discontinuities l_t by multi-level and binary MRFs O_t [13] and L_t [28],[37], respectively. We ensure the smoothness of $\mathbf{c}_t^p$ and o_t, and continuity of l_t by appropriate choice of Gibbs distribution parameters.

Motion trajectory model

As stated before, in order to uniquely characterize a MRF it is sufficient to specify parameters of its Gibbs distribution. These parameters must be compatible with some expected properties of the random field. Let the *a priori* distribution $p(\mathbf{c}_t^p | o_t, l_t, \mathbf{g}_{t_n})$ for the *motion trajectory model* be defined as follows:

$$p(\mathbf{c}_t^p | o_t, l_t, \mathbf{g}_{t_n}) = \frac{1}{Z_c} e^{-U_c(\mathbf{c}_t^p, l_t, \mathbf{g}_{t_n})/\beta_c}, \tag{3.27}$$

where Z_c and β_c are constants. The energy function U_c is defined as follows:

$$U_c(\mathbf{c}_t^p, l_t, \mathbf{g}_{t_n}) = \sum_{\theta_c \in \Theta_c} V_c(\theta_c, \mathbf{c}_t^p, l_t, \mathbf{g}_{t_n}). \tag{3.28}$$

θ_c is a clique for trajectories $\mathbf{c}_t$ while Θ_c is a set of all such cliques derived from the neighborhood system $\mathcal{N}_c$ defined over lattice Λ_c. V_c is a potential function essential to characterization of the properties of motion field $\mathbf{c}_t$.

To specify the *a priori* motion model, $\mathcal{N}_c$, θ_c and V_c have to be specified. For example, the first-order neighborhood system $\mathcal{N}_c^1$ (Figure 3.3(a)) and two-element horizontal and vertical cliques (Figures 3.3(b) and 3.3(c)) [40] can be chosen. To model the smoothness of trajectories $\mathbf{c}_t^p$, the potential V_c should be such that adjacent similar trajectories belonging to the same object give a small value of V_c (high probability), while dissimilar ones give a large value of V_c. Also, dissimilar adjacent trajectories located across an occlusion boundary should give a small V_c, since they belong to two differently moving objects. We define the potential function V_c as follows:

$$V_c(\theta_c, \mathbf{c}_t^p, l_t, \mathbf{g}_{t_n}) = (\mathbf{p}_i - \mathbf{p}_j)^T \Gamma(\mathbf{g}_{t_n})(\mathbf{p}_i - \mathbf{p}_j) \cdot [1 - l(<\mathbf{x}_i, \mathbf{x}_j>, t)], \quad \theta_c = \{\mathbf{x}_i, \mathbf{x}_j\} \in \Theta_c, \tag{3.29}$$

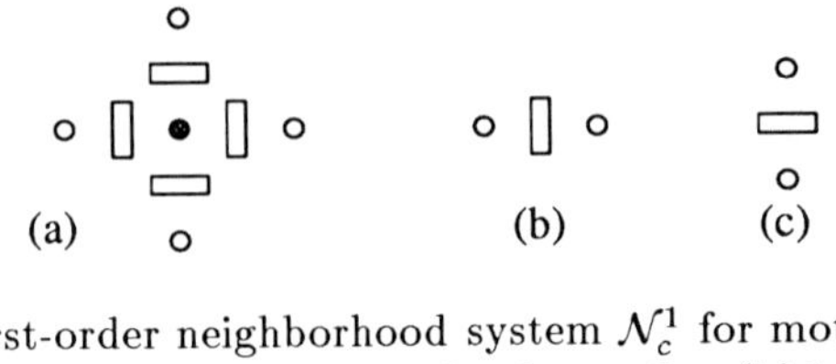

(a) (b) (c)

Figure 3.3: (a) First-order neighborhood system $\mathcal{N}_c^1$ for motion field $\mathbf{c}_t$ defined over Λ_c with motion discontinuities l_t defined over Ψ_l ; (b) horizontal clique; (c) vertical clique (empty circles: trajectory positions in $\mathcal{N}_c^1$; filled circle: central position not in $\mathcal{N}_c^1$; rectangles: positions of motion discontinuities).

where $\mathbf{p}_i$ and $\mathbf{p}_j$ are parameter vectors for trajectories at $(\mathbf{x}_i, t)$ and $(\mathbf{x}_j, t)$, respectively, and $\Gamma(\mathbf{g}_{t_n})$ is a positive-definite weight matrix depending on the observations. $(<\mathbf{x}_i, \mathbf{x}_j>, t) \in \Psi_l$ denotes the site of a motion discontinuity located between trajectory sites $(\mathbf{x}_i, t)$ and $(\mathbf{x}_j, t)$ which both belong to Λ_c. This potential captures smoothness of the random field $\mathbf{C}_t^p$; for $\mathbf{p}_i = \mathbf{p}_j$, $V_c = 0$.

According to (3.27) the probability of having a particular trajectory at location $(\mathbf{x}, t)$ depends on the occlusion and motion discontinuity fields as well as on the observations. The dependence on local discontinuities is expressed through the multiplicative term $1 - l(<\mathbf{x}_i, \mathbf{x}_j>, t)$ in (3.29) [28],[37]. For a line element "on" ($l(<\mathbf{x}_i, \mathbf{x}_j>, t) = 1$) no contribution (penalty) is added to the energy U_c. Since the potential V_c is non-negative, such a contribution would lower the probability of $\mathbf{c}(\tau; \mathbf{x}_i, t)$. Thus, a jump in trajectory parameters is not penalized if a motion discontinuity had been detected. Note that in equation (3.27) the energy U_c does not depend on the occlusion field o_t. We assume that sufficient information about occlusion boundaries is conveyed by the field l_t, and thus that o_t is not necessary here. To ensure that this assumption is valid, below we make the *occlusion model* dependent on motion discontinuities l_t.

The dependence of probability (3.27) on the observations is expressed through the weight matrix $\Gamma(\mathbf{g}_{t_n})$ in (3.29). This matrix permits different weighting of horizontal and vertical parameters of $\mathbf{c}_t^p$ as well as of lower and higher order parameters. If Γ is the identity matrix, the Euclidean norm results. In general, $\Gamma(\mathbf{g}_{t_n})$ does not have to be diagonal, and may include off-diagonal entries, thus causing cross-terms to appear in the potential function. Also, it may depend on the observations $\mathbf{g}_{t_n}$ to allow suitable adaptation of motion properties to local image structure. This kind of adaptive smoothness constraint for the case of linear trajectories (3.5) has been proposed in [47].

At this point we must specify the functional form of $\mathbf{c}_t^p$ in order to compute the potential (3.29). In the case of linear trajectory model (equation (3.5)) the parameter vector $\mathbf{p}_i$ takes the following form:

$$\mathbf{p}_i = \begin{bmatrix} v_x(\mathbf{x}_i, t) \\ v_y(\mathbf{x}_i, t) \end{bmatrix}, \tag{3.30}$$

where v_x and v_y are horizontal and vertical components of velocity $\mathbf{v}$, respectively [35]. Thus, Γ becomes a 2×2 matrix.

A natural extension of the linear model is a quadratic trajectory model accounting for acceleration of image points, which can be described by the equation

$$\mathbf{c}(\tau; \mathbf{x}, t) = \mathbf{x} + \mathbf{v}(\mathbf{x}, t) \cdot (\tau - t) + \mathbf{a}(\mathbf{x}, t) \cdot (\tau - t)^2, \quad \mathbf{x} \in \mathcal{V}(\tau; t). \tag{3.31}$$

This new model is based on two velocity (linear) variables $\mathbf{v} = [v_x, v_y]^T$ and two acceleration (quadratic) variables $\mathbf{a} = [a_x, a_y]^T$. Thus, each quadratic trajectory $\mathbf{c}_t^p$ is described by the following parameter vector:

$$\mathbf{p}_i = \begin{bmatrix} v_x(\mathbf{x}_i, t) \\ v_y(\mathbf{x}_i, t) \\ a_x(\mathbf{x}_i, t) \\ a_y(\mathbf{x}_i, t) \end{bmatrix}. \tag{3.32}$$

Γ is a 4×4 matrix. Higher-dimensional approximations to trajectories $\mathbf{c}_t^p$ can be obtained similarly.

Occlusion model

The state space for each occlusion label $o(\mathbf{x}, t)$ consists of a number of states depending on the cardinality of the set $\mathcal{I}_t$ and on the configurations chosen; in Table 3.1 two examples with 3 and 5 states, respectively, are given. Thus, as the *occlusion model* for o_t we use a discrete-valued scalar MRF model described by the following Gibbs distribution

$$P(O_t = o_t | l_t, \mathbf{g}_{t_n}) = \frac{1}{Z_o} e^{-U_o(o_t, l_t, \mathbf{g}_{t_n})/\beta_o}, \tag{3.33}$$

with Z_o and β_o being constants. The energy function U_o is defined as follows:

$$U_o(o_t, l_t, \mathbf{g}_{t_n}) = \sum_{\theta_o \in \Theta_o} V_o(\theta_o, o_t, l_t, \mathbf{g}_{t_n}), \tag{3.34}$$

where θ_o is an occlusion clique and Θ_o is the set of all occlusion cliques derived from the neighborhood system $\mathcal{N}_o$ defined over the lattice Λ_c of motion trajectories. Thus, it is natural to choose the first-order neighborhood system $\mathcal{N}_c^1$ and two-element horizontal and vertical cliques (Figure 3.3). Additionally, we use single-element cliques. Other configurations are possible as well.

The potential function V_o provides a penalty associated with an occlusion state; otherwise energy (3.17) could be reduced freely by a suitable choice of occlusion states. V_o is usually expressed in a tabulated form. It can be expected that a typical occlusion field consists mostly of patches of pixels labeled as visible, and some smaller clusters of pixels labeled as exposed or covered. To penalize the introduction of a label, in addition to the horizontal and vertical cliques from Figure 3.3, single-element cliques are used as well. For the 3-state occlusion field presented in Table 3.1(a), the choice of potential values is similar to the one used in [13] and is shown in Figure 3.4. To ensure that occlusion states get clustered, $V_o = 0$ (high probability) for adjacent identical labels, and a high value of V_o (low probability) for different labels should be specified (Figure 3.4(b)). The boundaries between different patches are expected to be occlusion boundaries, which as mentioned above should coincide with motion discontinuities. To achieve this goal, the dependence of the potential function V_o on field l_t is utilized. V_o is set to 0 whenever two different occlusion states are separated by a motion discontinuity ($l(<\mathbf{x}_i, \mathbf{x}_j>, t) = 1$). Finally, discontinuities in areas of identical occlusion labels are discouraged by assigning a high value to V_o (Figure 3.4(b)). The total potential for a given occlusion label is defined as a sum of potentials for single-element and two-element cliques $V_o = V_A + V_B$ (Figure 3.4).

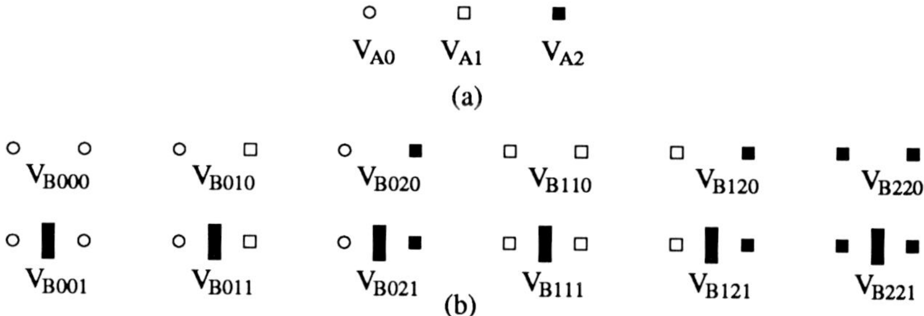

Figure 3.4: Costs associated with various configurations (up to rotation and permutation) of occlusion cliques for the 3-state occlusion field (Table 3.1(a)); (a) single-element cliques; (b) two-element cliques. Typically, $V_{A0} = 0 < V_{A1} = V_{A2}$, $V_{B000} = V_{B110} = V_{B220} = 0 < V_{B010} = V_{B020} \ll V_{B120}$ and $V_{B011} = V_{B021} = 0 < V_{B001} < V_{B111} = V_{B221} \ll V_{B121}$ (circle: moving/stationary pixel; open square: exposed pixel; filled square: covered pixel; filled rectangle: line element "on").

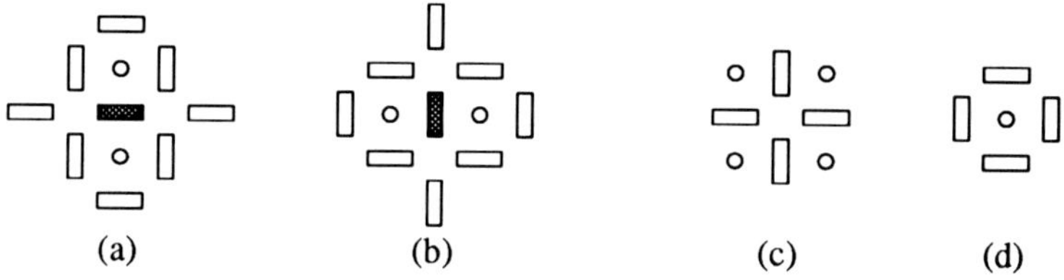

Figure 3.5: Neighborhood system $\mathcal{N}_l$ for motion discontinuity field l_t defined over Ψ_l ; (a) centered on a horizontal discontinuity; (b) centered on a vertical discontinuity; (c) four-element cross-shaped clique; (d) four-element square-shaped clique (empty rectangles: positions in $\mathcal{N}_l^2$; filled rectangle: central position not in $\mathcal{N}_l^2$; circles: positions of trajectories.)

Motion discontinuity model

The *motion discontinuity model* is based on binary MRF L_t, and is described by the Gibbs distribution

$$P(L_t = l_t|\mathbf{g}_{t_n}) = \frac{1}{Z_l} e^{-U_l(l_t,\mathbf{g}_{t_n})/\beta_l}, \tag{3.35}$$

with Z_l and β_l being the usual constants. U_l is the energy function defined as follows:

$$U_l(l_t, \mathbf{g}_{t_n}) = \sum_{\theta_l \in \Theta_l} V_l(l_t, \theta_l, \mathbf{g}_{t_n}), \tag{3.36}$$

where θ_l is a motion discontinuity clique and Θ_l is the set of all such cliques derived from neighborhood system $\mathcal{N}_l$ defined over Ψ_l. The potential function V_l provides a penalty associated with introduction of a discontinuity.

In order to model continuity of motion boundaries, a sufficiently large neighborhood system $\mathcal{N}_l$ for the "dual" sampling structure Ψ_l should be used [40].

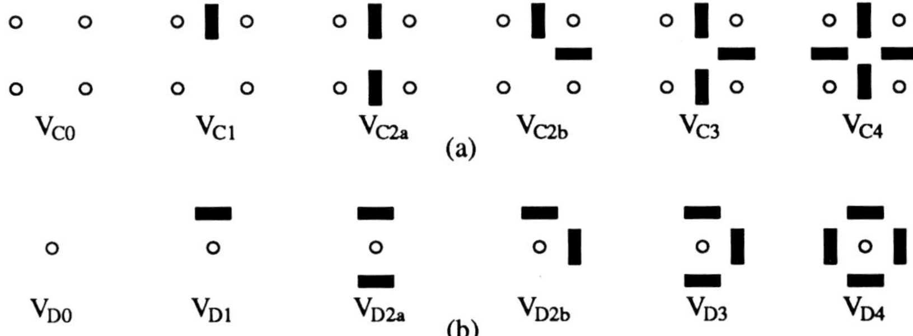

Figure 3.6: Costs associated with various configurations (up to rotation) of line cliques; (a) four-element cross-shaped cliques; (b) four-element square-shaped cliques. Usually, $V_{C0} = 0 < V_{C2a} < V_{C2b} < V_{C1}, V_{C3}, V_{C4}$ and $V_{D0} = 0 < V_{D1} < V_{D2b} < V_{D3} < V_{D2a} < V_{D4} = \infty$ (filled rectangle: line element "on", circle: position of a motion trajectory).

Example of such a neighborhood system is shown in Figure 3.5. In Figure 3.5(a) it is centered on a horizontal discontinuity, while in Figure 3.5(b) it is centered on a vertical discontinuity. To exercise control over straight lines, corners and intersections, typically four-element cliques are used (Figure 3.5(c) and 3.5(d)). The cross-shaped cliques (Figure 3.6(a)) with potential V_C discourage creation of unended and intersecting segments. The square-shaped cliques (Figure 3.6(b)) with potential V_D discourage formation of double lines and also inhibit generation of isolated trajectories ($V_{D4} = \infty$).

Note that the *a priori* probability of the line process (3.35) is conditioned on the observations $\mathbf{g}_{t_n}$. It means that image information should be considered when computing l_t. In general, a 3D scene giving rise to a motion discontinuity also contributes to an intensity edge. Only under specific circumstances does a motion discontinuity not correspond to an edge of intensity. Hence, we assume that the introduction of a motion discontinuity should coincide with an intensity edge. To enforce such a coincidence, we use single-element cliques with potential V_E assuming a high value whenever a motion discontinuity does not match an intensity edge. This can be done in two ways. We can detect intensity edges first, and then set potential V_E to a large value whenever motion discontinuity is attempted at locations where no intensity edge is present [28]. The other approach is to make V_E inversely proportional to a norm of the local image gradient [37]. The total potential for a given motion discontinuity is defined as a sum of potentials for single-element and four-element cliques $V_l = V_C + V_D + V_E$ (Figure 3.6).

3.2.6 A posteriori probability and cost function

In preceding sections we have defined the correspondence, trajectory, occlusion and motion discontinuity models through Gibbs distributions (3.14), (3.27), (3.33) and (3.35), respectively. We combine these probabilities using the Bayesian decomposition (3.6) to obtain the following Gibbs distribution:

$$p(\mathbf{c}_t^p, O_t = o_t, L_t = l_t | \mathcal{G}_t) = \frac{1}{Z} e^{-U(\mathbf{c}_t^p, o_t, l_t, \mathcal{G}_t)}, \tag{3.37}$$

where Z is a new normalizing constant incorporating the probability $p(\mathcal{G}_t^n | \mathbf{g}_{t_n})$ from (3.6), as well as Z_g, Z_c, Z_o, Z_l. The new energy function $U(\mathbf{c}_t^p, o_t, l_t, \mathcal{G}_t)$ is defined as follows

$$\begin{aligned} U(\mathbf{c}_t^p, o_t, l_t, \mathcal{G}_t) &= U_g'(\mathcal{G}_t, \mathbf{c}_t, o_t) + \lambda_c U_c(\mathbf{c}_t^p, l_t, \mathbf{g}_{t_n}) + \\ & \quad \lambda_o U_o(o_t, l_t, \mathbf{g}_{t_n}) + \lambda_l U_l(l_t, \mathbf{g}_{t_n}). \end{aligned} \tag{3.38}$$

The constituent energies are defined in (3.17), (3.28), (3.34) and (3.36) respectively, and $\lambda_c = 1/\beta_c$, $\lambda_o = 1/\beta_o$, $\lambda_l = 1/\beta_l$.

The *a posteriori* probability (3.37) may lead to different Bayesian solutions of the motion estimation problem. For example, it may be used in the Maximization of *A Posteriori* Probability [35] or in the Minimization of Expected Cost [36]. We will use the first criterion since for Gibbs-Markov models it leads directly to minimization of the following multiple-term cost function[5]:

$$\begin{aligned} (\widehat{\mathbf{c}_t^p}, \widehat{o}_t, \widehat{l}_t) = \min_{\{\mathbf{c}_t^p, o_t, l_t\}} \quad & U_g'(\mathcal{G}_t, \mathbf{c}_t, o_t) + \lambda_c U_c(\mathbf{c}_t^p, l_t, \mathbf{g}_{t_n}) + \\ & \lambda_o U_o(o_t, l_t, \mathbf{g}_{t_n}) + \lambda_l U_l(l_t, \mathbf{g}_{t_n}). \end{aligned} \tag{3.39}$$

The functional to be minimized consists of four terms: U_g' is a matching energy for all K components and describes the ill-posed problem of matching the data $\mathcal{G}_t$ by the motion field $\widehat{\mathbf{c}_t^p}$; U_c is responsible for conforming to the properties of the *a priori* trajectory model; U_o models occlusion areas; and U_l allows for occasional motion discontinuities. The four-term formulation of the energy function (3.38) can be viewed as regularization of the original correspondence problem.

In the formulation (3.39) the ratios of λ's play an important role weighting the confidence in the data and in the *a priori* models. A modification of any λ has an effect on the estimate; however the magnitude of this effect is highly dependent on the data itself. The parameters β_g, β_c, β_o and β_l, which characterize individual models, are difficult to compute. When MRFs are used in estimation of such observables as images or textures, β's can be estimated by analyzing a number of samples (training process), and then used to perform estimation on some other data. The success of the estimation is highly dependent on the similarity between the real data and the model (the training data). In the case of estimating an unobservable such as motion, it is not clear how to compute β's and thus they are usually chosen *ad hoc.*

3.2.7 Solution method

Minimization (3.39) is a very difficult task for two reasons. First, there are hundreds of thousands of unknowns to be found; for a typical 512×512 image, there

[5]Due to parametrization of motion trajectories the estimate $\widehat{\mathbf{c}}_t$ is equal to the field of trajectories determined by the parameter vector $\widehat{\mathbf{p}}$: $\widehat{\mathbf{c}_t^p}$.

are 262,144 vectors of trajectory parameters (minimum 2-D), 262,144 occlusion labels and 523,264 motion discontinuities. Secondly, the cost function under minimization is not convex. Thus, it is very likely that multiple minima exist. Additionally, the cost function is not differentiable since some energies are not expressed analytically but are obtained from look-up tables of potential values.

Consequently, a minimization method is sought which can handle hundreds of thousands of variables, which can find the global optimum and which can deal with non-differentiable functions. A method satisfying all three requirements is *simulated annealing* [32],[19] which is a stochastic search algorithm. This method is a computer simulation of the process of *annealing* of solids; the behavior of a solid is simulated by generating sample configurations from a Gibbs distribution with a suitable energy function divided by a "temperature" parameter T. The sample configurations are produced using *stochastic relaxation* such as the *Metropolis algorithm* [43] or the *Gibbs sampler* [19]. Initially T is chosen to be sufficiently high to generate a full range of configurations, even the unlikely ones. This assures avoidance of local minima and is equivalent to melting the solid. As the process evolves, T is very slowly reduced. It has been shown [19], that if the reduction of T is sufficiently slow, then the system attains (in a limit) the state of minimal energy.

Simulated annealing has been proposed as the solution method for MAP estimation of 2-D motion [40]. The method has been applied to the case of linear motion (equation (3.5)) without occlusions (λ_o=0) for scalar [40] and vector (K=3) [38] observations. In fact, two types of simulated annealing algorithms for the estimation of 2-D motion have been proposed: a discrete state space algorithm and a continuous state space one. In the first case, *a priori* precision (quantization step) is chosen and only vectors satisfying the precision constraint are generated. In the second case, a local approximation of the Gibbs distribution by a Gaussian leads to a Gauss-Newton iterative update with an additional T-dependent stochastic update. This equation is a variant of the diffusion equation proposed in [20] for global optimization. In the limit with $T \to 0$, the stochastic term disappears resulting in a deterministic Gauss-Newton algorithm.

By instantaneously "freezing" the system ($T = 0$), equivalent deterministic methods can be obtained. For example, the discrete state space simulated annealing results in pixel matching with smoothness constraints, while the continuous state space one leads to equations similar to those proposed in [27],[28].

If the point of departure is a four-term cost function like in (3.39), then a different strategy must be used. Note that trajectories $\mathbf{c}^p$ are described by continuous-valued parameters, while the occlusion labels and line elements are discrete-valued with a finite number of states. This suggests that different methods must be used in order to estimate $\mathbf{c}^p$ and o, l. This can be done in an interleaved fashion, i.e., while one is iteratively updated, the others are kept constant. For given estimates $\widehat{o}_t$ and $\widehat{l}_t$, in order to improve the estimate of $\mathbf{c}_t^p$, one iteration (full scan of a field) of minimization

$$\min_{\{\mathbf{c}_t^p\}} U_g'(\mathcal{G}_t, \mathbf{c}_t, \widehat{o}_t) + \lambda_c U_c(\mathbf{c}_t^p, \widehat{l}_t, \mathbf{g}_{t_n}), \tag{3.40}$$

is carried out. U'_g is still non-quadratic, but U_c is quadratic because $\widehat{l}_t$ is given and the potential (3.29) is quadratic (as explained in Section 3.2.5 dependence on o_t has been omitted). To solve (3.40), for example Gauss-Newton optimization can be used, as was done in [3],[39]. In this method, a sequence of local quadratic approximations of U'_g is carried out. Once $\mathbf{c}_t^p$ has been updated, it becomes a known estimate $\widehat{\mathbf{c}_t^p}$ and o_t is improved by executing one iteration of

$$\min_{\{o_t\}} U'_g(\mathcal{G}_t, \widehat{\mathbf{c}_t^p}, o_t) + \lambda_o U_o(o_t, \widehat{l}_t, \mathbf{g}_{t_n}). \tag{3.41}$$

Since the number of possible states for each occlusion label is small, usually an exhaustive search is used. Also, such methods as Besag's *Iterated Conditional Modes* [6] can be used. Finally, l_t can be re-iterated by executing

$$\min_{\{l_t\}} \lambda_c U_c(\widehat{\mathbf{c}_t^p}, l_t, \mathbf{g}_{t_n}) + \lambda_o U_o(\widehat{o}_t, l_t, \mathbf{g}_{t_n}) + \lambda_l U_l(l_t, \mathbf{g}_{t_n}). \tag{3.42}$$

Once all three fields have been updated, the process is repeated until no reduction of energies can be observed. This approach usually leads to good results.

The stochastic algorithms for linear motion with piecewise smoothness have been shown to perform very well on synthetic and natural images [40]. These algorithms are, however, very demanding computationally and require from one (for the continuous state space) to two (for the discrete state space) orders of magnitude more time than deterministic algorithms. In a direct comparison [39], the stochastic algorithms perform better than equivalent ("frozen") deterministic ones both in terms of the cost function value and subjective evaluation. The improvement is particularly pronounced for some cases of ambiguous data with synthetic motion [39]. For typical video images, however, the improvement is small.

Other methods have been proposed to minimize variants of energy (3.38) for the case of linear trajectories and scalar data. For example, the simplest cost function, using only matching and motion energies, proposed in [35], has been optimized using mean field annealing [1]. Also, method called Highest Confidence First (HCF) [11] has been used to optimize the three-term energy (with additional motion discontinuity energy) [9].

3.2.8 Estimation over a hierarchy of resolutions

Both stochastic estimation methods summarized in the preceding section as well their deterministic counterparts are inappropriate for the estimation of fast motion (large displacements). The computational complexity of discrete state space algorithms becomes prohibitively large because the state space for each trajectory grows very rapidly with the maximum allowed displacement. On the other hand, the continuous state space methods fail when certain underlying assumptions are violated. For example, to simplify U'_g (Section 3.2.7) small displacements are assumed so that local intensity linearity holds; this linearity does not hold for large displacements.

To deal with the above problems various multiresolution methods have been proposed for motion estimation. One class of such methods uses coarse-to-fine strategy to compute motion. First, a pyramid of observations is constructed by low-pass or band-pass filtering and subsampling (usually by 2 in both directions) [8]. Then, a pyramid of fields to be estimated is built by similar subsampling. Estimation starts at the top of the pyramid. Since only a fraction of the field must be computed and since distances between field elements (e.g., trajectories) are relatively large, the convergence is very rapid. Once an estimate is obtained, it is propagated to a lower level by an appropriate operator, such as interpolation. The process is repeated until an estimate is obtained at full resolution. At each level any of the methods discussed in the preceding section can be used. For example, a matching algorithm for the estimation of discrete displacements has been proposed in [2] (see Chapter 1 in this book as well). Also, gradient-based methods for the continuous state space have been suggested in [23], and [17].

A similar multiresolution approach has been proposed for Gibbs-Markov models of motion [36]. Markov models at different levels of hierarchy have been linked by experimentally established parameters. Stochastic algorithms for discrete and continuous state space, and piecewise-smooth motion model (including motion discontinuities) have been given in [40]. Since subsampling of the observations aims only at efficient storage, no subsampling has been applied to images, thus resulting in a constant-width observations pyramid. The filtering of the observations has been used, however, as it is essential for gradient-based methods (reduction of degree of violation of local intensity linearity assumption). The same filtering, although not essential, is helpful in discrete matching methods because it reduces ambiguity due to local detail. Unfortunately, the filtering may also introduce unwanted artifacts, such as confusion of objects with background, which usually leads to locally unreliable estimates.

The above arguments suggest that for discrete state space methods it may be possible to use an estimation pyramid and raw data with no filtering or subsampling. Such a method, called *multiscale* estimation has been proposed for globally smooth linear motion (displacements) based on Markov hierarchical model [25]. In this approach Markov models at higher levels of the hierarchy have been rigorously derived from the full resolution model. This method belongs to another class of coarse-to-fine algorithms where instead of subsampling, the size of a block in which all estimates are kept constant varies from level to level. In this way a hierarchy of detail scales of an estimated field is obtained. The method has been reported to give results very similar to those obtained by stochastic monoresolution techniques [35] and superior to results obtained by deterministic multiresolution methods using filtering and subsampling of observations [39].

3.3 Motion-Compensated Processing

This section presents a number of applications of 2-D motion estimates to motion-compensated processing of video signals for coding, enhancement or standards conversion. The first application is predictive coding, in which we encode information along motion trajectories, where redundancy is highest. The second application is in noise reduction, where low-pass filtering along motion trajectories can be used to reduce the level of additive noise with little effect on picture content. We then discuss motion-compensated sampling structure conversion, which is a problem of spatiotemporal interpolation. Examples are frame rate increase, interlace-to-progressive conversion, and general standards conversion (e.g., NTSC $\leftrightarrow$ PAL, or NTSC $\leftrightarrow$ HDTV).

3.3.1 Motion-compensated prediction

Predictive coding is the most widely used technique for temporal redundancy reduction, and is incorporated in some way in most existing or proposed image coding standards, from low resolution teleconferencing [42] to HDTV [21],[60]. Motion trajectory estimates are used to predict an image field from previous fields using samples along the same estimated trajectory. The prediction error is then encoded and transmitted. Motion-compensated predictive coding is normally accomplished using the DPCM structure with a configuration similar to that shown in Figure 3.7. In the DPCM encoder, the input g_r to the predictor is equal to the image reconstructed at the receiver (in the absence of transmission errors). Both the input g and the reconstructed signal g_r can potentially be used as input to the motion estimation process. If only g_r is used, the receiver can duplicate the same operations, and motion information need not be transmitted. However, that may compromise the accuracy of the motion trajectory estimates and reduce coding efficiency. Since it is known that motion fields can be coded efficiently [51], we can use the original image to get the best possible motion field estimate and transmit it as side information.

The goal of prediction is to determine an estimate $\hat{g}(\mathbf{x},t)$ of the image $g(\mathbf{x},t)$ on the image sampling lattice. Thus, we must estimate a field of motion trajectories terminating at $(\mathbf{x},t)$, $\mathbf{x} \in (\Lambda_g)_t$ and passing through M previous fields

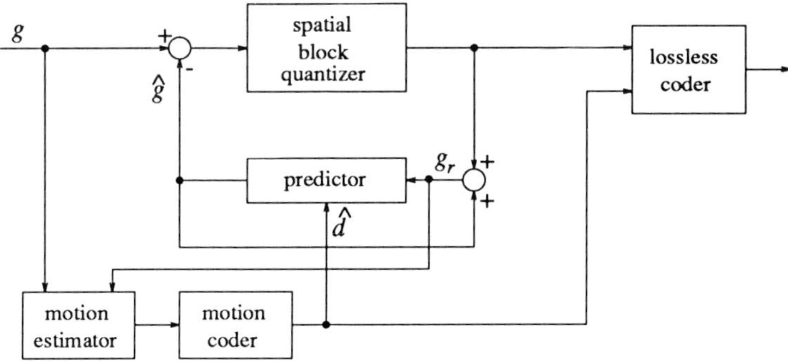

Figure 3.7: Block diagram of motion-compensated DPCM coder

using methods of the previous sections. The prediction is then given by

$$\widehat{g}(\mathbf{x},t) = \sum_{m=1}^{M} b_m \widetilde{g}_r(\widehat{\mathbf{c}^p}(t - mT_g; \mathbf{x}, t), t - mT_g), \tag{3.43}$$

where M is the predictor order and $\widehat{\mathbf{c}^p}(\tau; \mathbf{x}, t)$ is the trajectory terminating at $(\mathbf{x}, t)$ determined by the estimated parameter vector $\widehat{\mathbf{p}}(\mathbf{x}, t)$. In the case of first order prediction,

$$\widehat{g}(\mathbf{x},t) = b\widetilde{g}_r(\mathbf{x} - \widehat{\mathbf{d}}(\mathbf{x},t), t - T_g), \tag{3.44}$$

where b is usually taken to be unity due to the high correlation along motion trajectories and the fact that this gives perfect prediction in flat areas, where the viewer is most sensitive to distortions. For higher order prediction, the coefficients can be determined by estimating the correlation function along motion trajectories and solving the normal equations to get minimum mean square prediction error. Specifically, the coefficients b_m are chosen to minimize

$$\mathrm{E}\left[g(\mathbf{x},t) - \sum_{m=1}^{M} b_m \widetilde{g}_r(\widehat{\mathbf{c}^p}(t - mT_g; \mathbf{x}, t), t - mT_g)\right]^2. \tag{3.45}$$

By the orthogonality principle, the optimal coefficients satisfy the normal equations

$$\mathbf{Rb} = \mathbf{r} \tag{3.46}$$

where for $m = 1, 2, \ldots, M$

$$r_m = \mathrm{E}[g(\mathbf{x},t)\widetilde{g}_r(\widehat{\mathbf{c}^p}(t - mT_g; \mathbf{x}, t), t - mT_g)], \tag{3.47}$$

and for for $k, m = 1, 2, \ldots, M$

$$[R]_{km} = \mathrm{E}[\widetilde{g}_r(\widehat{\mathbf{c}^p}(t - kT_g; \mathbf{x}, t), t - kT_g)\widetilde{g}_r(\widehat{\mathbf{c}^p}(t - mT_g; \mathbf{x}, t), t - mT_g)]. \tag{3.48}$$

To a first approximation, the correlations can be estimated using the input signal along estimated motion trajectories (i.e., ignoring the spatial block quantizer in the DPCM loop) over a representative training set containing a wide variety of images and motion trajectories.

In many cases, it may be advantageous to use coefficients that depend on the motion field, since the correlation function along trajectories may depend on the motion (i.e., on $\widehat{\mathbf{p}}$). Interlace plays an important role in the form of the correlation function along a trajectory, since intrafield spatial aliasing may be present. With reference to Figure 3.8(a), for trajectory *i* the correlation with field t_{-2} may be greater than with field t_{-1}, while for trajectory *iii* the opposite is true. A prediction with motion-dependent coefficients takes the form

$$\widehat{g}(\mathbf{x},t) = \sum_{m=1}^{M} b_m(\widehat{\mathbf{p}}) \widetilde{g}_r(\widehat{\mathbf{c}^p}(t - mT_g; \mathbf{x}, t), t - mT_g). \tag{3.49}$$

In this case, we must solve the equation

$$\mathbf{R}(\widehat{\mathbf{p}})\mathbf{b}(\widehat{\mathbf{p}}) = \mathbf{r}(\widehat{\mathbf{p}}) \tag{3.50}$$

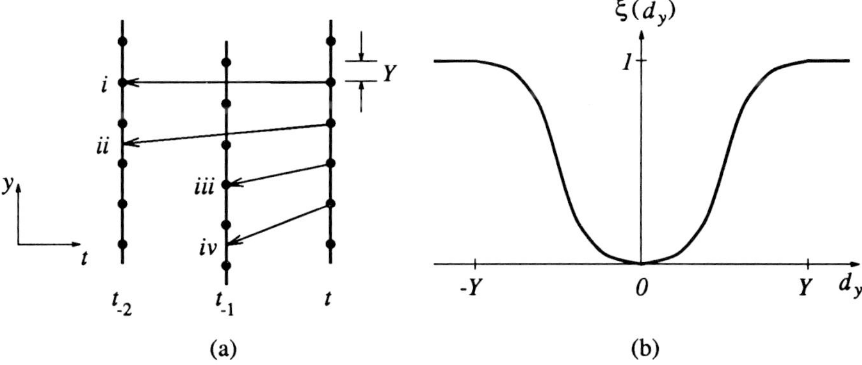

Figure 3.8: (a) Three fields used for prediction in an interlaced sequence. The field t is to be predicted from the two fields t_{-1} and t_{-2}. (b) The weighting function $\xi(d_y)$.

for each $\widehat{\mathbf{p}}$ of interest. In order to determine the correlations that specify $\mathbf{r}(\mathbf{p})$ and $\mathbf{R}(\mathbf{p})$, the parameter vectors can be clustered into classes within which the correlation function is similar, and the normal equations would then need to be solved for each of these classes.

Another approach to motion-adaptive prediction, that is based on *linear* trajectories and is applicable to second order prediction for interlaced sequences, is [12]

$$\widehat{g}(\mathbf{x},t) = \xi(\widehat{d}_y)\widetilde{g}_r(\mathbf{x}-\widehat{\mathbf{d}}(\mathbf{x},t),t-T_g)+(1-\xi(\widehat{d}_y))\widetilde{g}_r(\mathbf{x}-2\widehat{\mathbf{d}}(\mathbf{x},t),t-2T_g). \quad (3.51)$$

The coefficients depend only on the vertical component of $\widehat{\mathbf{d}}$, and enforce perfect prediction in flat areas. When the vertical component of $\widehat{\mathbf{d}}$ is zero, we choose $\xi = 0$ so that the prediction is based on the second previous field only, where correlation is highest. As $|d_y|$ increases, ξ increases until the point where $|d_y| = Y$ (Figure 3.8(a)), where we use $\xi = 1$, i.e., previous field prediction. A suitable function is (Figure 3.8(b))

$$\xi(d_y) = \begin{cases} \frac{1}{2}(1-\cos(\pi d_y/Y)) & 0 \leq |d_y| \leq Y \\ 1 & |d_y| > Y \end{cases} \quad (3.52)$$

This adaptive prediction can also be used to form the matching energy for the objective function used in the motion estimation phase

$$U_g = ||g(\mathbf{x},t) - \widehat{g}(\mathbf{x},t)||^2. \quad (3.53)$$

Since ξ is a continuous function of $\mathbf{d}$, it can be used without problems in the gradient descent algorithms described previously.

This method was shown to perform better than either fixed previous field or fixed second previous field prediction on a variety of image sequences [12].

3.3.2 Motion-compensated noise reduction

Random noise can be a significant impairment in video signals. Many techniques for image filtering to reduce perceived noise level have been studied, such as Wiener filtering or non-linear filtering (e.g., median filtering) [29], [52]. The major difficulty is to distinguish between noise and image information. If spatiotemporal motion trajectories are known, most random temporal variation along the trajectory can be attributed to noise. Thus, one-dimensional processing along motion trajectories is an attractive approach for noise reduction. We assume in the following that motion trajectory estimates are available for the noisy sequence. They may either be estimated from the noisy sequence by a robust motion estimation algorithm, or from the original uncorrupted sequence and assumed to be available as side information that was transmitted in a noiseless fashion (e.g., digitally, in a hybrid analog/digital transmission scheme [57]).

Let $g(\mathbf{x},t)$ be the noisy input image. The idea is to perform a one-dimensional filtering for each $(\mathbf{x},t)$ of the 1-D signal $s(\tau;\mathbf{x},t) = g(\mathbf{c}(\tau;\mathbf{x},t),\tau)$. If FIR filtering is used, this can be expressed as

$$g_r(\mathbf{x},t) = \sum_{m=-M_1}^{M_2} b_m \tilde{g}(\mathbf{c}(t-mT_g;\mathbf{x},t), t-mT_g). \tag{3.54}$$

As in the previous section, the b_m could be determined from knowledge of the correlation function of the image along the trajectory by solution of the normal equations. However, a fairly high filter order is required to obtain adequate noise reduction, which is likely to be too costly in terms of frame memories. Furthermore, it may be difficult to accurately specify motion trajectories over long time periods. A very attractive alternative is recursive filtering which uses only a few frames of delay.

An all pole motion-compensated recursive filter has the form

$$g_r(\mathbf{x},t) = \gamma g(\mathbf{x},t) + (1-\gamma)\sum_{m=1}^{M} b_m \tilde{g}_r(\mathbf{c}(t-mT_g;\mathbf{x},t), t-mT_g). \tag{3.55}$$

The transfer function of the 1-D filter operating along the motion trajectory is

$$H(z) = \frac{\gamma}{1-(1-\gamma)\sum_{m=1}^{M} b_m z^{-m}}. \tag{3.56}$$

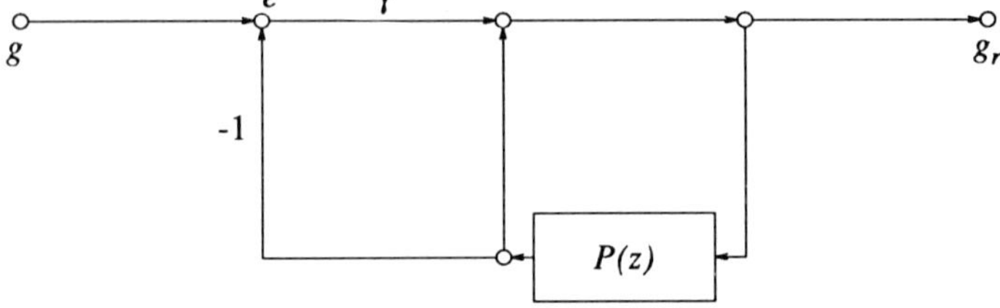

Figure 3.9: All-pole temporal recursive filter.

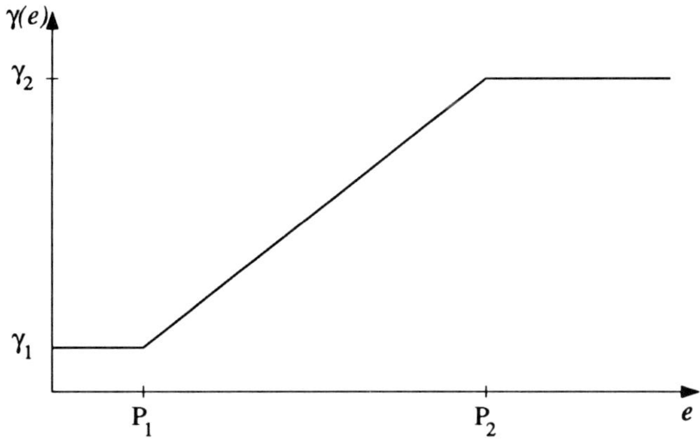

Figure 3.10: Nonlinear function relating multiplication coefficient γ to the prediction error.

This can be represented by the block diagram shown in Figure 3.9, where $P(z) = \sum_{m=1}^{M} b_m z^{-m}$ has the form of a predictor. Thus, this component can be designed using the methods of the previous section. For progressively scanned image sequences, a first order filter of the form

$$g_r(\mathbf{x},t) = \gamma g(\mathbf{x},t) + (1-\gamma)\tilde{g}_r(\mathbf{x} - \hat{\mathbf{d}}(\mathbf{x},t), t - T_g) \tag{3.57}$$

is suitable. This low-pass filter increasingly attenuates noise as the parameter γ approaches 0; it can be shown that the increase in SNR when the true image intensity is constant on the motion trajectory is given by $10\log_{10}(2-\gamma)/\gamma$ dB [16]. For $\gamma = 1$, there is no filtering. For interlaced sequences, a second order temporal filter is advantageous for the same reasons cited for the case of prediction. In this case, the noise reduction filter takes the form

$$\begin{aligned} g_r(\mathbf{x},t) &= \gamma g(\mathbf{x},t) + (1-\gamma)[\xi(d_y)\tilde{g}_r(\mathbf{x} - \hat{\mathbf{d}}(\mathbf{x},t), t - T_g) + \\ &\quad (1-\xi(d_y))\tilde{g}_r(\mathbf{x} - 2\hat{\mathbf{d}}(\mathbf{x},t), t - 2T_g)]. \end{aligned} \tag{3.58}$$

The motion-compensated filters described above perform well when motion is uniform and correctly estimated. In practice, due to effects such as occlusion or complex motion, the image intensity is not constant along the estimated motion trajectory. The filtering will then significantly impair the image. The value of the "prediction error" $e(\mathbf{x},t)$ can be used as an indicator of whether filtering should be performed. If $e(\mathbf{x},t)$ is relatively small (relative to the noise level), it can be assumed that the motion trajectory estimate is accurate enough and filtering should be performed. However, if $e(\mathbf{x},t)$ is large, we assume that filtering should be reduced or disabled. This can be achieved by replacing the multiplier γ with a memoryless nonlinearity having the desired effect: the output of the nonlinearity is $\gamma(e)\cdot e$ where $\gamma(e)$ has a form such as shown in Figure 3.10. The overall noise reduction system is then as shown in Figure 3.11.

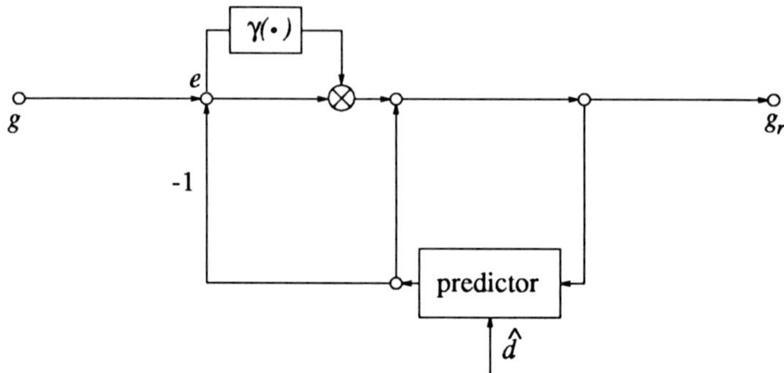

Figure 3.11: Block diagram of motion-compensated noise reducer.

A study of this system showed that on a variety of image sequences, a good choice of parameters was $P_1 = 10$, $P_2 = 20$, $\gamma_1 = 0.3$ and $\gamma_2 = 1.0$ when the noise standard deviation was 5.0. A clear subjective improvement in image quality was noted for all sequences [7].

3.3.3 Motion-compensated interpolation

Interpolation refers to the process of filling in missing samples in a signal based on neighboring samples. If the signal has been sampled respecting the multidimensional Nyquist criterion, the interpolation can be carried out by fixed interpolation filters with good results. However, video signals are often sampled in space-time in a way that the Nyquist criterion is violated, so that aliasing is present and fixed interpolation filters do not give good results. This is particularly true with respect to temporal sampling, as described in [15]. Motion-compensated interpolation along motion trajectories can overcome the limitations of fixed filtering.

Two cases arise in spatiotemporal interpolation. The first case is when no input data is available at the time instant t for which interpolation is being carried out. In other words, an image field is being generated at a time for which no input image field exists, as in Figure 3.2(a). This is the case for applications such as field rate standards conversion (e.g., between 50 Hz and 60 Hz), up-conversion from temporally subsampled signals, and field rate increase to reduce display artifacts or for slow-motion portrayal. In this case, purely temporal interpolation is involved. The second case is when input image samples exist at the given time (as in Figure 3.2(b)), but spatial aliasing may be present due to a non-orthogonal sampling structure, such as in interlaced sampling. In this case, temporal interpolation may give better results than spatial interpolation, and should be incorporated into the overall interpolation algorithm, perhaps jointly with spatial interpolation.

Motion-compensated interpolation takes the form

$$\widehat{g}(\mathbf{x},t) = \sum_{t_j \in \mathcal{I}_t^{\mathbf{x}}} b_j \widetilde{g}(\mathbf{c}(t_j;\mathbf{x},t),t_j), \tag{3.59}$$

where the trajectory covers only points visible at $(\mathbf{x}, t)$. Since the intensity varies slowly along the motion trajectory, fairly simple 1-D filters can be used, such as equal coefficients $b_j = 1/|\mathcal{I}_t^{\mathbf{x}}|$, or coefficients that decrease with distance from t. As with prediction and noise reduction, the coefficients can be made dependent on the motion trajectory parameters, and thus on position with respect to the sampling lattice. It should be noted that errors in motion estimates can lead to erroneously interpolated values that can seriously impair the picture, thus making highly reliable motion estimates a necessity for motion-compensated interpolation. If the field to be interpolated is available at the motion estimation phase (as in some coding systems where the motion field is transmitted) much better interpolation results can often be obtained [13].

In the second case of spatiotemporal interpolation, the motion-compensated interpolation should be combined with a spatial interpolation. As an example, consider the conversion of an image sequence from interlaced to progressive format. Referring to Figure 3.8, we see that if the vertical component of the motion is near zero, a motion-compensated temporal interpolation will be effective (where the horizontal motion is compensated). However, as the vertical velocity increases, the temporal interpolation is less effective and spatial interpolation might as well be applied. Thus the overall interpolation scheme is given by

$$\widehat{g}(\mathbf{x},t) = \xi \widehat{g}_T(\mathbf{x},t) + (1-\xi)\widehat{g}_S(\mathbf{x},t) \tag{3.60}$$

where $\widehat{g}_T(\mathbf{x}, t)$ is obtained by motion-compensated temporal interpolation, as in equation (3.59), and $\widehat{g}_S(\mathbf{x}, t)$ is obtained by spatial interpolation. The coefficient ξ can be based simply on d_y as previously, or it can be based on a more complex local analysis comparing the relative efficacy of temporal versus spatial interpolation [49].

The spatial interpolation can be a simple fixed vertical interpolation, such as cubic interpolation. However, because of the significant spatial aliasing present in single fields of interlaced sequences, more sophisticated spatial interpolation schemes may be advantageous. For example, directional interpolation based on local orientation analysis has been shown to provide a useful improvement in interpolation of diagonal edges, reducing the serration effect produced by pure vertical interpolation [50].

3.4 Summary

Estimates of 2-D motion are required in order to perform different types of temporal processing of time-varying images along trajectories of motion. In this chapter, we have presented a general Bayesian formulation for the estimation of motion trajectories from observed image sequences. By an appropriate parametrization of the trajectories, the formulation can handle nonlinear trajectories that result from accelerating motion. Occlusion effects such as newly exposed or disappearing image features are also included. General multichannel image observations (e.g., three color components) are assumed and incorporated into a vector image formulation. A maximum a posteriori probability (MAP) estimation criterion has been developed on the basis of a number of models: a model for the observation process, a structural model relating the unknown motion to the underlying image, and an a priori model of the motion field, including occlusion and discontinuity tags. Most estimation criteria that have been proposed can be seen to be special cases of this general formulation. Some of these special cases have been identified.

We have pointed out methods appropriate for minimization of the resulting cost function, specifically stochastic and deterministic relaxation methods. The importance of multiresolution and multiscale methods in efficient localization of a near global optimum has also been addressed.

Several applications of motion-compensated processing to the problems of prediction, interpolation and smoothing have been discussed. General methods for motion-compensated predictor design have been proposed, including some methods that can deal with the problematic effects of interlace. These predictors can be used in both motion-compensated coding and motion-compensated noise reduction. Finally, motion-compensated spatiotemporal interpolation, required for effective sampling structure conversion, has been addressed.

Many aspects of the general formulation presented remain to be investigated, and the relative importance of many of its features must be established. We believe that this formulation can be a useful framework for further research in advanced motion estimation and compensation techniques.

Appendix

In this appendix the definitions of a *Markov random field* and of the *Gibbs distribution* are given, and the relationship between them, the *Hammersley-Clifford theorem*, is described. For more detailed account, please consult [19],[5],[55].

Let sampling structure Ψ be a collection of sites in R^N [14]. A collection $\mathcal{N}$ of subsets of Ψ

$$\mathcal{N} = \{\eta(\mathbf{x}_i) : \mathbf{x}_i \in \Psi, \eta(\mathbf{x}_i) \subset \Psi, \forall i\} \tag{3.61}$$

is a *neighborhood system* on Ψ, if and only if, the *neighborhood* $\eta(\mathbf{x}_i)$ of site $\mathbf{x}_i \in \Psi$ satisfies both of the following conditions:

1. $\mathbf{x}_i \notin \eta(\mathbf{x}_i)$,

2. if $\mathbf{x}_j \in \eta(\mathbf{x}_i)$, then $\mathbf{x}_i \in \eta(\mathbf{x}_j)$ for any $\mathbf{x}_i \in \Psi$.

Examples of low-order neighborhood systems for orthogonal and non-orthogonal sampling structures Ψ over R^2 are given in Figures 3.3 and 3.5, respectively. For higher-order neighborhood systems see [19].

Random field over Ψ is a multidimensional stochastic process where each site in Ψ is assigned a random variable. A vector random field has a random vector (ensemble of random variables) assigned at each site in Ψ.

A *Markov random field* Υ is a random field with the following properties:

1. $P(\Upsilon = v) > 0, \quad \forall\, v \in \mathcal{S}$,

2. $P(\Upsilon_i = v_i | \Upsilon_j = v_j, \forall j \neq i) = P(\Upsilon_i = v_i | \Upsilon_j = v_j, \forall j \in \eta(i)), \quad \forall i, \forall v \in \mathcal{S}$,

where P denotes a probability measure and $\mathcal{S}$ is a state space. For a discrete $\mathcal{S}$, P is a probability for a given state, while for a continuous $\mathcal{S}$, P is replaced by the cumulative distribution F_Υ. If F_Υ is differentiable, the above property applies directly with the densities p replacing the probabilities P.

In order to define the *Gibbs distribution* the concepts of *clique* and *potential function* are needed. A *clique* θ defined over Ψ with respect to $\mathcal{N}$ is a subset of Ψ such that either θ consists of a single site or every pair of sites in θ are neighbors, i.e., $\mathbf{x}_i \in \eta(\mathbf{x}_j)\ \forall\{\mathbf{x}_i, \mathbf{x}_j\} \in \theta$. The set of all cliques is denoted by Θ. Examples of low-order cliques are shown in Figures 3.3 and 3.5.

Let v be a sample field from random field Υ defined over Ψ and over state space $\mathcal{S}$. A *Gibbs distribution* with respect to Ψ and $\mathcal{N}$ is a probability measure π on $\mathcal{S}$ such that

$$\pi(v) = \frac{1}{Z} e^{-U(v)/\beta}, \tag{3.62}$$

where β, Z are constants, and the *energy function* U is of the form

$$U(v) = \sum_{\theta \in \Theta} V(v, \theta). \tag{3.63}$$

$V(v, \theta)$ is called a *potential function*, and depends only on those samples from v which belong to the clique θ. Z is called a *partition function* and is a normalizing constant such that π is a probability measure. β is a parameter called *natural temperature*.

A MRF can be uniquely characterized by a finite-dimensional joint probability distribution, and thus all initial and transitional (conditional) probability distributions are needed. This approach is cumbersome, because the conditional probability distributions must satisfy certain consistency conditions [5], and because the computation of the joint distribution is usually difficult. Also, the relationship between the form of a conditional probability distribution and the characteristic properties of a sample field is not obvious. A clear and simple relationship can be provided, however, through the *Hammersley-Clifford theorem* [55],[5] which states that if Υ is a MRF on Ψ with respect to $\mathcal{N}$, then the probability distribution of its sample realizations is a Gibbs distribution with respect to Ψ and $\mathcal{N}$. This unique characterization of a MRF by a Gibbs distribution results in a straightforward relationship between qualitative properties of a MRF and its parameters via the potential functions V. Extension of the Hammersley-Clifford theorem to vector MRFs is straightforward (only a new definition of a state has to be provided).

Acknowledgement

Our work on motion estimation and motion compensation has been funded by the Natural Sciences and Engineering Research Council of Canada, by the Communications Research Centre and by the Canadian Institute for Telecommunications Research.

References

[1] I. Abdelqader, S. Rajala, W. Snyder, and G. Bilbro, "Energy minimization approach to motion estimation," *Signal Processing*, vol. 28, pp. 291–309, Sept. 1992.

[2] P. Anandan, "A unified perspective on computational techniques for the measurement of visual motion," in *Proc. IEEE Int. Conf. Computer Vision*, pp. 219–230, June 1987.

[3] C. Bergeron and E. Dubois, "Gradient-based algorithms for block-oriented MAP estimation of motion and application to motion-compensated temporal interpolation," *IEEE Trans. Circuits Syst. Video Technol.*, vol. 1, pp. 72–85, Mar. 1991.

[4] M. Bertero, T. Poggio, and V. Torre, "Ill-posed problems in early vision," *Proc. IEEE*, vol. 76, pp. 869–889, Aug. 1988.

[5] J. Besag, "Spatial interaction and the statistical analysis of lattice systems," *J. Roy. Statist. Soc.*, vol. B 36, pp. 192–236, 1974.

[6] J. Besag, "On the statistical analysis of dirty pictures," *J. Roy. Statist. Soc.*, vol. B 48, pp. 259–279, 1986.

[7] H. Boutrouille, "Réduction du niveau de bruit dans les séquences d'images video par filtrage compensé par le mouvement," Tech. Rep. 91–29, INRS-Télécommunications, Sept. 1991.

[8] P. Burt, "Fast filter transforms for image processing," *Comput. Vision, Graphics Image Process.*, vol. 16, pp. 20–51, 1981.

[9] R. Buschmann, "Improvement of optical flow estimation by HCF control and hierarchical block matching," in *Proc. IEEE Workshop Vis. Sig. Process. Comm.*, pp. 270–273, Sept. 1992.

[10] C. Cafforio and F. Rocca, "Methods for measuring small displacements of television images," *IEEE Trans. Inform. Theory*, vol. IT-22, pp. 573–579, Sept. 1976.

[11] P. Chou and C. Brown, "Multimodal reconstruction and segmentation with Markov random fields and HCF optimization," in *Proc. Image Understanding Workshop*, pp. 214–221, Apr. 1988.

[12] R. Depommier and E. Dubois, "Motion-compensated temporal prediction for interlaced image sequences," in *Proc. IEEE Workshop Vis. Sig. Process. Comm.*, pp. 264–269, Sept. 1992.

[13] R. Depommier and E. Dubois, "Motion estimation with detection of occlusion areas," in *Proc. IEEE Int. Conf. on Acoust. Speech Signal Process.*, pp. III.269–III.272, Mar. 1992.

[14] E. Dubois, "The sampling and reconstruction of time-varying imagery with application in video systems," *Proc. IEEE*, vol. 73, pp. 502–522, Apr. 1985.

[15] E. Dubois, "Motion-compensated filtering of time-varying images," *Multidim. Syst. Sig. Process.*, vol. 3, pp. 211–239, 1992.

[16] E. Dubois and S. Sabri, "Noise reduction in image sequences using motion-compensated temporal filtering," *IEEE Trans. Commun.*, vol. COM-32, pp. 826–831, July 1984.

[17] W. Enkelmann, "Investigations of multigrid algorithms for the estimation of optical flow fields in image sequences," *Comput. Vision, Graphics Image Process.*, vol. 43, pp. 150–177, 1988.

[18] W. T. Freeman and E. H. Adelson, "The design and use of steerable filters," *IEEE Trans. Pattern Anal. Machine Intell.*, vol. 13, pp. 891–905, Sept. 1991.

[19] S. Geman and D. Geman, "Stochastic relaxation, Gibbs distributions, and the Bayesian restoration of images," *IEEE Trans. Pattern Anal. Machine Intell.*, vol. PAMI-6, pp. 721–741, Nov. 1984.

[20] S. Geman and C.-R. Hwang, "Diffusions for global optimization," *SIAM J. Control and Optimization*, vol. 24, pp. 1031–1043, Sept. 1986.

[21] General Instruments Corporation, *DigiCipher HDTV system description*, Aug. 1991.

[22] M. Gennert and S. Negahdaripour, "Relaxing the brightness constancy assumption in computing optical flow," Tech. Rep. 975, MIT Artificial Intelligence Laboratory, June 1987.

[23] F. Glazer, *Hierarchical motion detection*. PhD thesis, Univ. of Massachusetts, Dept. Comp. Inform. Sci., Feb. 1987.

[24] D. Heeger, "Optical flow using spatiotemporal filters," *Intern. J. Comput. Vision*, vol. 1, pp. 279–302, 1987.

[25] F. Heitz, P. Perez, and P. Bouthemy, "Parallel visual motion analysis using multiscale Markov Random Fields," in *Proc. IEEE Workshop on Visual Motion*, pp. 30–35, Oct. 1991.

[26] E. Hildreth, "Computations underlying the measurement of visual motion," *Artificial Intell.*, vol. 23, pp. 309–354, 1984.

[27] B. Horn and B. Schunck, "Determining optical flow," *Artificial Intell.*, vol. 17, pp. 185–203, 1981.

[28] J. Hutchinson, C. Koch, J. Luo, and C. Mead, "Computing motion using analog and binary resistive networks," *Computer*, vol. 21, pp. 52–63, Mar. 1988.

[29] A. Jain, *Fundamentals of digital image processing*. Information and System Sciences Series, Prentice Hall, 1989.

[30] J. Jain and A. Jain, "Displacement measurement and its application in interframe image coding," *IEEE Trans. Commun.*, vol. COM-29, pp. 1799–1808, Dec. 1981.

[31] R. Keys, "Cubic convolution interpolation for digital image processing," *IEEE Trans. Acoust. Speech Signal Process.*, vol. ASSP-29, pp. 1153–1160, Dec. 1981.

[32] S. Kirkpatrick, C. Gelatt Jr., and M. Vecchi, "Optimization by simulated annealing," *Science*, vol. 220, pp. 671–680, May 1983.

[33] J. Konrad, *Bayesian estimation of motion fields from image sequences*. PhD thesis, McGill University, Dept. Electr. Eng., June 1989.

[34] J. Konrad, "Use of colour in gradient-based estimation of dense two-dimensional motion," in *Proc. Conf. Vision Interface VI'92*, pp. 103–109, May 1992.

[35] J. Konrad and E. Dubois, "Estimation of image motion fields: Bayesian formulation and stochastic solution," in *Proc. IEEE Int. Conf. on Acoust. Speech Signal Process.*, pp. 1072–1075, Apr. 1988.

[36] J. Konrad and E. Dubois, "Multigrid Bayesian estimation of image motion fields using stochastic relaxation," in *Proc. IEEE Int. Conf. Computer Vision*, pp. 354–362, Dec. 1988.

[37] J. Konrad and E. Dubois, "Bayesian estimation of discontinuous motion in images using simulated annealing," in *Proc. Conf. Vision Interface VI'89*, pp. 51–60, June 1989.

[38] J. Konrad and E. Dubois, "Use of colour information in Bayesian estimation of 2-D motion," in *Proc. IEEE Int. Conf. on Acoust. Speech Signal Process.*, pp. 2205–2208, Apr. 1990.

[39] J. Konrad and E. Dubois, "Comparison of stochastic and deterministic solution methods in Bayesian estimation of 2D motion," *Image & Vision Computing*, vol. 9, pp. 215–228, Aug. 1991.

[40] J. Konrad and E. Dubois, "Bayesian estimation of motion vector fields," *IEEE Trans. Pattern Anal. Machine Intell.*, vol. PAMI-14, pp. 910–927, Sept. 1992.

[41] D. Le Gall, "MPEG: A video compression standard for multimedia applications," *Communications ACM*, vol. 34, pp. 46–58, Apr. 1991.

[42] M. Liou, "Overview of the p x 64 kbits/s video coding standard," *Communications ACM*, vol. 34, pp. 59–63, Apr. 1991.

[43] N. Metropolis, A. Rosenbluth, M. Rosenbluth, H. Teller, and E. Teller, "Equation of state calculations by fast computing machines," *J. Chem. Phys.*, vol. 21, pp. 1087–1092, June 1953.

[44] A. Mitiche, Y. Wang, and J. Aggarwal, "Experiments in computing optical flow with the gradient-based, multiconstraint method," *Pattern Recognition*, vol. 20, no. 2, pp. 173–179, 1987.

[45] C. Moloney and E. Dubois, "Estimation of motion fields from image sequences with illumination variation," in *Proc. IEEE Int. Conf. on Acoust. Speech Signal Process.*, pp. 2425–2428, May 1991.

[46] D. Murray and B. Buxton, "Scene segmentation from visual motion using global optimization," *IEEE Trans. Pattern Anal. Machine Intell.*, vol. PAMI-9, pp. 220–228, Mar. 1987.

[47] H.-H. Nagel and W. Enkelmann, "An investigation of smoothness constraints for the estimation of displacement vector fields from image sequences," *IEEE Trans. Pattern Anal. Machine Intell.*, vol. PAMI-8, pp. 565–593, Sept. 1986.

[48] A. Netravali and J. Robbins, "Motion-compensated television coding: Part I," *Bell Syst. Tech. J.*, vol. 58, pp. 631–670, Mar. 1979.

[49] A. Nguyen and E. Dubois, "Adaptive interlaced to progressive image conversion," in *Proc. Int. Workshop on HDTV*, Nov. 1992.

[50] A. Nguyen and E. Dubois, "Spatial directional interpolation using steerable filters," in *Proc. Canadian Conf. Electr. Comp. Eng.*, pp. MA4.8.1–MA4.8.4, Sept. 1992.

[51] H. Nguyen and E. Dubois, "Representation of motion fields for image coding," in *Proc. Picture Coding Symposium*, pp. 8.4.1–8.4.5, 1990.

[52] A. Nieminen, P. Heinonen, and Y. Neuvo, "A new class of detail-preserving filters for image processing," *IEEE Trans. Pattern Anal. Machine Intell.*, vol. PAMI-9, pp. 74–90, Jan. 1987.

[53] T. Reuter, "Standards conversion using motion compensation," *Signal Processing*, vol. 16, pp. 73–82, 1989.

[54] M. Sezan, M. Ozkan, and S. Fogel, "Temporally adaptive filtering of noisy image sequences using a robust motion estimation algorithm," in *Proc. IEEE Int. Conf. on Acoust. Speech Signal Process.*, pp. 2429–2432, May 1991.

[55] F. Spitzer, "Markov random fields and Gibbs ensembles," *Amer. Math. Mon.*, vol. 78, pp. 142–154, Feb. 1971.

[56] C. Stiller and D. Lappe, "Gain/cost controled displacement-estimation for image sequence coding," in *Proc. IEEE Int. Conf. on Acoust. Speech Signal Process.*, pp. 2729–2732, May 1991.

[57] G. Thomas, "HDTV bandwidth reduction by adaptive subsampling and motion-compensated DATV techniques," *SMPTE J.*, vol. 96, pp. 460–465, May 1987.

[58] O. Tretiak and L. Pastor, "Velocity estimation from image sequences with second order differential operators," in *Proc. IEEE Int. Conf. Pattern Recognition*, pp. 16–19, July 1984.

[59] K. Wohn, L. Davis, and P. Thrift, "Motion estimation based on multiple local constraints and nonlinear smoothing," *Pattern Recognition*, vol. 16, no. 6, pp. 563–570, 1983.

[60] Zenith and AT&T, *Digital Spectrum Compatible: Technical details*, Sept. 1991.

4

Edge Based 3-D Camera Motion Estimation with Application to Video Coding

A. Zakhor and F. Lari

University of California, Berkeley CA, USA.

4.1 Introduction

Motion Compensation (MC) plays an important role in image compression applications such as video conferencing, video telephony, medical imaging, and CD-ROM storage. The basic idea is to take advantage of temporal redundancies between adjacent frames in an image sequence to reduce the information transmission rate. This is accomplished by estimating the displacement between frames of picture elements, which may be either uniformly sized blocks or individual pixels.

Ideally, the apparent motion in most moving sequences taken with a real camera can be attributed to either camera motion or the movement of the objects in a scene. The movement due to the camera is generally referred to as global motion, whereas the movement of the objects is called local motion. There are a number of advantages in separating these two classes of motion in MC algorithms. First, if there is no local movement in the scene and only the camera is moving, the dynamics of the resulting video sequence can be adequately modeled by only estimating camera motion parameters. Second, in more realistic situations where there is both local and global motion present in the video sequence, it is conceivable to remove a great deal of redundancy between successive frames by estimating the parameters related to camera motion. This is because the degrees of freedom of camera movement is small compared to the complex motion of the objects in typical scenes. Specifically, camera motion parameters can be effectively used to predict the stationary parts of the scene, thus saving the bandwidth needed to transmit motion vectors for them.

Most existing video coding techniques such as CCITT's Recommendation H.261, also referred to as $p \times 64$, and the MPEG standards only use local motion estimation to form a prediction of the current frame [1]. Even though global motion estimation is not officially part of these two standards, several authors have exploited camera motion estimation in the context of video sequence coding. For example, Adolph and Buschmann [2] propose a coder for television video signals at 1.15 mbps, where global MC is carried out before local one. In doing so, they assume the global motion to consist of only zoom and pan, and estimate it using a frame matching algorithm. Their experimental results indicate that for the given rate of 1.15 mbps the quantization step size for coding error frames can be reduced by a factor of three if global MC is exploited, and that the quality of the coded scenes is considerably improved if global MC is applied.

In addition to [2], Baker [3] has shown that a two stage global/local motion compensation approach improves motion prediction and reduces the amount of motion side information. Keesman [4], Hoetter [5] and Wu and Kittler [6] also show the advantages of global motion estimation schemes. Hoetter models zoom and pan, while Keesman, Wu and Kittler model rotation as well as zoom and pan parameters. The techniques used for estimating the global motion parameters vary considerably. For instance [4, 5, 6] use pel recursive algorithms, while [2, 3] use luminance block matching.

From the above survey, it is clear that most of the existing global MC techniques used for video compression applications only consider zoom, pan and possibly rotation as global motion parameters. Thus, an important parameter that is missing from these approaches is camera translation. This is not surprising since there are some intrinsic difficulties involved with using camera translation parameters in generating motion compensated frames. As we will see on section 4.2, this has to do with the fact that, unlike other global parameters such as zoom, pan and rotation, translation parameters require the depth map of the pixels in a scene before they can be used to derive the motion compensated frames.

In this chapter, we will consider a seven parameter camera model, including rotation and translation for global motion compensation of video sequences. In doing so, we will exploit the results on the problem of "structure and motion" recovery in the Computer Vision (CV) literature[7, 8, 9, 10], and "camera calibration" in the photogrammetry[11, 12, 13]. These problems appear in diverse applications such as autonomous navigation, stereo reconstruction, robot vision, object recognition, scene analysis and cartography [11, 14, 15]. The most basic formulation of "structure and motion" problem in CV consists of recovering translation and rotation parameters and structure of an object in three dimensional (3-D) space. This problem is in fact equivalent to the problem of recovering the rotation and camera translation parameters of a camera, and the 3-D depth map of a stationary scene from its video sequence. Indeed, this is the motivation behind applying results on "structure from motion" to global camera parameter estimation.

The basic approach to structure from motion in CV literature consists of three steps [10]: extracting feature points, establishing correspondence by matching the features, and finally computing the structure and motion parameters based on the feature matches. Our basic approach to global motion estimation in this chapter is similar to this. Since the number of matched features in our applications is large enough, we need not be concerned about uniqueness issues typically considered in CV applications [8]. Another inherent advantage of the large number of matches is that we can use simple linear algorithms, rather than iterative, or more complex nonlinear optimization techniques [8, 9, 10].

The two major steps in most existing global motion estimation algorithms, including ours are: (a) estimating local motion vectors ; (b) using the motion vectors in linear least squares estimation of global parameters. One of our goals in this chapter is to demonstrate that edge matching can be successfully used in step (a) of most global motion estimation algorithms. Specifically, we will show that edges alone are sufficient to determine the seven parameters in rotation/translation model as well as the parameters in the zoom/pan model. This can have important practical consequences since it implies that for camera motion estimation, the 8 bit luminance information in video sequences can be collapsed to one bit edge information without loss of performance.

Edge matching is a subclass of token tracking algorithms used for motion estimation. More generally, in token tracking schemes, distinctive image features are detected and their correspondences tracked from frame to frame in a sequence. The features, such as corners, blobs, and straight lines are assumed to arise from distinctive scene features. There are a number of motivations behind the use of edges as tokens for matching: First, it has been argued that the human visual system computes motion by temporal filtering of edge signal [16]. Second, it has been suggested that one of the advantages of using edges over raw irradiance schemes is that they are tied more closely to physical features [17].

In spite of these motivations, it has been found that the locations of edges by themselves are not sufficient for local motion estimation [18]. As a result, a number of other attributes such as orientation, strength, and curvature are used in conjunction with the location of edges in order to overcome the aperture ambiguity problem [19, 20, 21].

In this chapter, we will demonstrate that unlike local motion estimation, global estimation can be accomplished by matching the location of edges in consecutive frames of a video sequence. In doing so, we compare the performance of our edge based global motion estimation techniques with the well known Block Matching Algorithm (BMA), which is extensively used in today's video compression standards. Note that BMA can also be considered a subclass of token tracking and matching algorithms. Specifically, the particular token which is matched in BMA is the intensity of raw images.

Compared to intensity matching, the major payoff in using edge matching techniques is the simplicity involved in using one bit edge information rather than 8 bit intensity information: this simplicity can potentially have implications on both hardware and software implementation of global motion estimation techniques. In addition, since edges are more closely tied to physical features in a scene than individual pixel intensities, they are likely to be useful in other parts of typical video compression systems. Examples of these would be edge based vector quantization, edge based local MC, local MC in conjunction with global MC, segmented video coding, model based coding, etc [35].

To summarize, our goal in this chapter is two fold: First, we will show that the seven parameter camera model consisting of rotation and translation is applicable and useful in video compression applications. Second, we show that unlike local motion estimation, edges matching can be used to replace intensity based block matching for global motion estimation. The outline of the chapter is as follows: In section 4.2, we describe the global motion models used by our proposed algorithms presented in sections 4.3.1 and 4.3.2. In section 4.4, simulations are used to examine the performance of our global motion estimation algorithms in the context of video compression. Section 4.5 compares the computational complexity of our algorithms with traditional intensity based BMA algorithms, and section 4.6 includes conclusion.

4.2 Global Motion Models

In this section, we describe the specific model parameters for our global motion estimation algorithms. In general, camera movement can be decomposed into the following categories:

- Change of the camera focal length: zoom.
- Rotation around an axis normal to the camera axis: pan.
- Rotation around the camera axis.
- Translation along the camera axis.
- Translation in the plane normal to the camera axis.

Each of the above affects the temporal appearance of the resulting video sequence in a specific way. In this section, we describe the analytic relationship between the above parameters and pixel intensities of resulting video sequences of stationary scenes. These relationships can be exploited in video compression applications to predict future frames based on current frames and camera motion parameters.

As we will see in this section, the first three parameters, i.e. zoom, pan and rotation, by themselves are enough to determine the apparent intensity change in a video sequence. On the other hand, for translation along, or in the plane normal to the camera axis, the translation parameters by themselves are not sufficient, and the depth map of the scene under consideration is also needed to estimate the changes in a video sequence. To distinguish between these two cases, we consider two classes of models: the first class, discussed in section 4.2.1 includes zoom, pan and rotation, while the second class, discussed in section 4.2.2, considers translation and rotation.

4.2.1 Zoom, pan, and rotation model

A video camera uses central projection to map the 3-D space onto the image plane at the focal point[22]. Using the notation in Figure 4.1, the object space coordinates (x, y, z) of a point in the 3-D space and the image plane coordinates, (X, Y), of its image are related by the perspective transformation

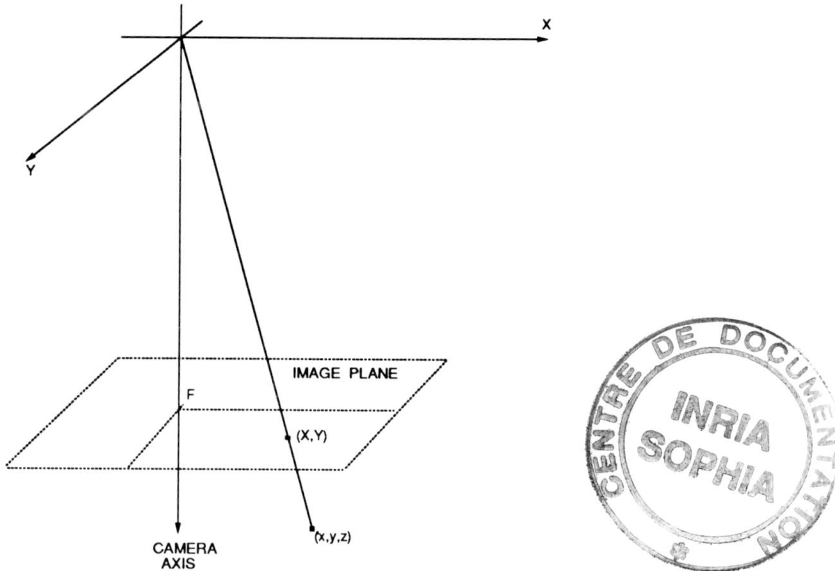

Figure 4.1: Projection of the point (x, y, z) in 3-D space onto (X, Y) in the camera image plane.

$$X = \frac{Fx}{z} \qquad Y = \frac{Fy}{z} \tag{4.1}$$

We now describe mathematical models for zoom and pan. A zoom is a change in the camera focal length. If X_1, Y_1 denote the image plane coordinates of a point (x, y, z)before zoom, and X_2, Y_2 the image plane coordinates of the same point after zoom, it is easily shown that

$$X_2 = \frac{F_2}{F_1} X_1 \qquad Y_2 = \frac{F_2}{F_1} Y_1 \tag{4.2}$$

where F_1 and F_2 are the focal lengths before and after the zoom. We define $c_1 = F_2/F_1$ to be the zoom parameter. A pan is a rotation of the camera around an axis parallel to the image plane. As shown in [3], if the rotation is small enough, the entire frame is displaced uniformly by the same vector:

$$\begin{pmatrix} X_2 \\ Y_2 \end{pmatrix} = \begin{pmatrix} X_1 \\ Y_1 \end{pmatrix} + \begin{pmatrix} t_x \\ t_y \end{pmatrix} \tag{4.3}$$

where t_x, t_y are the rotation angles around the x and y axes respectively. In arriving at the above equations, we implicitly assume that the rotation angles t_x

and t_y are small so that $cos(t_x) \approx= 1$, $cos(t_y) \approx 1$, $sin(t_x) \approx t_x$ and $sin(t_y) \approx t_y$ [5]. Furthermore, we assume the rotation to be small enough so that $x_{1,2}t_x << z$ and $y_{1,2}t_y << z$ [5]. This implies that for a given rotation angle and field of view, the approximation becomes increasingly more valid for points with larger depth, z.

In addition to rotation around an axis parallel to the image plane, we also consider rotation around the camera axis, i.e. z axis in Figure 4.1. A rotation of t_z around the camera axis can be described by:

$$\begin{pmatrix} X_2 \\ Y_2 \end{pmatrix} = \begin{pmatrix} \cos t_z & \sin t_z \\ -\sin t_z & \cos t_z \end{pmatrix} \begin{pmatrix} X_1 \\ Y_1 \end{pmatrix} \approx \begin{pmatrix} 1 & t_z \\ -t_z & 1 \end{pmatrix} \begin{pmatrix} X_1 \\ Y_1 \end{pmatrix} \tag{4.4}$$

where the above approximation assumes that the rotation angle t_z is small.

Combining zoom, pan and rotation models described above, we arrive at the following model[22]:

$$\begin{pmatrix} X_2 \\ Y_2 \end{pmatrix} = \begin{pmatrix} c_1 & c_2 \\ -c_2 & c_1 \end{pmatrix} \begin{pmatrix} X_1 \\ Y_1 \end{pmatrix} + \begin{pmatrix} c_3 \\ c_4 \end{pmatrix} \tag{4.5}$$

where c_1, c_2, c_3 and c_4 are given by:

$$c_1 = \frac{F_2}{F_1}$$

$$c_2 = t_z c_1$$

$$c_3 = F_2 t_y + t_z F_2 t_x$$

$$c_4 = F_2 t_x - t_z F_2 t_y$$

Thus, assuming there is no translation, the four parameters c_i completely describe all the motion in a video sequence of a stationary scene resulting from camera zoom, pan and rotation.

4.2.2 Rotation and translation model

In this section, we will consider a seven parameter model consisting of camera rotation and translation for global motion estimation. Before dealing with camera motion, consider the problem of recovering the motion parameters of a 3-D rigid body from the video sequence captured by a stationary camera. This problem is extensively discussed in the CV literature [7, 8, 9, 10]. It has been shown that any 3-D rigid body motion is equivalent to a rotation by an angle θ around an axis through the origin with directional cosines n_1, n_2, n_3 followed by a translation $(\Delta x, \Delta y, \Delta z)$,

$$\begin{pmatrix} x_2 \\ y_2 \\ z_2 \end{pmatrix} = R \begin{pmatrix} x_1 \\ y_1 \\ z_1 \end{pmatrix} + \vec{T} \tag{4.6}$$

where R is a 3×3 rotation matrix:

$$R \equiv \begin{pmatrix} r_1 & r_2 & r_3 \\ r_4 & r_5 & r_6 \\ r_7 & r_8 & r_9 \end{pmatrix} =$$

$$\begin{pmatrix} n_1^2 + (1 - n_1^2)\cos\theta & n_1 n_2(1 - \cos\theta) - n_3 \sin\theta & n_1 n_3(1 - \cos\theta) + n_2 \sin\theta \\ n_1 n_2(1 - \cos\theta) + n_3 \sin\theta & n_2^2 + (1 - n_2^2)\cos\theta & n_2 n_3(1 - \cos\theta) - n_1 \sin\theta \\ n_1 n_3(1 - \cos\theta) - n_2 \sin\theta & n_2 n_3(1 - \cos\theta) + n_1 \sin\theta & n_3^2 + (1 - n_3^2)\cos\theta \end{pmatrix} \quad (4.7)$$

$\vec{T}$ is the translation vector defined by:

$$\vec{T} \equiv \begin{pmatrix} \Delta x \\ \Delta y \\ \Delta z \end{pmatrix}$$

and (x_1, y_1, z_1) and (x_2, y_2, z_2) correspond to the coordinates of a point on the object before and after motion respectively [22]. It has been shown that once the elements of R are computed, n_1, n_2, n_3, θ can be found using a technique described in [8].

Our problem is slightly different from the above formulation in that the scene is assumed to be stationary while the camera is translating and rotating in the 3-D space. Thus, we must interpret the variables in the above formulation slightly differently. Specifically, we change the coordinate system so that the center of the camera is at origin, and therefore stationary at all times. Under these circumstances, the coordinates of the points in the 3-D scene will appear to be moving with respect to the camera. Under these conditions, equation 4.6 still holds, except for a change in the definition of the variables: the coordinates of a fixed point in 3-D space with respect to the camera is denoted by (x_1, y_1, z_1) before motion, and by (x_2, y_2, z_2) after motion, and its projection in the image plane moves from (X_1, Y_1) before motion to (X_2, Y_2) after motion. In this new interpretation, the rotation matrix R and translation vector T will correspond to the rotation and translation of the camera rather than the rigid body in the scene.

Based on equation 4.6, the seven parameters n_1, n_2, n_3, θ $\Delta x, \Delta y, \Delta z$ are sufficient for recovering (x_2, y_2, z_2) from (x_1, y_1, z_1). However, in our video application, the ultimate goal is to recover, estimate , or predict the *image plane* coordinates in frame $n + 1$ from those in frame n. Specifically, our modeling goal is to relate (X_1, Y_1) to (X_2, Y_2) in terms of camera motion parameters. This way, once the camera motion is estimated from the image sequence, it can be effectively used for predictive coding.

Since the image and 3-D space coordinates are related to each other through equation 4.2, using equation 4.6, we can relate image coordinates before and after motion. Using the third line of equation 4.6 to compute the ratio $\frac{z_1}{z_2}$, using equation 4.2 to express x_1, x_2, y_1, and y_2 in the first two equations of 4.6 in terms of X_1, Y_1, X_2 and Y_2, and assuming for simplicity $F = 1$, we obtain [8]:

$$X_2 = \frac{(r_1 X_1 + r_2 Y_1 + r_3)\, z_1 + \Delta x}{(r_7 X_1 + r_8 Y_1 + r_9) z_1 + \Delta z} \quad (4.8)$$

$$Y_2 = \frac{(r_4X_1 + r_5Y_1 + r_6)z_1 + \Delta y}{(r_7X_1 + r_8Y_1 + r_9)z_1 + \Delta z} \tag{4.9}$$

Thus, camera rotation and translation results in a relationship between image points (X_1, Y_1) and (X_2, Y_2) which is not only dependent on their coordinates, but also on the depth z_1 of the point in 3-D space whose projection in the image plane is at coordinates (X_1, Y_1). Note that if there is no translation, i.e. $\Delta x = \Delta y = 0$, then from equations 4.8 and 4.9, (X_2, Y_2) can be easily related to (X_1, Y_1) without requiring any depth information. Thus, it is the translation, but not the rotation that requires the depth information.

4.3 Algorithm Description

In this section, we will propose algorithms for each of the camera models discussed in section 4.2. Specifically, section 4.3.1 describes a global motion estimation and compensation algorithm based on zoom, pan and rotation, and section 4.3.2 deals with a seven parameter global motion estimation and compensation algorithm for camera translation and rotation. As will be seen, there is a great deal of similarity between the two algorithms.

4.3.1 Zoom, pan and rotation algorithm

In this section, we describe our edge based algorithm for finding zoom, pan and rotation parameters of the camera. As mentioned earlier, there are a number of existing zoom and pan detection algorithms [2, 3, 23], most of which consist of two steps: (1) estimate local motion vector for each block; (2) use the location of blocks in the current frame and their corresponding motion compensated location in the next frame in equation 4.5 in order to find the linear least squares estimate of zoom and pan parameters [3]. Since zoom and pan are smoothly varying processes, in many practical situations it is sufficient to use the motion vectors between every few frames in finding the least squares estimate of zoom and pan parameters. In addition, since displacement vectors between every few frames are likely to be larger than between consecutive frames, pel recursive algorithms [5, 6] are not suitable; block matching algorithms can be used, but since their search area has to be large, they are computation intensive.

In this section, we propose an edge based approach to global motion estimation of zoom, pan and rotation parameters to be used in predictive coding of video. One motivation behind our algorithm is to reduce the computational intensity of existing luminance based estimation algorithms. This will be discussed in more detail in section 4.5. The flowchart of our approach is shown in Figure 4.2. If A and B denote the two frames between which zoom and pan parameters are to be estimated, then the steps of our algorithms can be described in the following way:

1. Find the edges in frames A and B: We have chosen an edge detection algorithm described in section 4.5.1 for its speed and precision. As will

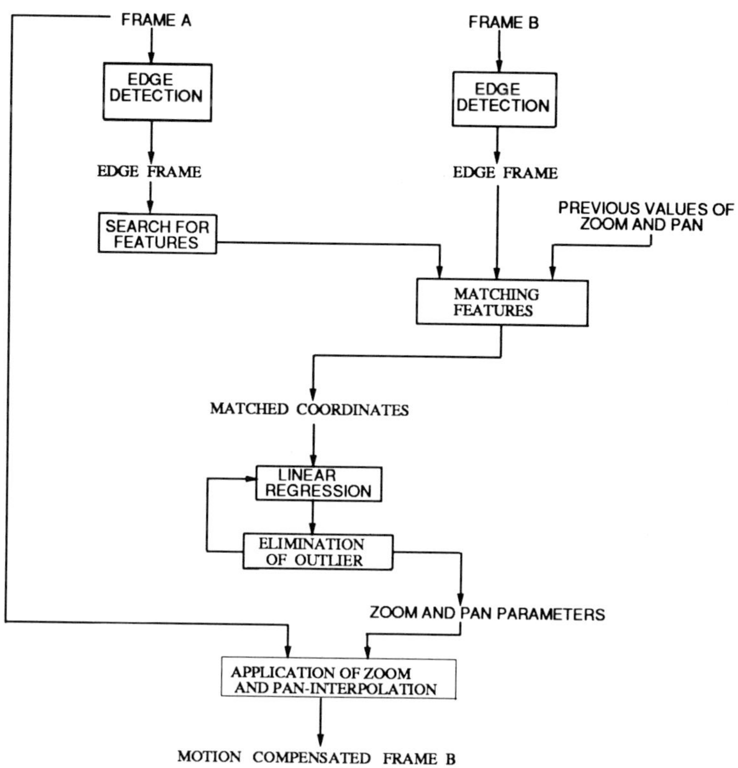

Figure 4.2: Flowchart of the zoom/pan algorithm of section 3.3.1

be seen, our edge detection algorithm consists of smoothing and finding directional derivatives [24, 25, 26].

2. Search for "features" in frame A: We refer to an 8×8 block as a "feature" if it contains more than 8 edge pixels. This is done in order to prevent false matches in the next step. To avoid aperture ambiguity, blocks with strictly vertical and horizontal edges are discarded. Note that the minimum distance between the coordinates of the location of two features is $(8, 8)$.

3. Search for perfect matches of the frame A features in frame B: Two edge features are said to be perfectly matched if they are identical in the binary domain. To maximize the computation speed, the search area is centered around the point the feature would have moved given the previously detected values of zoom and pan. If no previous information is known, the search begins around the feature position in frame A. If a match is found the coordinates of the feature in frame A, $(X_{i,A}, Y_{i,A})$ and the matched coordinates in frame B, $(X_{i,B}, Y_{i,B})$ are stored for use in the next step.

4. Find linear least squares estimates of c_1, c_2, c_3 and c_4 by fitting the model shown in equation 4.5 to the K pairs of coordinates obtained in the previous step. This can be accomplished by minimizing the following error expression

$$Error = \sum_{n=1}^{K} [(X_{n,B} - c_1 X_{n,A} - c_2 Y_{n,A} - c_3)^2 + (Y_{n,B} - c_1 Y_{n,A} + c_2 X_{n,A} - c_4)^2] \tag{4.10}$$

setting $\frac{\partial E}{\partial c_i}$ for $i = 1, 2, 3$ to zero, and solving a linear system of equations to estimate c_1, c_2, c_3, and c_4.

5. Remove "outlier" coordinates and repeat the previous step until the iterations converge: by outlier coordinates, we mean the ones whose residual errors are more than one standard deviation away from the average error of all the coordinate pairs. We consider the algorithm to have converged if the parameter estimates do not change substantially from one iteration to the next. For all the examples we present in this chapter, the above procedure converges in 2 to 3 steps.

6. Apply the global motion parameters obtained in the above step to the interpolated version of frame A in order to obtain "motion" compensated prediction for frame B. This is necessary because even though the motion vectors used in estimating global motion parameters have pixel accuracy, the effective motion vectors for each pixel does not necessarily have pixel accuracy. As a result, the displaced pixels according to the global parameters do not necessarily coincide with the sampling grid. Thus, we are forced to carry out interpolation to derive a motion compensated prediction of future frames. We adopt a very simple interpolation scheme in which the intensity value of a pixel (i, j) on the sampling grid is given by:

$$I(i, j) = \frac{\sum_{k \in A_{ij}} d_k I_k}{\sum_{k \in A_{ij}} (d_k + 1)} \tag{4.11}$$

where A_{ij} is a $2\Delta \times 2\Delta$ area centered around (i, j) pixel, Δ is the spacing on the sampling grid, d_k is the distance of the kth point with intensity I_k from the (i, j) pixel.

Note that the linear least squares approach of step 4 has already been reported in the literature [3, 6, 5, 4, 27]. Thus the major difference between our approach and existing global motion estimation techniques is the edge based feature matching, i.e. steps 1-3.

4.3.2 Rotation and translation algorithm

In this section, we will propose an algorithm for estimating rotation and translation parameters of a camera based on the model described in section 4.2.2. As mentioned earlier, camera translation parameter requires the knowledge of the depth map in order for it to be effectively used in predictively coding. Thus, our proposed algorithm in this section must recover both the seven parameters corresponding to translation and rotation of the camera and the depth map.

There are a number of algorithms in the CV literature for computing translation and rotation of rigid bodies [7, 8, 10]. The algorithm we propose in this section is heavily based on the approach proposed by Tsai and Huang in 1984 [8]. While the main objective in [8] is to derive conditions for unique recovery of rigid body objects in scene analysis, our objective is to use the camera parameters in predictive video coding.

Recall from section 4.2.2 that if the projection of a point, P in 3-D space onto the image plane is (X_1, Y_1) before camera motion and (X_2, Y_2) after camera motion, then the relationship between these coordinates is given by equations 4.8 and 4.9. Rewriting z_1 in each of these equations in terms of motion parameters and image plane coordinate, we obtain:

$$z_1 = \frac{\Delta x - \Delta z X_2}{X_2(r_7 X_1 + r_8 Y_1 + r_9) - (r_1 X_1 + r_2 Y_1 + r_3)} \tag{4.12}$$

from equation 4.8 and

$$z_1 = \frac{\Delta y - \Delta z Y_2}{Y_2(r_7 X_1 + r_8 Y_1 + r_9) - (r_4 X_1 + r_5 Y_1 + r_6)} \tag{4.13}$$

from equation 4.9. Equating the right-hand sides of the above equations, we obtain[8]:

$$[X_2 \quad Y_2 \quad 1] \quad E \begin{bmatrix} X_1 \\ Y_1 \\ 1 \end{bmatrix} = 0 \tag{4.14}$$

where the matrix E is referred to as the "essential matrix" and is given by:

$$E = \begin{pmatrix} \Delta z \cdot r_4 - \Delta y \cdot r_7 & \Delta z \cdot r_5 - \Delta y \cdot r_8 & \Delta z \cdot r_6 - \Delta y \cdot r_9 \\ \Delta x \cdot r_7 - \Delta z \cdot r_1 & \Delta x \cdot r_8 - \Delta z \cdot r_2 & \Delta x \cdot r_9 - \Delta z \cdot r_3 \\ \Delta y \cdot r_1 - \Delta x \cdot r_4 & \Delta y \cdot r_2 - \Delta x \cdot r_5 & \Delta y \cdot r_3 - \Delta x \cdot r_6 \end{pmatrix} \tag{4.15}$$

$$\equiv \begin{pmatrix} e_1 & e_2 & e_3 \\ e_4 & e_5 & e_6 \\ e_7 & e_8 & e_9 \end{pmatrix}$$

From equation 4.14 it is clear that the matrix E can be determined only to within a scale factor. Since elements of E are linear in $\Delta x, \Delta y$, and Δz, this implies that the translation parameters can be determined to within a scale factor. This can also be seen by noting that z_1 in equations 4.12 and 4.13 is linear in the translation parameters. This ambiguity is not merely a mathematical artifact; rather, it is inherent to the problem at hand, and can be physically explained by considering that an object at depth d_1 translating at T_1 appears the same to the camera as one at depth βd_1 translating at βT_1.

To avoid this ambiguity, without loss of generality, we set one of the e_i to 1. The only potential problem this might cause is in situations where the particular

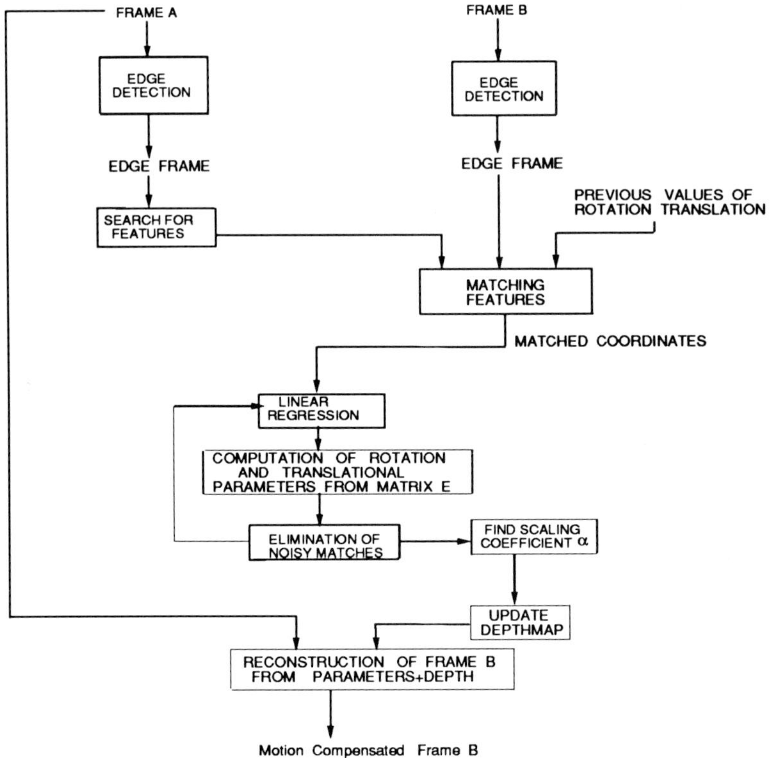

Figure 4.3: Flowchart of the rotation/translation algorithm in section 3.3.2.

e_i set to 1, is actually zero. However as we will see later, such problems can be easily detected numerically, and can therefore be avoided.

The flowchart of our proposed global MC algorithm is shown in Figure 4.3. Note that the flowchart implicitly assumes that the depth map for the first frame of the scene is available at the receiver. In spite of this, the depth map need not necessarily be *transmitted* to the receiver. For instance, if the first two frames of each scene are intra-frame coded, the transmitter and the receiver could compute the same depth map independently. At the end of this section, we will describe a way of determining the depth map for the first frame.

If A and B denote the two frames between which the global motion is to be estimated, then the steps of our algorithms shown in the flowchart of Figure 4.3 can be described in the following way:

1. Detect edges similar to step 1 in section 4.3.1.

2. Detect features similar to step 2 in section 4.3.1, except that the minimum distance between the location of features can be $(1,1)$ rather than $(8,8)$. This is because, even though the window size for detecting features is 8×8,

unlike in section 4.3.1, we allow them to overlap, thus resulting in a larger number of features and reduced noise sensitivity.

3. Feature matching similar to step 3 in section 4.3.1.

4. Find linear least squares estimates of e_i by fitting the model shown in equation 4.14 to the K pairs of coordinates obtained in step 3. This can be accomplished by minimizing the following error expression

$$\begin{aligned} Error \;=\; & \sum_{i=1}^{K}[(X_{i,B}X_{i,A}e_1 \;+\; X_{i,B}Y_{i,A}e_2 \;+ \\ & X_{i,B}e_3 \;+\; Y_{i,B}X_{i,A}e_4 \;+\; Y_{i,B}Y_{i,A}e_5 \;+ \\ & Y_{i,B}e_6 \;+\; X_{i,A}e_7 \;+\; Y_{i,A}e_8 \;+\; e_9)^2] \end{aligned} \tag{4.16}$$

   where $(X_{i,B}, Y_{i,B})$ denotes coordinates of a feature in frame B which is matched to a feature in frame A with coordinates $(X_{i,A}, Y_{i,A})$.

   As mentioned earlier, in solving the above linear least squares problem, we have to set one of the e_i to one in order to remove a scaling ambiguity. As an example, consider the case where e_8 is chosen to be one. Considering equation 4.15, this choice is only problematic when $e_8 \equiv \Delta y r_2 - \Delta x r_5 = 0$, i.e. in either of the following two situations: (a) the only non-zero translation is along the z direction, i.e $\Delta z \neq 0$ and $\Delta x = \Delta y = 0$; (b) translation along the x direction is zero and there is no camera rotation, i.e. $\Delta x = 0$, $\theta = 0$. We have found numerically that in physical situations where the choice of $e_i = 1$ is invalid, then the resulting translation parameters become unrealistically large, signaling an unappropriate choice of $e_i = 1$. Under these conditions, one simply switches from one e_i to another in order to remove the scaling ambiguity.

   Once, an appropriate parameter e_j has been set to one, we can find the remaining parameters by minimizing the error expression in equation 4.16, or equivalently by setting $\frac{\partial E}{\partial e_i}$ for $i = 1, ..9, i \neq j$ to zero, and solving a linear system of equations.

5. Once the matrix E is computed to within a scale factor, we find the seven rotation and translation parameters by computing the Singular Value Decomposition (SVD) of the 3×3 matrix E[8]. If the SVD of E is given by $E = U\Sigma V^T$, then there are two solutions for the rotation matrix, R, given by:

$$R = U \begin{pmatrix} 0 & -1 & 0 \\ 1 & 0 & 0 \\ 0 & 0 & s \end{pmatrix} V^T \tag{4.17}$$

   or

$$R = U \begin{pmatrix} 0 & 1 & 0 \\ -1 & 0 & 0 \\ 0 & 0 & s \end{pmatrix} V^T \tag{4.18}$$

where $s = \det(U)\det(V)$, and one solution for the translation vector, up to a scale factor α, given by:

$$\begin{pmatrix} \Delta x \\ \Delta y \\ \Delta z \end{pmatrix} = \alpha \begin{pmatrix} (-\phi_1^T\phi_1 + \phi_2^T\phi_2 + \phi_3^T\phi_3)^{\frac{1}{2}} \\ (\phi_1^T\phi_1 - \phi_2^T\phi_2 + \phi_3^T\phi_3)^{\frac{1}{2}} \\ (\phi_1^T\phi_1 + \phi_2^T\phi_2 - \phi_3^T\phi_3)^{\frac{1}{2}} \end{pmatrix} \tag{4.19}$$

where ϕ_i is the ith row of E for $i = 1,2,3$. Only one of the two solutions for R together with the appropriate sign for α, will yield positive z_1 and z_2. Since the object must be in front of the camera, and therefore have a positive depth, the solution is unique to within a positive scale factor.

6. Remove the outliers using the following steps: (a) recover rotation and translation parameters from the e_i; (b) use the matches in equations 4.12 and 4.13 to find z_1; (c) use equations 4.8 and 4.9 to compute new values for (X_2, Y_2); (d) find the error between these newly computed (X_2, Y_2) and those that were actually used in the matching process; (e) remove outliers by discarding matched features which have errors of more than one standard deviation.

7. Compute the scaling factor α: At this point, an important observation needs to be made. While both the depth z_1 and the coefficients $\Delta x, \Delta y, \Delta z$ are defined to within a common scale factor for any pair of frames, consecutive scale factors should be chosen in such a way that they result in consistent depths in consecutive frames. This implies that there is effectively only one degree of freedom for the entire sequence, as far as the scaling factor is concerned. Thus, changing the value of this degree of freedom, does not affect the relative depth of various frames, but only fixes their absolute depth.

   For the sake of argument, we assume that this degree of freedom has been fixed to a specific, but arbitrary value, so that the absolute depth map of frame $n-1$ is given by $\vec{z}_{n-1}$. Also, assume that step 5 has been applied to find the translation motion parameters between frames $n-1$ and n up to scaling factor, α_n. In addition to motion parameters, a by product of step 5, is α_n dependent depth map for frame $n-1$, which we will refer to as $\vec{z}_{n-1}(\alpha_n)$. Our approach is to choose α_n in such a way that $\vec{z}_{n-1}(\alpha_n)$ becomes as close as possible to its absolute depth map $\vec{z}_{n-1}$. This is done by solving another linear regression problem in which the depth of the pixels in $\vec{z}_{n-1}$ are matched to those in $\vec{z}_{n-1}(\alpha_n)$. Thus, we estimate

$$\alpha_n = \frac{\sum_{i \in K} \vec{z}^{\,i}_{n-1}}{\sum_{i \in K} \vec{z}^{\,i}_{n-1}(\alpha_n)}$$

   where the set K consists of the pixels for which depth is defined in both $\vec{z}_{n-1}$ and $\vec{z}_{n-1}(\alpha_n)$. Once α_n has been determined, the motion parameters between frames $n-1$ and n are uniquely determined, and can therefore be applied to the absolute depth map of frame $n-1$ in order to compute the absolute depth map of frame n through equation 4.6. The absolute depth map of frame n can then be used to determine subsequent scaling factors such as α_{n+1}.

8. Transmit the 7 parameters to the receiver so that the decoder can update the existing depth map based on the new set of rotation and translation camera parameters. Clearly, the depth map associated with the uncovered points in frame $n+1$ cannot be determined based on the depth map of frame n and the camera parameters.

9. Apply the rotation and translation parameters and the updated depthmap to frame A to obtain a global motion compensated prediction for frame B. Equations 4.8 and 4.9 are used to compute the new coordinates, (X_2, Y_2), of the point in frame B which corresponds to the point (X_1, Y_1) in frame A. In other words, the intensity value of pixel (X_2, Y_2) in frame B is assigned the same value as that of pixel (X_1, Y_1) in frame A, where the coordinates are related through equations 4.8 and 4.9. Since this mapping does not guarantee (X_2, Y_2) to fall on a rectangular sampling grid, interpolation is necessary. Our approach to interpolation is similar to the one described in section 4.3.1 for the zoom, pan, rotation algorithm.

10. Once the motion compensated version of B is computed, the error between the actual and compensated frame B is quantized and coded for transmission to the receiver. We will describe the particular quantization and coding scheme we consider in our experimental results in section 4.4.

Having described our basic algorithm, we now explain our approach to arriving at a depth map for the first frame. If we use the above algorithm to recover the translation and rotation parameters between frames 1 and 2, then equation 4.12 or 4.13 can be used to arrive at the depth z_1 for the pixel (X_1, Y_1) in frame 1. Since the depth map is only explicitly calculated for the first frame in a given scene, it is important for it to be dense; otherwise, the number of predicted pixels in motion compensated frames will not be dense, resulting in large error frames to be quantized and coded. However since the dense depth map need only be calculated once for each scene, this is only a one time cost.

4.4 Experimental Results

In this section, we will describe experimental results on the two algorithms described in section 4.3. Specifically, section 4.4.1 includes results on the algorithm of section 4.3.1, and section 4.4.2 includes results on the algorithm of section 4.3.2.

4.4.1 Zoom and pan results

We have applied the Zoom and Pan algorithm to three 720x480 pixel frame sequences. The "HDTV" sequence consists of 28 luminance frames with synthetic zoom and pan. The other two sequences are two pieces of the "Ping-Pong" sequence: "Ping-Zoom" with 40 frames of pure zoom along with random foreground motion, and "Ping-Pan" with 80 frames of pure pan along with random foreground motion. The zoom and pan parameters for "HDTV", "Ping-Pan", and "Ping-Zoom" are determined every 3, 10, and 3 frames respectively. Typical frames for the HDTV and PingZoom and their corresponding edges are shown in Figures 4.4, 4.5. Note that the edges corresponding to the pingpong table in the frames for the Ping-Zoom are jagged. This can be attributed to the fact that the edge detection algorithm was run on a frame, rather than a field and that there is a differential delay in between the even and odd fields causing the edge to appear jagged. Even though the found edges are imperfect, as we will see later, they do not affect the accuracy with which we can determine the zoom and pan parameters.

Figures 4.6, 4.7, and 4.8 show the predictive Mean Squared Error (MSE) for globally motion compensated frames obtained via traditional luminance based BMA and our edge based algorithm for the "HDTV", "Ping-Pan", and "Ping-Zoom" sequences respectively. By globally motion compensated frames via traditional BMA, we mean frames obtained by applying steps 4, 5, and 6 of our algorithms to the motion vectors obtained by full search luminance match. The solid curves in all the three figures denote the non compensated Frame Difference (FD) without any zoom or pan compensation. As seen, our edge based algorithm has comparable or better MSE performance than traditional intensity based BMA.

Few comments about the above results are in order: First, the sudden luminance change between frames 9 and 10 or the "HDTV" sequence does not affect the performance of our edge based technique, while the luminance method yields a large number of false matches resulting in substantially lower MSE performance. Second, we have found that iterations do improve the performance of our edge based techniques[1].

Visual results shown in Figure 4.9 fully confirm our results based on MSE plots. This figure shows the FD of intensity based BMA, our edge BMA based technique, and the non compensated FD for one frame of the PingZoom sequence. As seen, the edge based zoom and pan compensation results in approximately the same FD than conventional intensity based BMA. Furthermore, the stationary objects such as the picture on the wall, and the ping-pong table result in small FD, while the region corresponding to the player results in larger FD due to the players movement in addition to camera zoom. Similar results have been obtained for the Ping-Pan and HDTV sequence.

[1] Experiments show that this is also the case for traditional intensity based BMA based zoom and pan algorithms.

(a)

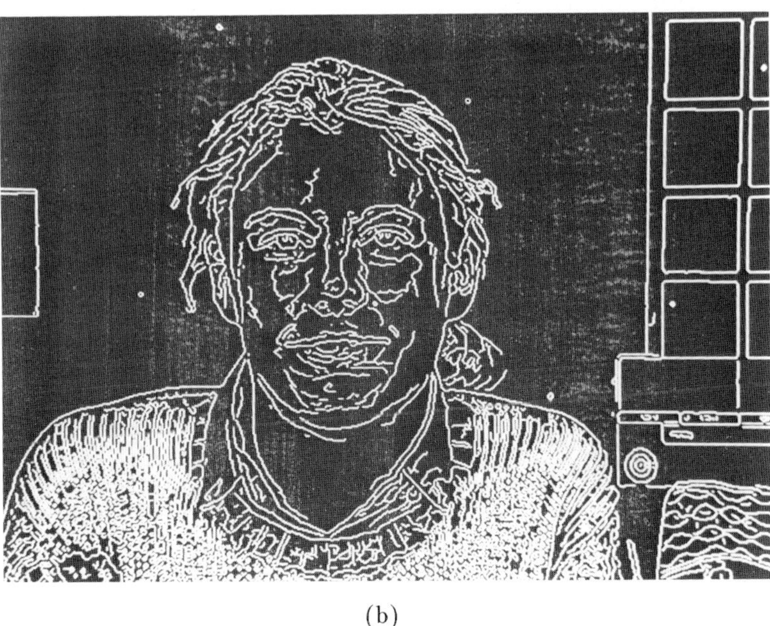

(b)

Figure 4.4: a) A typical frame of the "HDTV" sequence; b) edges of the image shown in (a).

(a)

(b)

Figure 4.5: a) A typical frame of the "Ping-Zoom" sequence; b) edges of the image shown in (a).

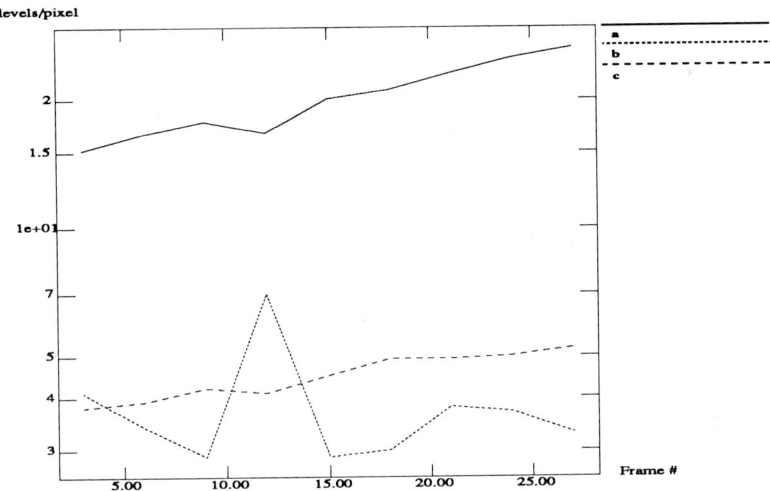

Figure 4.6: Zoom/pan compensated FD for the "HDTV" sequence: (a)non compensated FD ; (b)global motion compensated FD with luminance matching; (c) global motion compensated FD with edge matching.

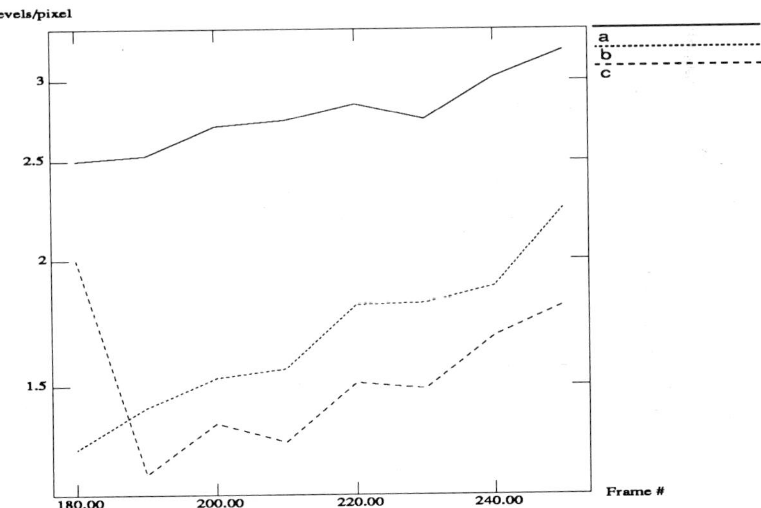

Figure 4.7: Zoom/pan compensated FD for the "PingPan" sequence: (a)non compensated FD ; (b)global motion compensated FD with luminance matching; (c) global motion compensated FD with edge matching.

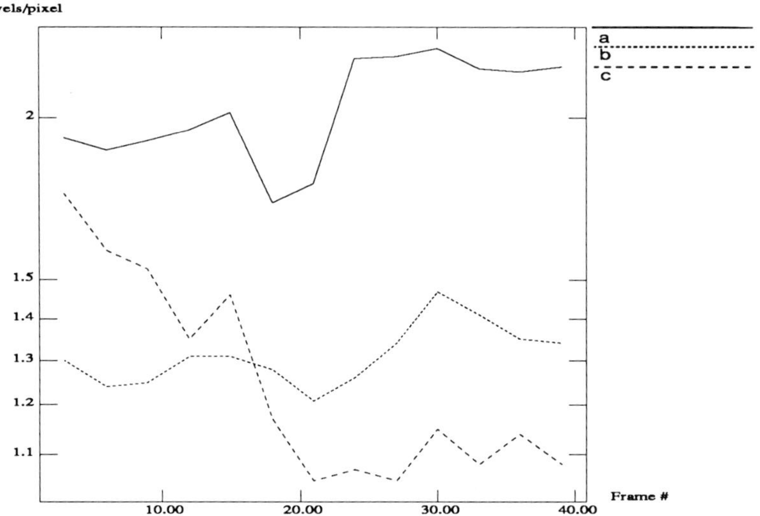

Figure 4.8: Zoom/pan compensated FD for the "PINGZOOM" sequence: (a)non compensated FD ; (b)global motion compensated FD with luminance matching; (c) global motion compensated FD with edge matching.

4.4.2 Rotation and translation results

We have applied our algorithm of section 4.3.2 to the first 40 frames of the "Flowers" sequence, which seem to only exhibit translational movement along the x direction. The dimensions of each frame is 486×720. This sequence exhibits a great deal of depth variation, featuring a tree close to the camera, a row of houses further away, and flowers in between. The first frame for this sequence and its corresponding edges are shown in Figure 4.10, and its depth map is shown in Figure 4.11. To visualize the depth map, we have heavily quantized it. As seen, the depth associated with the tree is smaller than that of the houses; in addition, the depth associated with the flowers is in between that of the houses and the tree. Finally, the sky has the largest depth as it is to be expected.

We compare the compression capability of our algorithm with that of an intensity based BMA with local motion compensation for each block. In our algorithm, the quantities to be quantized and coded for transmission are the rotation and translation parameters and the error frames. The corresponding quantities for a BMA based algorithm, with forward prediction only, is the motion vectors and the error frames. The error frames in both cases are quantized and coded in a fashion similar to MPEG [28]. Specifically, we apply DCT to 8×8 blocks of error frames, discard coefficients below a certain threshold, quantize with a uniform quantizer and variable length code the quantized coefficients in

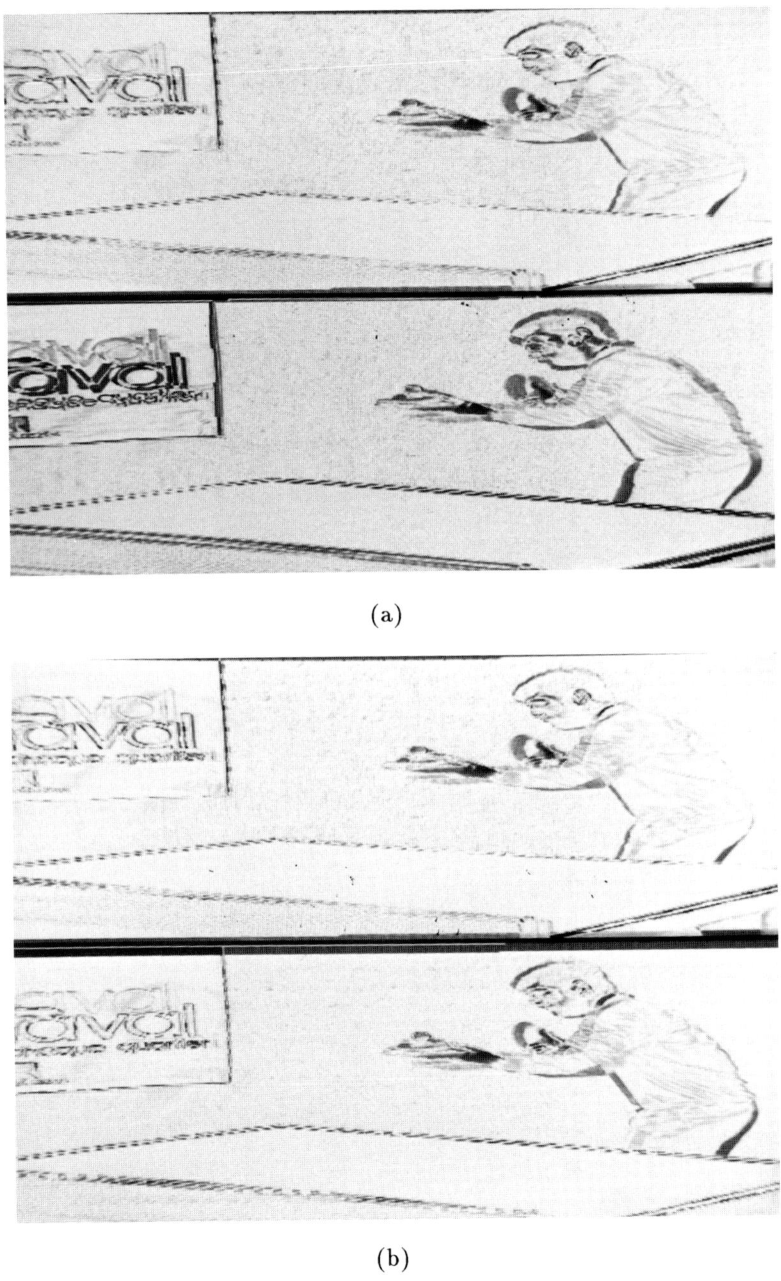

(a)

(b)

Figure 4.9: "Ping-Zoom" sequence: comparison of edge BMA based global motion compensated frame difference to (a) non compensated frame difference; (b) luminance BMA based global motion compensated frame difference.Edge based method is the top half for both part a and part b.

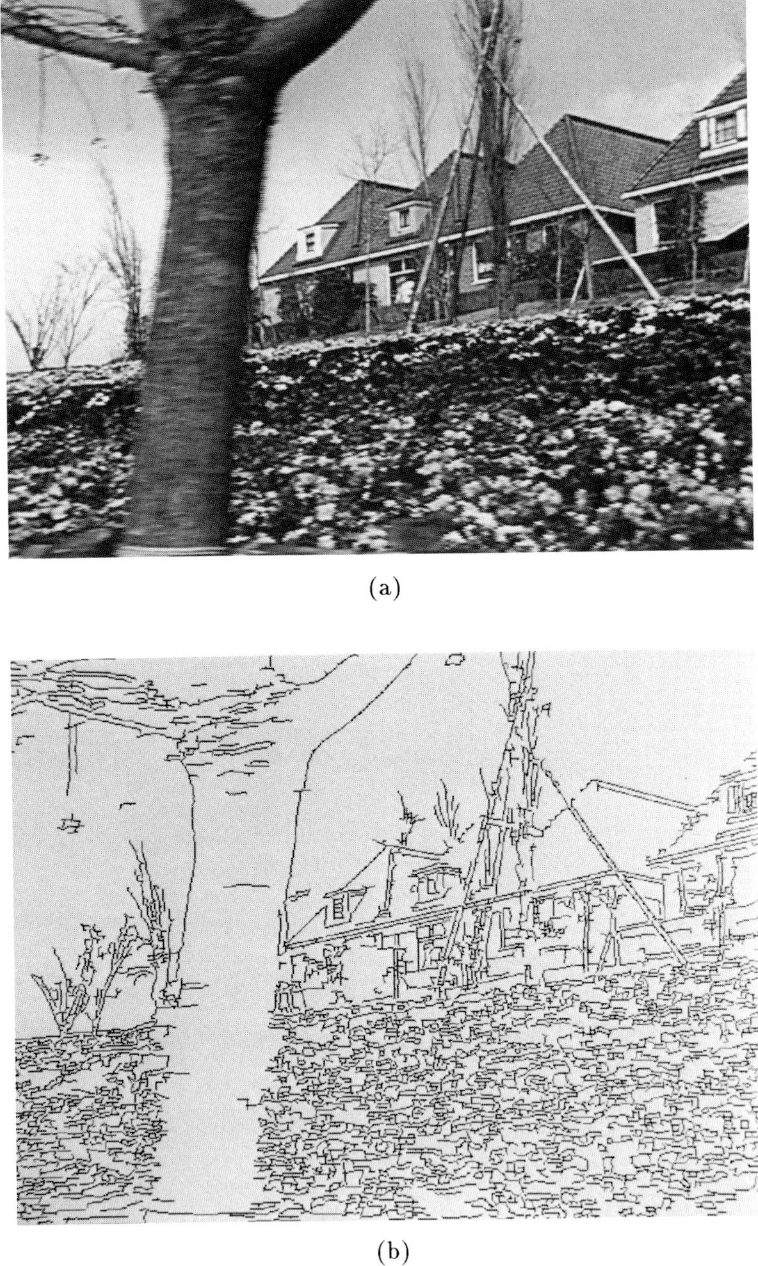

(a)

(b)

Figure 4.10: a) A typical frame of the "Flowers" sequence; b) edges of the image shown in (a)

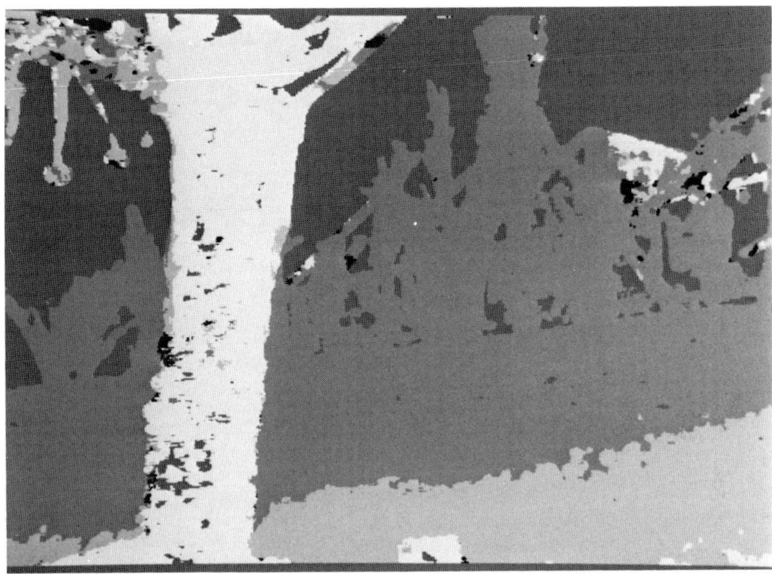

Figure 4.11: depth map of a typical frame of the "Flowers" sequence.

conjunction with the location of discarded coefficients. The motion vectors in the local motion estimation technique are also treated in a similar way to MPEG. Specifically, the motion vectors are first computed for each 16×16 block, and then differentially coded with variable length coder described in MPEG documents [28].

In applying our rotation/translation algorithm of section 4.3.2 to the flower sequence, the algorithm successfully finds all the rotation and translation parameters, except for Δx to be zero. The motion compensated residual error based on rotation/translation algorithm is found to have almost the same energy as that of local motion estimation based on BMA. Figures 4.12 and 4.13 show the resulting bit rates in bits per pixel, and distortion in intensity per pixel for (a) the edge based rotation, translation scheme of section 4.3.2; (b) intensity based BMA with local MC; (c) edge based zoom, pan, rotation technique of section 4.3.1. As seen, the bit rate of the three schemes is almost identical. The distortion for the rotation/translation model is similar to that of BMA, while that of zoom/pan/rotation model is slightly higher particularly for earlier frames. The visual impression of the reconstructed frames at the receiver seem to be more or less the same for all three techniques. It is interesting to note that even though the zoom/pan/rotation model is inappropriate for this sequence, its rate/distortion characteristics is comparable to those of the BMA based algorithm and rotation/translation model. This is true even when we remove the rotation part of the zoom/pan/rotation algorithm and apply only a zoom/pan model. This can be partially explained by considering that the computed pan parameter along the x direction is around 3 for most frames, which is approxi-

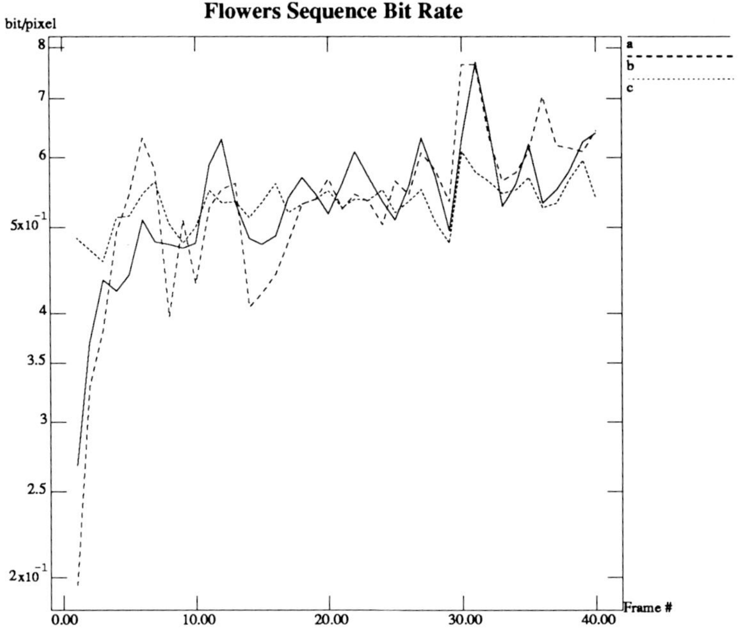

Figure 4.12: Bit rate for flowers sequence: (a) local motion estimation with intensity based BMA with quantization step set to 1.6 ; (b) edge based global motion estimation based on rotation/translation model with quantization step set to 1.8; (c) edge based global motion estimation based on zoom/pan/rotation model with quantization step parameter set to 2.0;

mately the average motion experienced by the most pixels in the scene, such as the pixels associated with houses.

Another observation to be made is that the peak bit rate exhibited by BMA based algorithm and the rotation/translation algorithm around frame 31 is absent in the zoom/pan algorithm. A similar trend was observed at other bit rates for these three algorithms.

Finally, we have found experimentally that the performance of the model based algorithms of section 4.3 remains the same if the edge matching part is replaced with traditional intensity matching. This is encouraging since it confirms our hypothesis that unlike local motion estimation, for global motion estimation, binary edge matching is sufficient and that 8 bit luminance information can be replaced with one bit edge information.

4.5 Computational Complexity

In this section, we compare the computational complexity of our techniques based on edges to that of intensity based BMA for both fixed and floating point arithmetic. In doing so, we only compare the complexity of our algorithm to

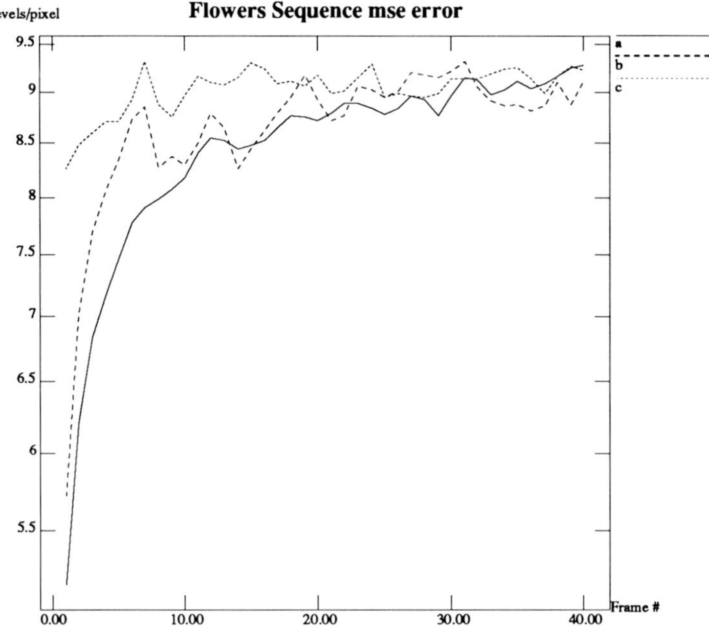

Figure 4.13: Mse for flowers sequence: (a) local motion estimation with intensity based BMA with quantization step set to 1.6 ; (b) edge based global motion estimation based on rotation/translation model with quantization step set to 1.8; (c) edge based global motion estimation based on zoom/pan/rotation model with quantization step parameter set to 2.0;

that of full search BMA, rather than hierarchical algorithms such as 3 step, logarithmic or conjugate gradient algorithms. There are two reasons behind this: First, in contrast with the hierarchical motion algorithms[29, 30], a major portion of our edge based approach does not require intermediate decisions and therefore does not require programmable hardware, and can potentially be implemented with highly parallel architectures. Specifically, as we will see, the major part of edge detection consists of convolution, which can be mapped into regular structures[31], while in hierarchical techniques the number of intermediate decisions *per block* is proportional to the logarithm of the dimension of the search area. The second disadvantage of hierarchical algorithms in intensity based local motion estimation algorithms is their inherent inaccuracy.

The outline of this section is as follows: in section 4.5.1, we describe the particular edge detection algorithm we used for the experimental results of the previous section and review its computational complexity. In section 4.5.2, we discuss the computational complexity of linear regression part of our algorithms and argue that the cost of regression can be neglected in comparison with edge detection for most practical purposes. In sections 4.5.3 and 4.5.4, we compare the computational complexity of our algorithms to that of full search intensity based BMA for floating and fixed point arithmetic respectively.

Finally, it is worthwhile to mention that the edge detection algorithm for our experimental results in section 4.4 uses fixed point arithmetic.

4.5.1 Edge detection

The edge detection algorithm we have used in section 4.4 is described in detail in [24, 25, 26]. In this section, we will briefly review the algorithm so that we can characterize its computational complexity.

The edge detection algorithm consists of two parts: smoothing, and taking directional derivatives. For smoothing purposes, we convolve our images with one dimensional Gaussians along both x and y directions. By separating the two dimensional convolution kernel into two one dimensional ones, we can save on the computations. The impulse response of the particular Gaussian filter we use has 8 taps and is of the form:

$$h(t) = \frac{1}{\sqrt{2\pi}} e^{-\frac{t^2}{2}} \tag{4.20}$$

To detect edges, we need to locate the maxima of the gradient or zeros of the second directional derivative along the gradient [24, 25, 26]. In doing so, we use a normalized symmetric exponential one dimensional filter implemented in the image processing software package KHOROS [32]. Specifically, the first and second directional derivatives of the image $I(n_x, n_y)$ are obtained by:

$$\begin{aligned}
I_x(n_x, n_y) &= I(n_x, n_y) * f_1(n_y) * f_2(n_y) * f_2(n_x) \\
I_{xx}(n_x, n_y) &= I(n_x, n_y) * f_1(n_y) * f_2(n_y) * [f_2(n_x) + f_1(n_x)] - 2I(n_x, n_y) * f_1(n_y) * f_2(n_y) \\
I_y(n_x, n_y) &= I(n_x, n_y) * f_1(n_x) * f_2(n_x) * f_2(n_y) \\
I_{yy}(n_x, n_y) &= I(n_x, n_y) * f_1(n_x) * f_2(n_x) * [f_2(n_y) + f_1(n_y)] - 2I(n_x, n_y) * f_1(n_x) * f_2(n_x)
\end{aligned}$$

where $*$ stands for convolution, and f_1 and f_2 are defined as:

$$f_1(n) = \begin{cases} 0 & n < 0 \\ a_0(1 - a_0)^n & n \geq 0 \end{cases} \tag{4.21}$$

$$f_2(n) = \begin{cases} 0 & n > 0 \\ a_0(1 - a_0)^{-n} & n \leq 0 \end{cases} \tag{4.22}$$

Since f_1 and f_2 are Infinite Impulse Response (IIR) filters, they can be implemented in a recursive fashion via a first order difference equation:

$$\begin{aligned}
I_1(n_x, n_y) &\equiv I(n_x, n_y) * f_1(n_x) \\
&= I_1(n_x - 1, y) + a_0[I(n_x, n_y) - I_1(n_x - 1, n_y)]
\end{aligned} \tag{4.23}$$

$$\begin{aligned}
I_2(n_x, n_y) &\equiv I(n_x, n_y) * f_2(n_x) \\
&= I_2(n_x + 1, y) + a_0[I(n_x, n_y) - I_2(n_x + 1, n_y)]
\end{aligned} \tag{4.24}$$

Based on the above, we can conclude that to find I_x and I_{xx}, we need to perform four convolutions. Specifically, we need two convolutions to compute

TASK	Multiply count	Add count
8 Tap 1-D filtering along x	8	8
8 Tap 1-D filtering along y	8	8
IIR filtering for I_{xx} and I_x		
Fixed point		11
Floating point	5	11
IIR filtering for I_{xx} and I_x		
Fixed point		11
Floating point	5	11

Table 4.1: Operation count for the edge detection algorithm

$\hat{I}(n_x, n_y) \equiv I(n_x, n_y) * f_1(n_y) * f_2(n_y)$, one convolution to form $\hat{I}(n_x, n_y) * f_2(n_x)$, and one convolution to form $\hat{I}(n_x, n_y) * f_2(n_y)$. Once these four convolutions are done, we need one addition [2] to compute $I_x(n_x, n_y)$ and two additions and a multiplication by 2 to compute $I_{xx}(n_x, n_y)$. Note that multiplication by 2 in fixed point arithmetic can be done at no almost no cost. Using equations 4.23 and 4.24, each convolution needs one multiplication and two additions. However, by choosing $a_0 = \frac{1}{2}$, the multiplications in fixed point computation can be done with shifting, and therefore their costs can be neglected.

Putting all of these together, we conclude that computing $I_x(n_x, n_y)$ and $I_{xx}(n_x, n_y)$ requires 5 multiplies and 11 adds per pixel for floating point implementation and 11 adds per pixel for fixed point arithmetic. Similar statments can be made for $I_y(n_x, n_y)$ and $I_{yy}(n_x, n_y)$. We can summarize the operation count per pixel as shown in table 4.1.

Once the gradient vector and the second derivative in the gradient direction for every point in the image is computed, the actual edge points are found by a series of thresholding operations [24, 25, 26]. The values of first and second hysteresis thresholds we have used for our various experiments are as follows: Ping-Pan $(1, 10)$, Ping-Zoom, HDTV and flowers $(1, 4)$.

In sections 4.5.3 and 4.5.4, we will use table 4.1 to compare the operation count of our algorithm with traditional intensity based techniques.

4.5.2 Linear regression

In this section, we argue that the computational complexity associated with linear regression can be neglected in comparison with edge detection. In doing so, we are primarily concerned with the complexity of computing the matrices involved in the linear regression, rather than that of solving the linear system of equations. The computation associated with solving linear systems of equation can be justifiably ignored because (a) the matrices under consideration are either 3×3 or 4×4 and therefore their solution can be computed in a closed form.

[2] We consider the complexity of a subtraction to be the same as an addition.

(b) unlike computing the matrices themselves, the cost associated with solving these systems is independent of the number of matched features; (c) on average, we solve only two systems per frame to remove outliers.

We begin with the zoom/pan/rotation algorithm of section 4.3.1, and then move on to the translation/rotation algorithm of section 4.3.2.

Zoom, pan and rotation algorithm

For this algorithm, we solve a linear least squares problem shown in equation 4.5 in order to determine c_i for $i = 1, 2, 3, 4$. Minimizing the error expression in equation 4.10 requires setting its derivatives with respect to c_i equal to zero, or equivalently solving the following system of equations:

$$A\vec{c} = \vec{d} \tag{4.25}$$

where $\vec{c}$ denotes the motion parameters, $\vec{d}$ is given by:

$$\vec{d} \equiv \begin{bmatrix} \sum_{n=1}^{K}[X_{1n}X_{2n} + Y_{1n}Y_{2n}] \\ \sum_{n=1}^{K}[X_{2n}Y_{1n} - X_{1n}Y_{2n}] \\ \sum_{n=1}^{K} X_{2n} \\ \sum_{n=1}^{K} Y_{2n} \end{bmatrix}$$

matrix A is given by:

$$A \equiv \begin{pmatrix} \sum_{i=1}^{K}(X_{1i}^2 + Y_{1i}^2) & 0 & \sum_{i=1}^{K} X_{1i} & \sum_{i=1}^{K} Y_{1i} \\ 0 & \sum_{i=1}^{K}(X_{1i}^2 + Y_{1i}^2) & \sum_{i=1}^{K} Y_{1i} & -\sum_{i=1}^{K} X_{1i} \\ \sum_{i=1}^{K} X_{1i} & \sum_{i=1}^{K} Y_{1i} & K & 0 \\ \sum_{i=1}^{K} Y_{1i} & -\sum_{i=1}^{K} X_{1i} & 0 & K \end{pmatrix}$$

and K denotes the number of matched features. To form matrix A and vector $\vec{d}$, there are 6 multiplications (mults) and 7 additions (adds) per feature match. In addition, the removal of outliers in step 5 of the algorithm, involves (a) applying the model shown in equation 4.5 which requires 4 mults and 4 adds per match; (b) computing two standard deviations which requires a total of 2 mults and 6 adds per match; (c) recomputing matrix A and vector $\vec{d}$ shown above which requires 6 mults and 7 adds per match. Thus, there are 13 operations per match for finding the parameters, and 29 operations per match for each iteration of removing the outliers. Assuming that our algorithm converges in two steps, i.e. the outliers are only removed once, there are a total of $29+13 = 42$operations per match. The average number of matches found for the three sequences considered in section 4.4.1 is one per 3500 pixels[3]. Putting all of these together, we get a total of .01 operations per pixel for the linear regression. This is negligible compared to the operations per pixels needed for computing the edge, regardless of whether fixed or floating point arithmetic is used.

Finally, if a global motion estimation algorithm uses intensity based BMA, the number of "matches" used in linear regression is by definition one per 64 for 8×8 blocks. This results in 0.65 operations per pixel, which can be neglected for most practical purposes.

Rotation and translation algorithm

Similar arguments can be applied to the rotation/translation algorithm of section 4.3.2. In that algorithm, the linear regression is carried out to minimize the error function of equation 4.16. For convenience, assume that the particular e_i chosen to be one is e_9. Minimizing the error expression in equation 4.16 requires setting its derivative with respect to c_i equal to zero, or equivalently solving the following system of equations:

$$Q\vec{e} = \vec{w} \tag{4.26}$$

where $\vec{e}$ denotes the vector containing the e_i parameters, $\vec{w}$ is an 8 dimensional vector given by:

$$\vec{w} = \sum_{n=1}^{K} [\rho\gamma \ , \ \dot{\rho}\gamma \ , \ \gamma \ , \ \dot{\gamma}\rho \ , \ \dot{\gamma}\dot{\rho} \ , \ \dot{\gamma} \ , \ \rho \ , \ \dot{\rho}] \tag{4.27}$$

and Q is an 8×8 matrix given by

$$\sum_{n=1}^{K} \begin{pmatrix} \rho^2\gamma^2 & \rho\gamma^2\dot{\rho} & \rho\gamma^2 & \rho^2\gamma\dot{\gamma} & \rho\gamma\dot{\gamma}\dot{\rho} & \rho\gamma\dot{\gamma} & \rho^2\gamma & \rho\gamma\dot{\rho} \\ \rho\gamma^2\dot{\rho} & \dot{\rho}^2\gamma^2 & \dot{\rho}\gamma^2 & \rho\gamma\dot{\gamma}\dot{\rho} & \dot{\rho}^2\gamma\dot{\gamma} & \dot{\rho}\gamma\dot{\gamma} & \dot{\rho}\gamma\rho & \dot{\rho}^2\gamma \\ \rho\gamma^2 & \dot{\rho}\gamma^2 & \gamma^2 & \dot{\gamma}\rho\gamma & \dot{\gamma}\dot{\rho}\gamma & \dot{\gamma}\gamma & \rho\gamma & \dot{\rho}\gamma \\ \rho^2\gamma\dot{\gamma} & \rho\gamma\dot{\gamma}\dot{\rho} & \dot{\gamma}\rho\gamma & \dot{\gamma}^2\rho^2 & \dot{\gamma}^2\dot{\rho}\rho & \dot{\gamma}^2\rho & \dot{\gamma}\rho^2 & \dot{\gamma}\rho\dot{\rho} \\ \rho\gamma\dot{\gamma}\dot{\rho} & \dot{\rho}^2\gamma\dot{\gamma} & \dot{\gamma}\dot{\rho}\gamma & \dot{\gamma}^2\dot{\rho}\rho & \dot{\gamma}^2\dot{\rho}^2 & \dot{\gamma}^2\dot{\rho} & \dot{\gamma}\dot{\rho}\rho & \dot{\gamma}\dot{\rho}^2 \\ \rho\gamma\dot{\gamma} & \dot{\rho}\gamma\dot{\gamma} & \dot{\gamma}\gamma & \dot{\gamma}^2\rho & \dot{\gamma}^2\dot{\rho} & \dot{\gamma}^2 & \dot{\gamma}\rho & \dot{\gamma}\dot{\rho} \\ \rho^2\gamma & \dot{\rho}\gamma\rho & \rho\gamma & \dot{\gamma}\rho^2 & \dot{\gamma}\dot{\rho}\rho & \dot{\gamma}\rho & \rho^2 & \rho\dot{\rho} \\ \rho\gamma\dot{\rho} & \dot{\rho}^2\gamma & \dot{\rho}\gamma & \dot{\gamma}\rho\dot{\rho} & \dot{\gamma}\dot{\rho}^2 & \dot{\gamma}\dot{\rho} & \rho\dot{\rho} & \dot{\rho}^2 \end{pmatrix}$$

with $\rho \equiv X_{1n}, \dot{\rho} \equiv Y_{1n}, \gamma \equiv X_{2n}, \dot{\gamma} \equiv Y_{2n}$. Since Q is a symmetric matrix, 40 mults and 44 adds are needed in computing it for each match, resulting in a total of 84 operations per match. In addition computing $\vec{w}$ requires 12 operations per match. In removing the outliers, we also need to re-compute matrix Q and vector $\vec{w}$. Before doing so however, we need to compute the

[3] As an example, of all the features in a 720×480 PingZoom frame, only 93 matches were found; the corresponding numbers for the PingPan and HDTV sequences are 113 and 81 respectively.

following: (a) find the singular value decomposition of the E matrix which takes 18 operations; (b) use the singular vectors in equations 4.17, 4.18 to compute the rotation parameters; this step requires 45 operation for 4.17 and 45 operations for 4.18; (c) use the e_i parameters in equation 4.19 to compute the translation parameters; this step requires 24 operations. (d) use the computed rotation and translation parameters in either equation 4.12 or 4.13 to compute z_1; this step takes 13 operations per feature match; (e) use equations 4.8 and 4.9 to compute (X_2, Y_2) coordinates for each match; this step requires 20 operations per match; (f) compute the standard deviation of the error using 8 operations; (g) re-compute matrix Q and vector $\vec{w}$ above using 96 operations per match.

Since steps (a), (b) and (c) are independent of the number of matched features, their contribution to the complexity of outlier removal can be ignored. Thus, the total number of operations per match needed for outlier removal is 136. Assuming the algorithm converges in two steps, i.e. we only remove the outliers once, there is a total of 232 operations per match. Furthermore, assuming there is approximately one match every 350 pixels, the number of operations per pixel becomes 0.66. This is again negligible compared to that needed for edge detection, regardless of whether fixed or floating point arithmetic is used.

Finally, if a global motion estimation algorithm uses intensity based BMA, the number of "matches" in linear regression is by definition one per 64 for 8×8 blocks. This results in $\frac{232}{64} = 3.6$ operations per pixel, which can be neglected for most practical purposes.

4.5.3 Floating point implementation

In this section, we compare the computational complexity of traditional full search BMA and that of our edge based technique for floating point implementation. For simplicity, we assume that the time/complexity required to match edges is negligible compared to the time/complexity required to find them. Furthermore, we assume that the amount of time needed for a floating point addition is roughly equal to that of a multiplication, and we will refer to either of those floating point operations as a flop.

Under these conditions, based on table 4.1, we conclude that the number of operations per pixel for our edge based technique is 64. Similarly, the number of flops for full search BMA is $2S_A$ since there are $2S_A$ additions. S_A in the above expressions stands for search area used for finding the matching vector. We now compare computational complexity of our algorithms in sections 4.3.1 and 4.3.2.

Based on the above reasoning, we conclude that for global motion estimation algorithms based on the zoom/pan/rotation model of section 4.2.1, the edge based algorithm results in lower computational complexity than full search intensity matching techniques, provided the search area is larger than 4×4. A similar statement can be made about global motion estimation based on rotation/translation model.

Since in most camera systems, the zoom and pan parameters vary slowly, it is enough to compute the parameters every P frames or so. As a result, we must use a fairly large search area. For instance in our experimental results in section 4.4.1, we subsample the frames by as much as 10 before computing the zoom and pan parameters, and as a result the search area for the PingPan and PingZoom sequences in section 4.4.1 was chosen to be 41×41. For this particular value of search area, the floating point implementation of the full search version of our algorithm requires 53 times fewer operations than full search BMA.

For the rotation/translation algorithm, the search area used for luminance matching was 17×17 resulting in a saving factor of 17.

4.5.4 Fixed point implementation

Unlike floating point implementation, the cost of multiplication in fixed point implementation is higher than that of addition. In general, several factors need to be taken into account in comparing the costs associated with various algorithms. These factors include, speed, chip area, power and can usually be traded off with each other depending on the architecture used. For example, parallel processing architectures allow one to trade off chip area (or hardware complexity) with speed. Thus, comparing costs associated with various algorithms is not an easy task.

In spite of these, it is possible to choose simple ways of defining complexity. Specifically, in comparing 8 bit fixed point implementations, we assume that (a) cost of an 8 bit by 8 bit multiplication is 4.5 times that of 8 bit addition [33], and (b) the cost of an 8 bit addition is 16 times that of one bit matchings [33]. The word cost in the previous sentence reflects either time or the hardware required to do a certain operation.

We now compare fixed point implementation of our edge based algorithm with that of intensity based full search BMA. The complexity of a full search block matching algorithm per pixel in units of cost of one bit matching is $S_A(2 \times 16)$. This is because full search matching requires 2 adds, and that the price of an 8 bit add is 16 times that of one bit matching with a logic gate. Thus, the cost associated with full search block matching is proportional to $32S_A$.

The complexity of an edge based full search matching algorithm is proportional to $(16 \times 4.5 \times 16)$ for the multiplications needed for edge detection, 38×16 for the additions needed for edge detection and S_A for one bit matching. Thus, the cost associated with edge based full search matching algorithm is $1760 + S_A$.

Comparing the above two techniques, we can conclude that for both global motion estimation based on zoom/pan/rotation and rotation/translation, as long as the search area is larger than 8×8, the cost associated with edge based matching is smaller than that of intensity based full search matching. For example, for search area size of 41×41 used in the examples of section 4.4.1, the intensity based matching is 13 times more expensive than edge based matching.

4.6 Discussion

We have used the results on recovery of 3-D rigid body motion to develop a new algorithm for global motion estimation. Based on this approach, global parameters consisting of camera translation and rotation are first estimated from a video sequence, and then used in conjunction with the depth map of a scene to form the MC prediction for future frames. Our approach is different from existing global motion estimation techniques in that camera translation, as well as rotation is modeled. The performance of our approach is comparable to intensity based local MC, at least in situations where the scene only consists of camera motion, such as in the flower sequence.

A salient feature in our algorithm is the use of edges in estimating camera parameters. There are several motivations behind edge based global motion estimation: First, the availability of VLSI edge detection chips and convolution chips [31, 34] make the possibility of using edges in motion estimation quite realistic and potentially rewarding. Second, since a number of edge based video coding schemes have been recently proposed[35], the additional use of edges for motion estimation is lucrative, particularly since it results in lower computational complexity as compared to traditional intensity based techniques.

During the course of this chapter, we have compared the complexity and performance of our edge based global MC algorithms to two different classes of MC algorithms: (a) full search intensity based BMA, with local MC ; (b) full search intensity based BMA with global MC. As far as performance goes, we have found that if the only motion in the sequence is due to the camera, then our edge based technique does as well as (a); An example of this is shown in the flower sequence curves in Figures 4.12 and 4.13, where the only variation in the video is due to camera translation. On the other hand, if there is additional object movement in the scene such as in PingZoom and PingPan, then our edge based technique does as well as any global MC technique such as (b). An example of this is shown in the plot of Figure 4.7 for camera pan and 4.8 for camera zoom. As for complexity, we showed in section 4.5 that our edge based global MC algorithms are less complex than both (a) and (b) for both floating point and fixed point implementations.

An important point to keep in mind is that, the degree to which global MC algorithms work is highly dependent on the relative motion of camera compared to the objects in the scene. Applying any global model estimation technique, including ours, to a sequence that does not correspond to any camera motion at all will be unsatisfactory. In fact, this is true for all model based techniques in image processing.

The work presented in this chapter can be potentially applied in many practical situations with video sequences resulting from stationary scenes and moving cameras. One application of such a scenario might be 3D video data bases in which 3D stationary objects are captured on a video sequence by a camera moving around them. In this situation, camera motion estimation techniques can not only be useful for efficient representation of the video sequence, but also for 3D interpolation of the views not directly captured by the video camera. Other interesting application might include 3D scene representation and reconstruction, and transmission.

4.7 Acknowledgement

This work has been sponsored in part by the National Science Foundation (NSF) Presidential Young Investigator (PYI) award MIP-9057466, in part by Office of Naval Research (ONR) young investigator award #N00014-92-J-1732, and in part by Digital Equipment Corporation and Eastman Kodak company. Techical interaction with Jonathan Zyngman of Teknekron Communications Systems, and John Liu of Sun Microsystems is greatfully acknowledged.

References

[1] J.Jurgen, "An abundance of video formats", *IEEE Spectrum*, March 1992

[2] D.Adolph, R.Buschmann, "1.15 Mbit/s coding of video signals including global motion compensation", *Signal processing: Image communication 3*, 1991, pp. 259-274

[3] Y.T.Tse, R.L.Baker, "Global zoom/pan estimation and compensation for video compression", *Proceedings of ICASSP 1991*

[4] G. Keesman, "Motion estimation based on a motion model incorporating translation, rotation and zoom", *Signal processing 4*, 1988, pp. 31-34.

[5] M. Hoetter, "Differential Estimation of the global motion parameters zoom and pan", *Signal processing*, Vol. 16, No. 3, March 1989, pp. 249-265.

[6] S.F.Wu and J. Kittler, "A differential method for simultaneous estimation of rotation, change of scale and translation", *Signal processing: Image communication 2*, 1990, pp. 69-80

[7] G.Adiv, "Determining three-dimensional Motion and Structure from Optical Flow generated by Several Moving Objects" *IEEE trans. on patt. an. and mach. int.*, Vol.PAMI-7, no.4, July1985, pp.384-401

[8] R.Y.Tsai, T.S.Huang, "Uniqueness and Estimation of Three-Dimensional Motion Parameters of Rigid Objects with Curved Surfaces", *IEEE Transactions on Pattern Analysis and Machine Intelligence*, Vol. PAMI-6, No. 1, Jan. 1984, pp. 13-26.

[9] T.S.Huang, A.N.Netravali, "3D Motion Estimation", Machine Vision for Three-Dimensional Scenes, Academic press, Inc., 1990

[10] J.Weng, N.Ahuja, T.S.Huang, "Motion and Structure from Point Correspondences with Error Estimation: Planar Surfaces", *IEEE trans. on sign. proc.*, Vol.39, no.12, Dec.1991, pp.2691-2717

[11] S. Y. Chen and W. H. Tsai, "A systematic approach to analytic determination of camera parameters by line features", *Image Recognition*, Vol. 23, No. 8, pp. 859-877, 1990.

[12] P.R. Wolf, "Elements of Photogrammetry", McGraw Hill, New York, 1983.

[13] M. A. Fischler and R. C. Bolles, "Random sample consensus: a paradigm for model fitting with applications to image analysis and automated cartography", *Commun. ACM,* Vol. 24, pp. 381-395, 1981

[14] D.Vernon and M.Tistarelli, "Using camera motion to estimate range for robotic parts manipulation", *IEEE trans. robotics and automation*, Vol.6, no.5, Oct.1990, pp.509-521

[15] H.H.Chen and T.H.Huang, "Matching 3-D line segments with applications to multiple-object motion estimation", *IEEE trans. on patt. an. and mach. int.*, Vol.12, No.10, October 1990, pp.1002-1008

[16] D. Marr and E.C. Hildreth, "Theory of edge detection", *Proc. of Royal Society London B., Vol. 207*, pages 187–217, 1980.

[17] E.C. Hildreth, "Computations underlying the measurement of visual motion", *Artificial Intelligence*, 23:309–354, 1984.

[18] David W. Murray and Bernard F. Buxton, "Experiments in the machine interpretation of visual motion", MIT Press, 1990, Cambridge, Mass.

[19] G. L. Scott, "Local and Global Interpretation of Visual Motion", Pitman and Morgan Kaufmann, London and Los Altos, 1987.

[20] L. A. Spacek, "Edge detection and motion detection", *Image and Vision Computing*, Vol. 4, No. 1, pp. 43-56, 1986.

[21] C. Wang H. Sun S. Yada and A. Rosenfeld, "Some experiments in relaxation image matching using corner features", *Pattern Recognition, Vol. 16, No. 2*, pp. 167-182, 1983.

[22] D.F.Rogers, J.A.Adams, "Mathematical Elements for Computer Graphics", New York: McGraw-Hill, 1976

[23] G.Kummerfeldt, F.May, W.Wolf, "Coding television signals at 320 and 64 kbit/s", *SPIE*, Vol.594 Image Coding, 1985, pp.119-128

[24] J. Shen and S. Castan, "An optimal linear operator for edge detection", *Proceedings of Computer Vision and Pattern Recognition*, Miami, FL, 1986.

[25] J. Shen and S. Castan, "Edge detection based on multi-edge model", *Proceedings of SPIE*, Cannes, 1987.

[26] J. Shen and S. Castan, "Further results on DRF method for edge detection", *9th International Conference on Pattern Recognition*, Rome 1988.

[27] P.J. Burt, J. R. Bergen, R. Hingorani, R. Kolczynski, W. A. Lee, A. Leung, J. Lubin and H. Shvaytser, "Object tracking with a moving camera", *Proceedings of IEEE Workshop on Visual Motion*, 1989, pp.2-12.

[28] "MPEG Video Committee Draft", December 1990.

[29] T. Komarek and P. Pirsch, "VLSI architectures for hierarchical block matching algorithms", *IEEE Int. Symp. Circuits and Systems*, New Orleans, LA, May 1990, pp. 45-48.

[30] T. H. Y. Meng and A. C. Hung, "Parallel array architectures for motion estimation", *Proceedings of Application Specific Array Processors*, September 1991.

[31] Zoran 2-D Convolver ZL33771, Zoran 1-D convolver ZR33891 and ZR33288.

[32] John Rasure and Danielle Argiro, "Visual Programming System and Software Development Environment for Data Processing and Visualization (Khoros)", User Manual, University of New Mexico, 1992.

[33] J. Rabaey, Private communication.

[34] C Lee, F. V. M. Catthoor and H. J. De Man, "An Efficient ASIC Architecture for al Time Edge Detection", *IEEE Transactions on CAS*, Vol. 36, No. 10, pp. 1350-1360, October 1989.

[35] M. J. Biggar and A. G. Constantinides, "Segmented video coding", *Proceedings of ICASSP 1988*, pp 1108- 1111.

5

Motion Compensation: Visual Aspects, Accuracy, and Fundamental Limits

B. Girod

Department of Computer Graphics, KHM Cologne, Germany

5.1 Introduction

As Einstein has observed, "mankind's great problem is a perfection of means but a confusion of ends." Several chapters in this book present means to estimate motion between two or more image frames in rather sophisticated ways. In this chapter, we will not consider motion estimation, but rather look at the ends of motion compensation, i.e., we will consider various ways that motion vectors are used in different applications.

With motion compensation, the first source of confusion can be terminology, and a word of clarification is needed. If processing of an image sequence combines data along an estimated motion trajectory, it is common to talk about "motion compensation". This is opposed to "motion adaptation," a term which is used for systems that switch between two modes, one for still image contents, one for moving image contents, but that do not take into account magnitude and direction of the motion vector. A motion adaptive system requires a "motion detector" (really a motion presence detector), while a motion compensating system requires a motion estimator.

In section 5.2, we will look at visual aspects of motion compensation. We will show that eye movements are very important for the spatiotemporal transfer function of human vision. It turns out that motion compensation in technical systems is important because there are smooth pursuit eye movements. In section 5.3, we consider motion compensating interpolation, both from non-interlaced and from interlaced grids. We look at de-interlacers with finite impulse response as well as frame-recursive de-interlacing. In section 5.4, we analyse the most common application of motion compensation today: motion compensating prediction for compression of video sequences. There are fundamental limits that cannot be overcome even by a perfect motion estimator.

Throughout this chapter, I emphasize the fundamentals more than details of a specific algorithm. A question repeatedly asked is what accuracy of motion compensation is required. Accurate motion measurements require more sophisticated and expensive methods than inaccurate ones. In compression applications, motion vector fields frequently are transmitted or stored as side information. Accurate motion vectors might require more bits for representation than available. Still, it turns out that in most situations fractional-pel accuracy is desirable. Fractional-pel accuracy requires spatial interpolation, and motion compensation will usually involve a 3D spatiotemporal filter.

5.2 Visual Aspects of Motion Compensation

The overwhelming majority of image sequences that are transmitted, processed, or stored today are ultimately viewed by humans. Any processing scheme therefore has to take into account the properties of human visual perception. If we are processing image sequences, we have to understand the spatiotemporal properties of human vision.

Motion compensation in technical systems makes sense for a very simple reason: our eyes move and compensate moving patterns in the real world, enabling us to perceive them easily. If, for some obscure reason, our eye balls were glued to our skull, motion compensation would be much less important in processing and compression of image sequences. Due to the limited temporal bandwidth of our visual receptors, everything moving in the world would appear blurred to us anyway, including moving television pictures. For frame interpolation, we would not object to motion blur introduced by filters without motion compensation. For low bitrate coders, we would simply blur moving areas, which makes them easy to encode even without motion compensating prediction. However, since the human visual system indeed compensates motion, we must account for this perceptual ability in designing any satisfactory system.

It seems only logical to start a chapter on motion compensation by considering motion compensation in the human visual system. In the following sections, we will briefly review the spatiotemporal frequency response of human vision and the influence of eye movements on this frequency response.

5.2.1 The spatiotemporal frequency response of the human visual system

A concept to which engineers and scientists readily relate is the frequency response of a system. Vision researchers have characterized the frequency response of the human visual system by measuring visibility thresholds for sine wave patterns. The threshold contrast sensitivity as a function of frequency is referred to as the modulation transfer function (MTF). The MTF provides some indication of the frequency response at suprathreshold levels as well, although in a nonlinear system like the human visual system the frequency response is amplitude dependent. For spatial sine wave patterns, the MTF has the characteristic of a roughly isotropic bandpass with a maximum at 2 to 4 cycles per degree (cpd) [1]. This bandpass characteristic is due to a lateral inhibition mechanism in the human retina [1], in combination with the optical limitations of the human eye [2].

Spatial MTF measurements are applicable and very relevant for still images, e.g., when dealing with printed material. However, they do not provide sufficient information for systems that display image sequences in rapid temporal succession, as, e.g., television or film. For these applications, we need the spatiotemporal frequency response of the human visual system. This can be measured using moving or temporally modulated spatial sine wave patterns. The spatiotemporal MTF of the human visual system was reported first by Robson in the 1960s [3], and later on by Kelly [4] over a wider range of conditions. A model fit to Robson's data is shown in Fig. 5.1 [5]. The spatiotemporal MTF is separable at high spatial and temporal frequencies, but non-separable at low frequencies.

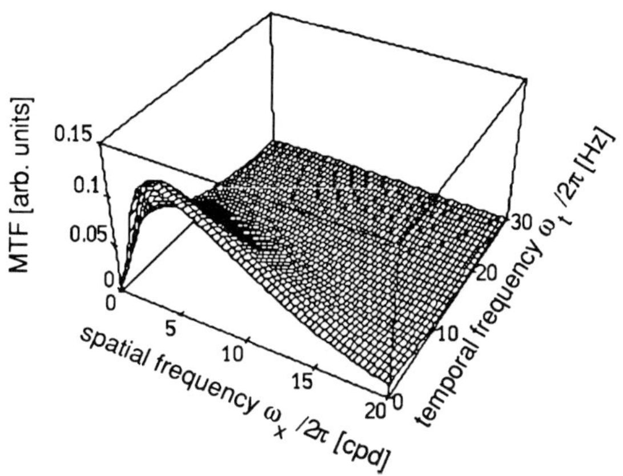

Fig. 5.1: Spatiotemporal modulation transfer function (MTF) of the human visual system. The figure shows a model fit to Robson's data [3] [5].

5.2.2 Spectrum of an image sequence containing translational motion

The simplest moving image sequence would be one that contains only translational motion at a constant velocity (v_x, v_y). Such a signal can be characterized as

$$l(x, y, t) = l(x - v_x t, y - v_y t, 0), \tag{5.1}$$

where $l(x, y, t)$ is the luminance signal as a function of the continuous spatial variables x and y, and continuous time t. The 3D Fourier transform of (5.1) yields

$$L(\omega_x, \omega_y, \omega_t) = L(\omega_x, \omega_y) \cdot \delta(\omega_x v_x + \omega_y v_y + \omega_t), \tag{5.2}$$

where $L(\omega_x, \omega_y)$ is the 2D Fourier transform of $l(x, y, t = 0)$. $\delta(\cdot)$ is the 1D Dirac impulse. According to (5.2), a signal with translational motion contains non-zero spectral components only within a plane in spatiotemporal frequency space. This plane is orthogonal to the augmented velocity vector $(v_x, v_y, 1)$. The region of support of $L(\omega_x,\omega_y,\omega_t)$ is shown in Fig. 5.2, assuming a spatial band-limitation

$$L(\omega_x,\omega_y) = 0 \quad \text{for} \quad |\omega_x| \geq \Omega_x \text{ or } |\omega_y| \geq \Omega_y. \tag{5.3}$$

A still image, with $v_x = v_y = 0$, contains non-zero components only in the plane $\omega_t = 0$. For translational motion, this plane is sheered in ω_t-direction [6] [7].

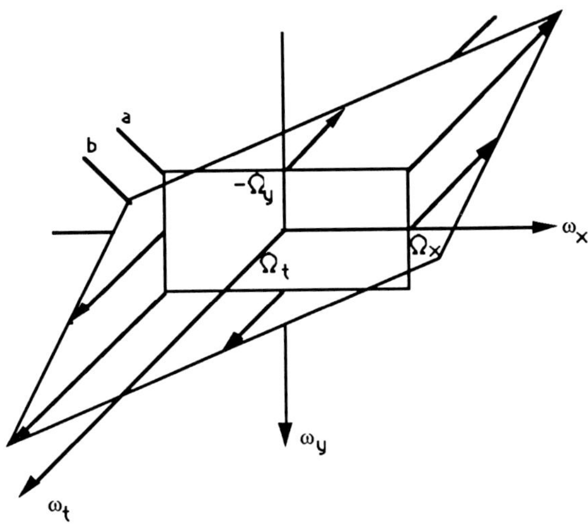

Fig. 5.2:(a) Spectrum of a band-limited image sequence with still image contents (b) with translational motion.

5.2.3 The importance of eye movements

As Fig. 5.2 illustrates, the spectrum of the luminance signal representing rapid motion will contain high temporal frequency components. Let us assume that we are viewing a display from a distance that is in reasonable correspondence with the spatial bandwidth of the system. To be more precise, say, $\Omega_x/2\pi = \Omega_y/2\pi \approx 20\,\text{cpd}$. If we now compare the spectrum according to (5.2) with the MTF shown in Fig. 5.1, we find that, even for moderate motion, the high spatial frequency components of the spectrum fall outside of the sensitive passband of the human visual system. The shape of the spatiotemporal MTF seems to imply that less spatial resolution is required when the picture contents move. In fact, various sampling or coding schemes for moving video signals have been proposed to take advantage of the shape of the spatiotemporal MTF, for example, by omitting spatially and temporally high frequency components altogether [8] [9] [10] [11] [12] [13].

A simple experiment that we carried out some years ago shows that motion dependent blur cannot be tolerated if impairments are required to be below visibility threshold

[14]. Motion was generated by artificially panning a still picture at three different velocities (0, 2, and 10 deg/sec). The picture contents were blurred by a 1D horizontal Gaussian filter which allowed a continuous adjustment of the degree of blur. The visibility threshold for just not visible blur was then measured in subjective tests. Surprisingly, blur visibility is not significantly affected by the motion speed. Even for the very rapid motion of 10 deg/sec, the visibility threshold is hardly higher than for a still picture. Subjects follow the picture contents by "smooth pursuit eye movements" and compensate its motion [14] [15] [16] [17] [18]. On the other hand, the spatiotemporal MTF, as shown in Fig. 5.1, is measured without eye movements. It cannot be applied without appropriate modification if the eye moves.

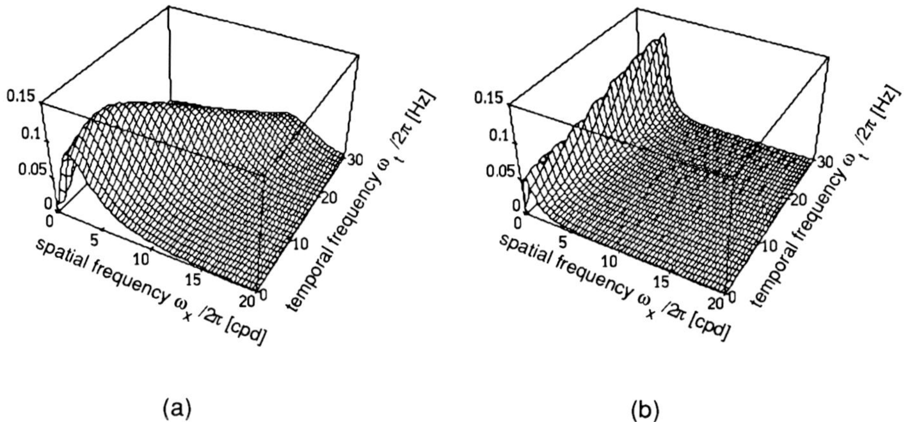

Fig. 5.3: Spatiotemporal modulation transfer function of the human visual system, as shown in Fig. 5.1, considering smooth pursuit eye movements with velocities of (a) 2 deg/sec (b) 10 deg/sec. Only one of the quadrants in the frequency plane that contain the peaks of the MTF is shown.

Smooth pursuit eye movements give rise to a Doppler effect. Assuming a smooth pursuit eye movement that shifts a still picture contents with a constant velocity of (v_x, v_y) across the retina, the frequency response of the human visual system can be easily transformed into the screen coordinate system using

$$\omega_x = \omega_x^*$$

$$\omega_y = \omega_y^*$$

$$\omega_t = \omega_t^* - \omega_x^* \cdot v_x - \omega_y^* \cdot v_y \tag{5.4}$$

[6] [7] [14] [19] . ω_x and ω_y are horizontal and vertical frequency, and ω_t temporal frequency in the screen coordinate system. ω_x^*, ω_y^*, and ω_t^* are the corresponding quantities in the eye coordinate system. The derivation of (5.4) is straightforward and very similar to the derivation of (5.2). Using this coordinate transformation, the spatiotemporal MTF (Fig. 5.1) has been plotted for smooth pursuit eye movements

of $v_x = 2$ deg/sec and $v_x = 10$ deg/sec (Fig. 5.3). The effective spatiotemporal frequency response of the human visual system is dramatically altered by eye movements. Through eye movements, we can perceive very high temporal frequencies at high spatial frequencies. According to [19] temporal frequencies up to 1000 Hz can be perceived. This effect does not contradict the flicker fusion effect that, depending on the brightness level, occurs at a temporal frequency below 80 Hz [20]. Flicker fusion is measured with uniform fields, i.e., $\omega_x = \omega_y = 0$.

Smooth pursuit eye movements have important consequences for the temporal bandwidth of display devices for moving images. Television phospors or liquid crystal materials, e.g., do not get away with a bandwidth of a few 10s of Hz, as Fig. 5.1 suggests. They need to display the full range, in the extreme up to 1000 Hz, to avoid visible blur of fast motion.

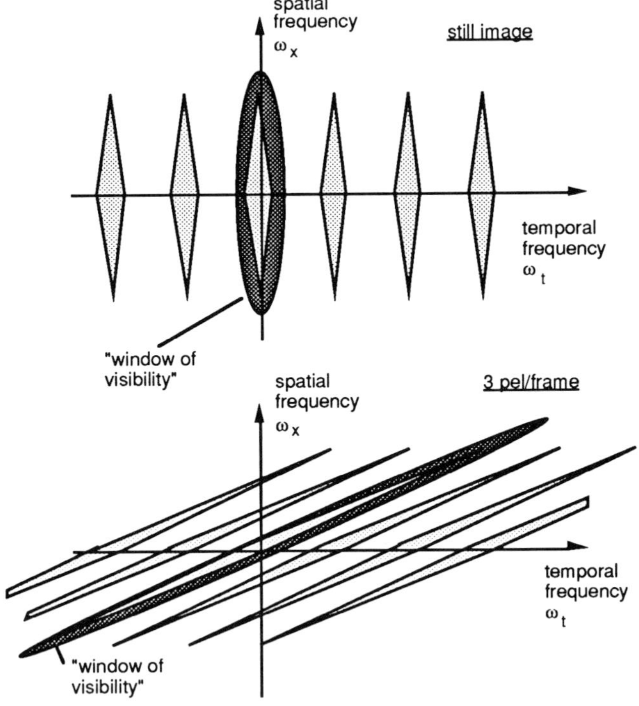

Fig. 5.4: Spectrum of a temporally sampled image sequence and the "window of visibility"

Temporal sampling of the video signal produces replications of the baseband at higher temporal frequencies. This is shown for velocities $v = 0$ and $v = 3$ pel/frame of the picture contents in Fig. 5.4. The spatiotemporal frequency response of the human visual system defines a "window of visibility" in spatiotemporal frequency space.

If the eye tracks the image contents, it fully passes the baseband of the signal and suppresses all temporal replications [6] [7] [14] [19] . The human visual system acts as a motion compensating interpolation filter (Fig. 5.4). It is sometimes argued that temporal aliasing sets in at a velocity of 1 pel/frame. This argument is valid only if a single pel is considered by itself. In spatiotemporal frequency space, there is no spectral overlap even at higher motion speeds. The baseband can be recovered as long as the image contents can be tracked.

For several applications, multidimensional filtering of the video signal with time as one dimension is desirable. Some of these applications will be discussed in section 5.3. Because of smooth pursuit eye movements, any type of temporal filtering has to be carried out with motion compensation in order to avoid visible blur of the picture contents. On the other hand, compensation is only required for motion that can be tracked by the human visual system. Today, we know too little about the limits of motion compensation by smooth pursuit eye movements, and little effort has been made to incorporate these limits into technical motion compensating filters.

5.3 Motion Compensating Interpolation

Frequently, "missing" luminance or chrominance values of a video signal have to be filled in from available samples that correspond to past or future time instances. This problem has to be solved, e.g., for the display of video sequences that have been encoded at a low frame rate. For 64 kbit/s transmission of video sequences, frame rates of 10 Hz or less are not uncommon. A related example is frame rate doubling for flickerfree display of television signals, say from 50 Hz to 100 Hz in the European countries. Another very demanding application is the conversion between interlaced 50 Hz and 60 Hz television sampling rasters, for example, between PAL and NTSC.

Today, the most common solution is the repetition of past samples at the same position, i.e., without regard of the motion occuring in the image sequence. The result is a jerkiness in the motion reproduction, that is more or less annoying depending on the initial frame rate. The jerkiness in a 10 Hz sequence shown with frame repeat is very objectionable even with only moderate motion. On the other hand, most people do not object strongly against the 24 Hz frame rate used with film. In the movie theater, each frame is projected twice, which leads to a considerable jerkiness and double contours, especially noticable with camera pans. When transfered to 50 Hz video, a motion picture is played back slightly faster (at 25 fps), and each frame constitutes two video fields. For a transfer of film to 60 Hz television, a frame repeat method known as "3/2 pull down" is used, where alternate film frames are displayed for three TV fields and two TV fields, respectively. The motion rendition in each of these examples can be improved by motion compensating interpolation techniques.

In the following sections, we analyze effects caused by spatiotemporal interpolation of missing television samples in the 3D frequency domain. Such an analysis is a very powerful way to understand effects of blurring and aliasing artifacts, and it is the basis for designing proper filters for high quality interpolators utilizing motion compensation. While section 5.3.1 will treat motion compensating interpolation starting from an orthogonal, non-interlaced sampling grid, section 5.3.2 will extend results to line-interlaced grids, as they are used in all television systems today.

5.3.1 Motion compensating frame interpolation

Sampling of a space-time continuous luminance signal $l(x, y, t)$ can be described by a multiplication with a periodic sampling grid, $grid(x, y, t)$, in the space-time domain

$$s(x,y,t) = l(x,y,t) \cdot grid(x,y,t). \tag{5.5}$$

In the 3D frequency domain, this corresponds to a 3D convolution with the Fourier transform of the sampling grid

$$S(\omega_x,\omega_y,\omega_t) = L(\omega_x,\omega_y,\omega_t) * GRID(\omega_x,\omega_y,\omega_t) \tag{5.6}$$

where $S(\omega_x,\omega_y,\omega_t)$ is the 3D Fourier transform of the sampled luminance signal $s(x,y,t)$. As $grid(x,y,t)$ consists of periodic repetitions of delta-functions, the same holds for its Fourier transform $GRID(\omega_x,\omega_y,\omega_t)$. Thus, the spectrum of the sampled signal contains replications of the signal baseband at higher spatial and temporal frequencies.

Conceptually, any sampling raster conversion can be interpreted as an interpolation to recover the space-time-continuous signal $l(x,y,t)$ and then resampling it to the new raster. The main problem is the interpolation of $l(x,y,t)$ by suppression of its spectral replications without loss of spatial resolution in the case of motion. This requires a motion compensating interpolation filter that takes into account the velocity of the scene contents.

For a further frequency domain analysis, it is useful to introduce the notion of an ideal motion compensating lowpass filter with the frequency response

$$H(\omega_x,\omega_y,\omega_t) = \begin{cases} 1, & \text{if } -\Omega_x < \omega_x < \Omega_x \\ & \text{and } -\Omega_y < \omega_y < \Omega_y \\ & \text{and } -V_x\omega_x - V_y\omega_y - \Omega_t < \omega_t < -V_x\omega_x - V_y\omega_y + \Omega_t \,; \\ & \\ 0, & \text{otherwise.} \end{cases} \tag{5.7}$$

The vector (V_x, V_y) will be called the nominal velocity of the motion compensating lowpass filter. Ω_x and Ω_y are its horizontal and vertical bandwidths, and Ω_t is its nominal temporal bandwidth. The passband of the ideal motion compensating lowpass filter has the form of a parallel-epiped (Fig. 5.5). The frequency response is separable into a lowpass filter operating along the nominal motion trajectory, a horizontal lowpass filter, and a vertical lowpass filter.

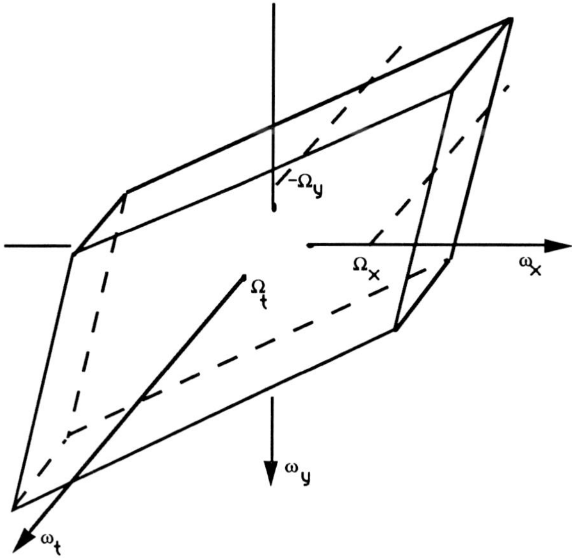

Fig. 5.5: Passband of an ideal motion compensating lowpass filter as defined by (5.7)

The spectrum of a signal which contains translational motion according to (5.1) and (5.2), and which is spatially bandlimited according to (5.3), passes the motion compensating lowpass filter (5.7) without any loss of spatial resolution if

$$\Omega_x |v_x - V_x| + \Omega_y |v_y - V_y| < \Omega_t \tag{5.8}$$

Thus, there exists a diamond-shaped "passband" of a motion compensating lowpass filter in the velocity-plane, which is shown in Fig. 5.6.

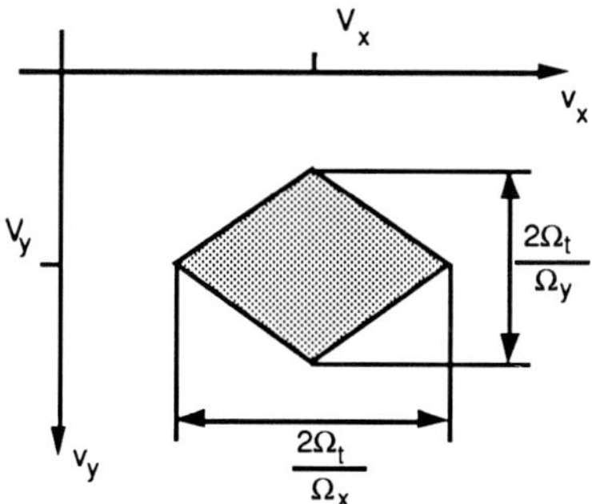

Fig. 5.6: Passband of a motion compensating lowpass filter in the velocity plane (5.8)

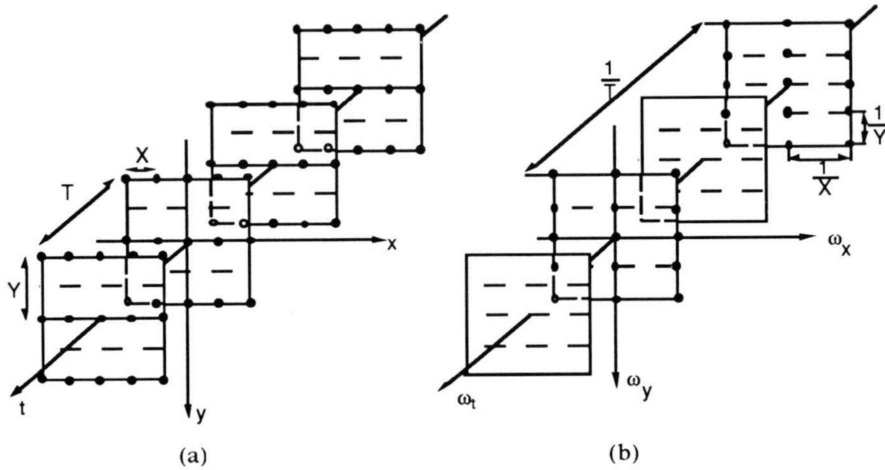

Fig. 5.7: (a) Non-interlaced sampling grid and (b) its 3D Fourier transform

Fig. 5.7a shows an orthogonal, non-interlaced, 3D sampling grid with horizontal sampling interval X, vertical sampling interval Y, and temporal sampling interval T. Its 3D Fourier transform is shown in Fig. 5.7b. The maximum spatial bandwidth of a motion compensating interpolation filter (5.7) that removes all baseband replications at higher spatial frequencies is

$$\frac{\Omega_x}{2\pi} = \frac{1}{2X}; \quad \frac{\Omega_y}{2\pi} = \frac{1}{2Y} \tag{5.9}$$

According to (5.8), there is no need to compensate the motion of the scene contents precisely. We can perfectly interpolate a signal containing any translational motion (v_x, v_y) according to (5.1) and (5.2) by covering the entire velocity plane by diamond shaped passbands as shown in Fig. 5.8. Within each passband, a fixed motion compensating lowpass filter with a nominal velocity of (V_x, V_y) corresponding to the passband center is used. The proper spacing of the passbands is determined by the ratios Ω_t/Ω_x and Ω_t/Ω_y. The widest spacing results for

$$\frac{\Omega_t}{2\pi} = \frac{1}{2T} \tag{5.10}$$

With bandwidths according to (5.9) and (5.10), the nominal velocities of the passbands can be chosen as

$$V_x = 2i \cdot \frac{X}{T}; \quad V_y = 2j \cdot \frac{Y}{T} \tag{5.11a}$$

and

$$V_x = (2m+1) \cdot \frac{X}{T}; \quad V_y = (2n+1) \cdot \frac{Y}{T} \tag{5.11b}$$

with integer i, j, m and n.

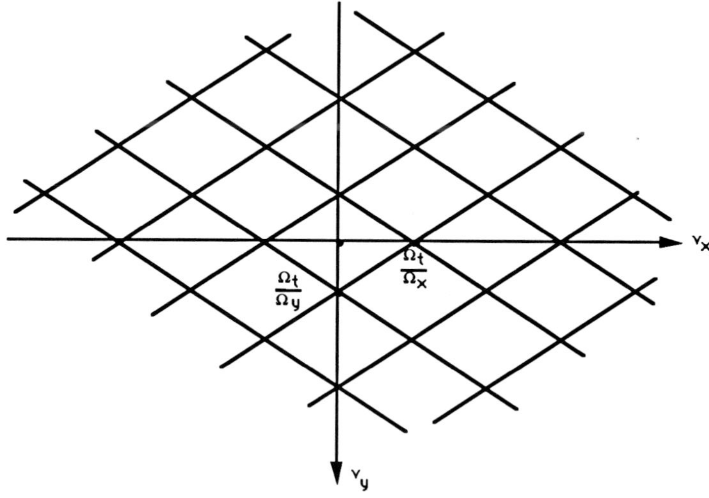

Fig. 5.8: Tessalation of the velocity plane by velocity passbands of motion compensating lowpass filters

Eq. (5.11) indicates that we can build a motion compensating interpolator that does not require spatial interpolation between samples. In practice, we would not attempt to realize a motion compensating interpolation filter with a steep temporal transition between passband and stopband. For such a filter, many frame stores would be required. In addition, we might encounter problems, if the motion changes from one frame to the next. The above analysis was based on the assumption of constant translational motion. For these reasons, we prefer short temporal filters with a wide transition band and a nominal temporal bandwidth smaller than indicated by (5.10). With such a temporal frequency characteristic, the velocity passbands (5.8) will be narrower and a closer spacing of the passbands in the velocity plane is required. Accordingly, the motion compensating interpolator will require fractional-pel accuracy of the motion compensation.

5.3.2 Motion compensating de-interlacing

Let us now extend the frequency domain analysis presented in the previous section to understand and solve the problem of motion compensating de-interlacing. A line-interlaced sampling grid and its Fourier transform are shown in Fig. 5.9. Again, X, Y, and T are the horizontal, vertical, and temporal sampling intervals. For an interlaced grid, motion compensating interpolation has to cope with an additional complication. The vertical resolution of the luminance signal $l(x, y, t)$ depends on the vertical velocity component v_y of the scene contents. If the scene does not contain vertical motion, i.e., $v_y = 0$, the vertical bandwidth is

$$\frac{\Omega_{y1}}{2\pi} = \frac{1}{Y}. \tag{5.12}$$

The resolution is twice as large as without interlacing. However, at critical velocities

$$v_y = \frac{(2k+1)}{2} \cdot \frac{Y}{T}, \quad k = \ldots, -2, -1, 0, 1, 2, \ldots \tag{5.13}$$

sampling lines will not interleave, as for $v_y = 0$, and the vertical bandwidth is only

$$\frac{\Omega_{y2}}{2\pi} = \frac{\Omega_{y1}}{4\pi} = \frac{1}{2Y}. \tag{5.14}$$

A good motion compensating interpolation scheme should reconstruct the signal containing motion at one of the critical velocities (5.13) with half vertical resolution Ω_{y2}, while keeping full vertical resolution over a wide range of other velocities.

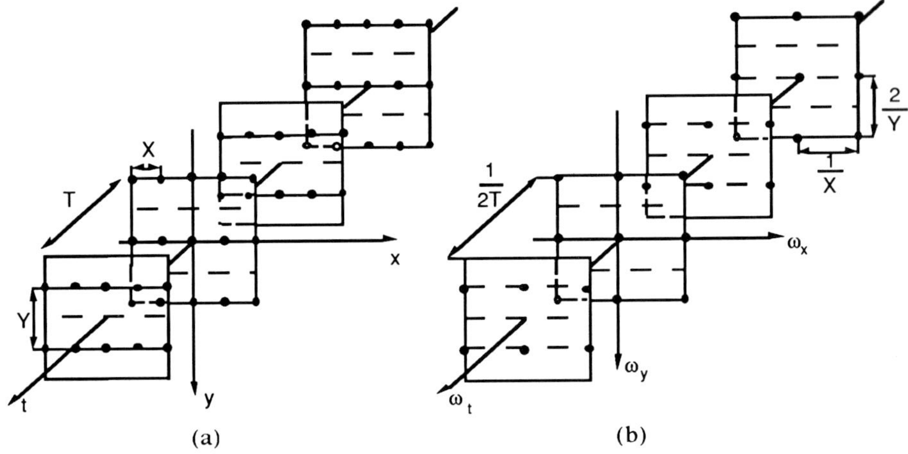

Fig. 5.9: (a) A line-interlaced sampling grid and (b) its 3D Fourier transform

A motion compensating interpolator that fulfills these requirements, can be constructed in a surprisingly simple fashion. To explain its principle, we again assume ideal motion compensating filters, as shown in Fig. 5.5. The vertical and the temporal nominal bandwidths of this filter are chosen as

$$\Omega_y = \Omega_{y1}, \quad \frac{\Omega_t}{2\pi} = \frac{1}{4T} \tag{5.15}$$

This filter may only be used at nominal vertical velocities of

$$v_y = k \cdot \frac{Y}{T}, \quad k = \ldots, -2, -1, 0, 1, 2, \ldots \tag{5.16}$$

since otherwise annoying aliasing patterns may result. The nominal velocities (5.16) are halfway between the critical velocities. Fig. 5.10 shows a subdivision of the velocity plane that corresponds to (5.16). Similar to the scheme described in the previous section, a fixed motion compensating filter is used within each of the rectangular regions of the velocity plane. The vertical resolution, however, varies within each of the rectangular regions.

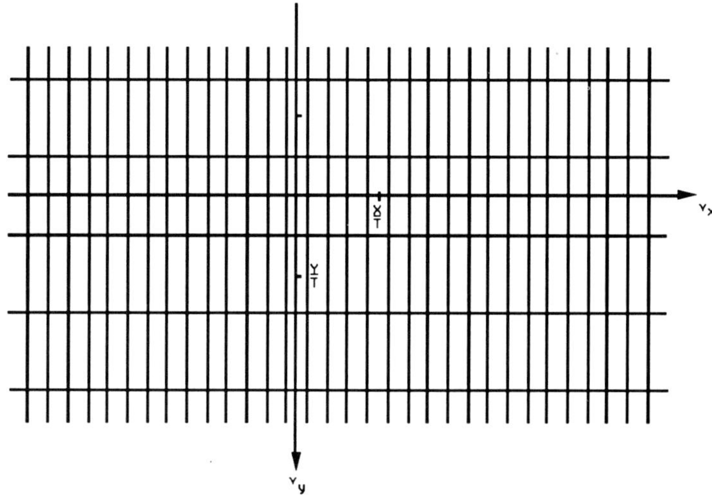

Fig. 5.10: Subdivision of the velocity plane for interpolation from a line-interlaced grid.

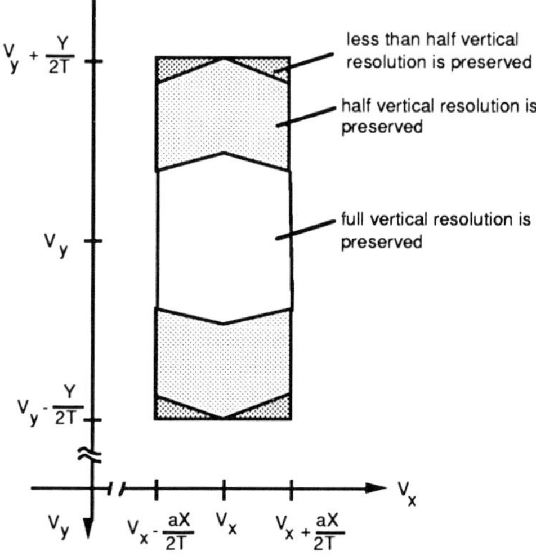

Fig. 5.11: Zones of different vertical resolution for each of the rectangular regions shown in Fig. 5.10.

We can easily calculate the velocity zones, where full or half vertical resolution is preserved for all horizontal frequencies (Fig. 5.11), by plugging Ω_{y1} and Ω_{y2} into (5.8). The parameter a in Fig. 5.11 determines the spacing of horizontal nominal velocities V_x. In order to maximize the vertical resolution for all velocities, a should be chosen as small as possible. Remarkably, horizontal motion has to be compensated accurately in order to achieve a high vertical resolution with the motion compensating de-interlacer described.

A motion compensating de-interlacer based on the above theory was first described by the author in [6] [7] and its performance was demonstrated for very critical moving zone plate patterns. The filters reported in [6] [7] require 11 field stores. Based on these ideas, a high quality standards converter was developed and demonstrated by researchers at the Heinrich-Hertz-Institut in 1988 [21]. Their converter used elaborate motion compensating filters to convert from 50 to 60 Hz line-interlaced HDTV signals. Superior picture quality was demonstrated with natural test scenes, using 6 polyphase filters with 5 field stores each.

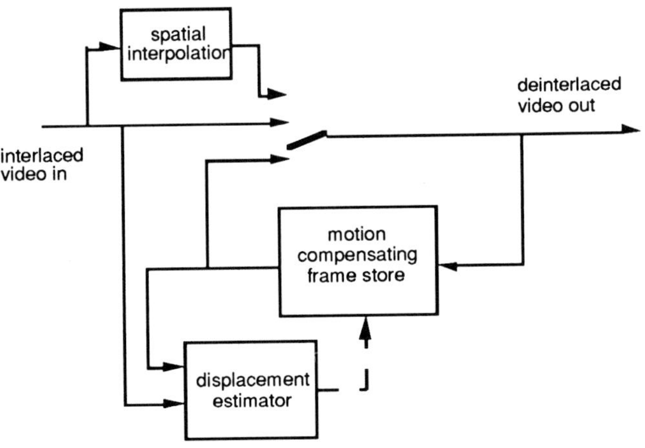

Fig. 5.12: Motion compensating temporally recursive de-interlacer

The above de-interlacers use linear phase FIR filters. Memory requirements and delay associated with such a scheme can be considerable. These drawbacks can be overcome with a temporally recursive de-interlacing scheme, as proposed in [22] (Fig. 5.12). It works under the assumption that we already have a de-interlaced, full-resolution frame available at the output that is used to de-interlace the next input field. The displacement between the contents of the frame store and the input field is measured with an appropriate algorithm. The contents of the frame store is then warped with this displacement field. In order to obtain the next ouput frame, either lines of the input field are passed straight through, or, for lines missing in the input, the motion compensated previous frame is filled in. If motion compensation fails, e.g., at the first picture of a scene or at critical velocities of the scene contents (5.13), the algorithm switches to an intrafield interpolation mode which provides a picture of half vertical resolution Ω_{y2}.

5.4 Motion Compensating Prediction

The most common application of motion compensation today is in predictive video compression algorithms. The idea of predictive coding dates back to 1952 [23]. More than twenty years later, after significant advances in computer power and memory technology, motion compensating prediction was seriously explored for video compression [24] -[45]. It turned out to be one of the keys to achieve the impressive compression ratios that allow the transmission of moving video at rates as low as 64 kbit/s with acceptable picture quality. A low bitrate video coder based on motion compensating prediction has now been standardized in CCIR Recommendation H.261 [46].

The majority of motion compensating predictors that are reported in the literature use motion compensation with integer-pel accuracy. In fact, CCIR Recommendation H.261 includes motion compensation with integer-pel accuracy. Only recently, it has been widely accepted that fractional-pel accuracy not only can significantly improve prediction, but is also feasible for a real-time coder. Early coding work that uses motion compensation with fractional-pel accuracy is reported, e.g., in [29], [41], [42], [44]. The recent ISO MPEG I standard for compression of moving video to rates at around 1 Mbit/s allows for half-pel accuracy [47]. Typically, fractional-pel accuracy is achieved by simple bilinear interpolation, which produces a spatially blurred prediction signal. It has been pointed out that the use of a spatial lowpass filter in the predictor can improve prediction in connection with integer-pel accuracy [24] [29] [48] [49]. Bilinear interpolation for fractional-pel accuracy introduces spatial lowpass filtering as a side effect, and benefits both from increased accuracy and from the filtering effect.

A theoretical framework has been developed for analysis of the performance of motion compensating prediction in terms of rate distortion theory [48]. It relates power spectral density of the prediction error to the accuracy of motion compensation. It is shown in [48] that with integer-pel accuracy of the displacement estimate the additional gain by motion compensating prediction over optimum intraframe encoding of the signal is limited to ~ 0.8 bit/sample in moving areas. Larger gains require fractional-pel accuracy.

In the following, we will examine the theoretical performance limits of motion compensating prediction, complemented by experimental results. In addition, we will consider the question of which interpolation kernel should be used for fractional-pel accurate motion compensation and what precision of displacement measurement and motion compensation is adequate.

5.4.1 Motion compensating hybrid coding schemes

The structure of many proposed state-of-the-art codecs resembles the motion compensating hybrid coding scheme shown in Fig. 5.13 [24] - [49]. This coding scheme combines differential pulse code modulation (DPCM) along an estimated motion trajectory of the picture contents with intraframe encoding of the prediction error e. The displacement estimate $(\hat{d}_x, \hat{d}_y)$ is transmitted in addition to the intraframe-encoded prediction error. At the receiver, the intraframe source decoder generates the reconstructed prediction error e', which differs from e by some source coding error. The transmitter contains a replication of the receiver in order to be able to generate the same prediction value $\hat{s}$.

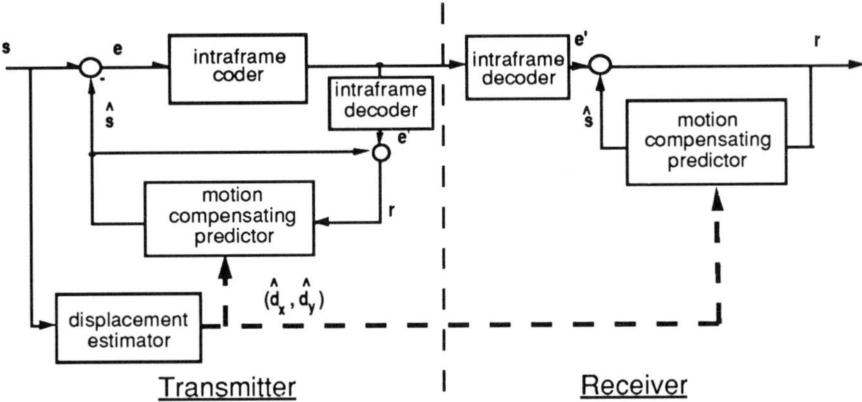

Fig. 5.13: Motion compensating hybrid coding scheme

In a simple scheme, the intraframe source encoder could be just a quantizer, such that the hybrid coder (Fig. 5.13) is a motion compensating DPCM coder [25], [26], [33], [37], [39]. More sophisticated schemes encode a spatial neighborhood of prediction errors e employing a blockwise transform, like the discrete cosine transform (DCT) [24], [28], [29], [31], [34], [35], [36], [38], [41], [42], a quadrature mirror filter bank [32], [40], vector quantization [27], [30], [43], or quadtree coding[45]. In each case, the intraframe encoder serves to remove spatial redundancy from the signal e and to adaptively reduce the spatial resolution, if the channel cannot accomodate a full-resolution signal. It has been pointed out by several authors that the motion compensated prediction error signal is only weakly correlated spatially [34] [44] [45] [48]. Thus, the potential for redundancy reduction in the intraframe source encoder is very small. This finding suggests that the prediction error variance σ_e^2 is a useful measure that is directly related to the minimum achievable transmission bit-rate for a given signal-to-noise ratio [48].

5.4.2 Theoretical performance limits of motion compensating prediction

Three factors limit the performance of a motion compensating predictor:

- The estimated displacement $(\hat{d}_x, \hat{d}_y)$ differs from the true displacement of the scene contents. To characterize this effect we introduce a displacement error $(\Delta d_x, \Delta d_y)$.
- There are signal components in the current frame that are not present in the previous frame, and vice versa. For analysis, we lump these signal components together as noise n with a power spectral density $\Phi_{nn}(\omega_x,\omega_y)$, and we assume that n is independent from the video signal s and the displacement error $(\Delta d_x, \Delta d_y)$.
- The motion compensated prediction image is somewhat blurry compared to the original image. This can be due to the kernel for sub-pel accuracy interpolation, and possibly to an additional smoothing filter introduced after the warping stage of the predictor. We summarize all spatial filtering effects of the predictor by a frequency response $F(\omega_x,\omega_y)$.

In order to analyse the performance of motion compensating predictors, we assume that the estimated displacement $(\hat{d}_x, \hat{d}_y)$ does not depend on (x, y). This allows us to treat motion compensating prediction with the theory of linear, space-invariant systems. The assumption of a constant displacement is not too much of a limitation. Many motion compensating predictors employ a constant estimated displacement vector within a block of, e.g., 16×16 samples, and our assumption holds within a block. Other schemes interpolate a sparse estimated vector field to obtain a dense, smooth field for motion compensation. With the assumption of constant $(\hat{d}_x, \hat{d}_y)$, we neglect effects caused by spatial changes in the estimated vector field.

With this approach, the power spectral density of the prediction error e can be shown to be

$$\Phi_{ee}(\omega_x,\omega_y) = \Phi_{ss}(\omega_x,\omega_y) \cdot \Big(1 + |F(\omega_x,\omega_y)|^2 - 2\Re\{F(\omega_x,\omega_y)P(\omega_x,\omega_y)\}\Big) + \Phi_{nn}(\omega_x,\omega_y)|F(\omega_x,\omega_y)|^2, \quad (5.17)$$

where $\Phi_{ss}(\omega_x,\omega_y)$ is the spatial power spectral density of the video signal s, $\Re\{\cdot\}$ denotes the real part of a complex number, and $P(\omega_x,\omega_y)$ is the 2-D Fourier transform of the probability density function (pdf) of the displacement error $(\Delta d_x, \Delta d_y)$. For details of the derivation of (5.17), the reader is referred to [48] and [49].

From (5.17), we can conclude that motion compensation works well for low spatial frequencies, but it is hard for high spatial frequencies. The spectrum of the motion compensated prediction error, $\Phi_{ee}(\omega_x,\omega_y)$, is shown in Fig. 5.14 for $\Phi_{nn}(\omega_x,\omega_y) = 0$. At high spatial frequencies, the power spectral density can be up to 3 dB higher than the signal power spectrum. In order to compensate high frequency components in the signal, very accurate motion compensation is required, otherwise the signal and its prediction are out of phase. For low frequency components, motion compensation can be less accurate. Since more energy implies a higher bitrate required for the encoding of the frequency components, it is advantageous to only encode the motion compensated prediction error at the low frequencies, and transmit the video signal itself at the higher spatial frequencies where motion compensation fails. This can be incorporated into the predictor by choosing an appropriate filter $F(\omega_x, \omega_y)$.

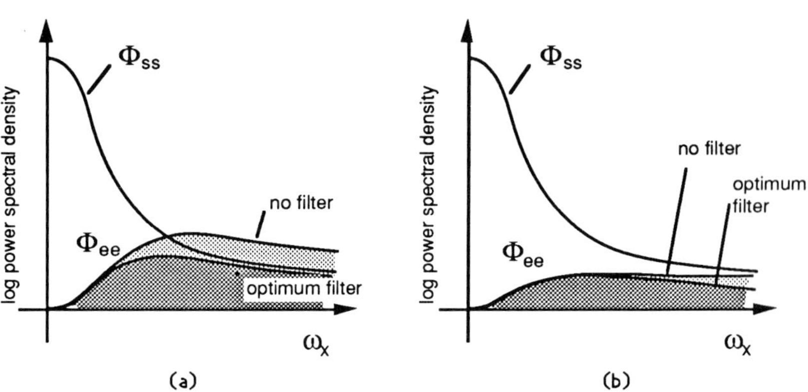

Fig. 5.14: Power spectrum of the motion compensated prediction error for (a) moderately accurate motion compensation (b) very accurate motion compensation

By differentiating (5.17) with respect to $F(\omega_x, \omega_y)$, it is readily shown that the mean squared prediction error is mimimized at each frequency if $F(\omega_x, \omega_y)$ is a Wiener filter with frequency response

$$F(\omega_x, \omega_y) = P^*(\omega_x, \omega_y) \cdot \frac{\Phi_{ss}(\omega_x, \omega_y)}{\Phi_{ss}(\omega_x, \omega_y) + \Phi_{nn}(\omega_x, \omega_y)}. \tag{5.18}$$

The superscript "*" denotes complex conjugation. Wiener filter (5.18) can be interpreted to consist of two stages, one of which is a Wiener filter with respect to the noise n, the other one is a Wiener filter that takes into account the inaccuracy of the displacement estimate.

Equation (5.17) allows us to study the influence of the displacement error pdf on the prediction error variance σ_e^2, which can be calculated on the basis of Parseval's relation

$$\sigma_e^2 = \frac{1}{4\pi^2} \int_{-\pi}^{\pi} \int_{-\pi}^{\pi} \Phi_{ee}(\omega_x, \omega_y)\, d\omega_x\, d\omega_y. \tag{5.19}$$

We have evaluated the prediction error variance assuming an isotropic signal power spectrum

$$\Phi_{ss}(\omega_x, \omega_y) = \frac{2\pi\sigma_s^2}{\omega_0^2} \cdot \left(1 + \frac{\omega_x^2 + \omega_y^2}{\omega_0^2}\right)^{-\frac{3}{2}} \tag{5.20}$$

[48], a flat noise power spectrum

$$\Phi_{nn}(\omega_x, \omega_y) = \frac{\sigma_n^2}{4\pi^2} \tag{5.21}$$

and an isotropic Gaussian displacement error pdf of variance $\sigma_{\Delta d}^2$

$$p(\Delta d_x, \Delta d_y) = \frac{1}{2\pi\sigma_{\Delta d}^2} \exp\left(-\frac{\Delta d_x^2 + \Delta d_y^2}{2\sigma_{\Delta d}^2}\right), \tag{5.22}$$

where σ_s^2 and σ_n^2 are signal and noise variances, respectively, and ω_0 has been set to correspond to a typical correlation of 0.93 between adjacent samples. Fig. 5.15 shows the influence of the displacement error variance $\sigma_{\Delta d}^2$ on the prediction error variance σ_e^2 for three variances σ_n^2 of noise. The curves compare a Wiener filter (5.18) with the case "no filter",

$$F(\omega_x, \omega_y) = 1. \tag{5.23}$$

The following observations are important:

- Prediction error variance is generally decreased by more accurate motion compensation.
- Beyond a certain "critical accuracy" the possibility of further improving prediction by more accurate motion compensation is small.
- The critical point is at a high displacement error variance for high noise variance and at a low displacement error variance for low noise variance.
- For low noise the Wiener filter is more effective for less accurate motion compensation than for accurate motion compensation.
- For high noise the Wiener filter is more effective for accurate motion compensation than for less accurate motion compensation.
- For accurate motion compensation the potential gain through the Wiener filter increases with noise level.

Fig. 5.15 also indicates the minimum displacement error variance that can be achieved for a given motion compensation accuracy in moving areas. Consider a perfect displacement estimator that always estimates the true displacement. Then, the displacement error $(\Delta d_x, \Delta d_y)$ is entirely due to rounding. In moving areas with sufficient variation of motion, the displacement error will be uniformly distributed between

$\pm\frac{1}{2}\beta X$ and $\pm\frac{1}{2}\beta Y$, where $\beta = 1$ for integer-pel accuracy, $\beta = 1/2$ for 1/2-pel accuracy, etc. For a grid with balanced horizontal and vertical resolution, $X = Y$, the minimum displacement error variance in moving areas is

$$\sigma_{\Delta d}^2 = \frac{(\beta X)^2}{12}. \tag{5.24}$$

It turns out that the precise shape of the displacement error pdf has hardly any influence on the variance of the motion compensated prediction error, σ_e^2, as long as the displacement error variance $\sigma_{\Delta d}^2$ does not change. A uniform pdf and a Gaussian pdf yield essentially the same variances σ_e^2.

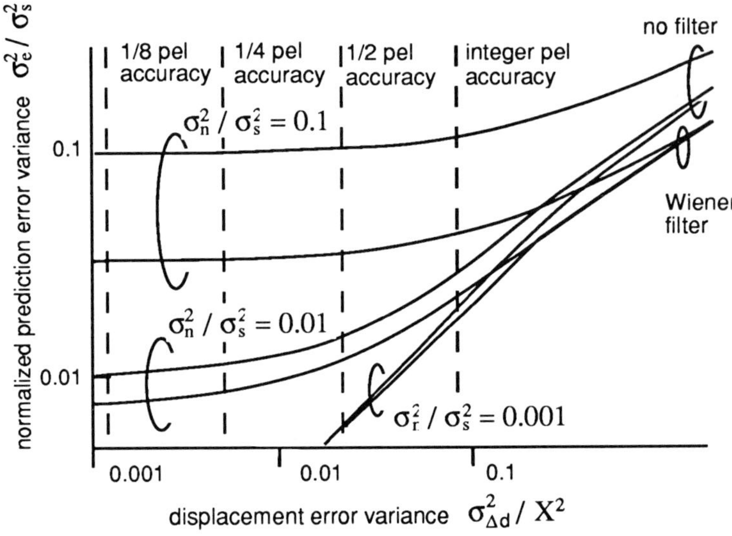

Fig. 5.15: Results of a model calculation showing the influence of motion compensation accuracy on prediction error variance for noisy signals. The potential gain of a spatial prediction filter is illustrated by the difference between "no filter" and the "Wiener filter" curves. The vertical dashed lines indicate the minimum displacement error variances that can be achieved with the indicated motion compensation accuracies in moving areas.

5.4.3 Experimental comparison of motion compensation accuracies

While the theory presented in the previous section provides important insights into the basic mechanism that governs the performance of motion compensating predictors, we still have to confirm the curves in Fig. 5.15 by experimental results obtained with natural image sequences. For this purpose, we would like to employ a displacement estimator as precise as possible, and use its displacement vector field at various degrees of accuracy.

For the experimental results presented in this section, we have used a three-stage displacement estimator, which is an extension of a technique presented in [50]. Its first stage estimates a histogram of displacements using a phase correlator in large overlapping measurement windows [51] [52]. The predominant peaks of the phase correlation surface are then passed on to a second stage as candidate vectors. The second stage performs an image segmentation by candidate vector assignment to smaller blocks. The first and the second stage each use integer-pel accuracy for the displacement vector. In the final third stage, the integer-pel vector is refined to the desired sub-pel accuracy by searching for the shift that minimizes the mean absolute displaced frame difference. The three-stage displacement estimator is described in more detail in [49].

With this displacement estimator, three different types of spatial prediction/ interpolation filters $F(\omega_x, \omega_y)$ were compared experimentally at different motion compensation accuracies.

1. *Sinc-interpolation* corresponds to the case $F(\omega_x, \omega_y) = 1$, or "no filter" in Fig. 5.15. Ideally, the spatial impulse response would be

$$f(x,y) = \frac{sin(\pi x/X) \cdot sin(\pi y/Y)}{(\pi x/X) \cdot (\pi y/Y)}. \tag{5.25}$$

   The sinc-interpolation kernel (5.25) has infinite extent. For our measurements, we approximated (5.25) by very long filters.

2. *Bilinear interpolation* uses an interpolation kernel

$$f(x,y) = \max\{0, 1 - |\frac{x}{X}|\} \cdot \max\{0, 1 - |\frac{y}{Y}|\}. \tag{5.26}$$

   Note that Sinc-interpolation and bilinear interpolation are identical for integer-pel accuracy of motion compensation.

3. *Wiener filters* were computed separately for each estimation accuracy and each source signal. (5.18) gives the Wiener filter as a function of signal and noise power spectra and displacement error pdf. Although this formulation is useful to understand MCP, it is not a useful formulation for filter design, since noise power spectrum and displacement error pdf are usually not known explicitly. We take an alternative approach here. Let us introduce an *unfiltered motion compensated previous frame* $c(x, y)$ that is obtained using the sinc-interpolation kernel (5.25). Our task is to find the filter that best predicts the current frame $s(x, y)$ when applied to $c(x, y)$. It is a well-known result from linear mean-square estimation [53] that the Wiener filter with frequency response

$$F(\omega_x, \omega_y) = \frac{\Phi_{sc}(\omega_x, \omega_y)}{\Phi_{cc}(\omega_x, \omega_y)} \tag{5.27}$$

minimizes the mean squared prediction error. In (5.27) $\Phi_{sc}(\omega_x, \omega_y)$ is the cross spectrum between $s(x, y)$ and $c(x, y)$, and $\Phi_{cc}(\omega_x, \omega_y)$ is the power spectrum of $c(x, y)$. Both spectra can be measured directly using standard methods of power spectral estimation.

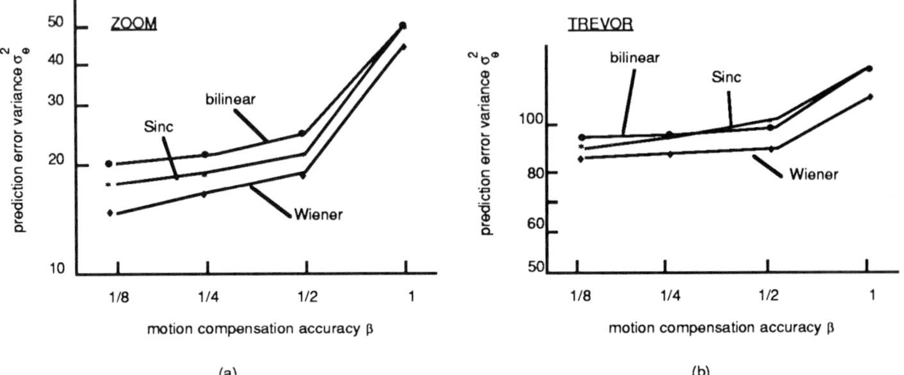

Fig. 5.16: Experimental comparison of different spatial prediction/interpolation filters for a broadcast TV signal ZOOM and a videophone signal TREVOR. For TREVOR, motion compensated prediction error variance is measured for moving areas only. Scaling of the axis "motion compensation accuracy β is such that $\beta = 1$ corresponds to integer-pel accuracy, $\beta = 1/2$ corresponds to $1/2$-pel accuracy, etc. [49]

The variance of the prediction error for the different spatial prediction/ interpolation filters is shown in Fig. 5.16 as a function of motion compensation accuracy. The prediction was based on the previous original picture rather than the previous reconstructed picture r, thus neglecting noise introduced by encoding of the prediction error e (Fig. 5.13). We show results for a broadcast TV signal ZOOM and for a videophone signal TREVOR (Fig. 5.17a,b). For TREVOR, only moving parts have been considered for both designing the Wiener filter and measuring the prediction error variance. Fig. 5.13 illustrates the following observations:

- Compared to integer-pel accuracy of MCP without filtering, prediction error variance can be reduced by more accurate motion compensation and Wiener filtering by up to 5.2 dB for ZOOM, and 1.8 dB for TREVOR.

- For TREVOR, bilinear interpolation is as good as Sinc-interpolation. This is typical for videophone signals. Motion compensation with high order spatial interpolation filters is advantageous only when practically all changes from one frame to the next can be motion compensated. This is the case for ZOOM.

CURRENT FIELD
PREVIOUS FIELD
FRAME DIFFERENCE
DISPLACEMENT VECTORS
INTEGER-PEL-ACCURACY
1/2-PEL-ACCURACY
1/4-PEL-ACCURACY
1/8-PEL-ACCURACY
(a) (b) (c) (d) (e) (f) (g) (h)

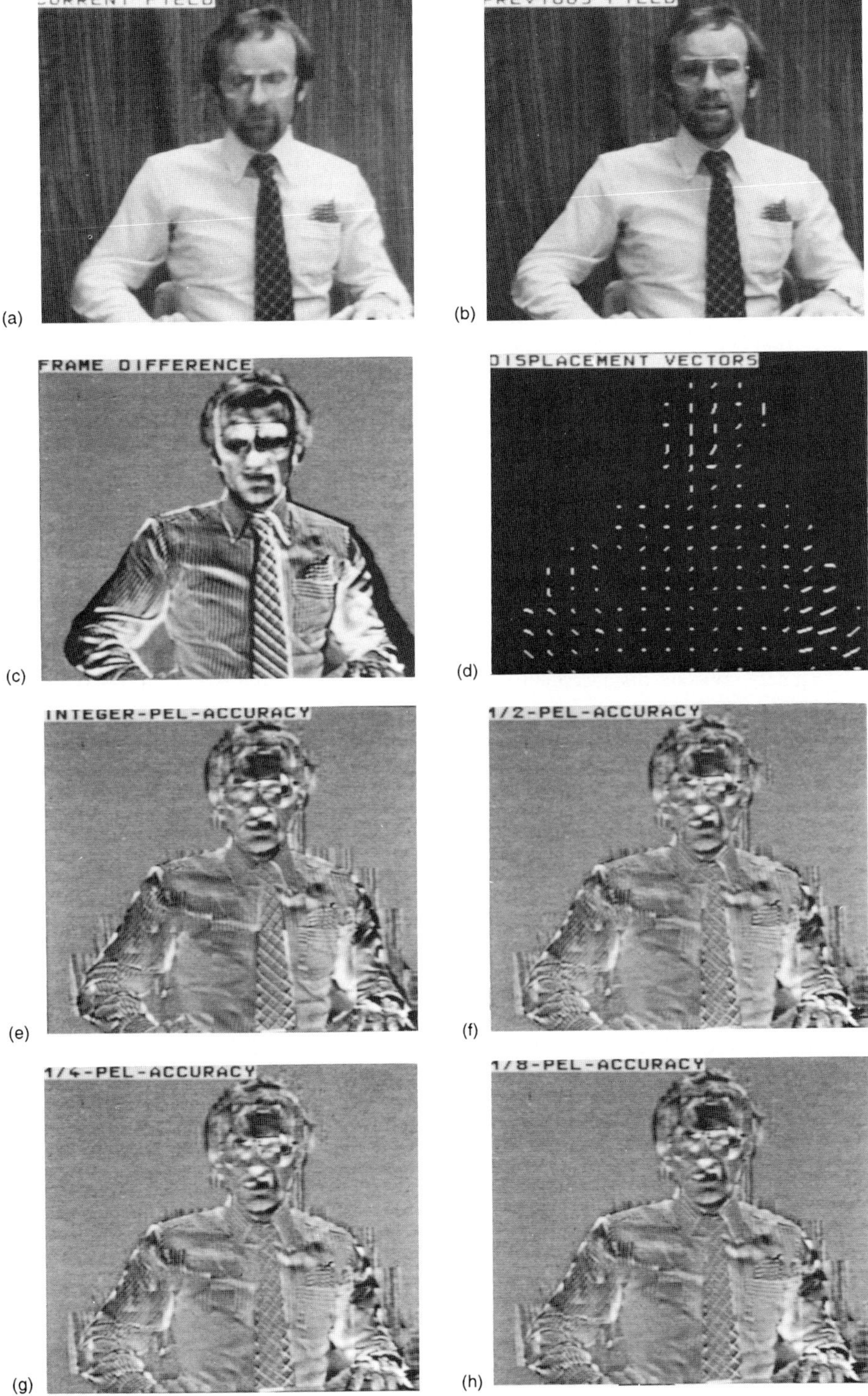

Fig. 5.17 (previous page): Picture pair TREVOR (a) current picture (b) previous picture (c) frame difference without motion compensation (d) displacement vectors (e) motion compensated prediction error with integer-pel accuracy (f) motion compensated prediction error with 1/2-pel accuracy (g) motion compensated prediction error with 1/4-pel accuracy (h) motion compensated prediction error with 1/8-pel accuracy.

- For broadcast TV signals like ZOOM, motion compensation with 1/4-pel accuracy is certainly sufficiently accurate for a practical coder.

- For videophone signals like TREVOR, 1/2-pel accuracy seems to be a desirable limit.

- The curves in Fig. 5.16 qualitatively correspond to the model curves in Fig. 5.15. The measurement results can be explained by the theory presented in Section III.

Fig. 5.17 shows the motion compensated prediction error $e(x, y)$ for TREVOR at different accuracies of motion compensation. For this example Sinc-interpolation was used. This series of pictures provides an intuitive explanation of the curves in Fig. 5.16. Once motion compensating prediction, say, with integer-pel accuracy, has been applied, the motion compensated prediction error contains mostly components that cannot be further reduced by more accurate motion compensation. The signal model "constant displacement within a block" is not sufficient to describe all signal changes occuring from one frame to the next. There is a variety of effects that limit the efficiency of MCP, most of which can be discovered in Fig. 5.17. A meaningful displacement does not exist where background is uncovered. There can be spatial resolution changes due to zoom, varying distance between camera and object, or temporal integration of the camera target. The apparent brightness of a surface in general varies when any of the angles between surface normal, light source, and observer changes, or when shadows are cast on the surface. Problems also occur at object borders, where displacement varies rapidly spatially, or with rotational movements. In addition to all this, there might be shortcomings of the camera that introduce noise and aliasing. All these effects contribute to the residual prediction error $e(x, y)$, that is encoded and transmitted in the motion compensating hybrid coder. In order to further reduce this residue in the future, we will have to use much more elaborate models than those underlying motion compensating prediction.

5.5. Conclusions

Motion compensation is one of the most important techniques in image sequence processing today. In this chapter, we have looked at the fundamentals that govern various motion compensating schemes, in particular, motion compensating filtering and interpolation, and motion compensating prediction for image sequence compression. We have considered visual aspects, accuracy of the motion compensation required, and performance limits.

Motion compensation is important because of smooth pursuit eye movements, which allow a human observer to track moving image contents. Eye movements have a dramatic effect on the spatiotemporal frequency response of the human visual system, and high temporal frequency components can be perceived. As a consequence, image sequences may not be bandlimited temporally without objectionable motion blur. Temporal filtering of image sequences always requires motion compensation.

Motion compensating temporal filtering is required for interpolation of missing television fields or frames, or in standards conversion. We have shown that, for interpolation from non-interlaced grids, motion compensating interpolation with integer-pel accuracy is theoretically sufficiently accurate with filters involving several frame stores. For short temporal filters, more accurate motion compensation is required. Motion compensating interpolation can also be used with line-interlaced sampling grids. It is desirable to preserve the full vertical resolution of the image except for critical velocities, where sampling lines in successive fields coincide relative to the picture contents and vertical resolution is reduced to one half. We presented the principles of an FIR de-interlacer and an elegant frame-recursive de-interlacer with only one frame store.

In the final section, we considered motion compensating prediction as is commonly used for image sequence compression. In particular, we studied the effects of fractional-pel accuracy on the efficiency of a motion-compensating predictor with various spatial interpolation/prediction filters. The motion compensated prediction error power spectrum can be related to the probability density function of the displacement error. Model calculations explain a critical accuracy beyond which the possibility of further improving prediction by more accurate motion compensation is small. Signal components that do not obey the paradigm of translational motion limit the performance of motion compensating prediction. The model calculations are confirmed by experimental data obtained with Sinc-interpolation, bilinear interpolation, and Wiener filtering for fractional-pel accuracy.

5.6 References

[1] T. N. Cornsweet, "Visual Perception," New York: Academic Press, 1970.

[2] F. W. Campbell, R. W. Gubisch, "Optical quality of the human eye," Journal of Physiology, vol. 186, 558 - 578, 1966.

[3] J. G. Robson, "Spatial and temporal contrast sensitivity functions of the visual system," Journal of the Optical Society of America A, vol. 56, 1141 - 1142, August 1966.

[4] D. H. Kelly, "Adaptation effects on spatio-temporal sine-wave thresholds," Vision Research, vol. 12, 89 - 101, 1972.

[5] B. Girod, "Ein Modell der menschlichen visuellen Wahrnehmung zur Irrelevanzreduktion von Fernsehluminanzsignalen," VDI-Verlag, Düsseldorf: VDI-Fortschritt-Berichte, Reihe 10, Nr. 84, 1988 (in German).

[6] B. Girod, R. Thoma, "Motion Compensating Field Interpolation from Interlaced and Non-Interlaced Grids", 2nd International Technical Symposium on Optical and Electro-Optical Applied Science and Engineering, Proc. SPIE Conf. B 594: Image Coding, Cannes, France, December 1985, 186 - 193.

[7] B. Girod, R. Thoma, "Motion-Compensating Conversion without Loss of Vertical Resolution in Line-Interlaced Television Systems", Proc. of the International EURASIP Workshop on "Coding of HDTV", L'Aquila, November 1986.

[8] G.J. Tonge, "Three-Dimensional Filters for Television Sampling," IBA Report, no. 117/82, June 1982

[9] G. J. Tonge, "The Television Scanning Process," SMPTE Journal, pp. 657 - 666, July 1984.

[10] W.F. Schreiber, et. al., "A Compatible High-Definition Television System Using the Noise-Margin Method of Hiding Enhancement Information," SMPTE Journal, vol. 98, pp. 873 - 879, Dec. 1989.

[11] W. F. Schreiber, A. B. Lippman, "Reliable Transmission of EDTV/HDTV Signals in Low-Quality Analog Channels," SMPTE Journal, vol. 98, pp. 5 - 13, July 1989.

[12] G. Karlsson and M. Vetterli, "Subband coding of video signals for packet switched networks," in Proc. Visual Communications and Image Processing II, SPIE vol. 845, pp. 446-456, 1987.

[13] A. B. Lippman and W. Butera, "Coding Image Sequences for Interactive Retrieval," Communications of the ACM, vol. 32, no. 7, pp. 852 - 860, July 1989.

[14] B. Girod, "Eye movements and coding of video sequences," SPIE vol. 1001, Visual Communications and Image Processing '88, 398 - 405, 1988.

[15] G. Westheimer, "Eye movement responses to horizontally moving stimulus," Arch. Ophthalmology, no. 52, pp. 932 - 941, 1954.

[16] D. A. Robinson, "The mechanics of human smooth pursuit eye movement," Journal of Physiology (London), no. 180, pp. 569 - 591, 1965.

[17] B. Girod, "What's Wrong With Mean Squared Error?" in A. B. Watson (ed.), Visual Factors of Electronic Image Communications, MIT Press, to be published 1992.

[18] B. Girod, "Psychovisual Aspects of Image Communication," Signal Processing, in print, 1992

[19] G. J. Tonge, "Time-sampled motion portrayal," in Proc. Second International Conference on Image Processing and its Applications, IEE Conf. Publications, London, June 1986.

[20] D. H. Kelly, "Theory of flicker and transient responses, I: uniform fields," Journal Optical Society of America, vol. 61, no. 4, pp. 537 - 546, April 1971.

[21] M. Ernst, T. Reuter, "Adaptive filtering for improved standards conversion," Proc. 2nd International Workshop on Signal Processing of HDTV, L'Aquila, Italy, Feb./March 1988.

[22] F. M. Wang, D. Anastassiou, A. Netravali, "Time-recursive motion compensated deinterlacing for IDTV and pyramid coding," Image Communication Journal, vol. 2, no. 3, October 1990, pp. 365 - 374.

[23] C. C. Cutler, "Differential quantization of communications," U.S. patent 2 605 361, July 1952.

[24] J. R. Jain and A. K. Jain, "Displacement measurement and its application in interframe image coding," IEEE Trans. Commun., vol. COM-29, pp. 1799-1804, Dec. 1981.

[25] T. Ishiguro and K. Iinuma, "Television bandwidth compression by motion compensated interframe coding," IEEE Communications Magazine, pp. 24-30, Nov. 1982.

[26] T. Koga, A. Hirano, K. Iinuma, Y. Iijima, and T. Ishiguro, "A 1.5 Mb/s interframe codec with motion compensation," in Proc. Int. Conf. Commun., Boston, MA, 1983, pp. 1161-1165.

[27] T. Murakami, K. Asai, A. Itoh, and E. Yamazaki, "Interframe vector coding of color video signals," in Proc. Int. Picture Coding Symp., Rennes, France, 1984.

[28] G. Kummerfeldt, F. May, and W. Wolf, "Coding television signals at 320 kbit/s and 64 kbit/s," SPIE vol. 594 Image Coding, pp. 119 - 128, Dec. 85.

[29] S. Ericsson, "Fixed and adaptive predictors for hybrid predictive/transform coding," IEEE Trans. Commun., vol. COM-33, pp. 1291-1302, Dec. 1985.

[30] M. J. Bage, "Interframe predictive coding of images using hybrid vector quantization,"

IEEE Trans. Commun., vol. COM-34, pp. 411-415, Apr. 1986.

[31] H. Brusewitz and P. Weiss, "A video conference system at 384 kbit/s," in Proc. Int. Picture Coding Symp., Tokyo, Japan, 1986.

[32] A. v. Brandt, "Motion estimation and subband coding using quadrature mirror filters," EUSIPCO '86, The Hague, The Netherlands, Sept. 1986.

[33] H. Murakami, S. Matsumoto, Y. Hatori, and H. Yamamoto, "15/30 Mbit/s universal digital TV codec using a median adaptive predictive coding method," IEEE Trans. Commun., vol. COM-35, no. 6, pp. 637-645, June 1987.

[34] M. Kaneko, Y. Hatori, and A. Kloike, "Improvements of transform coding algorithm for motion compensated interframe prediction errors – DCT/SQ Coding," IEEE J. Sel. Areas in Commun., vol. SAC-5, no. 7, pp. 1068 - 1078, Aug. 1987.

[35] P. Gerken and H. Schiller, "A low bit-rate image sequence coder combining a progressive DPCM on interleaved rasters with a hybrid DCT technique," IEEE J. Sel. Areas in Commun., vol. SAC-5, no. 7, pp. 1079 - 1089, Aug. 1987.

[36] Y. Kato, N. Mukawa, and S. Okubo, "A motion picture coding algorithm using adaptive DCT encoding based on coefficient power distribution classification," IEEE J. Sel. Areas in Commun., vol. SAC-5, no. 7, pp. 1090-1099, Aug. 1987.

[37] R.J. Moorhead II, S.A. Rajala, and L. W. Cook, "Image Sequence Compression Using a Pel-Recursive Motion-Compensated Technique," IEEE J. Sel. Areas in Commun., vol. SAC-5, no. 7, pp. 1100-1114, Aug. 1987.

[38] T. Koga and M. Ohta, "Entropy coding for a hybrid scheme with motion-compensation in subprimary rate video transmission," IEEE J. Sel. Areas in Commun., vol. SAC-5, no. 7, pp. 1166-1174, Aug. 1987.

[39] D.R. Walker, and K.R. Rao, "Motion-Compensated Coder," IEEE Trans. Commun., vol. COM-35, no.11, pp. 1171-1178, Nov. 1987.

[40] J. Biemond, B. P. Thieme, D. E. Boekee, "Subband-coding of moving video using hierarchical motion estimation and vector quantization," Proc. Int. Workshop on 64 kbit/s coding of moving video, Hannover, Germany, June 1988.

[41] B. Girod and F. Joubert, "Motion-compensating prediction with fractional pel accuracy for 64 kbit/s coding of moving video," Proc. Int. Workshop on 64 kbit/s coding of moving video, Hannover, Germany, June 1988.

[42] F. May, "Model based movement compensation and interpolation for ISDN videotelephony," 1988 IEEE Int. Symp. on Circuits and Systems (ISCAS 88), Espoo, Finland, June 1988.

[43] H. Hashimoto, H. Watanabe, and Y. Suzuki, "A 64 kb/s video coding system and its performance," Proc. SPIE Conf. Visual Commun. and Image Proc. '88, Cambridge, MA, SPIE vol. 1001, pp. 847 - 853, Nov. 1988.

[44] M. Gilge, "A high quality videophone coder using hierarchical motion estimation and structure coding of the prediction error," Proc. SPIE Conf. Visual Commun. and Image Proc. '88, Cambridge, MA, SPIE vol. 1001, pp. 864 - 874, Nov. 1988.

[45] P. Strobach, "Tree-structured scene-adaptive coder" IEEE Trans. Commun. vol. 38, no. 4, pp. 477-486, April 1990

[46] CCITT Recommendation H.261, "Video codec for audio visual services at px64 kbit/s," 1990.

[47] Committee Draft of Standard ISO11172, "Coding of moving pictures and associated audio," ISO/MPEG 90/176, December 1990.

[48] B. Girod, "The efficiency of motion-compensating prediction for hybrid coding of video sequences," IEEE J. Sel. Areas in Commun., vol. SAC-5, no. 7, pp. 1140-1154, Aug. 1987.

[49] B. Girod, "Motion-compensating prediction with fractional-pel accuracy," IEEE Transaction on Communications, to appear 1992.

[50] G. A. Thomas, "Television motion measurement for DATV and other applications," BBC Research Department Report no. 1987/11, Sept. 1987.

[51] C. D. Kuglin, and D. C. Hines, "The phase correlation image alignment method," Proc. of the IEEE 1975 Int. Conf. on Cybern. and Soc., pp. 163 - 165, Sept. 1975.

[52] B. Girod, and D. Kuo, "Direct Estimation of Displacement Histograms," Image Understanding and Machine Vision, OSA 1989 Technical Digest Series, vol. 14, pp. TuB3-1 - TuB3-4, July 1989.

[53] A. Papoulis, "Probability, Random Variables, and Stochastic Processes," McGraw-Hill, 1965.

6

Motion Field Estimators and their Application to Image Interpolation

S. Tubaro and F. Rocca

Politecnico di Milano, Dipartimento di Elettronica e Informazione

Milano, Italy

6.1 Introduction

In many video coding algorithms a subsampling of the input image sequence is considered. In fact to obtain low output data rates, with respect to the desired image quality, it is necessary to skip images at the transmitter, images that must be reconstructed at the receiver end. For example, this situation can occur when video services are provided over an ISDN network at rates under 384 Kbit/s, in the video-coding algorithm for interactive applications with digital storage media like the CD-ROM, proposed by the Moving Picture Experts Group (MPEG) [1] and for the image sequence transmission over ATM networks [2]. Moreover an alteration of the frame rate is required in a wide number of applications, such as for standard conversion (NTSC to PAL and viceversa), conversion from interlaced to progressive image format and slow play of sequences for more precise event understanding.

The recovery of the skipped or unavailable frames is a serious problem as simple reconstruction techniques like field/frame repetition or temporal linear interpolation result in jerkiness or blurring where there is movement in the video sequence, temporal luminance changes often occurring due to the presence of moving objects in the imaged scene.

An original frame from the sequence Trevor is shown in fig. 6.1a (the sequence is in Common Intermediate Format (CIF) [3]). Fig. 6.1b represents the same image reconstructed by a linear interpolation after a time subsampling of the sequence. In this case two out of every three frames are discarded.

Better results can be obtained by taking into account the motion of each object imaged in the scene. In fact, with this information, it is possible to estimate the exact position and orientation of each image element on the intermediate

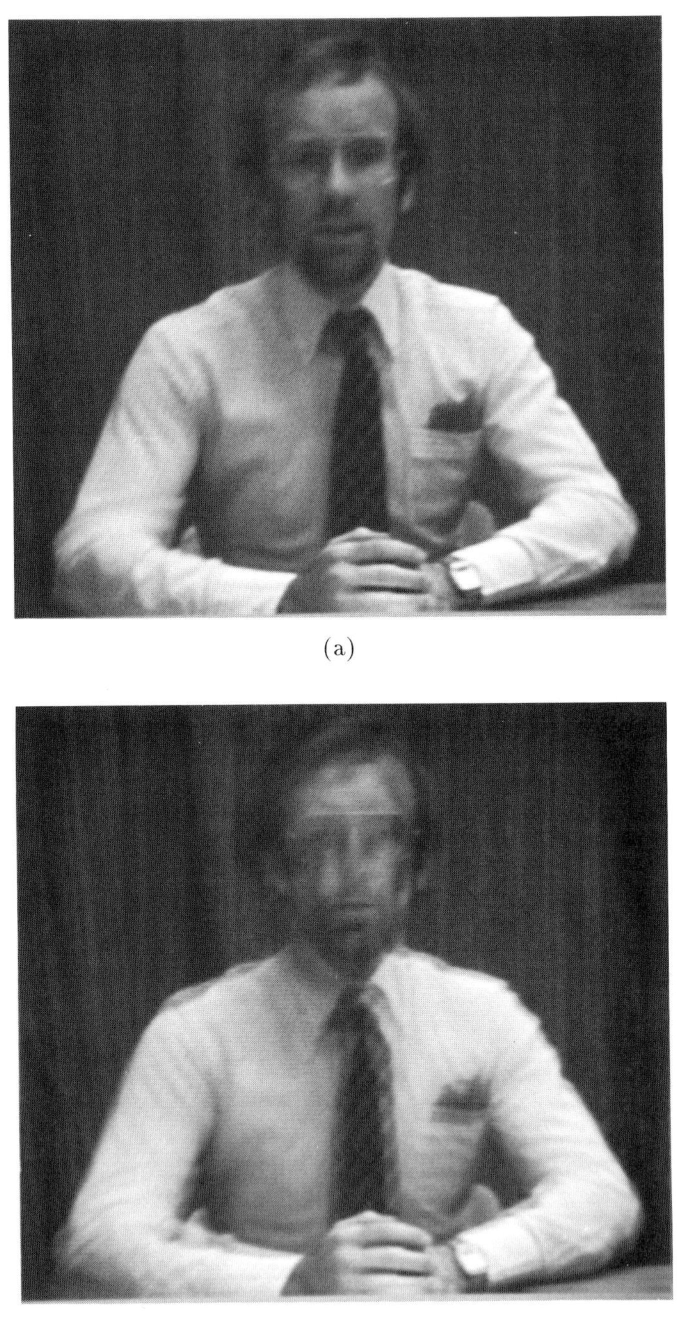

(a)

(b)

Figure 6.1: (a) An original frame from sequence Trevor. (b) The same image reconstructed by linear interpolation of the neighboring frames.

frames.

A common assumption in the motion compensated interpolation algorithms is that the 3D displacements of the imaged objects are rigid and that their projections on the image plane can be considered as purely translational. This is not always true, especially when complex motions are considered. Moreover during image acquisition the intrinsic and extrinsic parameters of the camera may vary, introducing temporal luminance differences that can be interpreted as motion of the image pixels. For example when the focal length of the camera changes (zooming), the image becomes warped, this effect cannot be interpreted on the basis of a translational motion field.

Clearly the 2D displacement field (also called optical flow field) estimated from two image frames represents the sum of the motion field due to the motion of the imaged object (local motion field) and of the field due to the change of the acquisition parameters (global motion field).

A brief description will be given in this chapter of the relations between the 3D object displacements, the intrinsic and extrinsic camera parameter variations and the optical flow field. The way in which global and local motion fields can be detected and used for image interpolation is also considered. Then the algorithms for motion compensated interpolation proposed in the literature are presented; these also take into consideration the use of image segmentation to minimize the problems related to occlusions among stationary and moving objects, and the limited resolution of the motion fields that must often be used.

Finally, we discuss some of the future research that is yet to be carried out are discussed.

6.2 A Model For Image Point Displacements

In this section we present a notation for describing the motion of a camera through an environment containing independently moving objects. We use a perspective projection to describe the relations between the 3D motion and the corresponding optical flow. In fact a pin-hole camera model describes a real image acquisition system very well [4].

Let (X, Y, Z) represent a Cartesian coordinate system which has its origin in the focus of the optical system of the camera (see fig. 6.2), let (x, y) represent the corresponding coordinate system on the image plane. The focal length f is assumed unknown and its variation represents a camera zoom.

The perspective projection at time t of the generic point $\mathbf{P} = (X, Y, Z)$ is $\mathbf{p}(t) = (x(t), y(t))$ where:

$$x(t) = f(t) * (X(t)/Z(t)) \tag{6.1}$$
$$y(t) = f(t) * (Y(t)/Z(t)) \tag{6.2}$$

The relative motion between the camera and a rigid object can be described by a translation $\mathbf{T} = (T_X, T_Y, T_Z)$ and a rotation $\mathbf{\Phi} = (\Phi_X, \Phi_Y, \Phi_Z)$ vector.

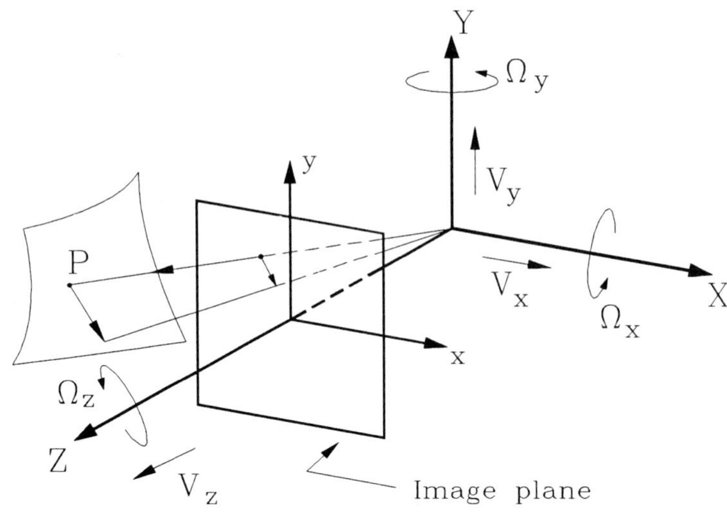

Figure 6.2: Camera model and coordinate systems.

After the motion the position $\mathbf{P}' = (X', Y', Z')$ of the generic point $\mathbf{P} = (X, Y, Z)$ at time $t + \Delta t$ is:

$$\mathbf{P}' = \mathbf{R} * \mathbf{P} + \mathbf{T} \tag{6.3}$$

where the rotation matrix $\mathbf{R}$ can be approximated, assuming small values of the rotation parameters, by [4]:

$$\mathbf{R} = \begin{pmatrix} 1 & -\Phi_Z & \Phi_Y \\ \Phi_Z & 1 & -\Phi_X \\ -\Phi_Y & \Phi_X & 1 \end{pmatrix} \tag{6.4}$$

The projection on the image plane of the points $\mathbf{P}$ and $\mathbf{P}'$ are respectively (x, y) and (x', y') where :

$$x' = \frac{\frac{x}{f} - \Phi_Z \frac{y}{f} + \Phi_Y + \frac{T_X}{Z}}{-\Phi_Y \frac{x}{f} + \Phi_X \frac{y}{f} + 1 + \frac{T_Z}{Z}} (f + \Delta f) \tag{6.5}$$

$$y' = \frac{\Phi_Z \frac{x}{f} + \frac{y}{f} - \Phi_X + \frac{T_Y}{Z}}{-\Phi_Y \frac{x}{f} + \Phi_X \frac{y}{f} + 1 + \frac{T_Z}{Z}} (f + \Delta f) \tag{6.6}$$

Δf represents the change in focal length of the acquisition system during the time Δt when the motion took place; the time dependency of the variables and parameters has been omitted for the sake of simplicity.

The instantaneous velocity of the point $\mathbf{p}$ can be obtained by differentiating (6.1, 6.2):

$$\dot{x} = u = \dot{f}\frac{X}{Z} + f[\frac{\dot{X}Z - \dot{Z}X}{Z^2}] \tag{6.7}$$

$$\dot{y} = v = \dot{f}\frac{Y}{Z} + f[\frac{\dot{Y}Z - \dot{Z}Y}{Z^2}] \tag{6.8}$$

Considering a translation velocity of the point $\mathbf{P}$ equal to $\mathbf{K} = (K_X, K_Y, K_Z) = (\dot{T}_X, \dot{T}_Y, \dot{T}_Z)$ and a rotation velocity $\mathbf{\Omega} = (\Omega_X, \Omega_Y, \Omega_Z) = (\dot{\Phi}_X, \dot{\Phi}_Y, \dot{\Phi}_Z)$ we have:

$$u = \frac{x\lambda}{f} - [\Omega_X \frac{xy}{f} - \Omega_Y f(1 + \frac{x^2}{f^2}) + \Omega_Z y] + \frac{K_X}{Z} f - \frac{K_Z}{Z} x \tag{6.9}$$

$$v = \frac{y\lambda}{f} - [\Omega_X f(1 + \frac{y^2}{f^2}) - \Omega_Y \frac{xy}{f} - \Omega_Z x] + \frac{K_Y}{Z} f - \frac{K_Z}{Z} y \tag{6.10}$$

where

$$\lambda(t) = \dot{f}(t) \tag{6.11}$$

These equations represent the relations among the motion parameters of the point $\mathbf{P}$ in the 3D space, the velocity of its projections on the image plane and the focal length of the camera.

The complexity of the two approaches (one based on the displacement of the image points, the other on their velocities) is very different: in fact the first approach requires the solution of more complex non linear equations whereas the second does not.

When the time interval Δt is small the acceleration effect can be neglected when estimating the image point displacement, thus the equations that describe the velocity field, (6.8) and (6.9), can be used to compute the displacements. Therefore what is done in an optical flow estimation and analysis is:

- the estimation of a sufficiently dense bidimensional velocity field through the observation of the temporal changes of the luminance profiles;
- the inversion of the equations (6.8), (6.9).

The second step is possible only if some assumptions on the imaged objects are made. Typical constraints are the modeling of the entire scene by a single rigid body [5], or by multiple rigid bodies with independent motions [6], [7] or the local description of the imaged objects by curved surfaces like a parabolic one that can be moved and deformed [8],[9]. Considering the relative motion between $\mathbf{P}$ and the camera, this entails three different situations:

1. Both the camera and scene are stationary, but objects in the scene move. The displacements on the image plane are due to the 3D motion of the imaged objects and to the change in the focal length of the lens.

2. The scene is completely stationary and the camera moves. The displacements are due to the motion of the camera and the changes in focal length of the lens.

3. The displacements are due to movement of both the camera and the objects in the scene.

For each of these situations a particular motion estimation and interpolation scheme has been carried out.

6.3 Motion Compensated Interpolation

In the previous section we have given a brief description of how the optical flow field is generated by the relative motion between camera and imaged objects.

For image interpolation we have at our disposal pairs of images and we must make the best estimation of all the motion and shape parameters needed to interpolate the unknown images between the two known ones(see fig. 6.3).

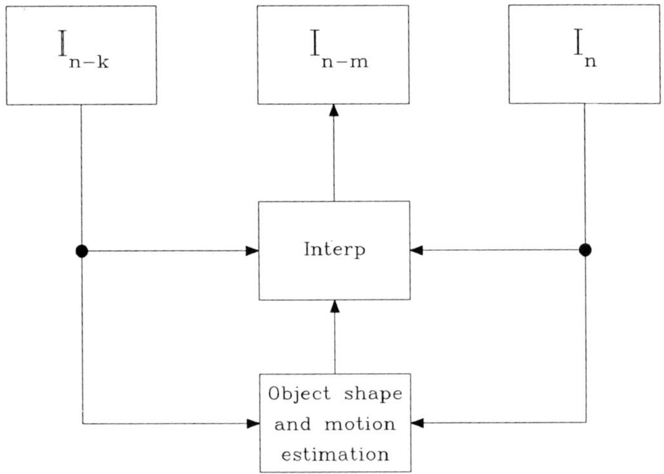

Figure 6.3: Image interpolation scheme based on parameter estimation.

As a first approach we can utilize the available optical flow field to directly describe how each image point moves from one known image to the other (see fig. 6.4) [10],[11],[12]. This is the same as modeling the imaged objects as planar surfaces moving in a plane parallel to the image plane. Considering real scenes this is a very crude hypothesis, but it is sufficiently accurate for the short time interval considered when two frames belonging to a subsampled image sequence are taken into account to interpolate some intermediate frames [13].

As already stated different systems require image interpolation. For example coding systems that use time subsampling techniques, require that the correct

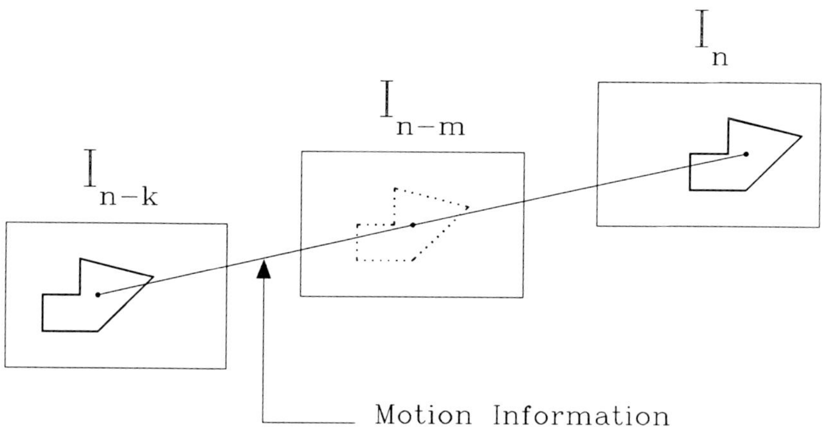

Figure 6.4: Use of the motion information for a direct image interpolation.

frame rate be recovered at the receiver side. During the coding process, when a progressive scan of the images is used, for every k frames, $k-1$ are completely discarded and are later interpolated at the decoder. The subsampling factor k normally varies from 2 to 4.

When an interlaced format is employed a certain number of fields are first discarded and then even and odd missing fields are interpolated separately . The time interval between two consecutive frames or consecutive homologous fields (even or odd) in the subsampled sequence varies from 80 to 160 ms (European standard).

When image interpolation is used for frame rate conversion or to play a sequence in slow motion the time interval between consecutive known frames reduces to 40 ms.

Although we took progressive image formats into account the presented results are also easily applicable to the case of interlaced sequences. We have considered only the luminance component of the images but the same processing can be applied to the chrominance for color sequences.

The interpolation algorithm uses the motion information to describe the trajectory of each point from one known image to the other (fig. 6.4); only if the estimated motion field is homogeneous and very close to the "physical" one there is the possibility of reassembling every part of the known images into the reconstructed images. Therefore an accurate motion estimation is of much greater importance than in the case of a motion compensated interframe coding scheme in which the motion information is used only to reduce the entropy of the frame luminance differences without considering any semantic meaning of this information.

Moreover the optical flow field must often be estimated using images taken at very different times ($\Delta t = 120, 160$ msec). In these cases the motion vectors that must be estimated are very large and some illumination changes or object distortions can be present between current and previous images. For different frames, the variations in projection shape of the same 3D object on the image plane are mainly due to changes in the acquisition parameters, or to the incomplete verification of the previously described hypothesis on the planarity of imaged 3D objects and their relative displacements and, at the end, to lens distortion [14],[15].

All these considerations indicate that when the motion information should be used for interpolation purposes some changes must be introduced in the algorithm normally used for motion estimation in interframe coding. In any case occlusions between stationary and moving objects can create serious problems to motion estimators. In fact the presence of a detail in one of the two available frames, but not in the other, means that it may be impossible to make an estimation of that detail's displacement [16]. Moreover the correspondent areas on the missing images cannot be reconstructed by an interpolation filter unless they are detected beforehand and then treated separately.

6.4 Motion Estimators For Image Interpolation

A lot of information concerning motion field estimation can be found in the literature. In general, as input signals the estimators have two images, the current and the previous one; first a hypothesis on the motion field is made, then the correspondent relative motion compensated frame differences and their first and second derivatives are used to verify/update this field.

Algorithms will only be briefly described here as they are dealt with in detail in other chapters of this book.

The most common motion estimators are block estimators. They subdivide the current image into a set of non-overlapping areas (usually squares), a single vector being calculated to give the best match between the considered block and those shifted in the previous images where only a limited search area is considered so as to reduce the computation load. The best match is carried out minimizing a suitable objective function, using either direct search techniques [17],[18] or gradient based descendent methods [19],[12]. In the second approach the objective function for the entire image is formulated as the weighted sum of a pixel based matching function and a function favoring smoothness of displacement field [20]. Both classes of methods are sometimes coupled with multiresolution techniques to avoid convergence to local minima [21],[22].

The block matching methods suffer from the obvious problem that the true motion field is not constant piecewise so that the final estimated motion field may be a poor approximation if the block size is too large. This is especially true near the boundaries of objects with different motion. This problem cannot be overcome by simply reducing the block size as the method can then lead to instability [12].

In a more sophisticated approach the motion field within a given block is described by a function of a few parameters that describe the optical flow in this area (translation, rotation, zooming, slight deformation) [6]. If the block dimension is very small (or even equal to one pixel) it is essential to introduce some constraints on the continuity of the motion so as to maintain the stability of the estimation process [12].

When the block dimension is equal to one pixel the relative algorithms are, in general, pel-recursive estimators because the motion vector of the current pel is an update of earlier motion information on the same or neighboring points [10]. This updating increases the capability of the algorithms to estimate, with great accuracy, large motion vectors. A very interesting approach is based on a multi-resolution iteration: starting from low resolution images the motion field is updated, and the image resolution increase. This cycle is repeated until a prefixed resolution is reached [22].

The optical flow field generated by pel-recursive algorithms is quite sensitive to noise since first and second partial derivatives of image brightness have to be estimated. In addition errors arise at object boundaries as these methods are based on certain continuity assumptions of the optical flow [6]; furthermore, computing a motion vector on a pel by pel basis can be quite a time-consuming task.

On the contrary block algorithms are, in general, less expensive in terms of computation load but they generate a low resolution motion field. Moreover the block boundaries and the boundaries of the moving objects do not coincide, thus disturbing artifacts are created.

A significant improvement in the motion estimation quality may be obtained if images are segmented according to their real content, and, for each semantic homogeneous region, a motion parameter set is carried out. In [9] a similar approach is presented. For each patch of the current image an estimation is made, a set of "mapping parameters" that take into account translation, rotation scale change and distortion.

Also in the case of squared block and simple translational motion estimation the segmentation can be used to improve the performance of the algorithms, especially for those blocks that include part of a moving object and part of a stationary background. In this case only points belonging to the moving part of the block should be used to calculate the matching function [23].

If some general illumination changes have occurred between the times when past and current images were acquired the motion estimation algorithm can take into account this luminance profile modification, estimating an additional parameter for either the block or pel algorithms [9].

Another class of motion estimators, called feature based algorithms, utilizes a discrete set of features, or brightness patterns, detected in the images. The objective is to derive the motion of the objects in the scene by the analyzing the motion of these features.

The first step is the extraction of these sets of relatively sparse, but highly discriminatory, two dimensional features such as object corners, surface edges and boundaries demarcating changes in surface reflectivity. Inter-frame correspondences are then established between these features, leading to the estimation of the feature displacements. At the end this information is used to compute the motion parameters of the imaged objects. Also in this case some constraints can be formulated, for example to take into account that the object motions are rigid [24].

When the interpolation algorithm is a part of a more complex video coding scheme, in which the motion field is estimated only at the transmitter side and used not only for interpolation but also, for example, in a motion compensated DPCM loop, the motion estimators are generally block-matching algorithms [25],[3],[26]. This is especially true when the computation bound and/or hardware complexity are highly important constraints.

If the motion field is not locally estimated but sent from the transmitter only a subsampled version of the entire optical flow field can be represented. Fortunately for natural and videoconference scenes, the motion field within an object varies only slowly. Abrupt changes take place only at the boundaries between static and moving objects or between objects with different motion. Therefore the use of a semantic segmentation, able to separate these various parts of the images, increases the effective resolution of the motion field. In practice some rules are used to manage the blocks that include parts of more than one moving object either for motion estimation or interpolation, and for interframe predictions [22],[23].

As stated before, when a high temporal subsampling factor is considered motion compensated interpolation requires the estimation of very large motion vectors. This is a problem, as the maximum allowable displacement can be greater than the dimension of the main lobe of the luminance autocorrelation function; therefore the objective function, evaluated by the block algorithms, presents more than one minimum. The value of the objective function in these minima greatly depends on the noise superimposed on the images and the effects of incorrect modeling, therefore erroneous matching can be carried out. For this reason the estimated motion fields present evident inconsistency and, moreover when a block matching procedure is employed, the use of fast search strategies like conjugate direction, logarithmic search [27],[28] lead to unsatisfactory results. Some regularization procedures embedded in the block matching algorithm, or carried out on the estimated motion field, must be employed to obtain reliable results.

6.4.1 Regularization of the motion field

Regularization is based on the assumption that the motion field can be modeled as a smooth field with abrupt changes corresponding to object boundaries [11].

In [23] as a matching function between the current block and the ones in the search area, a normalized cross-correlation function is used.

Blocks with a correlation coefficient, relative to the estimated displacement, above a high threshold are considered easily compensatable, and the displacement is recorded. For all the other blocks the motion vectors corresponding to correlation values above another threshold are selected first. From among these candidates the displacement chosen is the one most similar to those estimated for blocks in a window centered on the block under consideration. Finally the blocks for which the maxima of the computed correlation are too low are classified as barely predictable, and, once a motion vector has been chosen for all other blocks, the average displacement of the surrounding blocks is assigned to them.

In fig. 6.5a the motion field obtained by a conventional block matching algorithm is presented, while fig. 6.5b shows the results obtained with the regularization process previously described. In [10] the interdependence of adjacent vectors is modeled by a Gibbs/Markov random field [29]. Thus vectors leading towards a smooth vector field are preferred to those leading to a rough field.

The aim is not to minimize the Displaced Frame Differences (DFD), defined for an pixel area of K elements as:

$$DFD(V) = \frac{1}{K} \cdot \sum_{i=1}^{K} d^2(i, v(i)) \tag{6.12}$$

where $d(i, v(i))$ denotes the amplitude of the prediction error of the ith pel of the block after displacement with vector $v(i)$, but to maximize the a posteriori probability density:

$$p(V \mid I_n, I_{n-1}) \longrightarrow Max \tag{6.13}$$

where I_n and I_{n-1} are the present and past images and V the corresponding motion vector field.

Using Bayes rule we have:

$$p(I_n, I_{n-1} \mid V) \cdot p(V) \longrightarrow Max \tag{6.14}$$

Assuming that the amplitudes of the prediction error have a zero mean white gaussian distribution, and considering a small power of noise (P_n) due to the acquisition system, we have:

$$p(I_n, I_{n-1} \mid V) \propto \frac{1}{\sqrt{2\pi \cdot Max(DFD(V), P_n)}^K} \tag{6.15}$$

The interdependence between vector, modeled as a second order Gibbs random field, is given by:

$$p(V) \propto exp(-\sum_{j=1}^{M} C_i) \tag{6.16}$$

where M denotes the number of considered cliques (a clique is an ensemble of neighborhood pels) and C_i stands for the cost assigned to the ith clique. Considering only cliques of two pels, where one pel is one of the eight nearest neighbors of the other, the cost can be assigned as:

$$C = \frac{c}{l} \cdot \parallel v_1 - v_2 \parallel; \; c = cost. \tag{6.17}$$

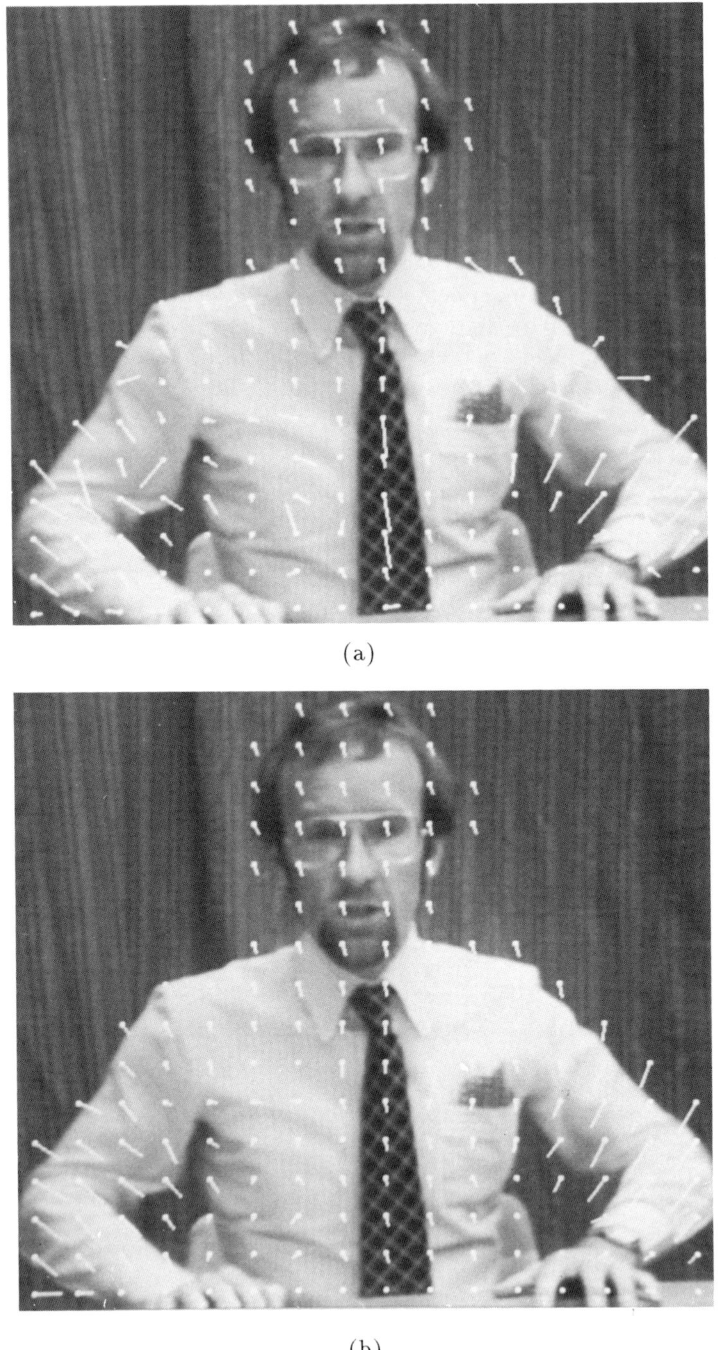

(a)

(b)

Figure 6.5: Motion fields obtained by a classical block matching procedure (a) and using regularized estimation (b).

with $\| \cdot \|$ denoting the norm of a vector and l the distance between the considered pel (1 or $\sqrt(2)$).

The evaluation of the maximum a posteriori estimation finally leads to:

$$K \cdot ln(Max(DFD(V), KP_n) + \sum_{i=1}^{K} \sum_{j=1}^{8} \frac{c}{2l} \cdot \| v(i) - v_j(i) \| \longrightarrow Min \quad (6.18)$$

$v_j(i)$ denotes the vector of jth neighbor of $v(i)$. The constant parameter c is determined by statistical tests.

To reduce the computation load a block of N pels can considered as macropels displaced by the same motion vector.

Starting with a rough vector field approximation the motion field is improved in a certain number of steps, according to the DFD/Gibbs criterion.

When a large subsampling factor (k) is used the estimation of the motion field through images belonging only to the subsampled sequence could be critical, thus the use of the information concerning skipped images would increase the performances of the estimators. The original block in the image I_n can be compared with all the blocks of the skipped images situated along the direction of the considered displacement vector and a global similarity function can be considered (see fig. 6.6a) [30].

In this way each part of the moving object is tracked also in the skipped frames and the problems related to multiple correspondence and object distortion are reduced. On the other hand, this algorithm calls for a greater computation load than other methods that consider only the frames n and $n - k$. To reduce the computational effort a suitable procedure for scanning the research area can be implemented, the aim being to improve the detection of small displacement vectors [31].

For example, in the search area, a spiral search path starting in the center of the area itself is considered (see fig. 6.6b). The process is stopped when, for a specific search position, the similarity function lies under a threshold adapted to the noise superimposed on the images. The displacement corresponding to this iteration is assigned to the current block. In this way the average computational effort is reduced, without introducing any deterioration in performance, to about a half that of a standard full search procedure.

6.4.2 Global pan and zoom estimation

In section 6.2 we saw how a change in the intrinsic and extrinsic parameters of the acquisition systems generates a motion of the image points. In particular the camera pan (rotation around x and y axes, see fig. 6.2) and zoom (change of the focal length) are the most common parameter changes in video conference application and broadcasting sequences. In this case equations (6.8), (6.9) become:

$$u = \frac{x\lambda}{f} - [\Omega_X \frac{xy}{f} - \Omega_Y f(1 + \frac{x^2}{f^2}) + \Omega_Z y] \quad (6.19)$$

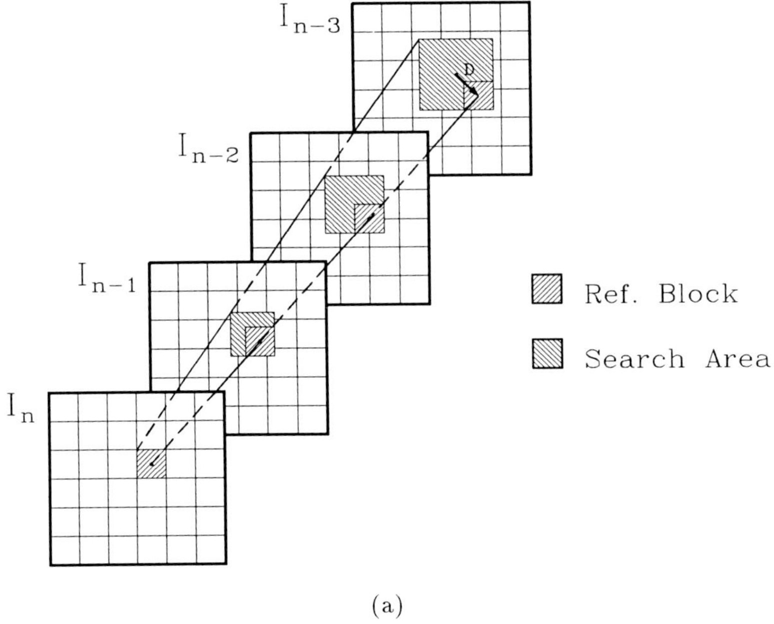

(a)

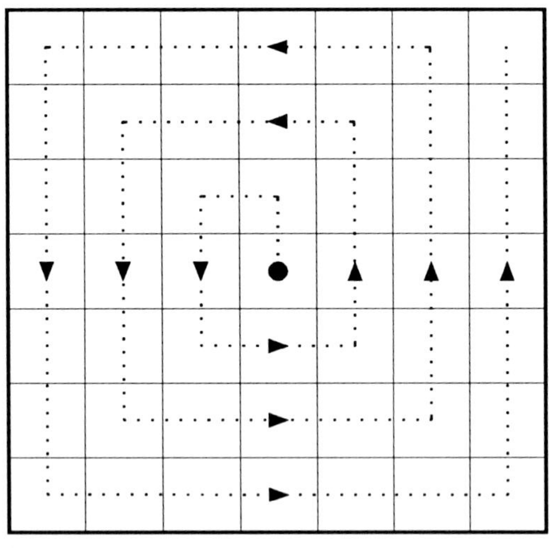

(b)

Figure 6.6: (a) Corresponding block for a multiple frame estimation algorithm. (b) Spiral search path employed for block matching.

$$v = \frac{y\lambda}{f} - [\Omega_X f(1 + \frac{y^2}{f^2}) - \Omega_Y \frac{xy}{f} - \Omega_Z x] \qquad (6.20)$$

$\lambda = f_n - f_{n-k}$ where f_n and f_{n-k} represent the focal lengths used in the acquisition of the image n and $n - k$.

The maximum rotation velocity of a camera around the x or y axis (see fig. 6.2) is about $12^o/s$ ($0.2rad/s$). With a time interval of 160 ms between two consecutive known images ($k = 4$) the maximum rotation angle that must be taken into account is approximately of $0.03rad$. Moreover considering the real dimension of the target surface of the camera and the focal length of the lenses we have that, in general, $x, y \ll f$ and therefore the following simplified relation holds:

$$u = \frac{x\lambda}{f} + \Omega_Y f \qquad (6.21)$$

$$v = \frac{y\lambda}{f} - \Omega_X f \qquad (6.22)$$

These formulas indicate that small panning causes the entire frame to be displaced uniformly by the same vector while zooming introduces a "stretching" motion of the image points.

In fig. 6.7, using equation (6.18) and (6.19), a synthetic motion field due to a simultaneous x and y camera pan is plotted to visualize the correctness of the approximated equations (6.20), (6.21).

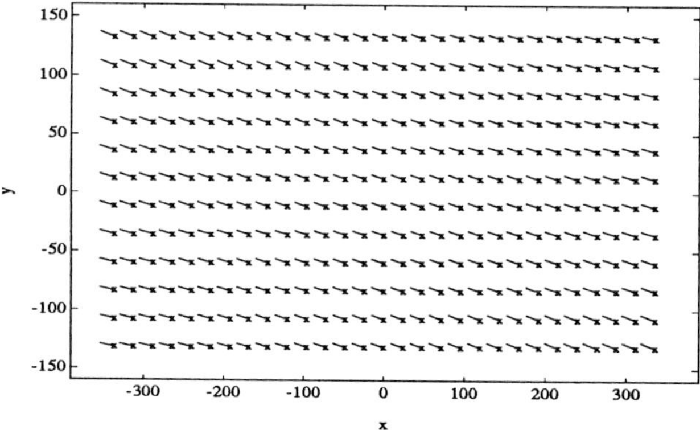

Figure 6.7: Synthetic motion field correspondent to a camera pan.

The estimation of the zoom and pan parameter is very important for image interpolation. In fact the motion field generated by varying the focus of the camera lens is difficult to model as a translational field like the ones estimated using a block matching algorithm. Therefore the estimated displacement field in the case of camera zooming is not very useful for interpolation. It is interesting to estimate from these "noisy" measurements the real pan and zoom parameters: using equations (6.20) and (6.21) a linear regression over the available motion field can be carried out to estimate these values [32]. Discarding the unreliable motion vectors and those due to the displacement of independent moving objects in the scene increases the performance of the estimation.

From equation (6.20) and (6.21) it can seen that u (the x component of the displacement) is constant in the y direction, whilst v (the y component of the displacement) is constant in the x direction. This suggests the discarding of the motion vectors whose x components are not in agreement with the others relative to blocks with the same y coordinate, and viceversa. In practice for each row and column of the block estimated motion field, the mean value of the y or x components is calculated, vectors whose corresponding components are quite different with respect to these values are discarded.

Fig. 6.8a shows the motion field estimated with a block algorithm in the presence of a significant pan and zoom. The images are taken from the sequence "Table Tennis". Fig. 6.8b plots the residual motion field after pan and zoom compensation. The figure indicates that it is possible to estimate the correct general motion parameters even in the presence of independent moving objects and partially incorrect motion estimation.

6.5 Semantic Segmentation Of Images

The knowledge of a good description of the imaged scene is an important aid in any field of image coding and processing. For example this knowledge allows the estimation of coherent motion information on all the points belonging to the same object and, furthermore, allows the adaptation of the processing to the semantic meaning of each segmented object.

In the literature several algorithms have been proposed for image segmentation, such as region growing, texture analyses [27], and so on.

When applied to image interpolation the segmentation must primarily take into account motion information. Pixels that probably have the same motion parameters should be put together to increase the reliability of the displacement estimation and to obtain more accurate interpolations [33],[4].

When we assume that the images are taken by a fixed camera, a first segmentation is the one that classifies each point of the current image as changed/unchanged with respect to its correspondent on the previous one [27]. In a more accurate scene description the changed regions of the images can be subdivided into moving areas and areas that represent unpredictable regions, like those due to uncovered background (see fig. 6.9) [34],[35],[36]. Therefore three different states can be assumed by each image pel:

1. Stationary background: represents the part of the current image that is unchanged with respect to the previous one, and in general represents the projection on the image plane of the fixed objects.

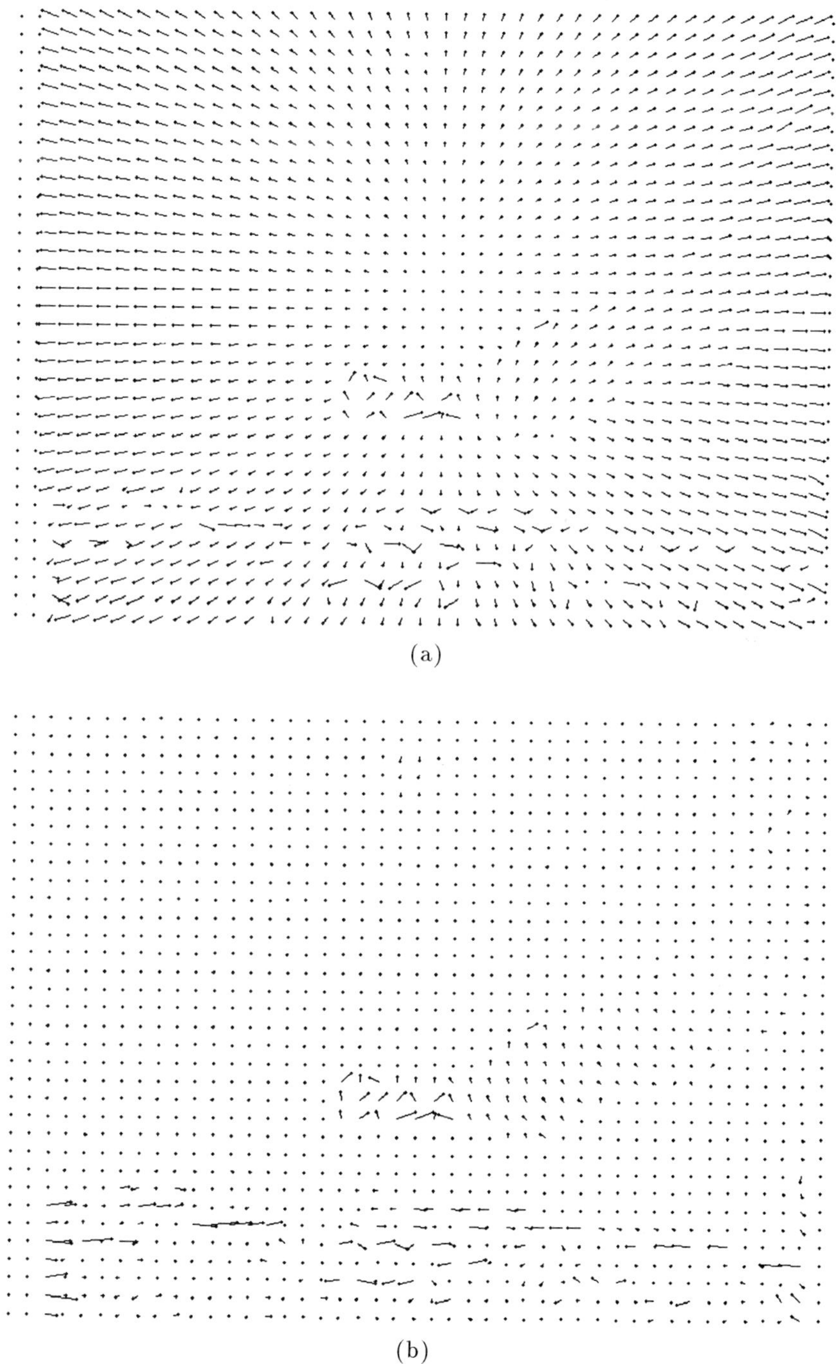

(a)

(b)

Figure 6.8: (a) Estimated motion field for a frame of the sequence Table Tennis. (b) Residual motion field after pan and zoom compensation.

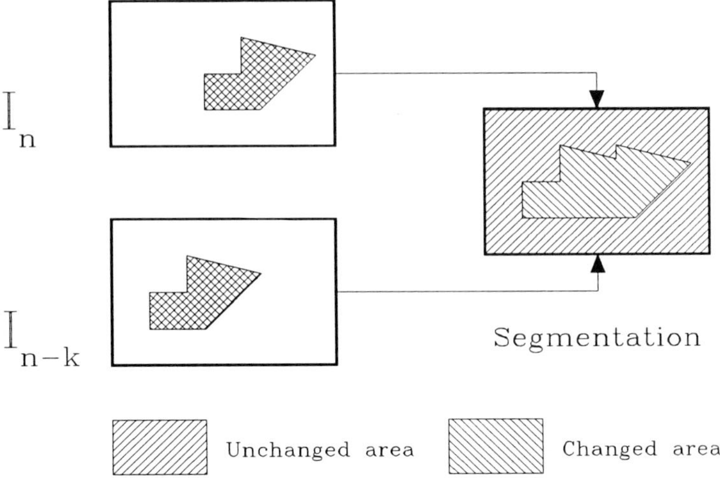

Figure 6.9: Image segmentation based on a changed/unchanged detector.

2. Moving object: represents the moving part of the images.
3. New scene: represents the part of the current image that is unpredictable with respect to the previous one, such as covered background, new imaged objects and so on.

Several algorithms have been developed to realize either a simple changed–unchanged image segmentation or a background–moving objects–new scene one.

In the first case the luminance differences between a pair of images is subdivided into two classes (changed/unchanged pixels) on the basis of their intensities. A simple threshold mechanism adapted to the noise level can be used. In a more sophisticated approach the segmentation is modeled as a bidimensional second order Markov process [37]. Given the conditional probability of observing a luminance difference at current pel and given its state (changed/unchanged), the segmentation process can be expressed in terms of a Maximum A-posteriori Probability (MAP) problem. In other words we must find the label set that maximize the product of the (a priori) probability of having the observed luminance differences and of the given segmentation. The former depends only on the values of the parameters of the Markov model, while the latter is expressed by the probability density function of luminance differences in still and changed areas.

The general bidimensional problem can be solved in a suboptimal way by either estimating an independent segmentation of each line of the image or using the one obtained for the previous line as a constraint for the current segmentation [23]. In either case the Viterbi algorithm [38] can be used to obtain the segmentation line by line.

When a three-state segmentation must be estimated it is necessary to subdivide the changed pixels between points belonging to the moving objects or to the covered-uncovered background. A simple way is the use of two consecutive changed/unchanged segmentations (between images I_{n-k}, I_n) and (I_n, I_{n+k}), and of a logical operator that separates the moving object shapes from the uncovered background (see fig. 6.10) [36].

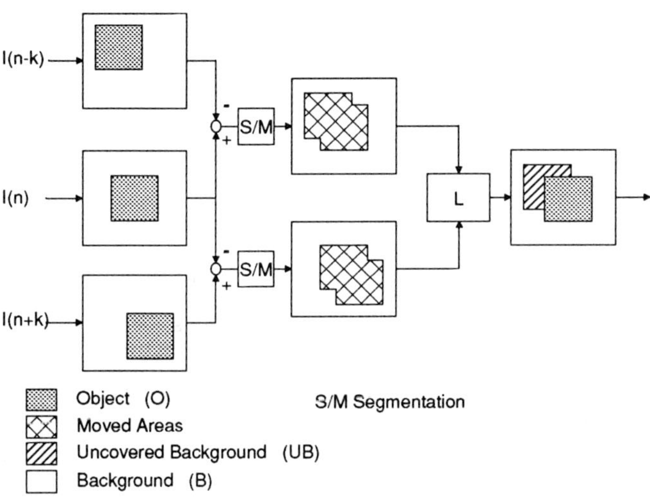

Figure 6.10: Image segmentation (background–moving objects–new scene) based on the comparison of two change–unchanged segmentations.

Figure 6.11 represents a three-state segmentation relative to a frame of the sequence Trevor obtained from two consecutive changed–unchanged segmentation obtained using a Viterbi algorithm. On the figure white areas correspond to moving object zones, gray areas to a new scene (uncovered background) and black areas to fixed background. Apart from a small spurious zone at the bottom right of the picture the segmentation appears to be very good.

When analyzing frame I_n it is also necessary to know the image I_{n+k} to carry out the complete segmentation. This significantly increases the coding delay when large subsampling factors are considered.

Another way to obtain segmentation information consists in the use of other

Figure 6.11: Three state segmentation obtained using a couple of two-state segmentation.

available information like the motion vectors and the segmentations relative to the previously considered images. Furthermore it is also possible to take a background memory into account so as to remember and update the fixed part of the imaged scene that can be covered or uncovered by the moving objects [36].

To perform the segmentation of the incoming image an *object mask* [39] is built up and improved from frame to frame using a recursive structure.

In fig. 6.12 a block diagram of the segmentation algorithm is given.

At the beginning the object mask is assumed empty. It is then built up and improved by three mechanisms:

1. the old mask is moved according to the estimated motion field;

2. the mask is improved with the help of the luminance differences, motion compensated or not, between the current image (I_n) and the previous one (I_{n-k});

3. the background memory is employed to discard pixels of the uncovered background included in the object-mask.

The first step for each frame is a *local segmentation* (changed/unchanged)

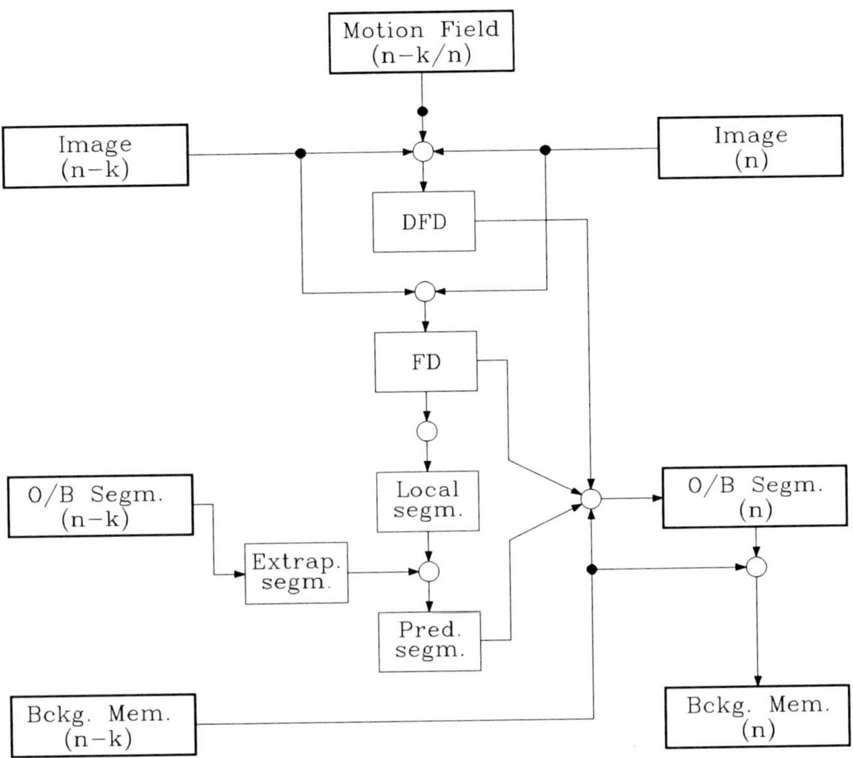

O/B Segm. = Object/Background Segmentation

Bckg. Mem. = Background Memory

FD = Frame Differences

DFD = Displaced Frame Differences

Figure 6.12: Block diagram of the recursive segmentation algorithm.

carried out considering the luminance differences between I_n and I_{n-k}. A threshold, adapted to the noise superimposed on the images, is employed for this segmentation. Furthermore a 3 by 3 median filter is used to smooth the mask's contour and to discard too small object regions.

As the second step, the object mask of the previous image is moved according to the estimated motion field to obtain a *predicted segmentation.* A logical operator is applied to the local and predicted segmentation to obtain a *merged segmentation*, the aim being to reduce, in the local segmentation, the "holes" generated by slow translations of nearly constant luminance areas.

Each point of the merged segmentation can assume only one of two different states: object or background, the transition between these states is controlled by the following rules:

a if a pixel of the local segmentation is declared changed its state in the merged one will be object;

b if a pixel of the local segmentation is declared unchanged but in the predicted one its state is object, in the merged one the pixel will be object;

c if the condition b happens more that h consecutive times, the state of the pixel in the merged segmentation will be background.

The conditions **b** and **c** are used to guarantee that when a moving object has a very small or even null displacement for a little time it is not immediately considered as a part of the background. The value of the parameter h depends on the considered subsampling factor k.

As the last step, the real object mask is obtained by considering the merged segmentation and the background memory content. If a point is classified as object in the merged segmentation, but its luminance value is very similar to that stored in the homologous position of the background memory, its state in the object mask is changed to background. Furthermore the pixels that are defined object, but present high motion compensated luminance differences, are considered not-well compensated and their state is changed to background because, with great probability, they are also part of the uncovered background. A final median filtering is carried out to regularize the shape of the object mask.

The object mask permits the segmentation of the current image in object and background. The uncovered background can be detected by comparing the merged segmentation and the object mask.

Like the object mask the background memory is filled and modified during the process. At the beginning the background memory starts as empty and no information is available on the fixed part of the imaged scene. When the object mask for the current frame is available, the luminance value of each background pixel (pixels not belonging to the object) is copied into the background memory if its corresponding location is empty. When the background memory is already filled, the current luminance value is compared with the stored one and, if there are significant differences, a replacement procedure is started. The pixel is marked as "possible changed background" and if, for the l successive frame, the differences between current and past images lie below a certain threshold, the considered pixel of the background memory is replaced with the luminance value relative to the last known image. Also in this case both the threshold and the l factor are adapted to the subsampling factor and noise level.

Fig. 6.13 shows the object mask, obtained using the recursive algorithm, of the same frame analyzed in fig. 6.11. Some little segmentation defects can be seen near the head of the speaker. These are due to the low luminance difference between the hair and the background, but the segmentation agrees with that

Figure 6.13: Object mask obtained using the recursive algorithm.

proposed in fig. 6.11.

The two classes of image segmenters presented here have different characteristics that it is useful to note.

The first class, based on the comparison of two consecutive changed/unchanged segmentations is based only on the observation of the interframe luminance differences. It introduces a delay because to segment the current image the next one is also required. Moreover if the segmenter is part of a coding system its results represent side-information that must be sent from the transmitter to the receiver. One advantage is that the segmentation can be used to increase the quality of the motion field estimation relative to the current frame, in fact it gives precise information on the shape and position of the moving objects.

On the contrary when the recursive segmentation is used the processing delay is reduced, and if it is employed in a coding system the segmentation can be independently estimated at the transmitter and receiver side without sending any side information if the quality of the coded images available at the receiver is good. A disadvantage is that the segmentation cannot be used for motion estimation because the displacement information relative to the current frame is used by the segmenter itself.

6.6 Use Of Motion Information For Image Interpolation

As previously indicated, when effective motion information is available it is possible to reconstruct, with great accuracy, intermediate frames between the current I_n and the previously available I_{n-k} ones.

Fig. 6.4 shows a simple interpolation algorithm: starting from a point on image I_n through the use of the correspondent motion vector its position on I_{n-k} and on the intermediate frames can be calculated. This is done under the hypothesis that motion speed is locally constant so the position occupied by an object pixel of I_n in frame I_{n-m} $(0 < m < k)$ can be obtained by multiplying the corresponding displacement vector by m/k.

If, on the intermediate frame, the estimated position no longer lies on the sampling grid, the nearest pel is selected. To this element is assigned the same luminance value of the starting point on I_n.

For "multiple selected" pel a simple average of the assigned luminance value can be carried out [12]. At the end unpredicted pels on the intermediate frames must be spatially interpolated from their neighbors.

A second motion field can estimated starting from image I_{n-k} and arriving at I_n [12]. This second estimation can be used to carry out further regularization of the motion field and to reduce artifacts and unpredicted pels on the intermediate frames. But if the first motion field is sufficiently accurate this should not be necessary. The literature proposes a great number of improvements of this simple algorithm for specific applications.

When the most important motion present on the scene is the general one (see section 6.4.2), there is available, after pan and zoom estimation, a different displacement vector for each image pel. When the time interval between two consecutive known images is not so high the pan and zoom motion are the most important part of the vector field [40]. This is very evident in sequences related to sporting events, such as football and soccer matches.

In this case it is very easy to implement a backward (from I_n to I_{n-k}) or forward (from I_{n-k} to I_n) interpolation because it is sufficient to change the sign of the pan and zoom parameters to obtain the two corresponding motion fields. This is very useful to interpolate image areas on the intermediate frames where there corresponds only one of the two known images. In fig. 6.14 the interpolation procedure is sketched in the case of pure zoom-out. Zoom-out stands for a focal length reduction of the camera during a scene.

Considering the case of a subsampling factor k=2 we have, by using the parameters describing the global motion from I_n to I_{n-2} and the luminance values of I_n, a first interpolation of the intermediate frame (I'_{n-1}). Then starting from I_{n-2} the process is repeated to obtain I''_{n-1}. At the end the defined part of I'_{n-1} is inserted in frame I_{n-1}. The other is taken from I''_{n-1} as indicated in fig. 6.14.

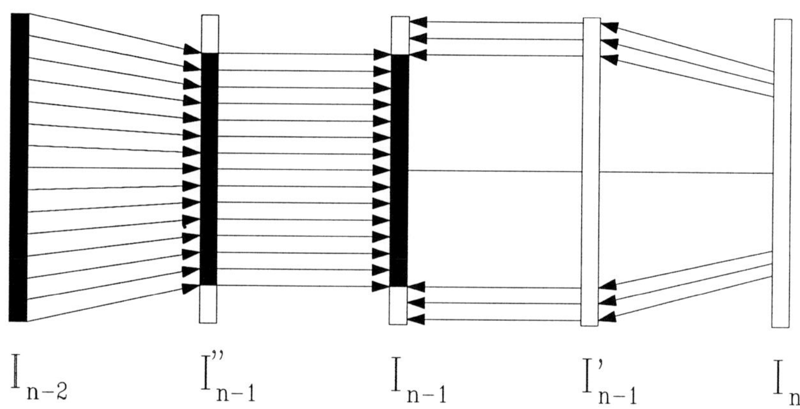

Figure 6.14: Motion compensated interpolation of an omitted frame in the case of pure zoom–out.

The performance of the interpolator is demonstrated in images of a soccer match where panning and zooming are done. First a time subsampling of the sequence with k=3 was carried out, then the missing images were reconstructed using either the motion information directly obtained by a block matching algorithm or the global motion parameters. In the two cases the Peak SNR (PSNR) of the interpolated images are plotted in fig. 6.15.

In fig. 6.16a a frame reconstructed from the motion information obtained by a block matching algorithm is presented. The artifacts due to incorrect motion estimation are evident. On the contrary in fig. 6.16b the same frame is reconstructed using the global motion parameter estimated by the block motion field. The quality of the interpolation is very good. This shows the accuracy and robustness of both the parameter estimator and the interpolation algorithm.

When the displacements of the independent moving objects are relevant an interpolation that uses only global motion parameters may be insufficient. In this case also the other motion present on the scene must be taken into account.

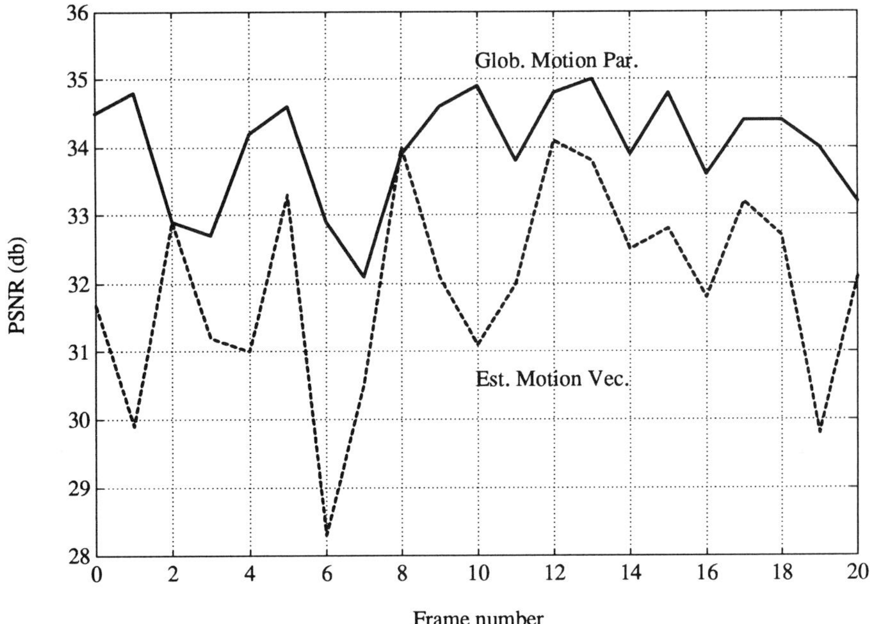

Figure 6.15: PSNR of interpolated images, using either the motion information obtained by a block matching estimator or the global motion parameters.

However, as previously indicated, when relevant global motion is present in the images, especially zoom, the performance of the block estimators is poor, and after global motion compensation the residual motion field is not very meaningful. In this case a two stage motion estimation can be useful (see fig. 6.17a) [32].

In the first stage, the pan and zoom parameters are estimated, then, using the previous frame, the global motion compensator constructs a globally compensated frame which is used, in the second stage, for a further motion estimation to model remaining object motions.

Fig. 6.17b shows the interpolation procedure that must be used when global-local motions are considered. The reconstruction of a missing image due to a time subsampling with k=2 is considered. The frames I_n and I_{n-2} are known and I_{n-1} must be interpolated. For the sake of simplicity a pure zoom out is considered. Using the parameter vector $(p_x, p_y, \Delta f)_{n,n-2}$, describing the global motion from I_{n-2} to I_n, the previous frame I_{n-2} is compensated obtaining I'_{n-2}. The local motion field is then employed to obtain I'_{n-1}.

In its real dimension the frame I_{n-1} can be reconstructed through the use of the relative global motion parameters. Near the boundaries there are some undefined regions that can be extrapolated from frames I_n and I_{n-2}, by taking

(a)

(b)

Figure 6.16: Interpolation of frame from a sequence relative to a soccer match. (a) Using the motion vectors obtained by a block matching estimator. (b) Using the global motion parameters.

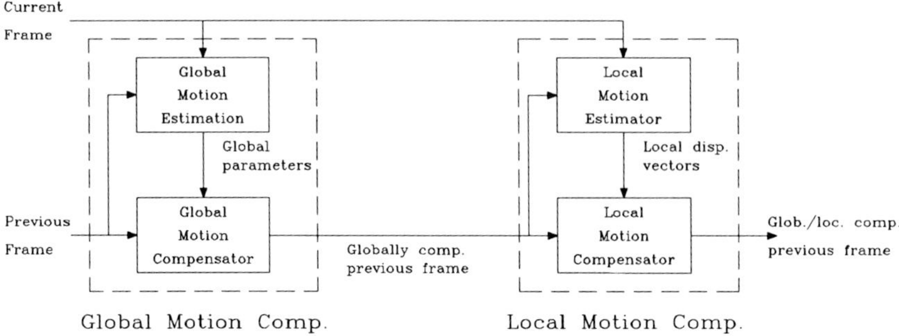

(a)

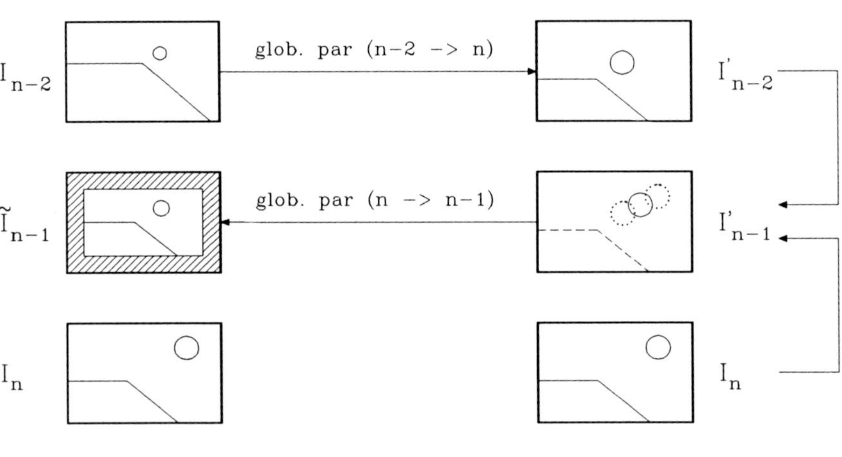

(b)

Figure 6.17: (a) Two-stage motion global-local motion estimator. (b) Interpolation scheme that uses global-local motion information.

into account the appropriate global compensation.

The result obtained from the proposed interpolation schemes is very interesting both for the consistency of the estimated global motion parameter and for the accuracy of the local motion field. Fig. 6.18 presents an interpolated frame from a subsampled (k=3) videoconference sequence with simultaneous panning,

Figure 6.18: Reconstructed image using global-local motion estimation and interpolation.

zooming and the independent motion of the speaker.

For the subsampling factor k=3 or k=4, the use of global/local motion interpolation gives good results, even with an average gain of 2-3 dB with respect to the case of simple motion compensated interpolation [30].

Another important situation in which interpolation is used regards videoconference application where the sequences are taken by a fixed camera. In this case a semantic segmentation of the input image can be carried out (see section 6.5). With this information a more precise interpolation algorithm can be implemented, especially with regard to the correct treatment of the unpredictable image areas (uncovered background, new scene and so on) [22],[30].

In fact, by having available the object mask of images I_n and I_{n-k} and the displacement field, the interpolator can first extrapolate the object position in the generic intermediate frame. However before doing so, tests are performed to verify whether a pel belonging to the object in frame I_n matches, according to its motion vector, a same-state pel in frame I_{n-k}. If not, local mean vector is tried in place of the original one, and if even this is not valid the considered point is classified as "uncompensatable".

After the object map has been determined the other semantic zones are extrapolated, as shown in fig. 6.19. Now that there is a complete scene segmentation the luminance values of the generic intermediate frame I_{n-m} can be interpolated according to simple rules depending on the state of the current pel: for background pels the weighted mean of values in homologous positions in I_n and I_{n-k} are taken. For covered and uncovered background pels the values are respectively from I_{n-k} and I_n. Assigned to the object points is a mean luminance value of the two corresponding points on the known images. Finally, "uncompensatable" points (that are normally less than 0.1 % of the total) are spatially interpolated from their neighbors.

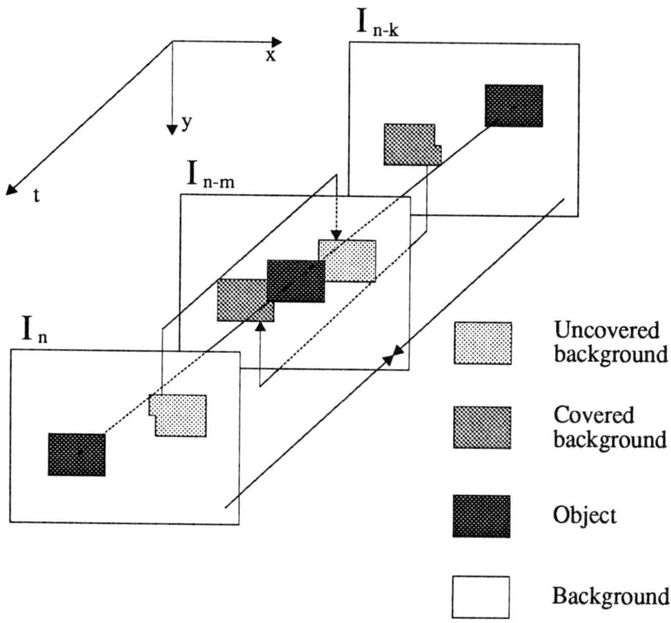

Figure 6.19: Use of semantic segmentation in the MCI algorithm.

The interpolation algorithm that uses semantic segmentation has been tested on the CIF sequence Trevor, with a time subsampling k=3. Fig 6.20 shows the PSNR of the reconstructed images in the case of linear interpolation and of the motion compensated one. Fig. 6.21 shows an interpolated frame when a MCI with semantic segmentation is used.

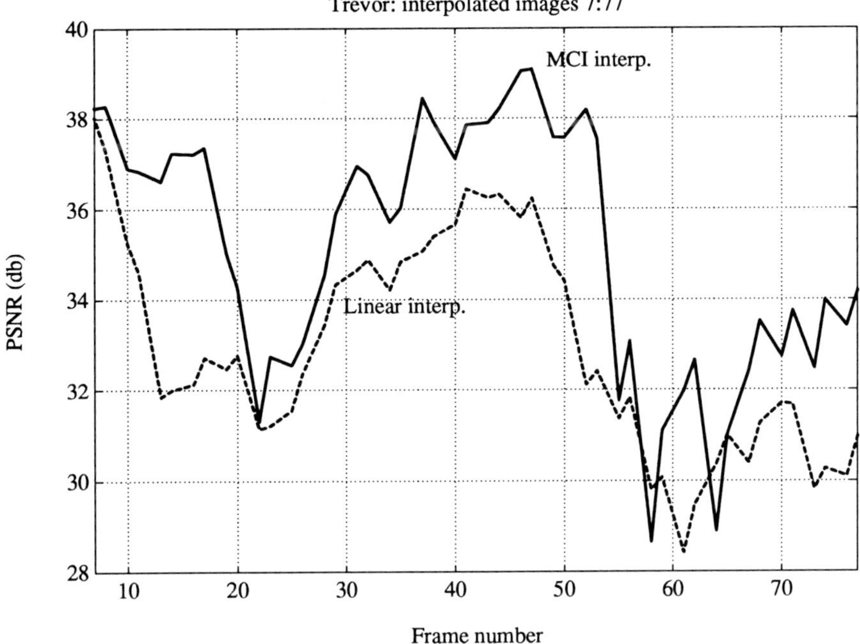

Figure 6.20: PSNR of reconstructed frames in case of linear interpolation and motion compensated one.

6.7 Conclusion

In this chapter the application of motion estimation to the problem of the interpolation of missing images has been presented.

The literature reports some very interesting algorithms and their application, not only in coding systems, but also anywhere there is an alteration in the frame rate of an image sequence. However there has been no unified approach to the problem and this has resulted in different algorithms being developed for different applications. Thus a generalization of the proposed algorithm could be necessary. In particular the extension of the use of the semantic segmentation and of background memories, even in the presence of global pan and zoom motions, would be interesting.

Moreover the use of a tridimensional interpretation of the imaged scene will be very important for interpolation in the future. Some work has already been done, such as the use of a stereo or multicamera system to extract depth information from multiple views of a scene; information of this type can be used for coding and interpolation purposes [41][42].

Figure 6.21: An interpolated image obtained with a MCI that uses semantic segmentation.

6.8 Acknowledgments

The authors would like to thank Ing. L. Mori of Italtel Sit-Central Research Laboratories (Milan), Ing. P. Migliorati of Cefriel Research Center (Milan) and Ing. L. Sorcinelli of Elettronica Industriale S.p.A. (Milan) for their help in the production of the simulation results presented here.

References

[1] A. Puri, R. Aravind, B.G. Haskell, R. Leonardi, *Video Coding with Motion-Compensated Interpolation for CD-Rom Applications*, Signal Processing: Image Communications, vol. 2, 1990, pp. 127–144.

[2] S. Tubaro, *A two layers video coding scheme for ATM networks*, Signal Processing: Image Communication 3, 1991, pp. 129–141.

[3] CCITT Recommendation H.261, *Video codec for audiovisual services at p*64kbit/s*, December 1990.

[4] G. Avid, *Determining Three-Dimensional Motion and Structure from Optical Flow generated by Several Moving Objects*, IEEE Trans. on PAMI, vol. PAMI-7, n. 4, July 1985, pp. 384–401.

[5] T.J. Broida, R. Chellappa, *Estimating the Kinematics and Structure of a Rigid Object from a Sequence of Monocular Images*, IEEE Trans. on PAMI, vol. 13, n. 6, June 1991, pp. 497–513.

[6] S.F. Wu, J. Kittler, *A Differential Method for Simultaneous Estimation of Rotation, Change of Scale and Translation*, Signal Processing: Image Communications, vol. 2, 1990, pp. 69–80.

[7] C. Bergeron, E. Dubois, *Gradient-Based Algorithms for Block-Oriented MAP Estimation of Motion and Application to Motion-Compensated Temporal Interpolation*, IEEE Trans. on CSVT, vol. 1, n. 1, March 1991, pp. 72–84.

[8] M. Subbarao, *Interpretation of Image Flow: A Spatio-Temporal Approach*, IEEE Trans. on PAMI, vol. PAMI-11, n. 3, March 1989, pp. 266–278.

[9] N. Diehl, *Object-Oriented Motion Estimation and Segmentation in Image Sequences*, Signal Processing: Image Communication, vol. 3, 1991, pp. 25–36.

[10] C. Stiller, *Motion Estimation for Coding of Moving Video at 8 kbit/s with Gibbs Modeled Vectorfield Smoothing*, Proc. of Visual communications and image processing '90, SPIE vol. 1360, Lausanne (CH), pp. 468–476.

[11] H.G. Musmann, M. Hötter, J. Ostermann *Object-Oriented Analysis-Synthesis Coding of Moving Images*, Signal Processing: Image Communications, vol. 2, 1989, pp. 117–138.

[12] C. Cafforio, F. Rocca, S. Tubaro, *Motion Compensated Image Interpolation*, IEEE Trans. on Comm., vol. 38, n. 2, February 1990, pp. 215-222.

[13] M. Hötter, *Object-Oriented Analysis-Synthesis Coding based on Moving Two-Dimensional Objects*, Signal Processing: Image Communica tions, vol. 2, 1990, pp. 409–428.

[14] R.Y. Tsai, *A versatile Camera calibration Techniques for High Accuracy 3D Machine Vision Metrology using Off-the-Shelf TV Cameras and Lenses*, IEEE Journal of Robotics and Automation, vol. RA3, n. 4, August 1987, pp. 323–344.

[15] D. Zhang, Y. Nomura, S. Fujii, *Error Analysis and Optimization of Camera Calibration*, Proc. of IEEE/RSJ International Workshop on Intelligent Robots and Systems IROS'91, November 3–5 1991, Osaka (Japan), pp. 292–296.

[16] D. Hepper, *Efficiency analysis and application of uncovered background prediction in a low bit-rate image coder*, IEEE Trans. On Comm., Vol. 38, NO. 9, Sept. 1990, pp. 1578–1584.

[17] T. Koga et al., *Motion Compensated Interframe Coding for Video Conferencing*, Proc. NTC, 1981, pp. G5.3.1–G5.3.5

[18] J. Jain, A. Jain, *Displacement Measurement and its Application in Interframe Image Coding*, IEEE Trans. on Comm., vol. 29, n. 12, 1981, pp. 1799–1808.

[19] A.N. Netravali, J. Robbins, *Motion Compensated Television Coding* Trans. on Inf. Theory, vol. 58, n. 3, 1979, pp. 631–670.

[20] B. Horn, B. Schunk, *Determining Optical Flow*, Artificial Intelligence, vol. 17, 1981, pp. 185–203.

[21] W. Enkelmann, *Investigations of Multigrid Algorithms for the Estimation of Optical Flow Fields*, Computer Vision, Graphics and Image processing, 1988, pp. 150-177.

[22] R. Thoma, M. Bierling, *Motion Compensated interpolation considering covered and uncovered background*, Signal Processing: Image Communications, vol. 1, n. 2, 1989, pp. 191–212.

[23] L. Mori, F. Rocca, S. Tubaro, *Motion compensated interpolation using foreground/background segmentation*, Proc. of Digital Signal Processing 91, Sept. 1991, Florence, Italy, pp. 379–384.

[24] J.K. Aggarwall, N. Nandhakumar, *On the Computation of Motion from Sequences of Images - A review*, Proceedings of the IEEE, vol. 76, n. 8, August 1988, pp. 917–935.

[25] G. Musmann, P. Pirsh, H.J. Grallert, *Advances in Picture Coding*, Proc. IEEE, vol. 73, April 1985, pp. 523– 548.

[26] D.J. Le Gall, *The MPEG video compression algorithm*, Signal Processing: Image Communications, vol. 4, n. 2, 1992, pp. 129–140.

[27] A.N. Netravali B.G. Haskell, *Digital Pictures, Representation and Compression*, Plenum Press, 1988.

[28] A. Puri, H.M. Hang, D.L. Schilling, *An Efficient Block Matching Algorithm For Motion-Compensating Coding,* Proc. ICASSP 87.

[29] H. Derin, *Segmentation of Textured Images Using Gibbs Random Fields*, Computer Vision, Graphics and Image Processing, vol. 35, 1986, pp. 72–98.

[30] P. Migliorati, L. Sorcinelli, S. Tubaro, *Motion Compensated Image Interpolation Using Semantic Segmentation*, Proc. Workshop on Intelligent Terminals, Cost 229, Working Group 3, 1992, V. Cappellini Editor, pp. 29–39.

[31] F. Lavagetto, L. Leonardi, *Block Adaptive Quantization of Multiple Frame Motion Field*, Proc. of Visual communications and image processing '91, SPIE vol. 1605, November 11/13, Boston (Mass.), pp. 534–543.

[32] Y. Tong Tse, R.L. Baker, *Global Zoom/Pan Estimation and Compensation for Video Compression*, Proc. ICASSP-91, pp. 2725–2728.

[33] M. Hötter, R. Thoma, *Image Segmentation Based on Object Oriented Mapping Parameter Estimation*, Signal Processing, vol. 15, 1988, pp. 315-334.

[34] S. Brofferio, F. Rocca, *Interframe Redundancy Reduction of Video Signals Generated by Translating Objects*, IEEE Trans. on Comm., vol. 25, April 1977, pp. 448–455.

[35] C. Lettera, L. Masera, *Foreground/background segmentation in videotelephony*, Signal Processing: Image Communications, vol. 1, n. 2, 1989, pp. 181–190.

[36] S. Brofferio, *An Object-Background Image Model for Predictive Video Coding*, IEEE Trans. on Comm., vol. 37, December 1989, pp. 1391–1394.

[37] F.R. Hansen, H. Elliot, *Image Segmentation Using Simple Markov Random Field Models*, Computer Vision, Graphics and Image Processing, vol. 20; 1982, pp. 101–132.

[38] G.D. Forney, *The Viterbi Algorithm*, Proc. of IEEE, March 1973, pp. 268–278.

[39] W. Guse, M. Gilge, B. Hurtgen, *Effective exploitation of background memory for coding of moving video using object mask generation*, Proc. of Visual communications and image processing '90, SPIE vol. 1360, Lausanne (CH), pp. 512–523.

[40] D. Adolph, R. Bushmann, *1.15 Mbit/s coding of video signals including global motion compensation*, Signal Processing: Image Communications, vol. 3, 1991, pp. 259–274.

[41] R. Skerjanc and J. Liu, *A Three Camera Approach for Calculating Disparity and Synthesizing Intermediate Pictures*, Signal Processing: Image Communications, vol. 4, n. 1, 1992, pp. 55–64.

[42] A. Tamtanoi, C. Labit, *3-D TV: joined identification of global motion parameters fro stereoscopic sequence coding*, Proc. of Visual communications and image processing 91: Visual Communication, pp. 720–731.

7

Subsampling of Digital Image Sequences using Motion Information

R.A.F. Belfor, R.L. Lagendijk and J. Biemond

Delft University of Technology, The Netherlands

7.1 Introduction

A digital image sequence consists of a set of pixels each describing the scene intensity at a specific location on a specific instance in time. In a natural scene these pixels are spatially and temporally correlated with each other. An image coding scheme utilizes these correlations in order to represent the image sequence more efficiently in this way allowing for a more cost-effective storage or transmission. In this chapter spatio-temporal subsampling will be discussed as a data reduction technique. Subsampling is in use as a data reduction method for the standardized HDTV transmission systems MUSE [1] and HD-MAC [2] which are completely based on this technique. Recent proposals use this technique in a combination with transform coding [3][4].

Subsampling, in its basic form, reduces the size of the set of pixels in such a way that the remaining pixels are sufficient to adequately described the original image sequence. It is obvious that a smaller set of pixels will require a smaller transmission bandwidth. In a spatial subsampling scheme the frames are treated independently, and only the correlation between spatially neighbouring pixels is used. In the extreme case that an image sequence consists of frames with a constant intensity value, only one pixel per frame is necessary to completely describe the image sequence.

Spatio-temporal subsampling also takes the temporal correlation into account. The temporal correlation of an image sequence is maximal if the content of the scene is not moving. A data compression scheme which is based on this approach is sub-Nyquist sampling [1][2][5][6], which is a simple way to compress image sequences. The main drawback of this method is that in order to maintain a good spatial resolution the temporal resolution has to be sacrificed. In order to achieve a good spatial resolution while maintaining the temporal resolution, the use of motion information is mandatory. With this information ordinary sub-Nyquist sampling can be extended to motion compensated sub-Nyquist sampling [7]. By using this technique the application range of sub-Nyquist sampling is no longer limited to stationary parts of the image sequence, but now consists of all the parts for which motion can be estimated accurately.

In this chapter the theoretical fundamentals of motion adaptive subsampling are described. Existing coding systems will not be described in full detail but will be treated as an illustration of the theory. For a detailed description the reader is referred to the specific references. The outline of the chapter reflects this objective. First the basics of multidimensional sampling in general and subsampling in particular will be reviewed. Next motion compensated interpolation which plays an important role in the remainder of this chapter will be discussed. After that the basics of sub-Nyquist sampling and motion compensated sub-Nyquist sampling will be presented. Also the way how to incorporate these techniques into a practical coding scheme will be shown. In the experimental results the different subsampling schemes will be compared with each other.

7.2 Spatio-Temporal Subsampling of Image Sequences

Sampling is a basic operation in image processing and image communication systems. An image sequence can be described by a continuous three dimensional intensity function. For the purpose of digital processing and transmission this three dimensional intensity function is sampled. Due to the limited bandwidth of communication channels and the limited capacity of storage devices in practice always a trade-off has to be made between the costs and the necessary sampling density. In subsampling schemes an effort is made to minimize the amount of pixels describing the original image sequence. A constraint in this process is that the quality of the reconstructed image sequence should match the quality necessary for the application. In this section subsampling is described in more detail. But first multidimensional sampling in general will be briefly considered [8][9], providing tools to describe the sampling and subsampling process.

7.2.1 Multidimensional sampling

The sampling of an image sequence is done by evaluating the value of the intensity function on a discrete set of points. In this section a method will be given to formally describe this set of points. Consider the D-dimensional non-singular matrix $\boldsymbol{S}$ †. Now the lattice L can be defined as the set of all linear combinations with integer coefficients n_i of the column vectors $\boldsymbol{s}_i$ of the matrix $\boldsymbol{S}$:

---

† Vectors are in bold lower case characters and matrices are in bold upper case characters.

$$L = \{n_1 s_1 + n_2 s_2 + \cdots + n_D s_D \mid n_i \in Z,\ i = 1,\cdots,D\} \tag{7.1}$$

So a lattice consists of a discrete set of points in R^D. Note that the matrix $\boldsymbol{S}$ does not uniquely describe the lattice L; all matrices whose column vectors are a linear combination of the vectors s_i describe the same lattice. An example of a two dimensional lattice is given in Figure 7.1(a).

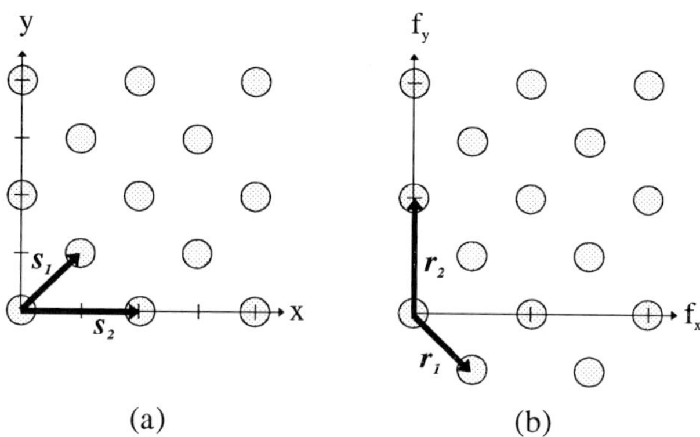

Figure 7.1: (a) Example of a two dimensional lattice with basis vectors $(1,1)^T$ and $(2,0)^T$. (b) Reciprocal lattice with basis vectors $(½,-½)^T$ and $(0,1)^T$.

The physical interpretation of lattices in image sequence processing is that the lattice points are the sampling points of the analog three-dimensional image intensity function $i_c(x,y,t)$:

$$i(x,y,t) = \begin{cases} i_c(x,y,t) & (x,y,t) \in L \\ 0 & otherwise \end{cases} \tag{7.2}$$

where $i(x,y,t)$ is the intensity function describing the sampled image sequence. The *sampling density* of a lattice is defined as the reciprocal of the absolute value of the determinant of $\boldsymbol{S}$. From this definition it is clear that the sampling density is unique for a particular lattice independent of the particular choice of basis vectors. An increase in the sampling density implies an increase in the number of samples used to represent the image.

The *reciprocal lattice* L^R of L is defined by the matrix $\boldsymbol{R}$ given by

$$\boldsymbol{R} = (\boldsymbol{S}^T)^{-1} \tag{7.3}$$

The reciprocal lattice of the example in Figure 7.1(a) is shown in Figure 7.1(b). From the theory of sampling it is known that the Fourier transform of a sampled function has infinite replica in the spectral domain. In [9] it is shown that if $i_c(x,y,t)$ is sampled according to the lattice L then the centre points of the replica are located in the Fourier space on the points of the reciprocal lattice L^R:

$$I(\omega_x,\omega_y,\omega_t) = \frac{1}{|\det(L)|} \sum_{(\Omega_x,\Omega_y,\Omega_t)\in L^R} I_c(\omega_x+\Omega_x,\omega_y+\Omega_y,\omega_t+\Omega_t) \tag{7.4}$$

The spectrum of the sampled signal is thus the superposition of shifted versions of the original spectrum.

Let L_o and L_s be lattices in R^D. L_s is a *sublattice* of L_o if every point of L_s is a point of L_o. The set

$$\{ \boldsymbol{c} + \boldsymbol{x} \,|\, \boldsymbol{x} \in L_s \} \quad , \quad \boldsymbol{c} \in L_o \tag{7.5}$$

is called a *coset* of L_s in L_o. So a coset is a shifted version of the lattice L_s. Two cosets are either identical or disjoint. The number of distinct cosets is given by the ratio of the sampling densities of L_s and L_o. This value is always an integer. This concept is illustrated in Figure 7.2. In this case L_o is the lattice defined by the unity matrix, and L_s is the lattice shown in Figure 7.1(a). In Figure 7.2 L_s is indicated by the light dots, which is according to the definition also a coset. The other coset is obtained by shifting L_s over the vector $(1,0)^T$, which are the dark dots in Figure 7.2. Note that the union of all cosets completely covers L_o. For more details about multidimensional sampling the reader is referred to [9].

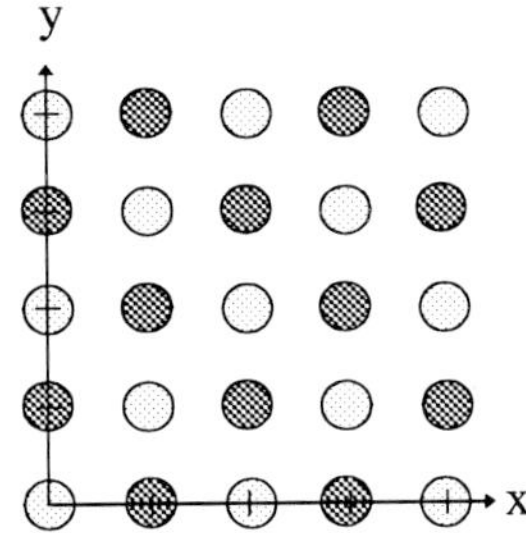

Figure 7.2: The cosets of a lattice. Dots with the same colour belong to the same coset.

7.2.2 Examples of sampling lattices

In this section some examples are given of three dimensional sampling lattices [9][11]. These are depicted in Figure 7.3. In Figure 7.3(a) an orthogonal lattice is shown. The basis vectors form an orthogonal basis, so this lattice can be represented by a scaled version of the identity matrix. If this sampling structure is used for an image sequence, this sequence is called *progressively* scanned.

Another scanning procedure which is often used is the *interlaced* scan. This lattice is depicted in Figure 7.3(b). Half the number of rows of the frame are discarded, hereby reducing the vertical resolution. The pattern in which the columns are discarded is shifted from frame to frame. After subsampling the individual images are called fields instead of frames. In practice interlaced sampling is used to increase the temporal resolution. While maintaining the same number of samples the field rate can be doubled with respect to the frame rate in the case of orthogonal sampling, hereby doubling the temporal resolution.

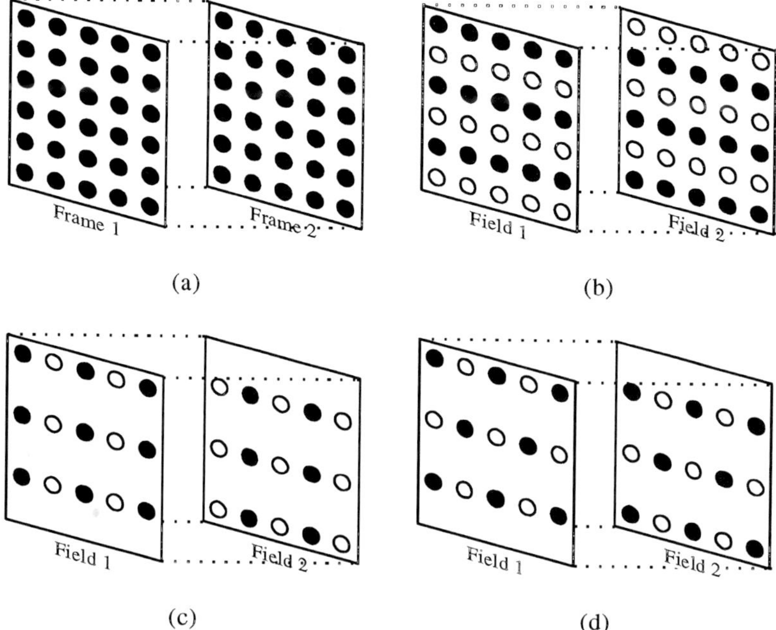

Figure 7.3: Examples of subsampling lattices. (a) Orthogonal (b) Interlace (c) Field quincunx (d) Line quincunx.

Starting from an interlaced input image sequence two other subsampling structure are possible to further reduce the sampling density. In Figure 7.3(c) the *field quincunx* sampling structure is shown. Half the pixels of a line are alternately discarded. The pattern in which the pixel are discarded is shifted from field to field. The consequence of this action is that the resolution in the vertical and horizontal direction is the same. In Figure 7.3(d) the *line quincunx* sampling structure is illustrated. Again half the pixels of a line are discarded but now the pattern is shifted from line to line. The vertical and horizontal resolution is also the same. The difference between field quincunx and line quincunx is that their temporal resolution is different.

7.2.3 Basic subsampling

In this section the basics of subsampling will be reviewed. Subsampling can be defined as the re-representation of an image sequence on a new sampling lattice with a lower sampling density then the original sampling lattice. A simple subsampling scheme is shown in Figure 7.4. The original image sequence is pre-filtered and subsampled prior to transmission. The pre-filter is surrounded by a dashed box because it will be shown that this action is not always necessary. At the receiver the band limited image sequence is recovered by using a spatial interpolation filter.

In order to describe the operations necessary prior to subsampling the concept of the *unity cell* is introduced. An unity cell $\mathfrak{P}$ of a lattice L is any set $\mathfrak{P} \subset R^D$ such that R^D

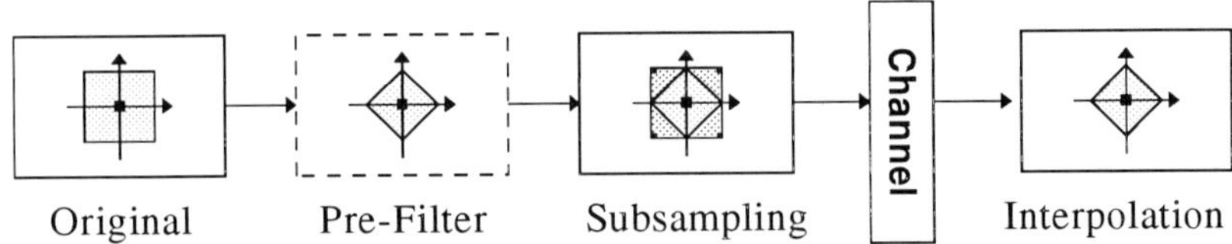

Figure 7.4: Basic subsampling scheme. The figures in the boxes illustrate the actions in the spectral domain.

is the disjoint union of copies of $\mathcal{P}$ centred on each lattice point:

$$(\mathcal{P} + x) \cap (\mathcal{P} + y) = \varnothing \quad x,y \in L, \quad x \neq y$$
$$\bigcup_{x \in L} (\mathcal{P} + x) = R^D \tag{7.6}$$

An unity cell is not unique. In Figure 7.5 two different examples of unity cells for the reciprocal lattice shown in Figure 7.1(b) are given. The concept of the unity cell is important for both the sampling and the subsampling of the image sequence. In order to avoid aliasing after the sampling of an image sequence, the spectrum of the image sequence should be confined to an unity cell. This should also be satisfied in the case of subsampling of the image sequence. This condition can be seen as an extension of the Nyquist sampling theorem [12] to multiple dimensions. To realize this condition a pre-filter can be used to reduce the original spectrum. If the original spectrum has already the shape of the unity cell, pre-filtering is not necessary.

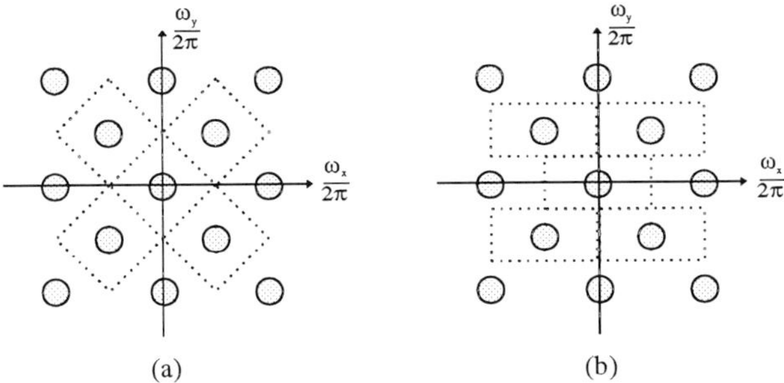

Figure 7.5: Examples of unity cells for the reciprocal lattice shown in Figure 7.1(b).

The shape of the unity cell controls the resolution of the subsampled image sequence. In Figure 7.5(a) a choice is made to reduce the diagonal frequencies in favour of the horizontal and vertical frequencies. This option is often chosen in subsampling schemes because studies have shown that the human eye is less sensitive for diagonal detail in comparison to horizontal and vertical detail [1]. In this case a diamond shaped pre-filter can be used prior to subsampling [13]. Another example is given in Figure 7.5(b), where horizontal frequencies are overemphasized in relation to vertical frequencies.

If the new sampling lattice is a subset of the original lattice, then the actual subsampling can be implemented by simply discarding pixels not present in the new lattice. If this is not the case, an intermediate sampling structure must be used, which has a relation with both the original and the new lattice [10]. The remaining pixels after subsampling are transmitted.

At the receiver the image sequence has to be converted to the original sampling lattice. This is done with an interpolation filter. This filter has to be designed in such a way that the extra replicas introduced by the subsampling process are cancelled. Another property of an interpolation filter is that it should not cancel the frequency components within the unity cell. Otherwise the interpolation would cause an additional loss of resolution. Therefore for the reciprocal lattice in Figure 7.5(a) a diamond shaped filter can also be used for the interpolation.

As can be seen from the theory, straightforward subsampling as discussed in this section will always result in an output image sequence with a lower spatial resolution than the resolution of the original image sequence, unless the original image sequence was oversampled. Therefore in the next sections more advanced subsampling strategies are presented which yields a better quality than basic subsampling at the cost of increased complexity. The schemes described will use motion information to improve the image quality. Therefore in the next section first the relation between motion and subsampling will be investigated.

7.3 Motion Compensated Interpolation

The main difference in a motion adaptive subsampling scheme as compared to traditional subsampling is that in the interpolation stages motion compensated interpolation is used instead of spatial interpolation. In this section the main features and advantages of motion compensated interpolation will be discussed. Fields where motion compensated interpolation is also applicable are deinterlacing [14][16][15], temporal interpolative coding [17] and frame rate conversion [18][19]. In order to describe motion compensated filtering in the spectral domain first the spectrum of a moving image sequence will be introduced.

7.3.1 The three dimensional spectrum of a moving image sequence

In this section a simplified but illustrative model for the three dimensional spectrum of a image sequence will be discussed [14]. If we assume only global motion, the image sequence can be modeled as a stationary scene $i(x,y,0)$ moving at a constant velocity $(v_x,v_y)^T$:

$$i(x,y,t) = i(x-v_x t, y-v_y t, 0) \tag{7.7}$$

According to the definition of the three dimensional Fourier transform, the spectrum $I(\omega_x,\omega_y,\omega_t)$ of this signal is given by

$$I(\omega_x,\omega_y,\omega_t) = \int_{-\infty}^{\infty}\int_{-\infty}^{\infty}\int_{-\infty}^{\infty} i(x,y,t)\exp(-j(\omega_x x+\omega_y y+\omega_t t))\,dx\,dy\,dt \tag{7.8}$$

Substituting (7.7) into (7.8) gives after some basic manipulations

$$\begin{aligned} I(\omega_x,\omega_y,\omega_t) &= \int_{-\infty}^{\infty}\int_{-\infty}^{\infty} i(x,y,0)\exp(-j(\omega_x x+\omega_y y))\,dx\,dy \cdot \int_{-\infty}^{\infty}\exp(-j(\omega_x v_x+\omega_y v_y+\omega_t)t)\,dt \\ &= I_0(\omega_x,\omega_y) \cdot \delta(\omega_x v_x + \omega_y v_y + \omega_t) \end{aligned} \tag{7.9}$$

The first part of this equation, $I_0(\omega_x,\omega_y)$ describes the two-dimensional spatial Fourier transform of the frame $i(x,y,0)$. This part of (7.9) is not a function of the temporal frequency, hence it is valid for every ω_t. The second part of the equation represents a delta function whose value is only non-zero when the argument is zero. Non-zero entries occur only when

$$\omega_x v_x + \omega_y v_y + \omega_t = 0 \tag{7.10}$$

which denotes a plane in the three dimensional Fourier space perpendicular to the vector $(v_x,v_y,1)^T$. Hence the resulting Fourier transform of a sequence containing global motion are the spatial frequency components intersecting the plane perpendicular to the motion vector. The temporal bandwidth W_t is linearly proportional to the velocity components v_x and v_y:

$$W_t = W_x v_x + W_y v_y \qquad (rad/s) \tag{7.11}$$

where W_x and W_y are the horizontal and vertical bandwidths.

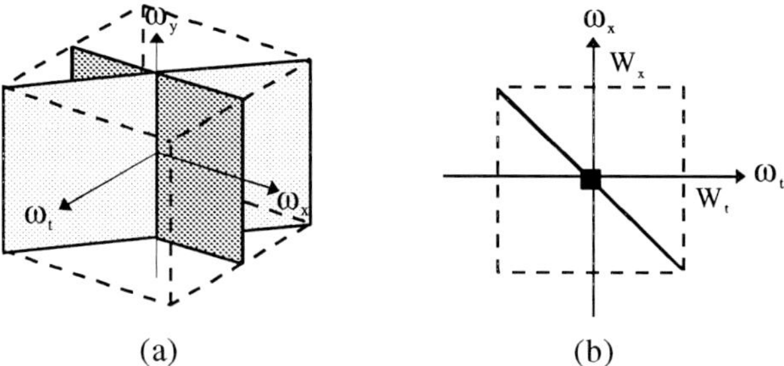

Figure 7.6: The effect of motion on the image spectrum ($v_x = 1$, $v_y = 0$). (a) Support of the three-dimensional spectrum. (b) Intersection of the spectral support with the plane $\omega_y = 0$.

Figure 7.5(a) illustrates Equation (7.9) for the case of $(v_x,v_y)^T$ is $(1,0)^T$. The planes indicate the support of the spectrum. The actual spectrum determines whether there are actual non-zero values. The dark plane indicates the support of the spectrum in the case of no motion, while the lighter plane indicates the support of the three dimensional Fourier transform $I(\omega_x,\omega_y,\omega_t)$. Due to the velocity the original image plane rotates, introducing temporal frequency components. Note that the intersection of the spectrum in Figure 7.5(a) with the $\omega_y = 0$ plane is a line. The graphical representation of the spectrum in this case is shown in Figure 7.5(b). To simplify the graphical representation in the following we shall restrict ourself to the two-dimensional case. Without loss of generality only the *x-t* dimensions will be considered.

7.3.2 Advantage of motion compensated interpolation

The advantage of motion compensated interpolation is illustrated in this section by an example. We assume a velocity of 1 pixel/frame in the horizontal direction. The original image sequence is sampled according to an orthogonal lattice. From this lattice half the number of columns is discarded without any offset, so the matrices $\boldsymbol{S}$ and $\boldsymbol{R}$ describing the sampling lattice and the reciprocal lattice are

$$\boldsymbol{S} = \begin{pmatrix} 2 & 0 & 0 \\ 0 & 1 & 0 \\ 0 & 0 & 1 \end{pmatrix}, \quad \boldsymbol{R} = \begin{pmatrix} \frac{1}{2} & 0 & 0 \\ 0 & 1 & 0 \\ 0 & 0 & 1 \end{pmatrix} \tag{7.12}$$

Extra replica will be introduced in between the original replica in the direction of the ω_x-axis. The upper part of Figure 7.7 illustrates a subsampling scheme in the spectral domain where spatial filtering is used at the receiver, whereas the lower part depicts the situation in which motion compensated interpolation is used. Only the baseband ranging from $-\pi$ rad/s to $+\pi$ rad/s is shown.

If spatial interpolation is used (Figure 7.7(a-c)) the high spatial frequency components cause aliasing, therefore these have to be eliminated prior to subsampling. In order to achieve this the input frames need to be spatially low-pass filtered (Figure 7.7(a)). After that the image sequence can be properly subsampled in the horizontal direction (Figure 7.7(b)). The dashed lines in the figure are the replica introduced by the horizontal subsampling of the image sequence. At the receiver again a spatial low-pass filter is used to eliminate the replica (Figure 7.7(c)). It is observed that both the spatial and temporal resolution is reduced. Spatial interpolation may be acceptable for image regions moving at high velocities, but in regions with a low velocity, where the eye is capable of following the motion, this causes visible blur [20][21]. In order to get a good image quality both the spatial and temporal resolution must be preserved.

In the case of motion compensated interpolation (Figure 7.7(d-f)) *no* pre-filtering is used prior to subsampling (Figure 7.7(d)). In the subsampling stage (Figure 7.7(e)) the replica enters the baseband, but they do not overlap. In Section 7.5.1 we will see that this is not the case for all possible velocities and subsampling lattices. Using the interpolation filter from Figure 7.7(c) in this case will give rise to aliasing, causing a degradation of the interpolated image sequence. Therefore at the receiver a motion compensated interpolation filter [14][22] can be used to interpolate the frame instead (Figure 7.7(f)). This filter temporally interpolates the image sequence in the direction of the velocity in this way adapting the interpolation filters passband to the shape of the spectrum of the subsampled sequence. Thus at the receiving end the choice of the unity cell is adapted to the motion. Aliasing can now be avoided in the reconstruction process without affecting the spatial resolution. It is however necessary that the receiver knows the local displacement.

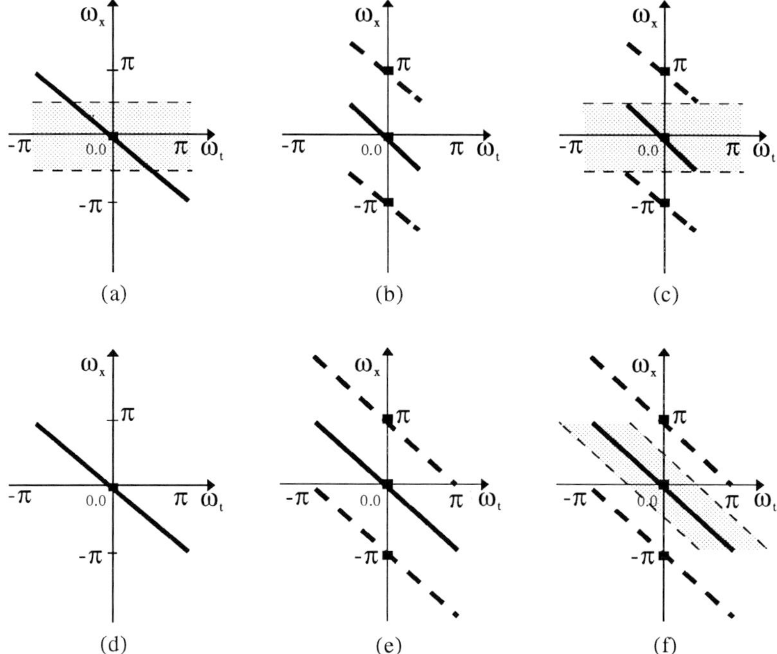

Figure 7.7: Spectral description of subsampling scheme. The lines show the support of the spectral components. (a-c) Spatial interpolation. (d-f) Motion Compensated interpolation.

From this discussion it is evident that motion compensated interpolation has the potential to improve the resolution after subsampling as opposed to using straightforward spatial interpolation.

7.3.3 Motion compensated interpolation in practice.

Motion compensated filtering is done by temporally filtering an image sequence in the direction of the motion. In the previous discussions we assumed a constant velocity over a number of frames. In [23] it is shown that this is not a necessary assumption and that it is sufficient to follow an object along a motion *trajectory*. The assumption made in this case is that the characteristics of the object do not change dramatically from frame to frame due to variations in the illumination, object deformations or occlusions.

The most simple temporal filter is a two taps filter with coefficients ½ and ½, which just averages the current and previous frame in the direction of the motion:

$$i_f(x,y,t) = \tfrac{1}{2} i(x,y,t) + \tfrac{1}{2} i(x-d_x(x,y,t), y-d_y(x,y,t), t-1) \qquad (7.13)$$

This filter has a poor frequency response, but is particulary interesting due to the fact that it imposes small memory requirements. If a longer filter would be used, the stop-band attenuation will increase. But in this case also the number of frames involved in the filtering process will increase, requiring more frame stores. Another disadvantage of long temporal filters is that the assumption that the object does not change will become less valid over a longer period of time. The result of motion compensated filtering also depends on the accuracy of the motion estimate. If the motion is estimated with a higher precision then the filter response can be more appropriately shaped to the actual frequency content of the image sequence. This aspect will be discussed in more detail in the Section 7.3.5.

In motion compensated interpolation schemes it is important to have a consistent vector field, as opposed to hybrid coding schemes where a best match vector field is advantageous for increasing coding performance. Also particular precautions have to be taken in regions where the motion estimation fails. This could be the case in occluded regions or regions moving at high velocity. Errors are also introduced when a region is moving with a velocity which is half the resolution of the estimated motion vectors. In this case the replica introduced by the subsampling are poorly attenuated because they are located in the transition area between passband and stopband of the temporal interpolation filter.

7.3.4 Nominal velocities

To cover the entire range of velocities it is not necessary to have a filter for each possible velocity. The two dimensional velocity plane depicting all the possible velocities can be subdivided into regular regions. Each region can be associated with one specific motion compensated interpolation filter. The set of velocities covering the entire velocity plane are called the *nominal velocities* $(v_{nx},v_{ny})^T$. In this section the range covered by each nominal velocity will be derived [14]. This is done using Figure 7.8 in which the solid line symbolizes the spectrum of an object moving with a nominal velocity, and the shaded area is the range of the filter covering the area around the nominal velocity.

In order to reconstruct the image sequence, the three dimensional baseband of the original sequence should be well inside the three dimensional passband of the temporal interpolation filter. The baseband of the original image sequence is given by

$$\omega_t = -v_x\omega_x - v_y\omega_y \qquad (-W_x \le \omega_x \le W_x\ ,\ \ -W_y \le \omega_y \le W_y) \qquad (7.14)$$

This plane is bounded in the ω_x-direction by the horizontal bandwidth W_x rad.'s and in the ω_y-direction by the vertical bandwidth W_y rad/s. As an example two possible planes are depicted in the upper part of Figure 7.8 as the dashed lines. The slope of these lines is controlled by the actual velocity.

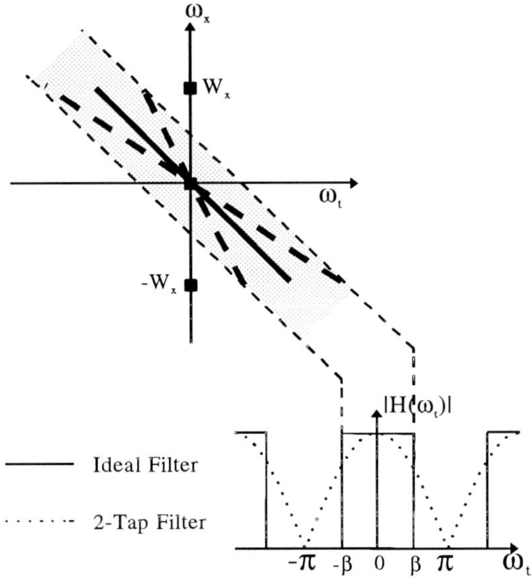

Figure 7.8: Accuracy of the motion compensated filter.

The boundaries of a one dimensional filter in the three dimensional space are planes. If the filter has a cut-off frequency at β rad/s, then the planes describing the boundaries are

$$v_{nx}\omega_x + v_{ny}\omega_y + \omega_t = \pm\beta \qquad (\beta>0) \tag{7.15}$$

The intersection of the filter boundaries and the baseband spectrum can be computed by substituting Equation (7.14) in Equation (7.15)

$$(v_{nx} - v_x)\omega_x + (v_{ny} - v_y)\omega_y = \pm\beta \tag{7.16}$$

In the case of a good reconstruction, intersections only occur at the boundaries of the baseband spectrum. Therefore $\omega_x = \pm W_x$ and $\omega_y = \pm W_y$ is substituted into Equation (7.16). By taking account the exact condition under which each intersection occurs, Equation (7.16) now becomes

$$|v_x - v_{nx}|W_x + |v_y - v_{ny}|W_y = \beta \tag{7.17}$$

This equation describes diamond shaped regions in the velocity plane. For an efficient subdivision of the velocity plane these regions should not overlap. Therefore the nominal velocities should be chosen at

$$v_{nx} = \frac{2\beta}{W_x}i + c_x\,, \qquad v_{ny} = \frac{2\beta}{W_y}j + c_y\,, \qquad i,j \in Z \tag{7.18}$$

where c_x and c_y are arbitrarily chosen constants.

Next this result will be illustrated with an example. Suppose the original image sequence was sampled on an orthogonal lattice according to the Nyquist criterium, hence $W_x = W_y = \pi$ rad/s. For the temporal interpolation filter a perfect low-pass filter is used, so $\beta = \frac{1}{2}\pi$ rad/s. Substituting these values into Equation (7.18) and including $(0,0)^T$ in the set of nominal velocities, yields for the nominal velocities

$$v_{nx} = i, \quad v_{ny} = j, \quad i,j \in Z \qquad (pixels/frame) \tag{7.19}$$

The dotted line in the lower part of Figure 7.8 is the transfer function of a simple two tap filter. In order to get a good reconstruction result β should be chosen smaller then in the case of a perfect lowpass filter. Note that the short filter has a pole for $\omega_t = \pi$ rad/s. Hence in the case that the object is moving with a velocity which exactly equals the nominal velocity the aliasing is completely cancelled, providing perfect reconstruction.

7.3.5 Necessary resolution of the motion estimate

In the previous section a relation for the nominal velocities has been derived. Since the velocity of a moving object is derived from the displacement in a fixed time interval, the accuracy of the velocity is coupled to the accuracy of the displacement estimate. From Equation (7.18) it follows that the necessary accuracy of the motion estimation is given by

$$\Delta d_x = \frac{2\beta}{W_x}, \quad \Delta d_y = \frac{2\beta}{W_y} \qquad (pixels) \tag{7.20}$$

where Δd_x is the accuracy in the horizontal direction and Δd_y the accuracy in the vertical direction. If the values from the example with the ideal motion compensated interpolation filter are substituted in Equation (7.20) then the resulting accuracy becomes

$$\Delta d_x = 1, \quad \Delta d_y = 1 \qquad (pixels) \tag{7.21}$$

The values prescribed by Equation (7.20) are lower boundaries for the motion accuracy. If it is possible to realize a motion estimator with a higher accuracy then Equation (7.20) can be solved for β instead. A lower value for $\Delta d_{x,y}$ will result in a lower value for β, giving opportunities for shorter temporal interpolation filters. This principle will be explained with an example.

In Figure 7.9 the effect of increasing the accuracy has on the two taps temporal filter is illustrated. Shown is the part of the filter characteristic, which is applied to the baseband spectrum, as a function of the velocity. In the case of perfect interpolation this should give a straight line for all values of ω_x. In Figure 7.9(a) integer accuracy is used. Figure 7.9(b) illustrates the characteristics when half pixel accuracy is used. Solving Equation (7.20) with $\Delta d_{x,y} = \frac{1}{2}$ pixel, under the same assumptions as the example in the previous paragraph, gives for β a value of $\frac{1}{4}\pi$. The nominal velocities are now given by

$$v_{nx} = \tfrac{1}{2}i, \quad v_{ny} = \tfrac{1}{2}j, \quad i,j \in Z \qquad (pixels/frame) \tag{7.22}$$

This example shows that an increase of the accuracy gives a better approximation of the straight line in the spectral domain, effectively improving the simple two tap filter.

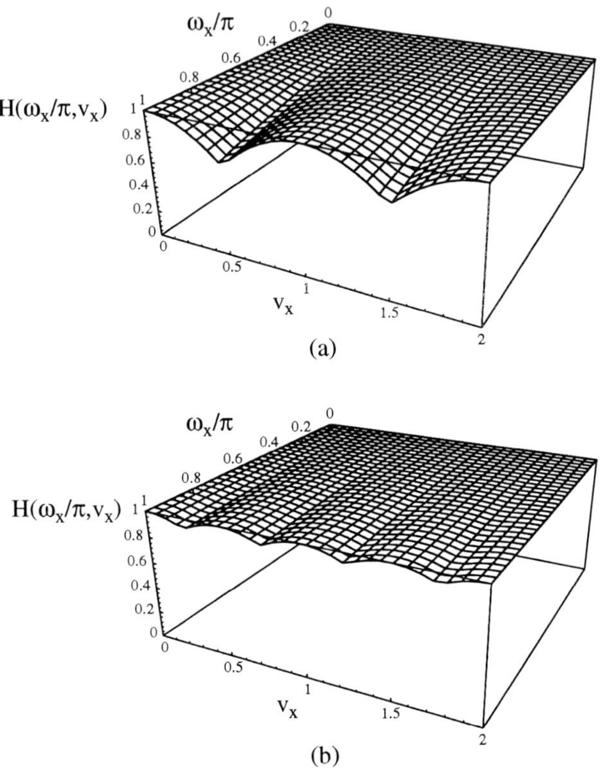

Figure 7.9: Effect accuracy on temporal filter. (a) $\Delta d_x = 1$ pixel, (b) $\Delta d_x = \frac{1}{2}$ pixel.

To apply motion compensated interpolation on a subsampled image sequence the estimated displacement should be defined on the sampling lattice. In order to meet this accuracy it is therefore necessary to expand the spatial resolution of the subsampled image sequence back to the spatial resolution on which the motion was estimated. In the case of fractional accuracy this means that the original image sequence has to be upsampled. If the upsampling filters involved in this process are not sufficiently long this may cause a loss of resolution.

7.4 Non-Adaptive Sub-Nyquist Sampling

Sub-Nyquist sampling is used to subsample an image sequence while still maintaining the full spatial resolution. It is currently used in the Japanese MUSE [1] and the European HD-MAC system [2]. Using this technique the frame is sampled with a frequency lower than the necessary Nyquist frequency, without taking extra precautions against aliasing. In this section sub-Nyquist sampling will be discussed in more detail.

7.4.1 Sub-Nyquist sampling

Consider the image sequence shown in Figure 7.10. The original image is defined on an orthogonal lattice L_o. L_{n-1} is the lattice used for the subsampled frame at $t = n$-1. The sampling density is halved compared to the original orthogonal sampling lattice. At $t = n$ the lattice

$$\begin{aligned} L_n &= \{x \mid x \notin L_{n-1}\} \quad , \quad x \in L_o \\ &= \{x + (1,0)^T \mid x \in L_{n-1}\} \end{aligned} \tag{7.23}$$

is used. The second line of this equation shows that this lattice is a coset of the lattice L_{n-1}. As was shown in Section 7.2.1, the union of these two lattices completely covers the entire image. So if these two lattices are combined, using a non-motion adaptive temporal filter the original frame can be reconstructed exactly at the receiver if the content of the scene is stationary. Hence this technique will result in an output frame with the full original spatial resolution. If the first sampling lattice contains more cosets, i.e. the sampling density is lower, then the number of frames involved in the interpolation will increase. A higher compression ratio can be achieved at the expense of a longer temporal interpolation filter.

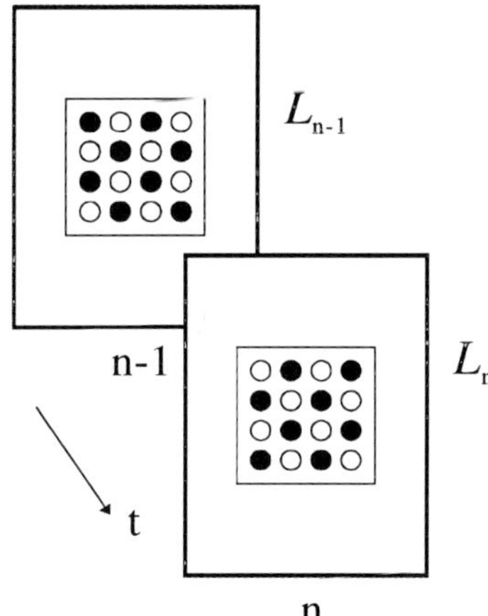

Figure 7.10: The principle of sub-Nyquist sampling.

The range of velocities for which sub-Nyquist sampling is possible, is not limited to completely stationary image sequences. Because a fixed temporal interpolation filter is used and no further motion information is taken into account, it is also applicable for image sequences containing little motion. The velocity range for sub-Nyquist sampling is closely related with the interpolation filter used. If a lowpass filter is used with cut-off frequency β then, using Equation (7.18), the range is given by

$$v_x = \pm\frac{\beta}{W_x}, \qquad v_y = \pm\frac{\beta}{W_y} \qquad (pixels/frame) \tag{7.24}$$

In the case that $W_x = W_y = \pi$ rad/s and a perfect low-pass filter with a cut-off frequency at $\frac{1}{2}\pi$, the range is $\pm\frac{1}{2}$ pixels/frame in both directions. If a lattice with a lower sampling density is used, the replica will come closer to the origin. In this case β must also decrease, lowering the velocity range.

Sub-Nyquist sampling can also be described in the spectral domain. This is done in Figure 7.11 for a completely stationary image sequence. The newly introduced replica are the dashed lines. They are located in the empty space between the original temporal replica. This is due to the fact that the sampling lattice is temporally shifted. The temporal space is empty because there is no motion causing the temporal bandwidth to be equal to zero. Therefore it can be said that in this case the temporal resolution is sacrificed in favour of the spatial resolution. If the sampling lattice has a lower sampling density, more replica are introduced and the area between the original replica will become more densely packed. Also show in the figure is the passband of the temporal filter. After temporal filtering only the replica from the original image sequence are retained.

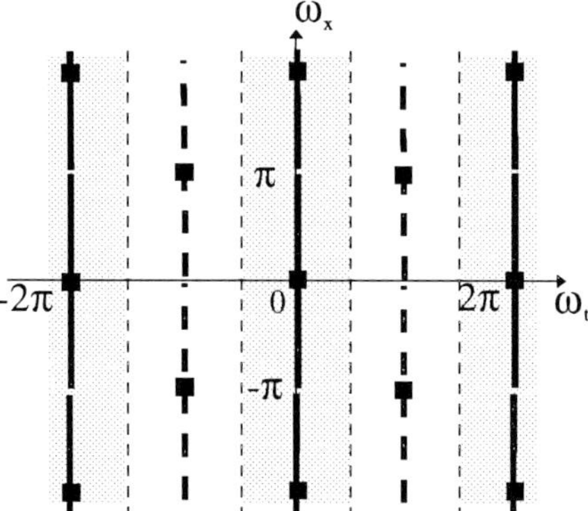

Figure 7.11: Sub-Nyquist sampling in the spectral domain in the case of no motion.

7.4.2 The practical implementation of a sub-Nyquist coding scheme

In this section some schemes for the practical implementation of sub-Nyquist sampling will be discussed. The schemes are simplified versions of the MUSE and HD-MAC. For details about these systems the reader is referred to the references [1][2]. The descriptions as given here are by no means complete and can by no means be used to faithfully compare the two systems.

The first system is a simplified version of the MUSE system. As we saw in Section 7.4.1 sub-Nyquist sampling can only be applied if there is no motion. Therefore in the case of motion a fall-back mode is necessary. How this is implemented is shown in Figure 7.12(a). The scheme consist of two branches. One branch consists of a spatial low-pass filter (LPF1) followed by spatial subsampling (SS). The second branch employs sub-Nyquist sampling (SNS). The spectrum is folded back two times, in such a way that the frequency components around the DC-level do not contain any aliasing. The motion detector (MD) controls the switching between the two branches. The decision which branch to use is made for every pixel. The choice which branch to use is *not* transmitted to the receiver. At the receiver a motion detection is made based on the part of the signal spectrum which is assumed not to contain any aliasing. Because the detection is based on a different signal then the detection made at the transmitter, the detection result may differ. At the receiver there are again two branches. One branch uses a spatial filter (LPF2) to interpolate the image spatially, whereas the second branch uses temporal filtering (TF) in order to preserve the spatial resolution.

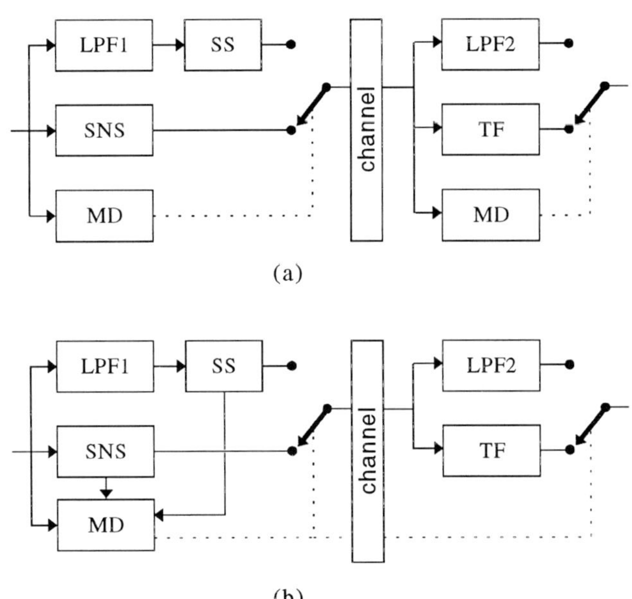

Figure 7.12: Implementations sub-Nyquist sampling: (a) Simplified MUSE coding scheme. (b) Simplified HD-MAC coding scheme.

The second system is a simplified version of the HD-MAC system and is illustrated in Figure 7.12(b). A motion detection is only done at the transmitter based on the original input image sequence and the result of both the subsampling branches. This information is transmitted to the receiver as side information. In order to reduce the amount of side information, motion detection is done on a block basis. As we see from this simple discussion HD-MAC will result in a simpler receiver due to the fact that no motion detection is necessary at the receiver. With the MUSE system the detection is done on a pixel basis which makes it easier to adapt to variations in the local displacement.

It should be noted that the actual HD-MAC system contains also a mode in between sub-Nyquist sampling and plain subsampling. This mode recovers some of the spatial resolution which is sacrificed in the non-adaptive subsampling case at the expense of temporal resolution. This loss of temporal resolution is solved by using motion compensated interpolation.

7.5 Motion Adaptive Sub-Nyquist Sampling

If an image sequence contains little or no motion then after subsampling the sequence can be reconstructed using Sub-Nyquist sampling. This was shown in the previous section. If the image sequence contains motion, direct combination of the pixels from the different subsampled frames is not possible. This problem can be partially solved by using motion information to interpolate the original frame. Unfortunately perfect reconstruction using motion compensated interpolation cannot be achieved in all situations. For a given subsampling lattice certain object velocities exist, called *critical velocities* [14], which do not allow for a perfect reconstruction of an object moving at that velocity. The condition under which these critical velocities occur will be discussed in the next section [24].

7.5.1 Critical velocities

In this section an expression will be derived for the critical velocities given a certain subsampling lattice. It is assumed that the original sequence was sampled on an orthogonal sampling lattice and that the subsampling factor is two. The raster points $(x,y,t)^T$ after subsampling can always be described as follows:

$$\begin{pmatrix} s_{11} & s_{12} & s_{13} \\ 0 & s_{22} & s_{23} \\ 0 & 0 & s_{33} \end{pmatrix} \cdot \begin{pmatrix} k \\ l \\ m \end{pmatrix} = \begin{pmatrix} x \\ y \\ t \end{pmatrix}, \quad k,l,m \in Z \tag{7.25}$$

The matrix $\boldsymbol{S}$ with elements s_{ij} determines the structure of the subsampling lattice. The upper triangular form imposes no constraint because every regular basis can be rewritten to this form. At $t = 0$ and $t = 1$ all possible sampling positions $\boldsymbol{x}_0$ and $\boldsymbol{x}_1$ are given by

$$\boldsymbol{x}_0 = \begin{pmatrix} s_{11}k_0 + s_{12}l_0 \\ s_{22}l_0 \end{pmatrix}, \quad \boldsymbol{x}_1 = \begin{pmatrix} s_{11}k_1 + s_{12}l_1 + \dfrac{s_{13}}{s_{33}} \\ s_{22}l_1 + \dfrac{s_{23}}{s_{33}} \end{pmatrix}, \quad k_0,k_1,l_0,l_1 \in Z \tag{7.26}$$

This is done by substituting the values for t and eliminating the variable m. If the velocity $\boldsymbol{v}$ is defined as the displacement of a pixel between two consecutive frames, then $\boldsymbol{v}$ is equal to

$$\boldsymbol{v} = \boldsymbol{x}_1 - \boldsymbol{x}_0 = \begin{pmatrix} s_{11}(k_1 - k_0) + s_{12}(l_1 - l_0) + \dfrac{s_{13}}{s_{33}} \\ s_{22}(l_1 - l_0) + \dfrac{s_{23}}{s_{33}} \end{pmatrix} = \begin{pmatrix} s_{11}i + s_{12}j + \dfrac{s_{13}}{s_{33}} \\ s_{22}j + \dfrac{s_{23}}{s_{33}} \end{pmatrix}, \quad i,j \in Z \tag{7.27}$$

Two conclusions follow from (7.27). In the first place, if a pixel moves at a velocity $\boldsymbol{v}$ that satisfies (7.27), and if the pixel also satisfies (7.25), i.e. its position is an element of the subsampled lattice, then we can always associate a certain pixel on the same subsampling lattice in any other frame with the pixel considered. Secondly, if however a pixels is moving with a velocity $\boldsymbol{v}$ and is *not* an element of the subsampled grid, then we can *never* associate any other pixel on the same subsampling grid in any other frame with this pixel, otherwise contradicting the first conclusion. The latter conclusion implies that if a pixel (or object) moves with a velocity satisfying (7.27), it can never be perfectly recovered using a motion compensated interpolation filter. Hence (7.27) gives an expression for critical velocities.

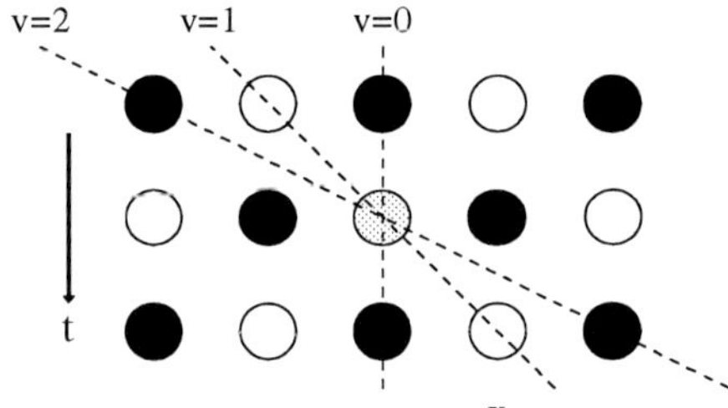

Figure 7.13: Illustration of critical velocities. In this figure $v_x = 1$ is a critical velcocity.

As an example consider the following matrix $\boldsymbol{S}$:

$$\boldsymbol{S} = \begin{pmatrix} 2 & 0 & 1 \\ 0 & 1 & 0 \\ 0 & 0 & 1 \end{pmatrix} \tag{7.28}$$

The subsampling associated with this matrix is the discarding of every other column in the image, and shifting this pattern for every frame. In Figure 7.13 a cross section showing only the x and t axis is depicted for this case. According to (7.27) the critical velocities $\boldsymbol{v}_c$ are:

$$\boldsymbol{v}_c = \begin{pmatrix} 2i + 1 \\ j \end{pmatrix}, \quad i,j \in Z \tag{7.29}$$

Figure 7.13 illustrates that it is indeed impossible to interpolate the dashed pixel for the velocity $v_x = 1$ irrespective of v_y, due to the fact that in this direction only discarded pixels are encountered.

The conclusion which can be drawn from this discussion is that every regular sampling lattice using spatial subsampling possess the problem of critical velocities. In the next section different solutions for this problem will be argued. It should be noted that the method described in this section does not necessarily gives *all* the critical velocities for an arbitrary subsampling factor. If a higher subsampling factor is used it may be possible to go from one sampling point to another sampling point while skipping some intermediate frames. The method described in this section can then be extended in a straightforward manner by taking more pairs of frames into account.

7.5.2 Possible solutions for critical velocities

In the previous section the problem of critical velocities was examined. In this section some possible solutions for this problem will be argued. These solutions are shown in the spectral domain in Figure 7.14. The subsampling structure used for the example in Figure 7.13 is also used, so again half the number of columns are discarded. It was shown in the previous section that the critical velocities depends on the velocity. Therefore in Figure 7.14 velocities of 1 pixels/frame in the horizontal direction (upper part) and 2 pixels/frame in the horizontal direction (lower part) are used in order to examine the implications of the different solutions for different velocities. The dashed lines are the replica introduced by subsampling the image sequence.

• In Figure 7.14(a) and Figure 7.14(d) the spatial resolution of all the moving regions is sacrificed in order to avoid critical velocities. This is done by low pass filtering the frame prior to subsampling. As we saw in Section 7.3.2 this may be objectional in the case of slowly moving objects.

• Another solution is shown in Figure 7.14(b) and Figure 7.14(e). The spatial resolution is only sacrificed if objects are moving with a critical velocity. This solution is generally used in the case of motion compensated deinterlacing [14][22]. For motion compensated subsampling this possibility may be rejected using the same argument as the first solution.

• The cause of the critical velocities is the fact that in the interpolation stages only discarded pixel are encountered in the direction of the motion. For this reason it is possible to solve this problem by using spatially adaptive subsampling [7]. The idea behind this solution is to shift the subsampling structure depending on the amount of motion. This is illustrated in Figure 7.15. Different subsampling lattices are used for different velocities. In Figure 7.14(c) we see that due to the horizontal shift in the subsampling lattice, the replica introduced by the subsampling are located in a different location causing no interference with the other replica. The advantage of this

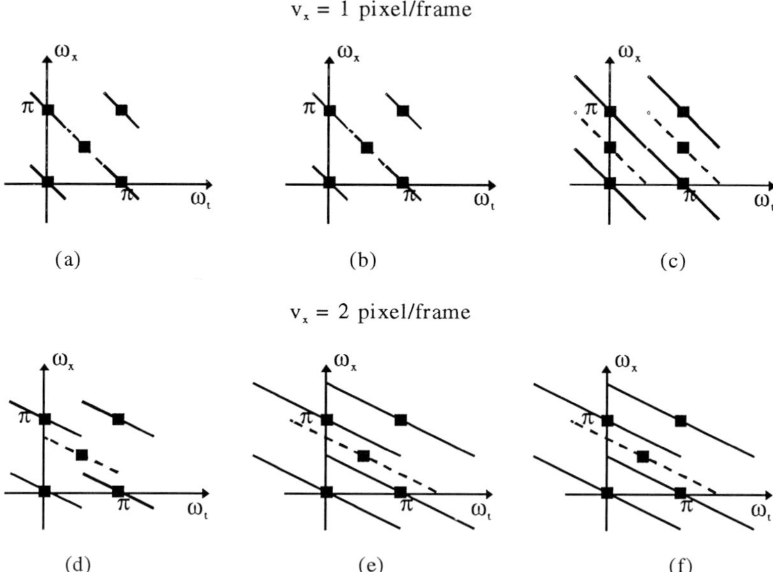

Figure 7.14: Possible solutions for critical velocities. In (a-c) the velocity is 1 pixel/frame. In (d-f) the velocity is 2 pixels/frame.

solution is that the image can be subsampled without sacrificing any spatial or temporal resolution. Therefore in the next sections this solution will be discussed in further detail.

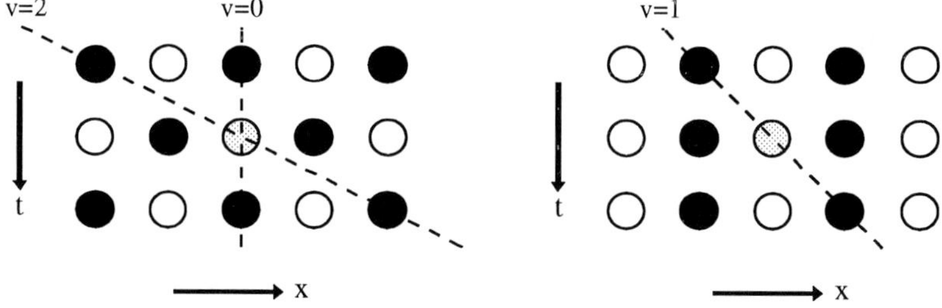

Figure 7.15: Subsampling structures (a) v_x is even. (b) v_x is odd.

7.5.3 Principle of motion adaptive sub-Nyquist sampling

In a motion adaptive subsampling scheme the sub-Nyquist principle can be extended. The pixels which are transmitted are (using the same notation as in Section 7.4.1):

$$L_n = \{\boldsymbol{x} \mid \boldsymbol{x} - \boldsymbol{d}(\boldsymbol{x},n) \not\in L_{n-1}\} \tag{7.30}$$

The vector $\boldsymbol{d}(\boldsymbol{x},n)$ is the estimated displacement vector for the pixel $(\boldsymbol{x},n)$. After motion compensated interpolation the union of L_n and L_{n-1} form L_o. This is due to the fact that no discarded pixel is present in the direction of the motion so it is always possible to reconstruct the current frame based on the previous frame and the subsampled current frame using motion compensated interpolation. Note that Equation (7.30) degenerates to Equation (7.23) in the absence of motion. Therefore this method can be called motion compensated sub-Nyquist sampling. Equation (7.30) however does not guarantees a constant data reduction as is always the case with non-adaptive sub-Nyquist sampling. Further, $\boldsymbol{d}(\boldsymbol{x},n)$ has to be transmitted as additional side information.

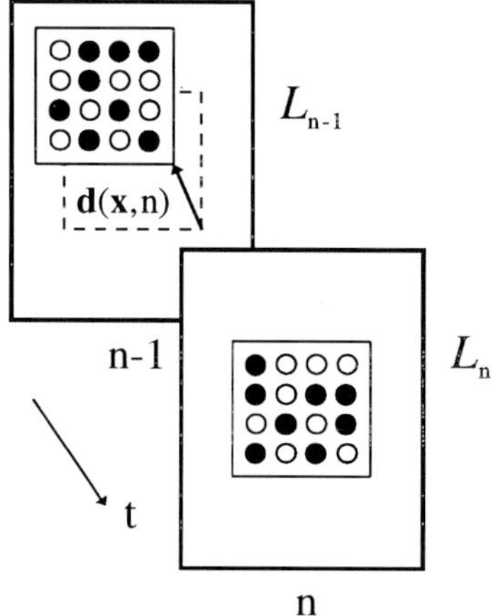

Figure 7.16: Motion adaptive subsampling.

Starting from a fixed subsampling grid, a recursive update is made of the subsampling structure.This principle is illustrated again in Figure 7.16. The subsampling structure of frame n is based on the subsampling structure of frame n-1 and the displacement vector $\boldsymbol{d}(\boldsymbol{x},n)$. If the pixel at $(\boldsymbol{x}-\boldsymbol{d}(\boldsymbol{x},n),n-1)$ of the previous frame was not an element of L_{n-1}, then the corresponding pixel $(\boldsymbol{x},n)$ in the current frame is transmitted. If the pixel was not discarded in the previous frame, then the corresponding pixel along the motion trajectory in the current frame is not transmitted.

One disadvantage of this method is that the spatial correlation structure can vary on a local basis due to the different subsampling structures possible. This could be disadvantageous in a coding scheme that uses spatial subsampling followed by DPCM or transform coding. This effect will be reduced if the motion estimator provides a consistent vector field. It can be easily verified that if the displacement vector $\boldsymbol{d}(\boldsymbol{x},n)$ is constant over a large region, all the blocks in this region will be subsampled with the same subsampling structure, resulting in a homogenous correlation structure provided that the previous subsampled frame also had a homogenous correlation structure.

7.5.4 Motion adaptive subsampling and fixed subsampling

In an HDTV application it may be advantageous to combine the motion compensated subsampling scheme with non-adaptive subsampling. The non-adaptive stage can be used as a first step to reduce the redundancy between the pixels. A simple scheme to combine these two data compression methods is shown in Figure 7.17(a). Without loss of generality again we only consider the *x-t* dimensions and assume a constant velocity of 1 pixel/frame in the horizontal direction. What is shown in each step of the figure is *one* line of two consecutive frames.

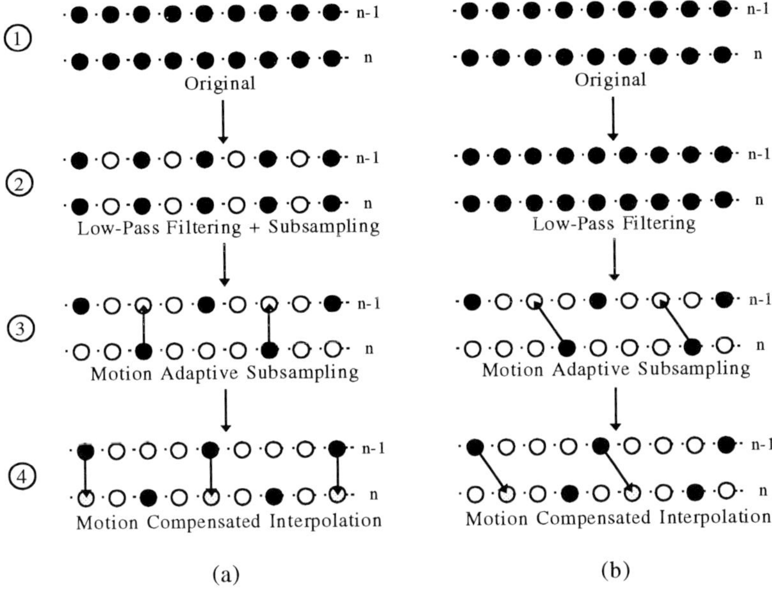

Figure 7.17: Two possible solutions to combine motion adaptive sub-Nyquist sampling with fixed subsampling.

The frame is first pre-filtered in the horizontal direction, reducing the horizontal bandwidth W_x from π rad/s to $\frac{1}{2}\pi$ rad/s. Due to this pre-filtering the frame may be subsampled by discarding half the columns of the image. These two operations are illustrated in the second step of Figure 7.17(a). By substituting $W_x = \frac{1}{2}\pi$ rad/s in Equation (7.18) and again assuming a perfect interpolation filter with a cut-off frequency at $\frac{1}{2}\pi$ rad/s the nominal velocities in the horizontal direction v_{nx} are now

$$v_{nx} = 2i, \quad i \in Z \qquad (pixels/frame) \tag{7.31}$$

Therefore a displacement of 1 pixel/frame has to be truncated to either 0 pixel/frame or 2 pixels/frame. In the third part of Figure 7.17(a) the displacement of 1 pixel/frame is assigned to a region covered by the nominal velocity of 0 pixel/frame. Using this nominal velocity, motion compensated sub-Nyquist sampling is applied on the subsampled grid, resulting in a overall data reduction with a factor four. This is shown in the third step of Figure 7.17(a). At the receiver the frame is interpolated with a motion compensated filter (step four in Figure 7.17(a)), followed by a spatial interpolation filter. A major disadvantage of this scheme is that a velocity of 1 pixel/frame falls in the transition band of the temporal interpolation filter, resulting in a poor attenuation of the replica and a distortion of the baseband. Another disadvantage is that no benefit is taken from the extra accuracy available in the displacement estimate, which is lost due to the necessary truncation.

A more attractive alternative is shown in Figure 7.17(b). It is based on the discussion in Section 7.3.5, which showed that a decrease of the nominal velocities improved the result after motion compensated interpolation. After the spatial low-pass filtering the frames are *not* immediately subsampled. On this sampling structure a velocity of 1 pixel/frame is still defined, and the nominal velocities are given by Equation (7.19). So truncation of the velocity is not necessary. Now the subsampling following the spatial low-pass filter is combined with the motion adaptive sub-Nyquist sampling. The algorithm is modified in such a way that after interpolation a sequence of three consecutive discarded pixels may not exist, in this way satisfying the sampling theorem. This is illustrated in the third step of Figure 7.17(b). Because the frames are not yet subsampled a velocity of 1 pixel/frame can be used, exploiting the full accuracy of the displacement estimate. If now motion compensated interpolation is used at the receiver (step four in Figure 7.17(a)), a velocity of 1 pixel/frame will fall in the middle of the pass-band of interpolation filter at the receiver, achieving a better suppression of the replica. This scheme can be extended to two dimension in a straightforward manner. Only the shape and the size of the region which should not contained discarded pixels must be adapted to the sampling lattice used.

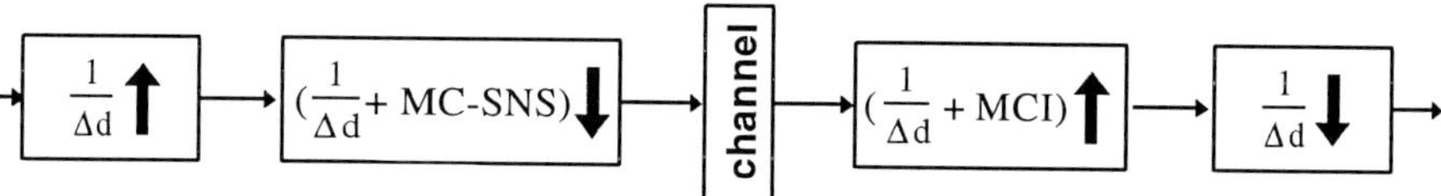

Figure 7.18: Motion compensated sub-Nyquist sampling combined with fractional accuracy of the motion estimate.

The above scheme also presents a way to use motion compensated sub-Nyquist sampling using motion vectors estimated with a higher accuracy. The modifications which have to be made to the original algorithm are illustrated in Figure 7.18. If the accuracy of the motion estimate is Δd pixel, then first the original sequence is upsampled with a factor $1/\Delta d$. On this sampling lattice this motion accuracy is defined on a pixel basis. This concept was previously discussed in Section 7.3.5. At this higher resolution the motion compensated sub-Nyquist sampling (MC-SNS) is combined with a downsampling of a factor $1/\Delta d$. The downsampling with a factor $1/\Delta d$ cancels the upsampling with a factor $1/\Delta d$, so the net result of these operations is the subsampling factor of the motion adaptive sub-Nyquist sampling. At the receiver these operations are inverted, replacing the subsampling with motion compensated interpolation (MCI) and an upsampling. The incorporation of the fractional accuracy means that the motion compensated interpolation filter can more adequately be shaped to the actual frequency content, resulting in better suppression of the replica. This may be especially useful in the case of short interpolation filter. In Figure the improved filter characteristics were shown in the case of an accuracy of a ½ pixel. This accuracy will require an upsampling with a factor two.

The same argument as above may be followed for the use of interlaced image sequences with motion compensated subsampling. Instead the first upsampling stage has to be replaced with a deinterlacing scheme.

7.6 A motion adaptive sub-Nyquist coding scheme

In this section motion adaptive sub-Nyquist sampling as discussed in the previous section will be integrated in a coding scheme. An important aspect is how the failure of the motion estimation can be detected and what actions should be taken. Therefore first the errors resulting from motion adaptive subsampling will be modeled. The distribution of the errors will be compared with a scheme where only one fixed subsampling pattern is used, using spatial interpolation at the receiver. This scheme will be used as a fall-back mode if the motion extimation fails. Also an experimental validation of the model will be given.

7.6.1 Error analysis

First the errors in case of motion adaptive subsampling will be discussed. Two distinctive situations will be considered: the case when an accurate motion estimate is made and the situation when an wrong motion estimation is made. It will be shown that these situations will introduce different types of errors. In a practical situation the intensity function describing an image sequence will not only contain information about the actual scene, but also includes degradations introduced during the image acquisition process. This factor can be modeled by an additive noise term:

$$i(\boldsymbol{x},t) = \tilde{i}(\boldsymbol{x},t) + n(\boldsymbol{x},t) \tag{7.32}$$

where $\tilde{i}(\boldsymbol{x},t)$ is the actual intensity, $i(\boldsymbol{x},t)$ the observed intensity and $n(\boldsymbol{x},t)$ the additive noise term. If an object is translating in the image plane according to the vector $\tilde{\boldsymbol{d}}(\boldsymbol{x},t)$, then $\tilde{i}(\boldsymbol{x},t)$ can be modeled by

$$\tilde{i}(\boldsymbol{x},t) = \tilde{i}(\boldsymbol{x}-\tilde{\boldsymbol{d}}(\boldsymbol{x},t),t-1) \tag{7.33}$$

The *estimated* motion vector is the vector $\boldsymbol{d}(\boldsymbol{x},t)$.

If an accurate motion estimate is made, $\tilde{\boldsymbol{d}}(\boldsymbol{x},t)$ will be equal to $\boldsymbol{d}(\boldsymbol{x},t)$. Then in practice $\tilde{i}(\boldsymbol{x}-\tilde{\boldsymbol{d}}(\boldsymbol{x},t),t-1)$ will be approximated by $i(\boldsymbol{x}-\boldsymbol{d}(\boldsymbol{x},t),t-1)$, which will result in a reconstruction error value $e(\boldsymbol{x},t)$ of

$$\begin{aligned} e(\boldsymbol{x},t) &= i(\boldsymbol{x},t) - i(\boldsymbol{x}-\boldsymbol{d}(\boldsymbol{x},t),t-1) = n(\boldsymbol{x},t) - n(\boldsymbol{x}-\boldsymbol{d}(\boldsymbol{x},t),t-1) \\ &= \Delta n(\boldsymbol{x},t) \end{aligned} \tag{7.34}$$

where $\Delta n(\boldsymbol{x},t)$ denotes the part of the error introduced by noise. So in the case of a correct motion estimate the error introduced is noise coloured by the motion vector field. The variance of the reconstruction error will be of the same order as the noise variance in the original image sequence, causing no clearly visible errors.

If an incorrect motion estimate is made, Equation (7.33) no longer hold. The reconstruction error is now given by

$$\begin{aligned} e(\boldsymbol{x},t) &= \tilde{i}(\boldsymbol{x},t) - \tilde{i}(\boldsymbol{x}-\boldsymbol{d}(\boldsymbol{x},t),t-1) + n(\boldsymbol{x},t) - n(\boldsymbol{x}-\boldsymbol{d}(\boldsymbol{x},t),t-1) \\ &= \Delta\tilde{i}(\boldsymbol{x},t) + \Delta n(\boldsymbol{x},t) \end{aligned} \quad (7.35)$$

An extra error term $\Delta\tilde{i}(\boldsymbol{x},t)$ is introduced as compared to Equation (7.34). If $\tilde{i}(\boldsymbol{x}\text{-}\boldsymbol{d}(\boldsymbol{x},t),t\text{-}1)$ differs significantly from $\tilde{i}(\boldsymbol{x},t)$ this term will dominate, introducing conspicuous errors. These are due to the fact that the variance of the error now depends on the actual image sequence contents. If the intensity value changes dramatically from one frame to another in the direction of the estimated motion vector the $\Delta\tilde{i}(\boldsymbol{x},t)$ term increases, causing an increase of the error variance. From this discussion it is clear that the errors made in the case of a good motion estimate differ significantly from the errors made in the case of a bad motion estimate.

If non-adaptive subsampling is used the outcome of the interpolation is the input image sequence $i(\boldsymbol{x},t)$ convolved with a low-pass filter $h(\boldsymbol{x})$. Now the error becomes

$$e(\boldsymbol{x},t) = i(\boldsymbol{x},t) - h(\boldsymbol{x}) * i(\boldsymbol{x},t) \quad (7.36)$$

where the asterisk ($*$) symbolises the convolution operator. Substituting (7.32) yields

$$e(\boldsymbol{x},t) = (1-h(\boldsymbol{x})) * \tilde{i}(\boldsymbol{x},t) + (1-h(\boldsymbol{x})) * n(\boldsymbol{x},t) \quad (7.37)$$

Hence the reconstruction error consists of the high frequency content of the image sequence signal and the noise contribution, causing a blurry image sequence with reduced noise content. These results will next be verified experimentally.

7.6.2 Experimental validation

The distribution of the absolute error value was determined for the *CALTRAIN* sequence. The scene contains a camera pan over a scene consisting of a moving train and calender. In Figure 7.19 the distribution of reconstruction error of a fixed subsampling scheme (QNX-SS) and a motion adaptive subsampling scheme (MA-SS) is shown. In the fixed subsampling case a quincunx pattern is used. The figure as shown consists of three part. At top the complete distribution is shown. The two bottom graphs both show a specific part of the horizontal axis. The left part shows the behaviour of the distribution for small error values whereas the right part shows the behaviour for large error values.

As can be seen from the figure the two distributions differ significantly. The motion adaptive scheme (solid line) has a high probability for low error values . This corresponds with Equation (7.34) which described the characteristics of the reconstruction error when an accurate motion estimation is available. The reconstruction error is $\Delta n(\boldsymbol{x},t)$, which has a variance in the same order as the noise in

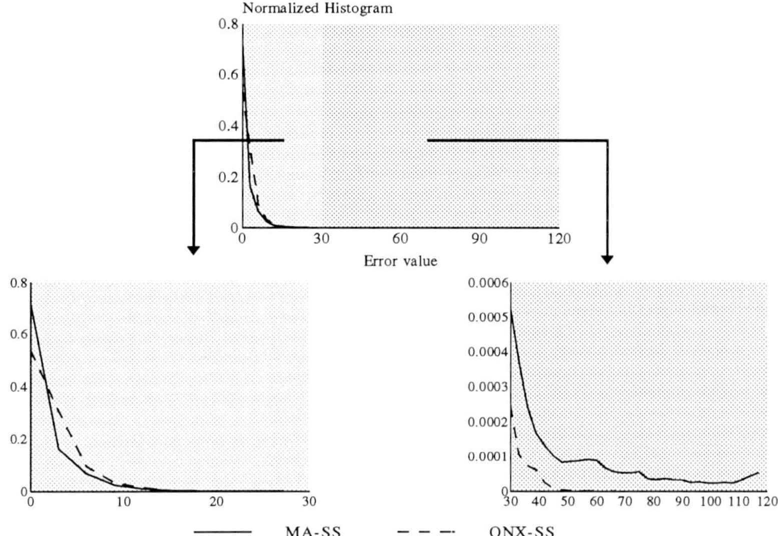

Figure 7.19: Error distribution for *CALTRAIN* sequence.

the original image sequence. However there is also a non-zero probability for large error values. This corresponds with a poor motion estimate as modeled in Equation (7.35). In this case the error consist of $\Delta\tilde{i}(\boldsymbol{x},t)$ and $\Delta n(\boldsymbol{x},t)$. This part of the distribution will cause clearly visible artifacts in the reconstructed frames due to the high error values introduced by $\Delta\tilde{i}(\boldsymbol{x},t)$.

The non-adaptive subsampling scheme (dashed line) has a lower probability for small error values but zero probability for high error values. This will result in a completely blurred reconstructed image but without clearly visible isolated artifacts due to the zero probability for large error value. Experiments on different image sequences show similar results. Hence Figure 7.19 provides a experimental validation of the theoretical analysis given in the previous section. In a practical coding scheme a combination of these two distributions should be made, using a threshold on the motion adaptive branch. In the case of high error value the fixed subsampling branch is chosen. This will give rise to a high resolution image in the case of a good motion estimate without clearly visible degradations in the case of motion estimation failure.

7.6.3 System overview

In this section a system incorporating motion adaptive sub-Nyquist sampling is described. We saw in the previous section that large errors in the motion adaptive branch were due to failure in the motion estimate, causing noticeable artifacts in the reconstructed image. Therefore when the motion estimate fails special precautions have to be taken in order to provide an acceptable image quality while still realizing the same data reduction. How this affects the overall system is shown in Figure 7.20. The system consists of two branches:

• The lower branch is used if a correct motion estimate (ME) is made, using motion adaptive subsampling (MA-SS) to subsample the image. At the receiver motion compensated interpolation (MCI1) is used in the reconstruction process.

• The upper branch is used in the case of failure of the motion compensated branch. In this case low-pass filtering (LPF1) and a non-adaptive quincunx subsampling structure is used (QNX-SS). This sampling structure is chosen due to the preservation of horizontal and vertical resolution. At the receiver spatial interpolation (LPF2) is used to reconstruct the frame.

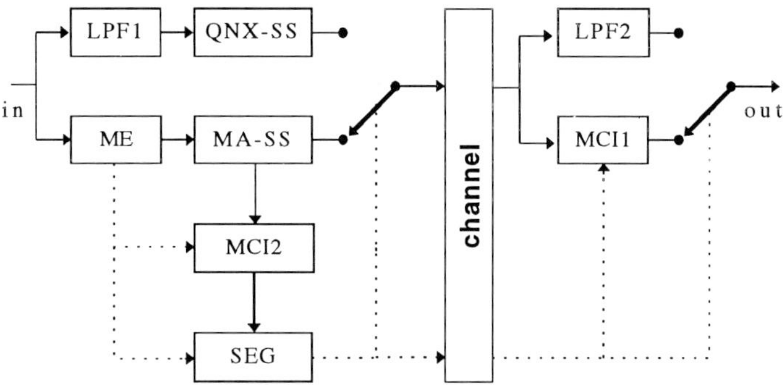

Figure 7.20: System overview.

The reconstruction quality of the motion adaptive branch controls the switching process between these two modes. This guarantees that always the highest possible resolution is chosen, regardless of the spectrum of the current frame. The error analysis in the previous section validates this principle. In order to be able to make a decision the image is also reconstructed at the transmitter side (MCI2). The decision is made on a block basis using the same blocks as the motion estimation. Smaller blocks will improve the reconstruction quality but will also increase the amount of side information. If the blocks of the motion estimation coincides with the segmentation blocks then the segmentation information can be multiplexed into the motion information due to the fact that in the case of non-adaptive subsampling no motion information is necessary.

7.7 Experimental Results

In this section the experimental result will be discussed. The experiments consists of a comparison between the different subsampling schemes discussed in this chapter. Fixed subsampling schemes, sub-Nyquist sampling and motion compensated sub-Nyquist sampling are investigated and compared with each other. Also motion compensated sub-Nyquist sampling with fractional motion estimation accuracy is considered.

7.7.1 Implementation details

For the experimental results three different schemes are used. In this section several details of these systems are described.

The first system uses fixed spatial subsampling as a data reduction method. A block diagram of this system was already given in Figure 7.4. A quincunx subsampling structure is used, yielding a data reduction of two. The pre-filter and interpolation filter which are used are described in [?]. In [?] is also described how the different filtering operations should be implemented.

The second system is a sub-Nyquist sampling scheme. Again a quincunx subsampling structure is used, but without any spatial filtering operations. Instead a temporal interpolation filter with two filter taps is used. With this system also a data reduction of two is achieved. The input image sequence is compensated for the global pan in order to make the image sequence appear to be stationary. This is especially useful if the entire sequence is moving. In order to evaluate the performance of the method in the case that the image sequence is moving, no fall-back mode is used.

The third system employs motion compensated sub-Nyquist sampling. Two variants of this system are considered. One uses motion vectors with pixel accuracy, whereas the other one uses motion vectors with half pixel accuracy. For motion estimation hierarchical blockmatching is used [25][26]. This scheme provides a consistent vector field, which is advantageous in a motion compensated interpolation application. Some modifications were made to this algorithm in order to incorporate fractional motion accuracy. In Section 7.5.4 it was shown that to utilize fractional accuracy, spatial upsampling is necessary. For this purpose separable maximally-flat filters were used [11][27]. One of the properties of these filters is that for a filter $h(n)$ with $2N+1$ filter taps:

$$h(0) = \tfrac{1}{2} \;\wedge\; h(n) = 0\,, \quad n = \pm 2, \pm 4 \cdots, \pm N \tag{7.38}$$

If this filter is used on an alternating input source of zeroes and original pixel values, the original input samples are not changed by the filtering process hereby preserving the original resolution as best as possible. A temporal interpolation filters with two taps was used. In Figure 7.9 the frequency characteristics of this filter was shown for the different motion estimation accuracies. In order to evaluate the performance in the case of a bad motion estimate, no fall-back mode is used.

7.7.2 Comparison between different subsampling schemes

In this section the schemes as described in the previous section are compared with each other. The image sequence used to compare these methods is the *CALTRAIN* sequence. In Figure 7.21(a) the second frame of the sequence is shown and in Figure 7.21(b) the corresponding estimated vector field.

This sequence is chosen to compare these different methods based on the following features:

- The sequence contains a considerable amount of high frequency content.
- Almost the entire sequence is moving in a translational manner due to a camera pan.
- There are also objects moving in a non-translational manner (the ball).
- The scene includes areas which are occluded due to the motion (the train).
- The sequence also contains objects moving in a translational manner with a non-integer speed (the calendar).

A completely stationary scene would obviously favour the sub-Nyquist sampling based techniques whereas a scene containing complex motion would favour fixed subsampling.

In Figure 7.21(c) the stretched difference between the original image and the reconstructed image using fixed subsampling is shown. As can be expected the difference image in Figure 7.21(c) shows that the errors are mainly concentrated around the edges of the scene causing a noticeable loss of resolution. The image detail shown in Figure 7.22(a) also shows this.

In Figure 7.21(d) the difference in the case of sub-Nyquist sampling is shown. As can be seen sub-Nyquist sampling works well for a large region of the frame but fails if the actual motion deviates from the global pan.

In the next experiment motion compensated sub-Nyquist sampling was used. These results are shown in Figure 7.21(e) and Figure 7.21(f). In Figure 7.21(e) pixel accuracy is used for the motion estimate and in Figure 7.21(f) half pixel accuracy is used. The results show that motion compensated sub-Nyquist sampling works better then non-adaptive sub-Nyquist sampling. The area with low error values now consists of the regions which move according to the global pan and regions for which an accurate motion estimate is possible. The results also show that if fractional accuracy is used for the motion estimate the result also improves. In Figure 7.22(b) an image detail is shown for the integer motion estimation accuracy case. This detail clearly shows the improvement compared to fixed subsampling.

Note that Figure 7.21(c-f) again illustrate the difference in the error distribution as was discussed in section 7.6.1. In case of a good motion estimate only the noise component is present whereas in the case of a bad motion estimate more clearly visible artifacts are introduced. This is also reflected in the actual value of the errors.

In Table 7.1 the signal to noise ratio and the mean square error for the different coding methods are shown. As can be expected motion adaptive sub-Nyquist sampling with half pixel motion estimation accuracy results in the smallest error value due to the fact that the temporal correlation can be exploited more efficiently. However the total error values do no differ significantly from the quincunx subsampling case. Besides the arguments discussed in Section 7.6 this has also another reason. The spatial high frequency contents of an image sequence usually represents a small

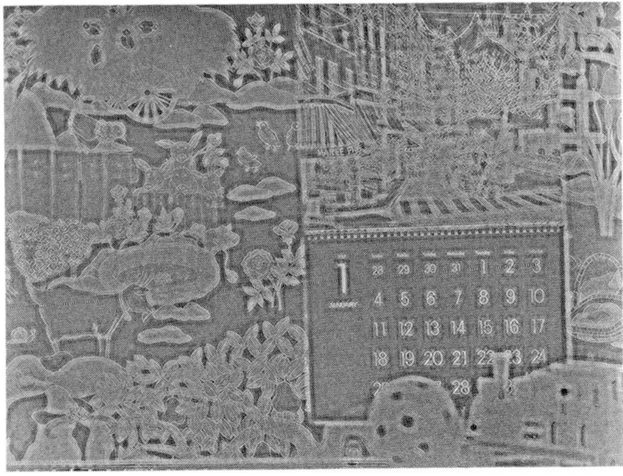

Figure 7.21: From top to bottom: (a) Original Image. (b) Estimated motion vectors. (c) Stretched reconstruction error using quincunx subsampling.

Figure 7.21(continued): From top to bottom: Stretched error using (d) sub-Nyquist sampling (e) motion adaptive sub_Nyquist sampling with pixel motion vector accuracy (f) motion adaptive sub-Nyquist sampling with half pixel motion vector accuracy.

Figure 7.22: Detail of reconstructed image. From left to right: (a) Quincunx subsampling (b) Motion adaptive sub-Nyquist sampling.

Table 7.1: Signal to noise ratio and mean square reconstruction error of the different coding methods.

CODING METHOD	SNR (dB)	MSE
Quincunx Subsampling	18.9	21.9
Sub-Nyquist Sampling	16.7	36.5
Motion Adaptive Sub-Nyquist Sampling (pixel accuracy)	18.3	25.6
Motion Adaptive Sub-Nyquist Sampling (half pixel accuracy)	19.2	20.9

portion of the total spectral energy. Therefore preserving the high spatial contents is not proportionally reflected in the mean square error or signal to noise ratio.

7.8 Conclusions

In this chapter motion adaptive subsampling schemes are discussed. Major topics to this respect are the multidimensional spectrum and motion compensated interpolation. The different trade-offs concerning temporal filters are discussed. Quantitative expressions are given to describe the relation between motion compensated interpolation and velocity. It is argued that increasing the accuracy of the motion estimation is more beneficiary then increasing the length of the temporal filter. This is due to the fact the assumptions of signal stationary along the motion trajectory is less valid over a longer number of frames.

Sub-Nyquist sampling has proven to be a valid method as to increase the performance of a subsampling scheme. Due to the fact that this method is only applicable for stationary or very slowly moving regions of a scene an extension is made to motion compensated sub-Nyquist sampling. This extension provides a high quality for a wider range of input sources. The price of this improvement is the increase complexity and vulnerability to transmission errors. Using motion compensated sub-Nyquist sampling requires a motion *estimation* instead of a motion *detection* necessary in the case of non-adaptive sub-Nyquist sampling. Generally motion estimation schemes are more complex then motion detection schemes. In order to prevent errors in the motion detection or motion estimation causing a complete breakdown of the algorithm, in both cases a fall-back mode is necessary. This mode should incorporate a scheme that provides an acceptable quality.

The schemes in this chapter mainly uses the temporal correlation in a image sequence. The objective of these algorithms is to provide a high resolution, regardless of the fact if the actual image content requires this high resolution. Therefore these methods could be further extended to incorporate spatial frequency adaptivity [28][29]. This extension will probably increase the data reduction factor due to the fact that in this case the subsampling can be better suited to the local signal statistics, and not only the temporal but also the spatial correlations can be exploited.

References

[1] Y. Ninomiya, "An HDTV Broadcasting System Utilizing a Bandwidth Compression Technique-MUSE", *IEEE Transactions on Broadcasting*, Vol. BC-33, No. 4, pp. 130-160, December, 1987

[2] F.W.P. Vreeswijk and M.R. Hagiri, "HDMAC Coding for MAC Compatible Broadcasting of HDTV Signals", *Signal Processing of HDTV, II: Proceedings of the third International Workshop on HDTV*, pp. 187-194, Elsevier Science Publishers, September, 1989

[3] G. Schamel, "Spatio-Temporal Subsampling and Transform Coding of HDTV Signals", *Image Comunnication*, Vol. 2, No. 3, pp. 305-318, October, 1990

[4] W. Vos, *A Combination of Spatio-Temporal Subsampling and Subband Coding of Image Sequences*, M.Sc. Thesis, TU Delft, The Netherlands, April, 1992 (in Dutch)

[5] G.J. Tonge, "Image Processing for Higher Definition Television", *IEEE Transactions on Circuits and Systems*, Vol. CAS-34, No. 11, pp. 1385-1398, November, 1987

[6] G. Schamel, "Pre- and Postfiltering of HDTV Signals for Sampling Rate Reduction and Display Up-Conversion", *IEEE Transactions on Circuits and Systems*, Vol. CAS-34, No. 11, pp. 1432-1439, November, 1987

[7] R.A.F. Belfor, R.L. Lagendijk and J. Biemond, "Motion Adaptive Sub-Nyquist Sampling of HDTV", *Signal processing VI: theories and applications: Proceedings of EUSIPCO-92*, Vol. I, J. Vandewalle (ed.), pp. 291-294, August, 1992

[8] N.T. Gaarder, "A Note on the Multidimensional Sampling Theorem", *Proceedings of the IEEE*, pp. 247-248, February, 1972

[9] E. Dubois, "The Sampling and Reconstruction of Time-Varying Imagery with Application in Video Systems", *Proceedings of the IEEE*, Vol. 73, No. 4, pp. 502-522, April, 1985

[10] T. Reuter, "Mehrdimensionale Abtastratenumsetzung", *AEÜ*, Vol. 40, No. 4, pp. 219-224, 1986

[11] G.J. Tonge, *The Sampling of Television Images*, Independent Broadcasting Authority, Rep. No. 112/81, May, 1981

[12] A.J. Jerri, "The Shannon Sampling Theorem-Its Various Extensions and Applications: A Tutorial Review", *Proceedings of the IEEE*, Vol. 65, No. 11, pp. 1565-1596, November, 1977

[13] B. Girod and W. Geuen, "Vertical Sampling Rate Decimation and Line-Offset Decimation of Colour Difference Signals", *Signal Processing*, Vol. 16, No. 2, pp. 109-127, February, 1989

[14] B. Girod and R. Thoma, "Motion-compensated Field Interpolation from Interlaced and Non-Interlaced Grids", *Proceedings of SPIE conference on Image Coding*, Vol. 594, M. Kunt and T.S. Huang (ed.), pp. 186-193, 1985

[15] P. Haavisto, Y. Neuvo and J. Juhola, "Motion Adaptive Scan Rate Up-converion", *Multidimensional Systems and Signal Processing*, Vol. 3, No. 2/3, pp. 113-130, May, 1992

[16] F.-M. Wang, D. Anastassiou and A.N. Netravali, "Time-Recursive Motion Compensated Deinterlacing", *Proceedings of the Third International Workshop on HDTV*, September, 1989

[17] ISO-IEC JTC1\SC2\WG11, *Test Model 2 for MPEG 2*, July, 1992 , edition MPEG 92/245

[18] M. Ernst, "Motion Compensated Interpolation for Advanced Standard Conversion and Noise reduction", *Proceedings Fourth International Workshop on HDTV*, September, 1991

[19] T. Reuter, "Standard Conversion using Motion Compensation", *Signal Processing*, Vol. 16, No. 1, pp. 73-82, January, 1989

[20] D.H. Kelly, "Motion and vision. II. Stabilized spatio-temporal threshold surface", *Journal Optical Society of America*, Vol. 69, No. 10, pp. 1340-1349, October, 1979

[21] U. Gölz and R. Schäfer, "Considerations on the Possibility to Exchange Temporal against Spatial Resolution in Image Coding", *Image Communication*, Vol. 2, No. 1, pp. 39-51, May, 1990

[22] M. Ernst and T. Reuter, "Adaptive Filtering for Improved Standards Conversion", *Proceedings Second International Workshop on HDTV*, pp. 449-458, Elsevier Science Publishers, February/March, 1988

[23] E. Dubois, "Motion-Compensated Filtering of Time-Varying Images", *Multidimensional Systems and Signal Processing*, Vol. 3, No. 2/3, pp. 211-239, May, 1992

[24] R.A.F. Belfor, R.L Lagendijk and J. Biemond, "Motion Compensated Subsampling of HDTV", *Proceedings of SPIE conference on Visual Communications and Image Processing '91: Visual Communication*, Vol. 1601, Kou-Hu Tzou and Toshio Koga (ed.), pp. 274-284, November, 1991

[25] M. Bierling, "Displacement Estimation by Hierarchical Block-Matching", *Proceedings of SPIE Conference on Visual Communications and Image Processing '88*, Vol. 1001, T. Russell Hsing (ed.), pp. 942-951, November, 1988

[26] R. Thoma and M. Bierling, "Motion Compensating Interpolation Considering Covered and Uncovered Background", *Signal Processing: Image Communications*, Vol. 1, No. 2, pp. 191-212, October, 1989

[27] C. Gumacos, "Weighting Coefficients for Certain Maximally Flat Nonrecursive Digital Filters", *IEEE Transactions on Circuits and Systems*, Vol. CAS-25, No. 4, pp. 234-235, April, 1978

[28] R. Kishimoto and N. Sakurai, "A "High-Efficiency TCM" Bandwidth Reduction for High-Definition TV", *Signal Processing of HDTV: Proceedings of the Second Workshop on Signal Processing of HDTV*, L. Chiariglione (ed.), pp. 129-136, Elsevier Science Publishers, 1988

[29] M. Tanimoto, A. Yamada and K. Shibata, "A New TAT Scheme for Higher Compression of HDTV", *Signal Processing of HDTV, II: Proceedings of the Third International Workshop on HDTV*, L. Chiariglione (ed.), pp. 235-241, Elsevier Science Publishers, 1990

This work was supported by NATO grant 5-2-05/CRG 900834.

8

Image Sequence Coding Using Motion Compensated Subband Decomposition

A. Nicoulin, M. Mattavelli, W. Li, A. Basso, A.C. Popat [†], M. Kunt

Laboratoire de traitement des signaux
Ecole Polytechnique Fédérale de Lausanne
1015-Lausanne, Switzerland

8.1 Introduction

Over the last two decades, the well known Descartes' phrase "I think therefore I am" became "I communicate therefore I am". The importance of communication is recognized at all levels, political, social, economical, even if sometimes one forgets to think. Audiovisual signals are the physical support of the information vital for communication.

Introduced in the fifties, analog TV system are definitely the most successful achievement of engineers for one way communication. There are more TV receivers in the world than telephone sets. Increasing demand for communication channels and the introduction of digital signal processing techniques are at origin of bandwidth reduction efforts which increase at a never seen before spread. The pace jumped even higher with the introduction of the HDTV, as notion for the public, new hope for the industry, new toy for politicians and pride for continents. The situation must be clear at all levels as well: one cannot satisfy the ever reincreasing demand for communication if the channel is not efficiently used. Superfluous information must be eliminated as automatically as possible to avoid expansive waste in channel capacity. Polluting channels with useless signals is as dangerous as polluting nature with chemicals.

As a results of increasingly large efforts, today a number of techniques are available for image sequence compression and coding. Technological constraints of the early seventies did not give much freedom to the imagination of designers. Simple techniques gave simple performances. Today we may be in a similar situation but at a different scale where old constraints do not exist anymore. That

[†] A.C. Popat is presently at the MIT Media Laboratory, Cambridge, MA 02139.

is how transform coding, viewed as an academical curiosity in the seventies, became today an industry standard with the discrete cosine transform (DCT), so that 10 to 1 compression is now taken for granted. A plethora of compression systems have been designed around DCT. They have the essential merit of existence with some performances. However they suffer from being closed systems with highly limited adaptivity. Furthermore they impose an external geometrical structure to the incoming data. They have no chance to success even for survival at the multimedia era. Systems that are designed today must be open systems using generic coders. Any dependence from resolution or technology is a definitive sign of failure.

The system described in this chapter has been designed with these observations in mind. Parallel to this effort newer versions and newer systems are under investigation for the next generation. The system relies heavily on by now well known principles, such as subband decomposition, motion compensation and entropy coding. It become specific by the special way these principles are implemented.

A general overview of the video sequence compression scheme is given in Section 8.2. A brief discussion about each functional block is also given. The description of design techniques, the optimization of design parameters, the discussion of optimality requirements of separate processing or of the cascade of several processing stages is developed in details in the following sections. Section 8.10 reports some simulation results about the performances of the system for the compression of standard digital TV test sequences. Section 8.11 concludes with a discussion of the main characteristics, strong and weak points of the scheme, and indicates some direction of research that could lead to more advanced and better performing new schemes.

8.2 Overview of the Coding System

This section gives a global overview of a compression system to encode color television digital sequences at a compression ranging from around 16:1 to 80:1. This corresponds to a bit rate in the range of 2–10 Mbit/sec for normal resolution CCIR 601 input format.

The major characteristics of the system are: (1) a motion compensated spatiotemporal subband decomposition, (2) a quasi-optimal uniform quantization strategy of the subband samples that avoids explicit bit allocation, (3) the use of a nonstationary model of the subband sample source that generates the probabilities for an arithmetic coder, exploits nonstationarities and allows a coding rate below the long term average entropy, (4) a feedback free rate control based on a statistical estimation of the quantization step size.

Although the system is designed for progressive sequences, in the case of interlaced sequences, a preprocessing stage is used to convert them into progressive sequences prior to encoding. The block diagrams of the encoder and the decoder are given in Fig. 8.1 and Fig. 8.2. The processing of the luminance Y is equal to the processing of the chrominance components U and V. The only difference is that motion estimation is performed only on Y and the resulting motion vectors are used to compensate the motion also the U and V components.

The digital video sequence is partitioned in blocks of M consecutive frames, called *Group Of Pictures* (GOP) . Each GOP is coded separately. The energy compaction and decorrelation along the temporal axis is obtained by means of a motion compensation scheme, followed by a temporal M-point DCT. The correlated motion vectors corresponding to blocks of pixels within each frame are transmitted prior adaptive arithmetic coding. A detailed description of the employed motion compensation scheme is given in section 8.4.

Spatially, a separable subband decomposition is applied to each frame of a GOP after the M-point DCT. Requirements and design of filter banks for image compression are discussed in section 8.3. The goal of subband analysis is to transform the source into an alternative representation, in such a way that most of the source energy is concentrated in a small fraction of the samples, while preserving the total energy. The transformed samples are called subband samples. The spectral components close to the DC spatiotemporal frequency after subband decomposition are removed using a second level of subband analysis, and constitutes the *DC component.* Due to the high level of the remaining correlation after subband analysis in the DC component, it is DPCM coded.

The unequal distribution of energy among subband samples can be exploited for the purpose of compression by allocating a larger fraction of the available bits to encode the high-energy samples and a smaller fraction to encode the low-energy samples [1]. This bit allocation procedure leads to an extremely efficient use of the available channel capacity.

Under entropy constraint, uniform scalar quantization is asymptotically optimum. An important characteristic of the described system is that the same step size can be used in quantizing all samples in a GOP. It will be shown that this strategy automatically results in optimal rate allocation with respect to a mean-squared error (MSE) criterion. If instead, a weighted MSE criterion related to perceptive human visual properties is adopted, then it is shown that it results in nearly optimal rate allocation. Section 8.5 discusses in details the quantization operation, and Section 8.6 discusses the rate allocation problem.

The resulting non-DC quantized subband samples, denoted x_n, exhibit a highly nonstationary behavior. Therefore, a dedicated statistical model of the source x_n for entropy coding has been designed. A multialphabet arithmetic coder is employed to code the samples. The model to generate the probabilities supplied to the arithmetic coder and decoder is time varying and is based on an estimation of the local variance σ_n^2. The subband samples x_n are assumed to follow a zero-mean Laplacian statistics with variance σ_n^2 that varies slowly spatially and spectrally. An estimate of this variances σ_n^2 is performed locally. These local variances estimates are compressed by vector quantization (VQ) and transmitted as side information. VQ allows an efficient compression of the data without excessive distortion when, like in the case of subbands, statistical dependencies across subbands are present. Section 8.7 describes the statistical model, and Section 8.8 deals with the encoding of side information.

In the case of fixed-rate transmission, variable-rate coding requires buffers at the transmitter and receiver. Usually, a feedback mechanism on the quantization stage is employed to control buffer occupancy. Such mechanisms work by periodically adjusting the short-term average bit rate, on the basis of measured short-term average rate and instantaneous buffer occupancy [2, 3, 4]. Section 8.6 shows how to determine quantizer step sizes for each GOP by a statistical operation, and to control the bit rate without buffers and feedback on quantization.

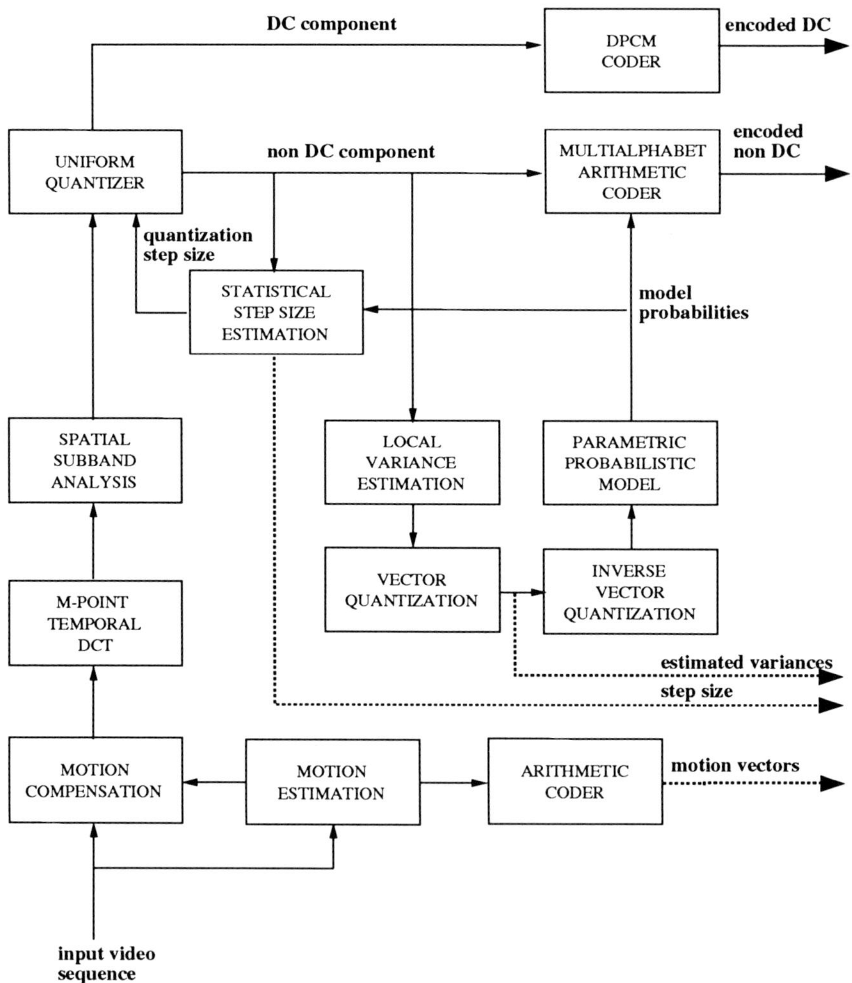

Figure 8.1: Encoder block diagram.

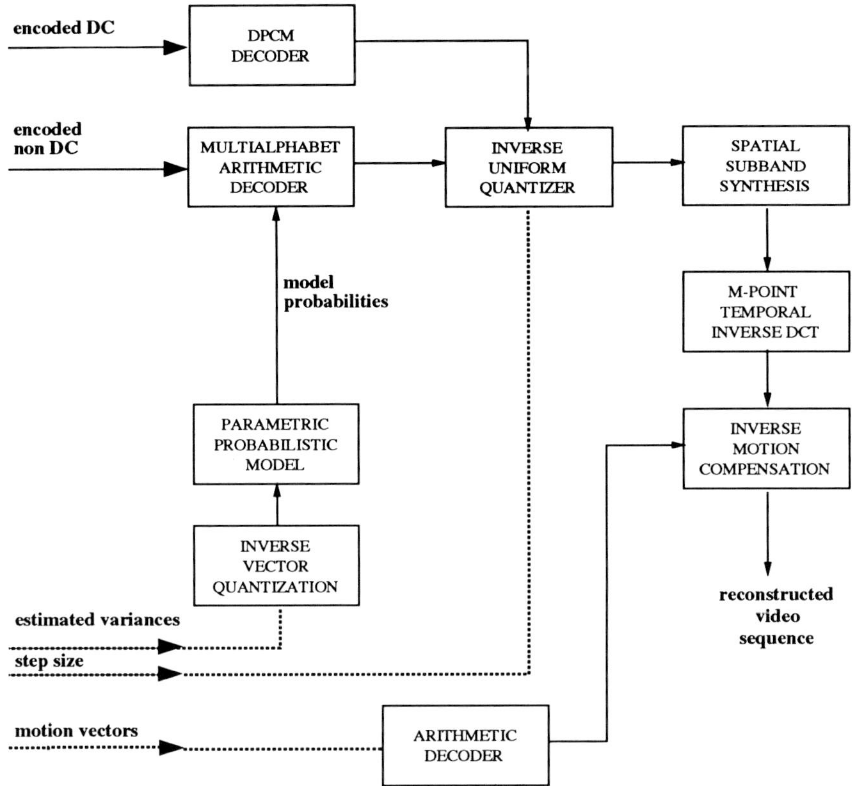

Figure 8.2: Decoder block diagram.

8.3 Filter Bank Design for Image Compression

Theory of multirate filter banks has considerably progressed in recent years. It has found many applications in various fields of communications such as speech processing, antenna systems, spectrum analysis, and image compression [5, 6, 7, 8]. Each application requires subband filters with dedicated characteristics, which are reached by different filter bank design techniques.

Subband decomposition filters for natural images in a compression scheme should be designed in accordance with the following requirements: quasi-perfect reconstruction, good simultaneous spatial and spatiofrequency localization, and confinement of the DC component in the lowest subband. For simplicity of the VLSI implementation, filter banks with small number of taps and finite wordlength precision are preferred. In this section, these properties are briefly discussed, and a technique to design such filter banks is presented.

Consider a one dimensional general filter bank analysis/synthesis system. Let $x(n)$ be the input signal of the filter bank, $y(n)$ the output of the filter bank, $h_k(n)$ the impulse response of the kth analysis filter, $g_k(n)$ the impulse response of the kth synthesis filter, N_k the length of the kth subband filter, M_k the subsampling/upsampling factor for the kth subband, and ψ_k the subsampling offset for the kth subband. Note that each subband in the system can have different subsampling/upsampling factors thus having different bandwidths and each analysis/synthesis filter can have different lengths.

Because of the subsampling/upsampling operations, the overall system is, in general, a periodically time-varying system, where the period is the least-common-multiple of the all subsampling/upsampling factors M_k. Let P denotes this period. There exist P different impulse responses, which can be shown to be:

$$t_p(n) = y(n+p)\mid_{x(n)=\delta(n-p)} = \sum_{k=0}^{K-1} \sum_{i \in \Gamma_p} g_k(i) h_k(n-i) , \tag{8.1}$$

where

$$\begin{aligned} \Gamma_p = \quad & \{ i : 0 \leq i < N_k \text{ and} \\ & (n+p-i) \in \{\psi_k, \psi_k \pm M_k, \psi_k \pm 2M_k, \cdots\}\} . \end{aligned} \tag{8.2}$$

From the above equation, the necessary and sufficient conditions for perfect reconstruction can be easily derived [9, 10]:

$$t_p(n) = \delta(n-D), \text{ for } p = 0, 1, \cdots, P-1 , \tag{8.3}$$

where D is the delay. The reconstruction error is defined as the square error between the P impulse responses $t_p(n)$ and the ideal responses $\delta(n-D)$:

$$E_R = \frac{1}{P} \sum_{p=0}^{P-1} \sum_{n=0}^{2N_k-2} [t_p(n) - \delta(n-D)]^2 . \tag{8.4}$$

Secondly, a numerical measure of localization is defined. Due to the non-flatness of the local spectra in natural images, their energy can be compacted by decomposing the source into spectrally disjoint subbands. Typical images are characterized by a spatial localization of the energy. By means of subband analysis, energy compaction can be achieved without losing localization property. However, because the subband analysis filters have certain extend, subband spectral analysis necessarily reduces this spatiotemporal energy compaction. Thus, it is advantageous to have analysis filters which are simultaneously compact spatially and spectrally. Such filters are known as *jointly localized filters.*

The joint localization for a Finite Impulse Response (FIR) filter whose impulse response is $h(n), n = 0, 1, ..., N-1$, is defined as:

$$E_{s,\omega} = \alpha_s E_s + \alpha_\omega E_\omega , \tag{8.5}$$

where

$$E_s = \frac{1}{E} \sum_{n=0}^{N_k-1} (n - n_c)^2 h^2(n) , \tag{8.6}$$

and

$$E_\omega = \frac{1}{E}\int_{\omega=0}^{\pi}(\omega-\omega_c)^2|H(\omega)|^2 d\omega \tag{8.7}$$

are respectively spatial localization and spectral localization measures. The weighting factors α_s, α_ω are both positive. Furthermore, the normalization factor is denoted by E, and the spatial central point of the filter and the central frequency of the passband are denoted by n_c and ω_c. Note that the localization measure is influenced both by the filter length N_k and the impulse response $h(k)$.

For simplicity, it is preferable to implement a separable spatial subband analysis. Furthermore, the subband analysis should be invertible and orthogonal (energy-preserving) to provide joint localization and preserve overall sampling rate.

The joint localization of the entire filter bank is defined as the sum of the joint localization measure for each analysis filter. The joint localization of the synthesis filter bank is entirely determined by that of the analysis bank, as the space-inverse constraint is imposed. This measure is given by:

$$E_L = \sum_{k=0}^{K-1}\left[\alpha_{s,k}E_{s,k} + \alpha_{\omega,k}E_{\omega,k}\right], \tag{8.8}$$

where $E_{s,k}$ and $E_{\omega,k}$ are respectively the spatial and spectral localization measures for the kth subband, and $\alpha_{s,k}$ and $\alpha_{\omega,k}$ are positive factors. Different trade-offs can be achieved by tuning these weighting factors.

The spatial inverse constraint is imposed for both theoretical and practical reasons. It was shown [11] that perfect reconstruction and orthogonality of a filter bank leads to this constraint. In practice, the designed filter banks will have poor magnitude-frequency response if this constraint is not imposed. The constraint can be expressed mathematically as follows:

$$g_k(n) = M_k \cdot h_k(N_k - 1 - n), \tag{8.9}$$

for $k = 0, ..., K-1$, and $n = 0, \cdots, N_k - 1$. The multiplication by subsampling factor M_k guarantees that each analysis filter has unit gain in its pass-band.

In addition, since the DC component contains most of the energy, the DC component must be strictly confined to the lowest subband to avoid aliasing of the DC component when quasi-perfect reconstruction is required. The condition of perfect reconstruction is not mandatory, since subband samples are anyway quantized to achieve compression.

The confinement of DC component in the lowest subband is stated as:

$$E_{DC} = \sum_{k=1}^{K-1}\left(\sum_{n=0}^{N_{h,k}-1} h_k(n)\right)^2, \tag{8.10}$$

where $h_k(n)$ is the impulse response of the kth analysis subband.

Finally a global compound error criterion is defined as:

$$E_{TOTAL} = \alpha_R E_R + \sum_{k=0}^{K-1} [\alpha_{s,k} E_{s,k} + \alpha_{\omega,k} E_{\omega,k}] + \alpha_{DC} E_{DC} \,, \tag{8.11}$$

where α_R and α_{DC} are positive factors. After having assigned the various weighting factors, a gradient based algorithm can be used to find the optimum coefficients with respect to this criterion. More details about the design procedure can be found in [9, 10]. Simulated annealing algorithms can be applied to design the filter banks with filter coefficients in finite wordlength or with powers-of-two coefficients [12, 13] without loosing optimality.

As an example, a 6-band parallel-structured filter bank, with each analysis filter having 12 taps has been designed. Figure 8.3 shows frequency responses of all analysis filters. The coefficients values of the analysis filter bank is given in Appendix B. The synthesis filter bank can be calculated using Eq.(8.9).

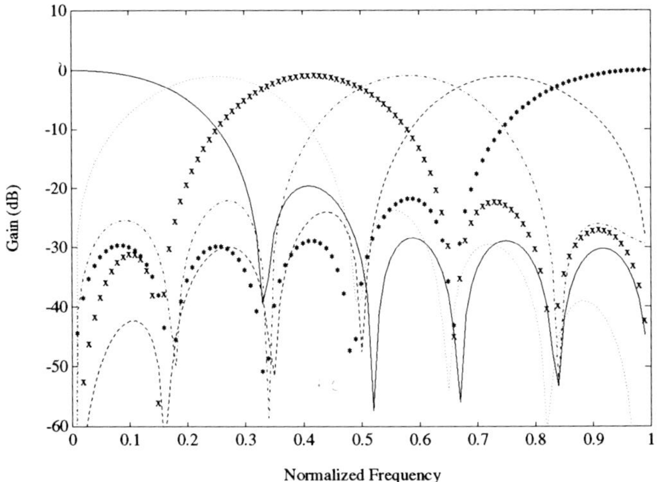

Figure 8.3: Frequency response of a 6-band analysis filter bank. Each filter has 12 taps.

8.4 Decorrelation by Motion Compensation and Spatiotemporal Subband Decomposition

Temporal decorrelation of natural image sequences is considerably different from spatial decorrelation. The main consideration about natural sequences is that they can be modeled as an ensemble of still and moving areas. An energy compaction technique similar to the spatial case is efficient for still areas but not much for moving ones. Many techniques for the estimation and compensation of motion have been developed and studied over the past years [14, 15,

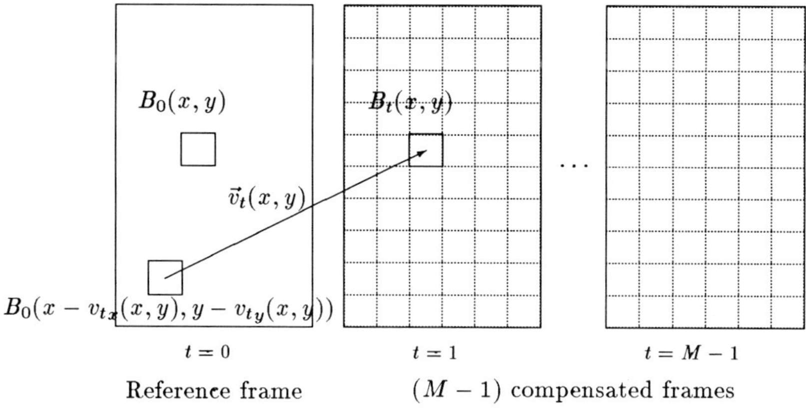

Figure 8.4: Motion compensation scheme in a GOP of M frames.

16]. Their application yields considerable improvements in the performances of video codecs. Usually, motion compensation is associated with temporal DPCM because it can be easily incorporated in the algorithms known as "hybrid schemes" [17].

This section describes in details the spatiotemporal decorrelation scheme used in the system. It is based on motion compensation, temporal subband decomposition and spatial subband decomposition. It differs from some classical schemes that are based on intraframe, predicted and interpolated frames.

Let consider a GOP of M frames (see Fig. 8.4). The first frame of each GOP, denoted $(t = 0)$, is called the reference frame. The $(M - 1)$ other frames are referred as compensated frames and are partitioned in blocks. A motion vector relative to the reference frame, is associated to each block in all the compensated frames. The technique used for the estimation of the motion vectors is a full-search block matching algorithm. Computationally more efficient algorithms can be found in literature [16, 18] and are not discussed in this chapter. They reach almost the same performances for the minimization of the block matching errors.

Motion compensation using block-based algorithms has some limitations due to the assumptions of rigid and translational motion and constant light condition along the motion trajectory. Therefore we associate to each pixel in the compensated frames the error given by the block matching. Before the temporal subband decomposition, all blocks $B_t(x, y)$ of the compensated frames are replaced by the blocks $\hat{B}_t(x, y)$ obtained from the algebraic summation of the reference $B_0(x, y)$ from the reference frame at the same spatial location, and the relative error block $e_t(x, y)$ (see Fig. 8.4). Noting $\vec{v}_t(x, y)$ the motion vectors

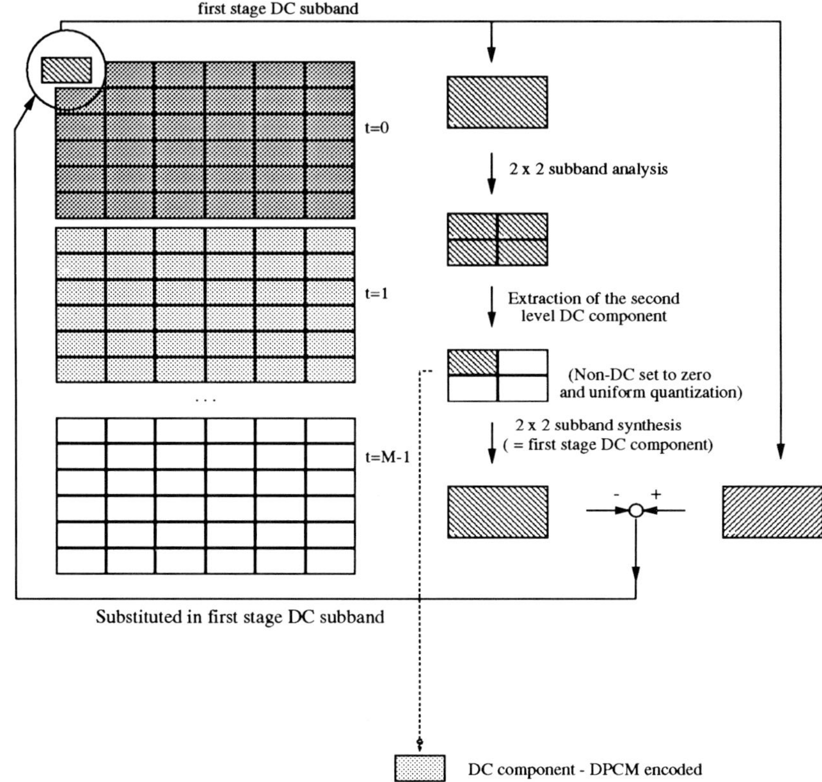

Figure 8.5: Spatiotemporal subband decomposition for a GOP of M frames with second level DC extraction.

associated to pels in block $B_t(x, y)$, we have:

$$\hat{B}_t(x,y) = B_0(x,y) + e_t(x,y) , \quad (8.12)$$

$$e_t(x,y) = B_t(x,y) - B_0(x - v_{tx}(x,y), y - v_{ty}(x,y)) . \quad (8.13)$$

After the temporal subband decomposition all the energy of still or well compensated areas will be concentrated in the low temporal frequency subband ($t = 0$). Energy in the high temporal frequency subbands, ($t = 1, \cdots, M-1$), will appear only in badly compensated areas.

This transformation originates a source of subband pixels characterized by a high degree of local nonstationarity. The description of an efficient modeling and coding of this particular data source will be the main focus of Section 8.7.

The determination of the number of frames M to compose each GOP will result from a compromise between two conflicting requirements: efficiency of the algorithm for the compression and simplicity of the hardware implementation. The higher the number of frames, the better the energy compaction. This holds particularly for still or slowly moving areas, but it vanishes in presence of sudden scene changes or uncovered backgrounds. Moreover, much more complex motion estimation algorithms, that enlarge the compensation range, are necessary to reach efficiency in presence of a large number of frames. A high number of frames requires a complex codec in terms of frame memories both on coder and decoder sides. In addition, the coding-decoding delay can become unacceptable for certain applications such as videoconference.

The straightforward application of spatial subband decomposition to the temporal axis is not reasonable because of the partitioning into GOPs of the input sequence. Filtering applied over the edges of the block will result in large amount of high frequency energy unless cumbersome mirroring modifications of the algorithms are used. This disadvantages are very clear if we consider the temporal filtering of a GOP of M frames by a L taps filter. M must be kept around 10 or below to reduce complexity and delay, while the number of taps of well designed filters usually can range from a very few up to 18-20. It may even happen to have a number of samples less than that of the filter tap. In conclusion, we have to renounce to all the desirable properties of well known subband filter banks, and we have to limit our attention to non-overlapping L-taps filter banks such as L taps DCT with the number of taps equal to the number of frames M composing a GOP.

After temporal subband decomposition, a two-dimensional separable parallel spatial subband decomposition is applied to the GOP. The filter bank is provided in Appendix B.

For natural images, the spectral region close to DC-spatiotemporal-frequency exhibits statistical properties significantly different from the other spectral regions, and is treated separately. For this reason, the DC component is removed from the lowest spatiotemporal-frequency subband using a pyramidal approach. In particular, the reconstructed DC component is extracted by lowpass filtering, subsampling, quantizing, interpolating, and finally subtracting from the original lowest subband (see Fig. 8.5). The quantized DC is transmitted separately since it has a very small number of samples.

Fig. 8.6 shows a GOP after spatiotemporal subbanddecomposition and DC extraction of the video sequence Table Tennis (see Fig. 8.14 (a)). The compaction of energy is clearly visible. Most of the energy is concentrated in the lowest temporal and spatial subbands. Energy in subbands of the compensated frame, denoted ($t = 1$), corresponds to moving objects. The frequency decomposition of the textured background into horizontal, vertical and diagonal components is clearly visible in the reference frame ($t = 0$).

In Fig. 8.7 the energy distribution among spatiotemporal subband samples x_n is displayed. In Fig. 8.7 (a), a scheme without motion compensation is applied, while in Fig. 8.7 (b), the described motion compensation method is used. It can be noticed that most of the energy of subbands of compensated frames ($t = 1$) and ($t = 2$) is compacted in the reference frame ($t = 0$) when motion compensation is applied.

Subband samples are rescaled by a factor 7, and a grey level of 128 is added for display purpose.

Figure 8.6: Spatiotemporal subband decomposition of a GOP of the sequence *Table Tennis*. Only the 2 first frames from a GOP of $M = 3$ frames are shown.

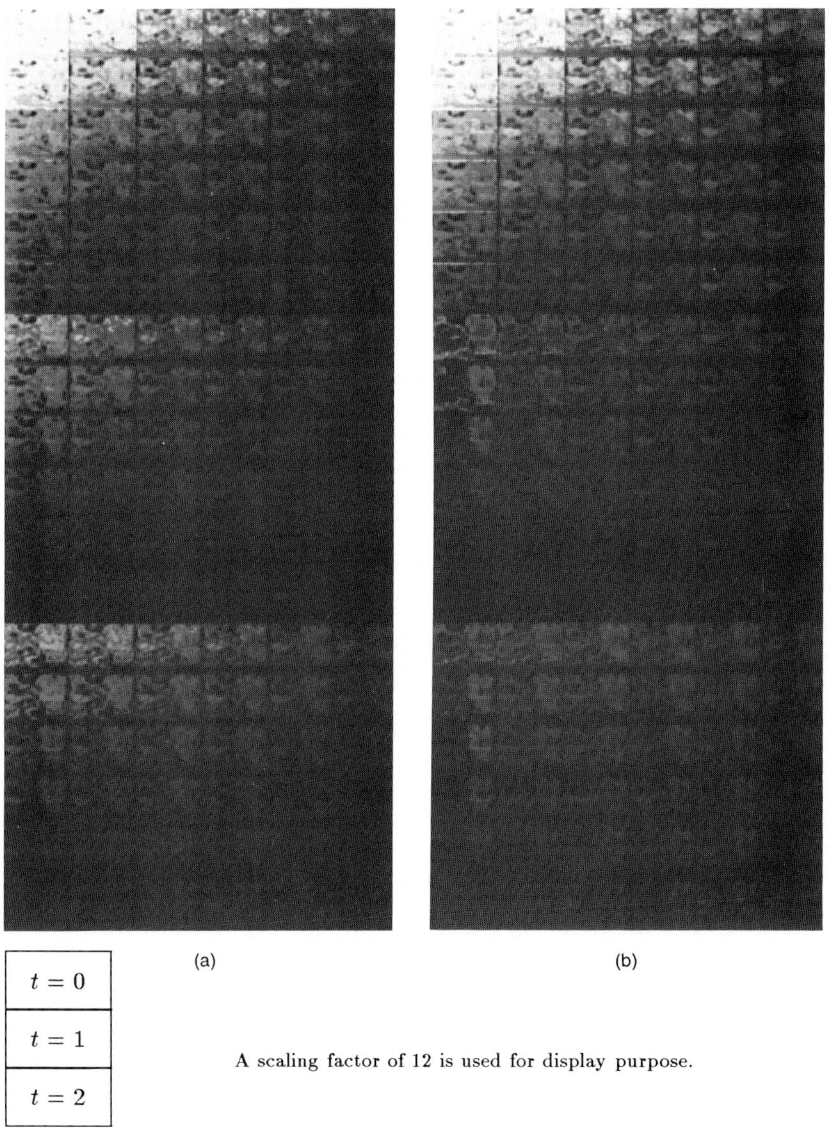

Figure 8.7: Energy distribution among spatiotemporal subband samples in a GOP of $M = 3$ frames of sequence *Mobile Calendar*: (a) without motion compensation, (b) with motion compensation.

8.5 Scalar Quantization with Multialphabet Arithmetic Coding

After the subband decomposition, quantization is necessary to reduce the entropy of the subband samples x_n in order to achieve compression of the original data. This operation approximates subband sample values by a finite set of allowed values. Quantization introduces distortion in the data. These distortions are much more important than thoses introduced by filter banks that perform quasi-perfect reconstruction.

In order to exploit the remaining statistical dependencies among samples, block quantization should be considered because it asymptotically reaches the rate distortion limit. This approach has been investigated using vector quantization [19, 20], but the results are not satisfactory [21], and the complexity of the scheme is too high for real time video applications. On the other hand, scalar quantization with entropy coding, though theoretically inferior than block quantization, is efficient in the case of memoryless sources. Due to the low level of correlation and statistical relation among the samples x_n after subband decomposition, they can be modeled as generated by a memoryless nonstationary source. It is well known that uniform scalar quantization of a memoryless source is asymptotically optimum under entropy constraint [22]. Section 8.6 considers the problem of rate allocation among the different subbands, and establishes an equation to calculate the optimum quantization step size Δ_n for scalar quantization of each subband.

After scalar quantization, entropy coding of the quantized samples must be performed [23, 24, 25]. The challenge in entropy coding comes about because the quantizer must have a large number of levels to avoid overload, while the entropy of the output can be relatively small. Usually, to reach the limit given by the entropy of the quantized samples with regard to a given model of the source, an encoding scheme is employed wherein successive quantizer outputs are concatenated to form *runs*, which are then Huffman coded. This technique reaches the entropy of the source model in some particular cases when each probability of the extended alphabet symbol is a negative integer power of 2, which is in fact a condition difficult to satisfy.

In contrast, arithmetic coding can work arbitrarily close to the entropy of the modeled source with very small constraint on its probability density function (pdf). Moreover, its efficiency is independent from the order in which samples are encoded, because source letters need not to be concatenated for the encoding to be efficient [26, 27]. A suitable arithmetic coding procedure is summarized in Appendix A. The problem of defining a model which considers the structural characteristics of the discrete source to be coded in order to reduce the entropy is presented in section 8.7.

Figure 8.8 illustrates the performances of arithmetic-coding, after uniform scalar quantization, applied to a memoryless Laplacian source. Each point is obtained by averaging performance over 50,000 samples of a simulated Laplacian source. For reference, the rate-distortion function computed via the Arimoto-

Blahut algorithm [28], the performances of uniform scalar quantization with ideal entropy coding (obtained numerically), and that of the Lloyd-Max quantization are also shown. Note the near-MSE-optimality of arithmetic-coded, uniform scalar quantized samples.

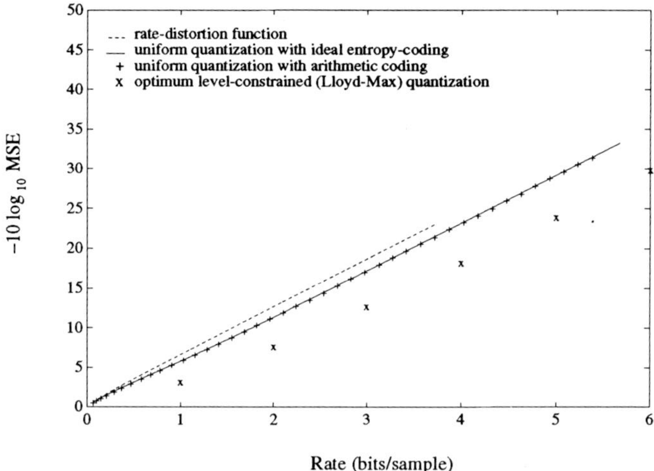

Figure 8.8: MSE-performance of arithmetic-coded uniform scalar quantization of a Laplacian memoryless source, relative to other techniques.

8.6 Automatic Rate Allocation

Ideally, rate allocation should be based on a psycho-visual subjective distortion measure. Because of the lack of satisfying human visual perception models, MSE or weighted MSE is typically used. In systems that employ Lloyd-Max (i.e. non-entropy-coded) quantizers, rate allocation must be carried out explicitly [29]. In contrast, we show here that rate is allocated implicitly in a system that uses quantization with entropy coding, by adjustment of the quantization resolution.

This section shows that nearly optimal (in a weighted MSE sense) rate allocation results when the step sizes are inversely proportional to the square roots of the corresponding distortion weighting factors. In the case of unweighted MSE, the optimal strategy is to use the *same* step size for all subband samples. The result is derived using familiar approximations to quantizer rate-distortion performance.

Explicit rate allocation

The rate allocation problem can be stated in the following way. Given a composite source having N components, each with variance ${\sigma_n}^2$ and perceptual-based weight factor w_n, and a given set of quantizers with known rate-distortion performance functions $D_n(r_n, \sigma_n)$, choose rates r_n that minimize :

$$D_{\text{tot}} = \sum_{n=1}^{N} w_n D_n(r_n, \sigma_n) \tag{8.14}$$

subject to:

$$r_n \geq 0 \qquad \text{for } n = 1, ..., N, \tag{8.15}$$

and to:

$$\sum_{n=1}^{N} r_n \leq R\,, \tag{8.16}$$

Where R is the total bit rate available. Clearly, an optimal rate allocation would satisfy (8.16) with equality. In the MSE case, the rate-distortion performance of a quantizer can be approximated over a range of r_n as:

$$D_n(r_n, \sigma_n) \approx \epsilon^2 2^{-2r_n} \sigma_n^2\,, \tag{8.17}$$

where the constant ϵ depends on the source pdf and the type of quantizer employed [22]. Substituting this approximation into Eq.(8.14), it can be shown that D_{tot} is a convex and separable function. Moreover, the constraints on r_n define a convex feasible set. Consequently, any locally optimal point in the feasible set is also globally optimal [30]. The method of Lagrange multipliers can be used to show that optimal values for r_n are given by the well-known formula:

$$r_n = \frac{R}{N} + \frac{1}{2} \log_2 \frac{w_n \sigma_n^2}{[\prod_{i=1}^{N} w_i \sigma_i^2]^{1/N}}\,. \tag{8.18}$$

If all the rates obtained using Eq.(8.18) happen to be nonnegative, then the explicit rate allocation problem is solved. Otherwise, some sort of iterative procedure is usually employed to obtain suitable values for r_n. In particular, the convexity of the feasible set and the separability and convexity of the function to be minimized allow a *greedy* algorithm to be used to obtain a globally optimal rate allocation [31, 32]. Such procedures tend to be computationally very demanding.

Near-MSE-optimality of implicit rate allocation

When uniform entropy-coded scalar quantization is used, rate and distortion are both determined by the quantization step size. The rate allocation problem thus becomes a step size determination problem.

Using approximation (8.17), the equation (8.18) can be rewritten as:

$$r_n = -\frac{1}{2} \log_2 \frac{D_n}{\epsilon^2 \sigma_n^2}\,. \tag{8.19}$$

It can be noticed from Fig. 8.8 that arithmetic-coded quantization is well approximated by Eq.(8.19) at all rates. In the high-rate region, the MSE for a uniform quantizer having step size Δ_n can be approximated as [22]:

$$D_n = \frac{\Delta_n^2}{12} . \tag{8.20}$$

Assuming that Eq.(8.20) is valid at all rates, (8.19) and (8.20) can be used to rewrite the overall rate constraint (8.16) in terms of the step size as:

$$\prod_{n=1}^{N} \Delta_n \geq C \prod_{n=1}^{N} \sigma_n , \tag{8.21}$$

where C is a constant.

Accordingly, the allocation problem becomes the minimization problem of

$$D_{\text{tot}} = \sum_{n=1}^{N} w_n \Delta_n^2 , \tag{8.22}$$

subject to (8.21). Note that Eq.(8.21) defines a convex region of vector space and (8.22) is a convex function, so that local optimality still implies global optimality. The solution, obtained by assuming equality in Eq.(8.21) and applying the method of Lagrange multipliers, is [33]:

$$\Delta_n = \frac{\Delta_c}{\sqrt{w_n}} , \tag{8.23}$$

where Δ_c is a constant. Eq.(8.23) specifies that all step sizes should be equal. Note that by formulating the problem in this way, the nonnegativity constraint (8.15) has become unnecessary.

Usually, the weighting factors w_n are determined empirically from subjective evaluation. An example of a weighting factor matrix used in the described system is given in appendix C. Different other techniques have also been proposed in the literature to design such matrices [33]. They are based on mathematical models of human perceptual relevance of subband coefficients.

8.7 Statistical Modeling of Quantized Subband Samples for Entropy Coding

The compression performance of an entropy coding scheme depends primarily on how good the statistical model is. The difficulty of modeling is to extract and exploit the relevant structural characteristics of the source. Unfortunately, it has been shown that the optimum modeling problem is an undecidable problem [34, 35]; it cannot be formulated in term of minimization of a functional.

Samples x_n generated by the spatiotemporal motion compensated subband decomposition exhibit a great deal of nonstationarity. For a nonstationary source, the pdf changes all the time. Therefore, the global pdf estimate neglects the local variation of the source, and thus will not lead to the lowest possible entropy. In the described system, we consider the samples x_n as being generated by a *composite source* which is made up of a number of subsources [36]. Since the composite source model is designed to take advantage of the nonstationary behavior of the source, the construction of the composite source model consists in the decomposition of the pdf of the source alphabet into different pdfs according to the local statistics.

In this system, we consider the subband samples x_n as a memoryless source with a pdf that changes slowly spatially (inside each subband) and spectrally (among the subbands). Experiments suggest modeling x_n as a zero-mean Laplacian random variable (see Fig. 8.9), with variances σ_n^2 :

$$p_{\sigma_n^2}(x_n) = \frac{1}{\sqrt{2}\,\sigma_{x_n}} \exp[\frac{-\sqrt{2}\,|x_n|}{\sigma_{x_n}}] . \qquad (8.24)$$

The local variance σ_n^2 of the pdf is estimated, coded, and transmitted as side information. In this way, the model pdf constitutes the conditional pdf of the source when the local variance σ_n^2 is known. Assuming that the local variance σ_n^2 varies slowly, it is estimated by squaring, lowpass filtering, and subsampling the subband samples x_n (see Fig. 8.11). Then the reduced amount of side information σ_n^2 is clustered by means of vector quantization [37]. Section 8.8 described in details this operation.

Fig. 8.9 reports histograms of quantized samples x_n conditioned on the local estimated variance σ_n^2. Statistical tests have shown a very good accordance with Laplacian functions with adequate variances σ_n^2.

According to the definition of adaptivity in [35], this scheme is not adaptive, because all the parameters of the model are not estimated on both encoder and decoder sides using the statistic of the previously transmitted samples, but are estimated knowing the current sample that has to be coded, and transmitted as side information.

Fig. 8.10 shows the local variance estimates σ_n^2 (a) and the real bit cost of each entropy coded sample x_n using the described model (b). We can notice that high σ_n^2 value areas (white pixels in the figure) result in a model of high entropy and high bit cost. Low σ_n^2 values areas (black pixels) correspond to very low bit cost. In a few cases the model fails, and small areas show a high bit cost when σ_n^2 values are small. Using this model, the available bits are automatically allocated in accordance with Equ. (8.19), which has been shown to be the optimal bit allcation strategy.

8.8 Encoding of Side Information

The estimated variances σ_n^2, used by the model to select a probability table for the arithmetic coder and decoder, need to be transmitted to the decoder. A direct encoding of this parameter would be very expensive in term of bit cost. In this section, we describe a method to compress drastically this information by taking advantage of its structural content and statistical dependencies among

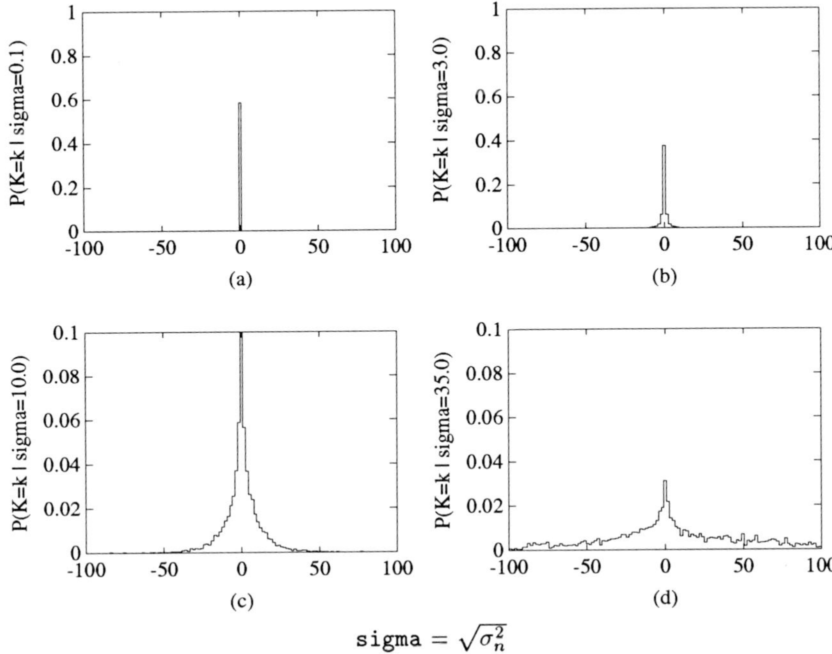

Figure 8.9: Histograms of quantized samples x_n conditioned on four different locally estimated standard deviation σ_n. Coordinates: x axis: quantized sample x_n values, y axis: relative occurancy of x_n.

local variance estimates. A block diagram of the local variance estimation and compression scheme is provided in Fig. 8.11.

Images can be modeled as slowly varying or textured regions, separated by edges. After the subband decomposition stage, edges are represented by samples with an energy proportional to the inverse square of the spatial frequency f^{-2}. Slowly varying areas are splitted in subbands where only the lowest frequency one contains energy. Finally, textured areas are also characterized by their local frequency content. In all these cases, there is a great deal of correlation in the local variance measure across the subbands. Therefore, VQ seems to be a natural way to compress this information, because VQ is able to take advantage of statistical dependencies among vector components[38, 19, 20]. A vector is built with each local variance σ_n^2 of a GOP coming from the same spatial location *across* the subbands (see Fig. 8.12). A nonlinear companding function prior and after VQ allows to have relative accuracy on the distortion of compressed σ_n^2: higher (lower) error is introduced on high (low respectively) σ_n^2 values. The binary logarithmic function is used for this purpose.

In VQ, a k-dimensional vector is quantized as a single entity. The coding process of the vector consists in finding the best match in a codebook containing

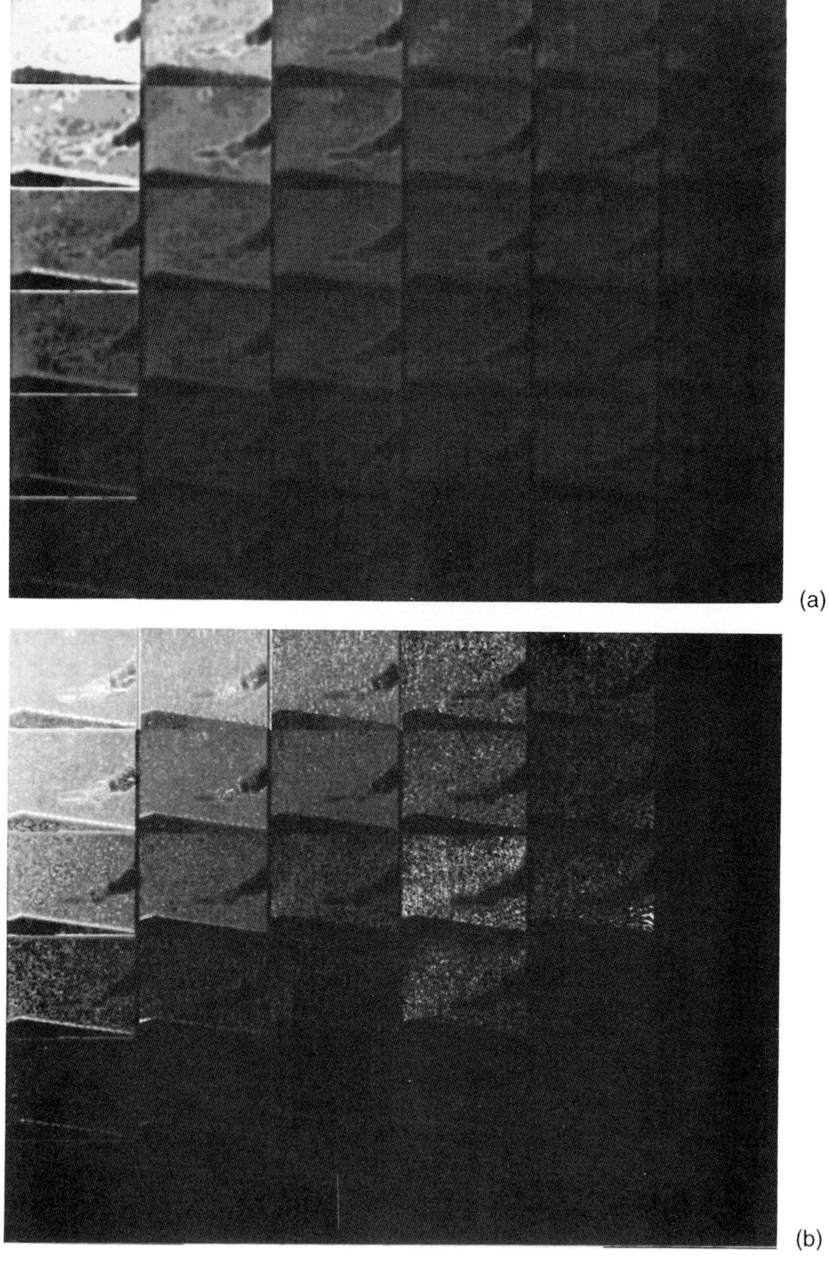

Figure 8.10: Modeling of a GOP of the sequence *Table Tennis*: (a) local variance estimates σ_n^2, (b) number of bits flushed by the arithmetic coder. Only the first frame of the GOP is shown.

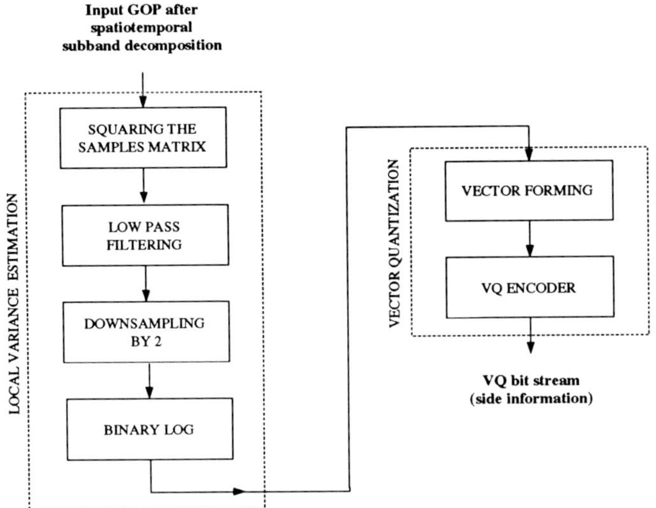

Figure 8.11: Local variance σ_n^2 estimation and compression algorithm.

2^m k-dimensional code vectors [39]. A word of m bits is sent to the receiver, as index for the code vector selected as a representative of the input vector. The receiver has a copy of the codebook used for transmission and the output vector is generated by table lookup [39, 40]. The VQ bit rate is given by :

$$R_{VQ} = \frac{m}{k} \quad \text{[bit/spl]}. \tag{8.25}$$

In the final setup of the system, described in Section 8.10, a 108-components vector (a 6 by 6 subband analysis for a GOP of 3 frames results in 108 subbands) has been employed. Simulations have shown that a relatively small codebook size of 256 is a good compromise between computational complexity and reconstruction quality. In this case, $R_{VQ} = 0.074$ bits/spl. Local errors on the received values of σ_n^2 due to VQ distortion will have a very small impact on the quality of a reconstructed sequence. In fact, local mismatch of the model will be spred over the entire GOP by leading to scalar quantization step sizes slightly higher for the GOP.

The codebook used in simulations has been generated using the LBG algorithm [41]. A full search strategy is used to select the reconstruction vector. A set made of 3 GOP coming from different standard sequences have been employed for the training of the VQ.

8.9 Statistical Feedback-Free Rate Control

Section 8.6 established equation (8.23) for the quantization step sizes Δ_n in terms of perceptually based weight factors and a constant Δ_c. The present section addresses the problem of determining the value Δ_c for transmission of

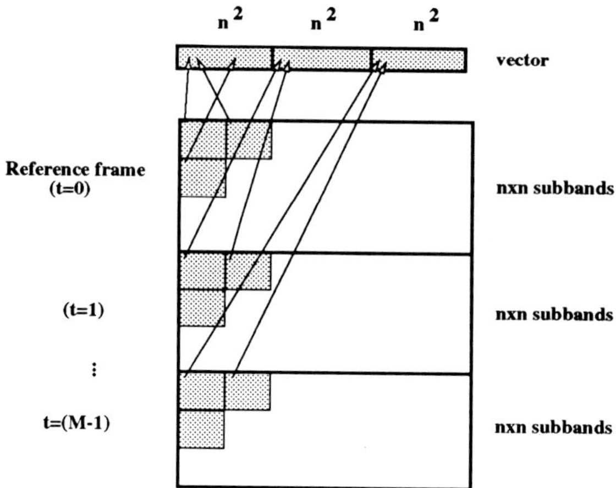

Figure 8.12: Vector quantization across the subband of a GOP: strategy to built the vectors.

the variable length code over a synchronous digital channel of a given capacity R.

Classical approaches [2, 3, 4] use a buffer combined with an adaptive feedback scheme in which the quantization parameter Δ_c is changed successively according to the occupancy state of the buffer. This regulation algorithm implies additional delays and has to prevent underflow or overflow of the buffer. All these schemes present the disadvantage of sudden variation of image quality, even in the same frame. This is particularly critical in presence of scene changes in the video sequence.

The system described here avoids the additional delay due to the buffer, and uses the same constant Δ_c for the entire GOP. The parameter Δ_c is determined based on a subset of subband samples randomly selected in each GOP. More precisely, a subset of random samples x_n, their corresponding model parameter σ_n^2, and their corresponding relative perceptually based weight factors w_n, are extracted from each GOP. Then a simple recursive converging method is used. It adapts Δ_c step by step as the random subset is scanned [42]. While subsets are scanned, the coding cost of x_n with a given Δ_c, using σ_n^2 and w_n, is calculated. This coding cost is compared with the channel capacity R, and Δ_c is updated by adding or subtracting δ_c. This factor δ_c is function of the number of samples x_n already scanned from the subset [42].

As the size of the random set increases, the accuracy of the estimation of Δ_c improves, according to the Tchebycheff inequality [43]. Example graphs of the evolution of the estimate of Δ_c are shown in Fig. 8.13 for different target bit rates and different sequences. Experimentally, convergence is obtained after approximatively only 1/10 of a GOP of 3 frames have been processed.

Due to the residual estimation error of Δ_c, the actual number of bits produced per GOP will not be exactly equal to R. A simple technique involves

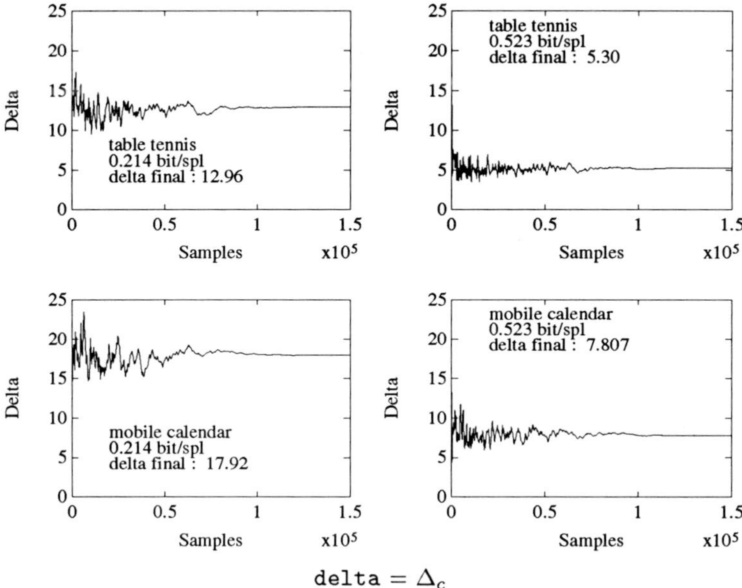

Figure 8.13: Evolution of step size estimate for two different luminance video sources and two different average bit rates, based on randomly scanned subband samples of a GOP.

either zero-padding the bit stream or truncating the source sequence, as needed, to obtain exactly R bits. In the case that zero-padding is necessary, a small (negligible) fraction of the available channel rate is wasted. In the case of truncation, the final samples in the sequence are lost, and are reconstructed by the receiver as zero. Note, however, that there is no loss of synchronization, since exactly R bits are always associated with a GOP. This approach may be enhanced by slightly overestimating Δ_c, to make zero-padding occur more frequently than the truncation. Another solution to the problem is to encode the entire GOP using several different values of Δ_c, choosing the one which gives the greatest number of bits not exceeding R. This option is also simple thanks to the computational simplicity of arithmetic coding (see appendix A).

8.10 Experimental Results

Simulation results of the system described in this chapter are presented in order to evaluate its performances. The configuration of the experimental setup was the following : GOP of $M = 3$ frames; full search block matching motion estimation algorithm based on a 16x16 block size and maximum search displacement of 16 pels; temporal 3-points DCT, uniform separable 6x6 subband decomposition using filter bank given in Appendix B. Luminance Y and the two chrominance U and V components are processed separately in the same manner. Motion vectors, which are estimated on the luminance component, are used on the chrominance as well.

Coding simulations on standard digital CCIR 601 test sequences[1] at a bit rate of 9 Mbit/sec, have been carried out. This corresponds to a compression ratio of 18.5 to 1. The sequences *Table Tennis*, *Flower Garden*, and *Mobile Calendar* have been selected due to their different characteristics (see Fig. 8.14). In particular, *Table Tennis* includes large displacement and zooming, while *Flower Garden* contains panning and high activity areas, and *Mobile Calendar* contains highly saturated and detailed moving areas.

Figure 8.14: First frames of three CCIR 601 test sequences: (a) *Table Tennis*, (b) *Flower Garden*, (c) *Mobile Calendar*.

Due to the rate regulation method of this coder, the bit rate is constant for each GOP. The repartition of the bits between Y, U, V, and the side information inside each GOP is fairly constant. In average, the allocation of the bit rate among different multiplexed component is given in Table 8.1.

Although the subjective evaluation of the quality of a compressed sequence is the only relevant test, numerical evaluation of the peak signal-to-noise ratio (PSNR) between the original image X and the reconstructed image $\hat{X}$ is

[1]Spatial resolution of 576 lines by 720 columns per frame, 25 frames/second, 4:2:2 subsampling pattern for chrominance components.

YUV component	Rate allocation	Bit rate [Kbit/sec]	percentage [%]
Y, U, V	motion vectors	96.32	1.13
Y	DC	192	2.13
Y	VQ	192	2.13
Y	non-DC x_n	7078.0	78.6
U	DC	48	0.53
U	VQ	48	0.53
U	non-DC x_n	624.53	6.94
V	DC	48	0.53
V	VQ	48	0.53
V	non-DC x_n	625.15	6.95

Table 8.1: Rate Allocation among the different rate components.

calculated using:

$$\mathrm{PSNR} = 10\log_{10}\frac{255^2}{E\{[X-\hat{X}]^2\}} . \tag{8.26}$$

In Fig. 8.15, the PSNR for Y, U and V components over 20 frames of the sequences *Flower Garden*, *Table Tennis*, and *Mobile Calendar* are reported. The behavior of the PSNR shows some regular periodicity inside each *GOP*. The reference frame results always 0.5–1 dB higher than the two compensated frames. The range of these oscillations remains very small when compared with the difference of intraframe and predicted or interpolated frames in the so called "hybrid schemes". This consideration is confirmed by the subjective good quality observed in the reconstructed sequences. In fact the coded sequences, are characterized by a complete absence of artifacts visible by non-expert observers and by a constant level of quality.

8.11 Conclusions

An example of a video signal compression scheme employing well known principles such as subband decomposition, motion compensation and entropy coding has been presented. The difficulty in achieving high performances in terms of channel occupancy, image quality and flexibility in a codec, is to link separate optimal processes in a complete coding scheme. It is important to notice how the use of these techniques, which themselves do not constitute a complete system, can lead to different implementation philosophies and solutions. Any system has weak parts that could be bottlenecks for the efficiency of the whole system. The identification of these bottlenecks is not simple and is a challenging task for further investigations.

In summary the main characteristics of the system are: (1) a motion compensated subband temporal scheme, followed by a uniform spatial subband stage, (2) a quasi-optimal uniform quantization strategy of the subband that avoids

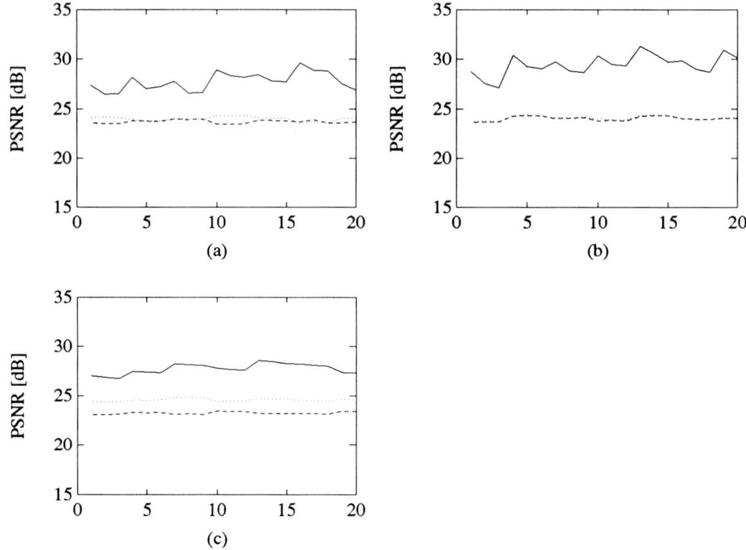

Figure 8.15: PSNR versus the frame numbersm for the Y (solid line), U (dotted), and V (dashed) components of CCIR-601 test sequences coded at 9 Mbits/sec: (a) *Flower Garden*, (b) *Mobile Calendar*, (c) *Table Tennis.*

explicit bit allocation, (3) a model for the arithmetic coder that, by the use of vector quantized side information, exploits nonstationarities and allows a coding rate below the long term average entropy, (4) a statistical estimation of the quantization step sizes that leads to a feedback-free rate control.

Although the simulations have shown good results, motion compensated temporal DCT seems to be a possible bottleneck for the system. A scheme based on intraframe, prediction and interpolation could probably result simpler to implement and give better performances. The time-varying model that builds the probability distribution for the arithmetic coder is also worth of further improvements. This could be done by increasing the accuracy of the statistical model by a different use of the vector quantized side information, such that the efficiency of the entropy coding would be increased. Another important consideration is that the system relies on progressive scanned inputs, which are certainly the future for video signals. As a consequence the performances of the coder for interlaced input video signal depend on the upconvertion algorithm employed.

Besides the relative progress that could be achieved by the mentioned improvements, a forward jump in the performances can be foreseen for the future. All the techniques used by now for video coding are based on the statistics of video signals. In the so called "second generation" coding schemes [44], instead the attention will be focused on global and semantic representation of the image content. This approach will certainly lead to more compact and efficient data compression systems that will introduce the still undiscovered era of the multimedia world. With all these efforts and the ones to come, it is hoped that one could say "I think therefore I communicate".

Acknowledgment

The authors wish to thank their colleagues at LTS-EPFL-Lausanne, P. Cicconi, F. Dufaux, T.Ebrahimi, I. Moccagatta, E. Reusens, and B. Rouchouze for their contributions, discussions and collaborations during research work in digital video field. We would also like to thank G. Auric for his continuous support for the computer environment.

Appendix A: A Suitable Multialphabet Arithmetic Code

Let $\mathbf{u}^N = [u_1, \cdots, u_N]$ denote a sequence of N letters to be encoded, where each letter u_i is a random selection from the alphabet $A = \{a_1, \cdots, a_K\}$. The sequence $\mathbf{u}^N$ can be regarded as a random selection from the *superalphabet* $A^N = \{\mathbf{a}_1^N, \cdots, \mathbf{a}_{K^N}^N\}$, where each $\mathbf{a}_l^N$ represents one of the K^N possible sequences of N letters. Let $U_i \in A$ be an experimental outcome of the letter u_i, and let $\mathbf{U}^N = [U_1, \cdots, U_N]$ be an experimental outcome of the random sequence $\mathbf{u}^N$. Here, the experimental outcomes $\mathbf{U}^i$ and $\mathbf{U}^j$, $0 < i, j \leq N$, always refer to the *same* experiment, so that, for example, $\mathbf{U}^i$ is simply obtained by appending U_i to $\mathbf{U}^{i-1}$.

Define $P_i(U_i)$ to be the nonzero probability that, in a particular experiment, the ith letter in $\mathbf{u}^N$ assumes the value U_i.

Conceptually, idea behind arithmetic coding is to associate with each sequence $\mathbf{a}_l^N$ an interval $\mathcal{J}(\mathbf{a}_l^N)$ of width $P_{\mathbf{u}^N}(\mathbf{a}_l^N)$, such that $\mathcal{J}(\mathbf{a}_l^N) \cap \mathcal{J}(\mathbf{a}_m^N) = \emptyset$ for all $m \neq l$, and $\bigcup_{l=1}^{K^N} \mathcal{J}(\mathbf{a}_l^N) = [0, 1)$.

To encode a particular sequence, a suitable number is selected from the corresponding interval, and its binary representation is used as the code bit sequence. Shannon demonstrated that such a technique can be very efficient in term of bit usage [45]. Others have shown that it can be carried out in a letter-by-letter fashion, using fixed-precision arithmetic [26]. Recently, an arithmetic coding technique that is well-suited to the present application has been described by Popat [46]. The technique, based on an algorithm devised by Rissanen and Langdon [27], can be summarized as follows.

Let ρ denote the precision in bits of the hardware. For nonnegative real x, let $\lfloor x \rfloor_\rho$ denote the result of truncating the binary representation of the fractional part of x to the most significant ρ bits, where the ith most significant bit corresponds to 2^{-i}. To ensure decodability, the conditional probabilities $P_i(a_k)$ must be rounded off to ρ-bit values $Q_i(a_k)$, in such a way that their sum is unity, and each is greater than or equal to $2^{-\rho+1}$. This rounding process has a strong influence on coding efficiency, and should be carried out using an optimization procedure[2].

Let X_i and W_i represent the contents of two ρ-bit registers, with initial value $X_0 = 0$ and $W_0 = 1 - 2^{-\rho}$. To encode the letter U_i for $i = 1, \cdots, N$, the registers are modified according to the recursion formulas,

$$X_i = 2^{s_i}[X_{i-1} + \lfloor W_{i-1} \sum_{j<k:a_k=U_i} Q_i(a_j)\rfloor_\rho], \tag{8.27}$$

$$W_i = 2^{s_i} \lfloor Q_i(a_j) W_{i-1} \rfloor_\rho, \tag{8.28}$$

where s_i is the number of positions that $\lfloor Q_i(a_j)W_{i-1}\rfloor_\rho$ can be shifted left without overflow. After these two operations are carried out, X_i is replaced with its fractional part, while the s_i bits comprising its integer part are transmitted as code bits. The minimum value of W_i is $2^{-\rho}$, so that the maximum number of bits that can be emitted in encoding any single letter is $\rho - 1$. It can be shown that the resulting code bit stream can be sequentially decoded using similar recursive fixed-precision procedures. A special procedure is employed to control the propagation of carry bits out the fractional part of X_i.

Note that the summation in 8.27 can be precomputed and stored in a table, so that encoding is computationally inexpensive. In particular, encoding each letter requires two ρ-bit integer multiplications, one ρ-bit integer addition, and one multiple-position shift operation. Thus, arithmetic encoding is less computationally expensive as a simple filtering operation. It can be shown that decoding requires one shift operation, and not more than $(\lceil \log_2 K \rceil + 1)$ ρ-bit multiplications and additions per letter. The need to control propagation of carry bits adds a few more bit operations to the encoding process. However, these can be omitted when the only purpose is to determine the number of bits required to encode a given sequence $\mathbf{u}^N$.

A more detailed description of this and other arithmetic coding techniques can be found in the cited literature.

[2]In the present application, rounded-off quantizer output probabilities are precomputed for 100 different value of Δ/σ in logarithmic increments, then stored in a table.

Appendix B: Coefficients of the 6 Bands, 12 Taps Analysis Filter Bank

k	0	1	2	3	4	5
$h_k(0)$	-0.027083	-0.032590	-0.013897	0.011789	0.030195	0.025120
$h_k(1)$	-0.003115	-0.035379	-0.082497	-0.082443	-0.035201	-0.001274
$h_k(2)$	0.050237	0.058134	0.021327	-0.025214	-0.056913	-0.047025
$h_k(3)$	0.116349	0.171900	0.184521	0.186820	0.171801	0.112529
$h_k(4)$	0.169597	0.200116	0.079225	-0.083244	-0.200547	-0.168234
$h_k(5)$	0.194238	0.089688	-0.188560	-0.184203	0.090602	0.198806
$h_k(6)$	0.194238	-0.089688	-0.188560	0.184203	0.090602	-0.198806
$h_k(7)$	0.169597	-0.200116	0.079225	0.083244	-0.200547	0.168234
$h_k(8)$	0.116349	-0.171900	0.184521	-0.186820	0.171801	-0.112529
$h_k(9)$	0.050237	-0.058134	0.021327	0.025214	-0.056913	0.047025
$h_k(10)$	-0.003115	0.035379	-0.082497	0.082443	-0.035201	0.001274
$h_k(11)$	-0.027083	0.032590	-0.013897	-0.011789	0.030195	-0.025120

Appendix C: Quantization Matrix

The following table provides the Δ_n used for each subband. The upper part of the table refers to reference frame. An hyphen (–) means that the corresponding subband are not transmitted.

1	1	2	4	7	25
1	2	3	5	9	–
2	3	6	9	12	–
4	6	9	13	20	–
7	9	12	20	–	–
25	–	–	–	–	–

1	1	2	3	5	20
1	2	3	5	9	–
2	3	5	8	10	–
3	5	8	10	18	–
5	9	10	18	–	–
20	–	–	–	–	–

1	1	2	3	5	20
1	2	3	5	9	–
2	3	5	8	10	–
3	5	8	10	18	–
5	9	10	18	–	–
20	–	–	–	–	–

References

[1] A. Segall. Bit allocation and encoding for vector sources. *IEEE Trans. on Inform. Theory*, IT-22(2):162–169, March 1976.

[2] N. Farvardin and J. W. Modestino. Adaptive buffer-instrumented entropy-coded quantizer performance for memoryless sources. *IEEE Trans. on Inform. Theory*, 11(1):9–22, January 1986.

[3] N. Farvardin and J. W. Modestino. On the overflow and underflow problems in buffer-instrumented variable-length coding of fixed-rate memoryless sources. *IEEE Trans. on Inform. Theory*, 32(6):839–845, November 1986.

[4] Coded representation of picture and audio information : Test model 2. Technical Report ISO-IEC/JTC1/SC29/WG11, Motion Picture Expert Group, August 1992.

[5] M. Vetterli. Multi-dimensional sub-band coding: Some theory and algorithms. *Signal Processing*, 6(2):97–112, 1984.

[6] M. Vetterli. A theory of multirate filter banks. *IEEE Transactions on Acoustics, Speech, and Signal Processing*, 35(3):356–372, March 1987.

[7] P. P. Vaidyanathan. Quadrature mirror filter banks, m-band extensions and perfect-reconstruction techniques. *IEEE ASSP Magazine*, 4(3):4–20, July 1987.

[8] P. P. Vaidyanathan. Multirate digital filters, filter banks, polyphase networks, and applications: A tutorial. *Proc. of IEEE*, 78(1):56–93, January 1990.

[9] A. C. Popat, W. Li, and M. Kunt. Numerical design of parallel multiresolution filter banks for image coding applications. In *Proc. SPIE Int. Symp. on Optical Applied Science and Engineering*, volume 1605, July 1991.

[10] W. Li, A. Basso, A. Nicoulin, and M. Kunt. Design of Perfect Reconstruction, Finite Wordlength, Non-uniform, Parallel-Structured Filter Banks. In *Proc. EUSIPCO'92, Brussels*, volume 3, pages 1345–1348, August 1992.

[11] J.W. Woods. *Subband image coding.* Kluwer Academic Publishers, Boston, 1991.

[12] W. Li, A. Basso, A.C. Popat, A. Nicoulin, and M. Kunt. Design of parallel multiresolution filter banks by simulated annealing. In *Visual Communications and Image Processing '91*, volume 1605, pages 124–136, November 1991.

[13] N. Benvenuto, M. Marchesi, and A. Ucini. A general optimization algorithm to design fir filters with power-of-two coefficients. In *Proc. EUSIPCO-90*, pages 565–568, September 1990.

[14] C. Cafforio, F. Rocca, and S. Tubaro. Motion compensated image interpolation. *IEEE Trans. on Communications*, 38:215–222, February 1990.

[15] T.S. Huang. *Image Sequence Processing and Dynamic Scene Analysis.* Springer-Verlag, Berlin, 1983.

[16] J.K. Aggarwal and N. Nandhakumar. On the computation of motion from sequences of images - a review. *Proceedings of the IEEE*, 76:917–935, August 1988.

[17] R. Forchheimer and T. Kronander. Image coding – from waveforms to animation. *IEEE trans. on Acoust., Speech and Signal Processing*, 37(12):2008–2023, December 1989.

[18] F. Dufaux and M. Kunt. Multigrid based motion estimation for interframe image sequence coding. In *EUSIPCO'92*, volume 3, pages 1323–1326, Brussels, Belgium, August 1992.

[19] P. H. Westerink, D. E. Boekee, J. Biemond, and J. W. Woods. Subband Coding of Images Using Vector Quantization. *IEEE Trans. on Communications*, 36(6):713–719, June 1988.

[20] P. H. Westerink, J. Biemond, and D. E. Boekee. Sub-band coding of images using predictive vector quantization. In *Proc. IEEE Int. Conf. Acoust., Speech and Signal Processing*, pages 1378–1381, April 1987.

[21] A. Nicoulin and M. Mattavelli. A statistical model for coding subband images using vector quantization and arithmetic coding. In *Visual Communications and Image Processing '92*, Boston, November 1992. Accepted paper.

[22] N. S. Jayant and P. Noll. *Digital Coding of Waveforms – Principles and Applications to Speech and Video.* Prentice-Hall, Englewood Cliffs, New Jersey, 1984.

[23] T. Berger. Minimum entropy quantizers and permutation codes. *IEEE Trans. on Inform. Theory*, 28(2):149–157, March 1982.

[24] N. Farvardin and J. W. Modestino. Optimum quantizer performance for a class of non-gaussian memoryless sources. *IEEE Trans. on Inform. Theory*, 30(3):485–497, May 1984.

[25] H. Gish and J. N. Pierce. Asymptotically efficient quantization. *IEEE Trans. on Inform. Theory*, 14(5):676–683, September 1968.

[26] G. G. Langdon. An introduction to arithmetic coding. *IBM J. Res. Develop.*, 28:135–149, 1984.

[27] J. Rissanen and G. G. Langdon. Arithmetic coding. *IBM J. Res. Develop.*, 23:149–162, March 1979.

[28] R. E. Blahut. Computation of channel capacity and rate-distorsion functions. *IEEE Trans. on Inform. Theory*, 18, July 1972.

[29] P. H. Westerink, J. Biemond, and D. E. Boekee. An Optimal Bit Allocation Algorithm for Sub-band Coding. In *Proc. IEEE Int. Conf. Acoust., Speech and Signal Processing*, pages 757–760, April 1988.

[30] D. G. Luemberger. *Optimization by Vector Space Methods.* Wiley, New York, 1969.

[31] Y. Shoham and A. Gersho. Efficient bit allocation for an arbitrary set of quantizers. *IEEE Transactions on Acoustics, Speech, and Signal Processing*, 36:1445–1453, September 1984.

[32] T. Ibaraki and N. Katoh. *Resource Allocation Problems.* MIT Press, Cambridge, Massachussetts, 1989.

[33] B. Macq. *Perceptual transforms and universal entropy coding for an integrated approach to picture coding.* PhD thesis, Université Catholique de Louvain, Louvain-la-Neuve, Belgium, 1989.

[34] A. Kolmogorov. Three approaches to the quantitative definition of information. *Prob. Peredach Inform.*, 1(1):3–11, 1965. Russian.

[35] J. Rissanen and G.G. Langdon. Universal modeling and coding. *IEEE Trans. on Information Theory*, 27(1):12–23, January 1981.

[36] T. Berger. *Rate Distortion Teory.* Prentice-Hall, Englewood Cliffs,New Jersey, 1971.

[37] A. Basso, W. Li, A. Nicoulin, and M. Kunt. Side information compression in subband coding of video. To be published in the *ECCV 92*, Paris, September 1992.

[38] J. Makhoul, S. Roucos, and H. Gish. Vector Quantization in Speech Coding. In *Proceedings IEEE*, volume 73, pages 1551–1587, November 1985.

[39] A. Gersho. On the structure of vector quantizers. *IEEE Trans. on Inform. Theory*, 28(2):157–166, March 1982.

[40] A. Gersho and R. M. Gray. *Vector Quantization and Signal Compression*. Kluwer Academic Publisher, Boston, Dordrecht, London, 1992.

[41] Y. Linde, A. Buzo, and R. M. Gray. An Algorithm for Vector Quantization Design. *IEEE Trans. on Communications*, 28(1):84–95, January 1980.

[42] A. Nicoulin. Adaptative subband coding for digital video. Master's thesis, Swiss Federal Institute of Technology, Lausanne, Switzerland, June 1991.

[43] A. Papoulis. *Probability, Random Variables, and Stochastic Processes*. McGraw-Hill, Inc., New York, 1965.

[44] M. Kunt, A. Ikonomopoulos, and M. Kocher. Second generation image coding techniques. *Proceedings of the IEEE*, 73(4):549–575, April 1985.

[45] C. E. Shannon and W. Weaver. *The Mathematical Theory of Communication*. University of Illinois Press, Urbana, 1963.

[46] A. C. Popat. Scalar Quantization with Arithmetic Coding. Master's thesis, Massachusetts Institute of Technology MIT, U.S.A., June 1990.

9

Vector Quantization for Video Data Compression

R.M. Mersereau, M.J.T. Smith, C.S. Kim
F. Kossentini, K.K. Truong

Georgia Institute of Technology, Atlanta, GA, USA

9.1 Introduction

Visual signals, even more than audio signals, typically contain an exorbitant amount of information. It is not uncommon for the visual component of a motion picture or television signal to contain several orders of magnitude more data than the audio component. During this age of visual information transmission and storage, efficient methods for compressing video have become particularly important.

Video sequences have many properties that can be exploited to reduce the volume of data. Many methods for exploiting these properties have been discussed in previous chapters. Vector quantization (or VQ) is one distinctive method that holds unusual promise and is the common theme of this chapter. It can be applied to coding the pixels in the video sequence as a self-contained compression method or in combination with other schemes.

Recent advances in vector quantization have inspired several new approaches to the video coding problem. In the remainder of this chapter we shall briefly develop the general concepts of VQ and discuss some recent VQ-oriented video compression research that has been under study in the digital signal processing laboratory at Georgia Tech. The first part of the chapter focuses on exploiting the spatial redundancies inherent in the image sequence while the later part addresses both spatial and temporal redundancies.

There are many applications in which coding visual information is important: HDTV, high quality teleconferencing, picture phone systems, etc. In this chapter the focus is on operation at bit rates in the range of 64 kbits/second and below.

9.2 Vector Quantization

Vector quantization is a simple concept that can be viewed as the extension of scalar quantization to a higher dimensional space. Inputs to the quantizer, $Q(\cdot)$, are vectors of dimension k that are mapped to a finite set of N code vectors. The

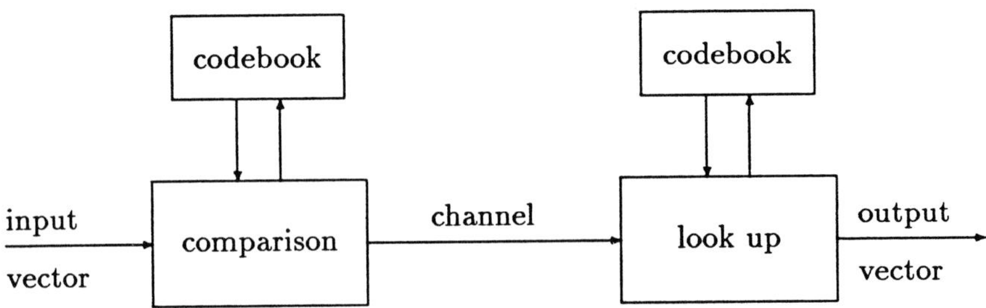

Figure 9.1: Block diagram of a VQ encoder (left) and decoder (right).

quantizer partitions the k-dimensional space into N cells and represents each cell by a binary index or number which is transmitted. The receiver uses the binary index to look up the corresponding code vector in the codebook and inserts it into the reconstructed image. A block diagram of a basic vector quantizer is shown in Figure 9.1. In the context of video frame encoding, the input vectors are often formed by selecting contiguous blocks of samples from a frame and arranging the pixels to form vectors.

An comprehensive treatment of VQ can be found in [1, 2]. Several points are worth highlighting before discussing video coding. First, VQ inherently introduces errors in the compression process. When a vector $\mathbf{x}$ is quantized to $\mathbf{y}_i = Q(\mathbf{x})$, a distortion $d(\mathbf{x}, Q(\mathbf{x}))$ can be calculated that represents the dissimilarity between the vectors $\mathbf{x}$ and $Q(\mathbf{x})$. This distortion is used to design VQ codebooks and measure their quality. The most commonly used distortion measure is the *mean squared error* defined by

$$d(\mathbf{x}, \hat{\mathbf{x}}) = ||\mathbf{x} - \hat{\mathbf{x}}||^2 = \sum_{n=1}^{k} (x_n - \hat{x}_n)^2, \tag{9.1}$$

where $\mathbf{x}$ and $\hat{\mathbf{x}}$ are the vectors in question, $||\cdot||$ denotes the Euclidean norm, and x_n and $\hat{x}_n$ are elements of $\mathbf{x}$ and $\hat{\mathbf{x}}$, respectively.

The vector quantizer is said to be an optimal (minimum distortion) quantizer if the codebook minimizes

$$D = E\{d(\mathbf{x}, Q(\mathbf{x}))\} = \int d(\mathbf{x}, Q(\mathbf{x})) f(\mathbf{x}) d\mathbf{x} \tag{9.2}$$

where $f(\mathbf{x})$ is the probability density function (pdf) of the input vectors. As with scalar quantization, there are two necessary conditions for optimality. The first is that each code vector $\mathbf{y}_i$ should be chosen to be the centroid of its k-dimensional cell. The second is that the quantizer should make its decisions using a minimum distortion or *nearest neighbor* rule:

$$Q(\mathbf{x}) = \mathbf{y}_i \quad \textit{iff} \quad d(\mathbf{x}, \mathbf{y}_i) \leq d(\mathbf{x}, \mathbf{y}_j), \quad i \neq j, \quad 1 \leq j \leq N. \tag{9.3}$$

Since the probability distributions of many practical sources, such as image sequences, are not known, the codebook design process typically uses the relative frequencies of representative vectors taken from a large training set.

The most widely known design technique is the generalized Lloyd algorithm (GLA) or LBG algorithm, as it is often called [2]. It is an iterative method that attempts to satisfy the nearest neighbor rule and centroid condition simultaneously. Its performance is generally very good; however, globally optimal solutions are not guaranteed. Thus, several techniques such as simulated annealing and stochastic relaxation have been proposed to help force convergence to a globally optimal codebook. Iterative methods of this type tend to be computationally intensive. Non-iterative design techniques such as the pairwise nearest neighbor (PNN) algorithm [3], and the self-organizing feature map (SOFM) algorithm [4, 5], have also been considered and have been shown to yield comparable performance results [6].

9.2.1 Design considerations for VQ systems

Well-known results from information theory can be used to show that VQ can achieve performance that approaches the optimal rate-distortion bound as the vector dimension increases. This is the upper bound on performance of any coding system. It provides a theoretical way to increase coding performance that makes VQ attractive. However, there are limitations and tradeoffs with respect to the complexity of the encoder, the codebook storage, and the encoding time that ultimately restrict the vector size.

The encoder complexity is a critical factor for real-time video compression systems because it determines the time required to compress the signal. It is directly related to the bit rate, the codebook size, the vector dimension, and the way in which the codebook is structured. For unstructured codebooks where exhaustive searching is performed, $2^r k$ distortion calculations are required for each input vector, where r and k are the bit rate and vector dimension, respectively. Tree-structured codebooks [1] can achieve an $O(\log_2 N)$ search time with an average distortion that is almost as good as that achievable with an exhaustive search. However, the required codebook memory is essentially doubled.

Codebook memory is often the constraint that limits the vector dimension and thus indirectly restricts the operating bit rate, r, which is defined as

$$r = \frac{\log_2 N}{k},$$

where N is the number of vectors in the codebook and k is again the vector dimension. The total required storage for the codebook is $kN = k2^{rk}$ samples. The codebook memory grows exponentially with the bit rate and the vector dimension. To circumvent this memory problem structurally constrained codebooks have been proposed, including product code VQ, lattice VQ, gain/shape VQ, and mean extraction VQ [1, 2]. In a later section, residual vector quantization (RVQ), a special class of product code VQ, is discussed in some detail. This technique is able to reduce both the memory and computation costs [7] and, as shown very recently, is also able to achieve good performance for coding frames of video sequences.

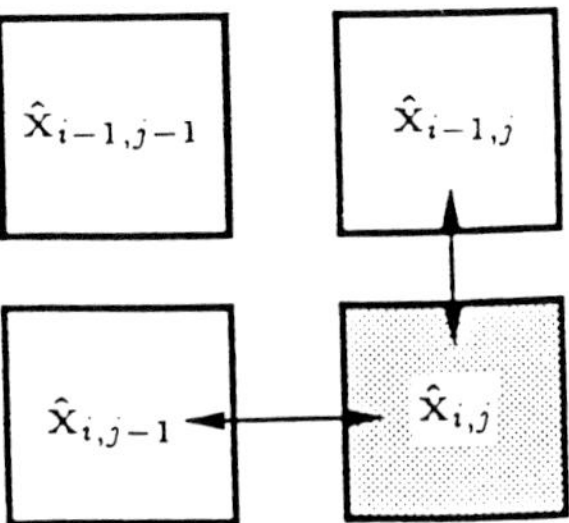

Figure 9.2: State specifications for FSVQ.

9.3 Exploiting Spatial Redundancies in Video

Image sequences are highly correlated spatially. Several different VQ-based approaches that exploit these spatial or intraframe redundancies are discussed in this section: FSVQ, residual VQ, cache VQ, and subband residual VQ. These approaches can be applied directly to the image sequence on a frame-by-frame basis or they can be applied in conjunction with motion compensation to exploit both temporal and spatial redundancies as discussed in Section 9.4.

9.3.1 Finite state vector quantization (FSVQ)

Classical vector quantization is effective in exploiting spatial redundancies within an image block, but it cannot capture redundancies between adjacent blocks. FSVQ and cache VQ, which will be described shortly, are two methods that attempt to take advantage of inter-block correlation.

FSVQ uses the information in neighboring blocks to predict which of several customized codebooks should be used to code the current block. This form of VQ was originally introduced by Foster, Gray, and Dunham [8] and later modified for image coding by Aravind and Gersho [9]. It can take advantage of both linear and non-linear dependencies present in an image frame.

FSVQ [9] is a multicodebook scheme that employs a state variable to select the codebook to be used to encode a particular input vector. Once the specific codebook has been chosen, the codeword is selected to minimize the distortion in the usual manner. The distinguishing feature of FSVQ is its use of states in the coding procedure or, equivalently, its use of multiple codebooks. The state specification is accomplished by classifying each of two previously coded neighboring vectors into one of a small number of classes. These *class* labels are used to specify the codebook to be used to code the present vector.

The *state specification* is illustrated in Figure 9.2. The blocks shown in the figure denote vectors, each of which represents a square $M \times M$ block of pixels where $M^2 = k$, the vector dimension. The image is partitioned into contiguous $M \times M$ blocks $\mathbf{x}_{i,j}$ where the subscripts indicate that the block is at the ith horizontal block position and the jth vertical block position in the image. The classes of the coded vectors $\hat{\mathbf{x}}_{i-1,j}$ and $\hat{\mathbf{x}}_{i,j-1}$ are used to determine the state $s_{i,j}$. The state $s_{i,j}$ for vector $\mathbf{x}_{i,j}$ is defined by

$$s_{i,j}=[\text{class}(\hat{\mathbf{x}}_{i-1,j}),\ \text{class}(\hat{\mathbf{x}}_{i,j-1})].$$

Thus, state determination relies on the classes of the two adjacent vectors as shown in Figure 9.2. The process assumes that vectors are coded sequentially from left to right across each row starting from the top. Once the initial states for the top vector row and leftmost vector column are specified, no side information needs to be sent. This is because the state specification is based on the classes of the coded vectors, which are known to both the encoder and the decoder systems. The initialization requires that some side information be sent; however, this can be reduced by using the knowledge of the class of the left adjacent vector for coding the top vector row and of the upper adjacent vector for coding the leftmost vector column. In general, the bit-rate cost associated with initialization is very small. It is worth mentioning that additional blocks surrounding the current block may be used in the state determination process to obtain a more accurate prediction. However, the number of states increases dramatically if this is done. Consequently, only two adjacent blocks are considered in this discussion.

Two factors influence the selection of the vector classes for the FSVQ system: on average, the mean value of the image vectors varies slowly; and edges should be continuous across vector boundaries. The vector classes are chosen to consist of shade classes and edge classes. The distinction among the shade classes is based on their average gray level, while the edge classes are formed on the basis of edge orientation.

Edge classes are defined using edge-oriented template vectors composed of "+1's" and "−1's". The +1's and −1's within the template are selected to enhance horizontal, vertical, and diagonal edge features. The classification first subtracts the mean from each input vector and then computes the inner product of that vector with the template. This inner product is computed for each template vector and the template producing the largest inner product determines the vector class. In the event that the largest inner product falls below a preset threshold, the vector is classified as a shade vector. Shade vectors may be further subdivided into classes based on the mean value of the vector, should that be desired.

FSVQ requires many codebooks. Therefore, a very large training data set is needed to populate all of the given states. Once the training data have been obtained for each state codebook, classical codebook design algorithms such as the LBG algorithm can be used.

9.3.2 Residual VQ

Residual vector quantization (RVQ), also known as cascaded VQ and multistage VQ, is a constrained VQ whose structure reduces both the search complexity and memory requirements. RVQ employs multiple structurally identical VQ stages; the first quantizer operates on the input vector while the subsequent quantizers operate on the residual differences between the input vectors and the quantized vectors from each stage. A two-stage RVQ is shown in Figure 9.3 for illustration.

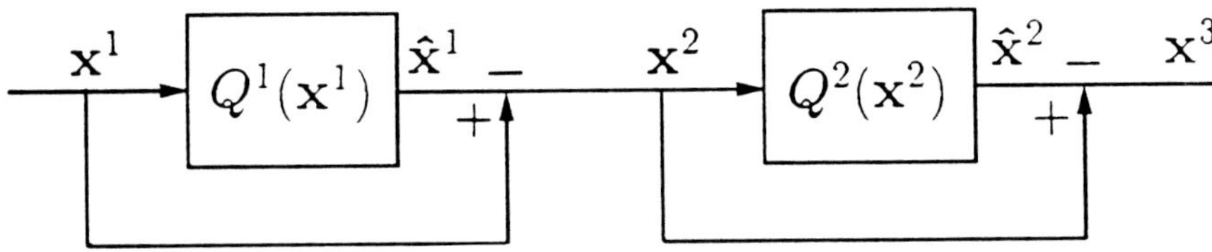

Figure 9.3: Two stage RVQ.

In this case, the input vector $\mathbf{x}^1$ is first quantized using the vector quantizer Q^1. The residual vector $\mathbf{x}^2$ is then formed by subtracting the quantized vector $\hat{\mathbf{x}}^1$ from the original vector $\mathbf{x}^1$. This residual is then used as the input to a second vector quantizer Q^2 whose quantized output is $\hat{\mathbf{x}}^2$. The final quantized value of $\mathbf{x}^1$ is simply the sum of the two vectors $\hat{\mathbf{x}}^1$ and $\hat{\mathbf{x}}^2$. A general multistage RVQ consisting of P stages is capable of uniquely representing

$$N = \prod_{i=1}^{P} N_i$$

vectors while only storing

$$N = \sum_{i=1}^{P} N_i$$

code vectors. Thus, a multistage RVQ is capable of achieving a huge potential savings of memory over unconstrained VQ. In addition, encoding a vector using RVQ can also be very efficient since it only requires on the order of

$$N = \sum_{i=1}^{P} N_i$$

distortion calculations for each vector that is encoded. This makes it possible to employ large vector blocks (e.g. 16×16 and 32×32) in the RVQ design.

The price paid for the computational and memory efficiency of RVQ is an expected reduction in performance. It is therefore important to design the RVQ codebooks very carefully. Juang and Gray [7] proposed an RVQ design algorithm that sequentially applies the LBG algorithm to the stage residuals of the training sets. Thus, their procedure first designs an LBG codebook for the first stage, computes a residual training set, designs an LBG codebook for the next stage, computes a residual training set, and so on. In this way each stage is locally optimal given all previous stage codebooks. However, a degradation in quality can be observed, the severity of which increases with an increasing number of stages [10]. The disparity in performance between unconstrained VQ and RVQ that was reported in [10] is a result of both the structural constraints and the RVQ design algorithm. Significant improvement in RVQ image frame coding performance has been obtained [11] using a new algorithm originally introduced by Barnes [12] and more recently discussed in [13] and [14]. Although a complete discussion of the design algorithm would take us too far afield, let it suffice to say that the design procedure attempts to jointly optimize each stage codebook to minimize the reconstruction error over all training data.

The RVQ encoding structure can be viewed as a series of paths through the constituent stages. These paths may be viewed as a tree structure for conceptual convenience. To our benefit, several well-known tree search strategies can therefore be used to search the RVQ codebook. The simplest is 1-path or sequential search. The disadvantage of this is that the selected equivalent code vectors are not, in general, closest to the input vectors. This is due to entanglements in the RVQ which can lead to large quantization errors.

Multipath searching (or M-search) can noticeably reduce these errors. Let M be the number of search paths (which is typically a small number) and N be the stage codebook size. The M-search algorithm proceeds one level deeper into the RVQ tree by extending all branches from the M saved nodes, and only saving the best M of these branches for the next step. Specifically, the encoder compares the input vector to all N code vectors of the first stage codebook. If N is less than M, then all code vectors of the first stage are retained as candidate path segments. Otherwise, only the best M matches are retained as candidates. For each remaining candidate, the next stage codebook is searched and only the best M matches of the MN combinations are retained. This procedure continues until the last stage of the RVQ codebook is reached, and then the code vector of the best path among the final M saved paths is used.

The M-search RVQ algorithm attempts to find the closest code vector to the input vector without exhaustively searching the equivalent codebook. For large block RVQ systems, exhaustive search is precluded by its exorbitant computational load. However, the performance of RVQ systems can be improved by using M-search encoders at the expense of a moderate increase in computation. Thus, by increasing M, the performance of M-search RVQ can come arbitrarily close to that of exhaustive search RVQ.

9.3.3 Cache VQ

Like FSVQ, cache VQ [15] is a sub-optimal coder that reduces inter-block correlation. This is particularly important for video coding where studies have suggested that motion vectors exhibit strong correlation between neighboring blocks or between consecutive frames [16]. The statistics of the motion vectors, however, vary from frame to frame and, therefore, a coder with fixed probabilities, such as an entropy coder [17], may not be efficient. Cache VQ, on the other hand, is adaptive and it can exploit statistics that vary with time. A side benefit of cache VQ is that it can also substantially reduce the average encoding time. Like RVQ, it can also be adjusted in real time to alter the tradeoff between the coding distortion and the average bit rate.

The name *cache VQ* was motivated by the conceptual similarities between it and a cache memory. Cache VQ consists of a high rate, high quality codebook which can be designed using any of the conventional methods. This large codebook is augmented by a small codebook that retains a reasonably small number of frequently used codewords from recently coded blocks. When a new vector needs to be coded, the cache is searched first, the best match in the cache is identified and, if the coding error is less than a threshold (an event known as a *hit*), the codeword is based on the address of the code vector within the cache. When the cache codebook cannot be used (an event known as a *miss*), the transmitted codeword is based on the address of the best match from the large codebook. If the large codebook is of size N and the cache is of size L, the effective bit rate, r_e, is

$$r_e = (1 + h \log_2 L + (1-h) \log_2 N)/k,$$

where h is the probability that a good fit can be found in the cache. This depends both upon the threshold value and on the cache update strategy. If a good fit is found in the cache, the large codebook is not searched, which reduces the average encoding time.

Several considerations affect the choice of a cache update strategy. First, the substitution must be determinable at both the encoder and decoder, preferably without the transmission of side information. Second, the size of the cache should be kept as small as possible. Finally, the updating should be such that the hit probability, h, for succeeding image blocks should be as high as possible. This can result in a VQ with a low effective bit rate.

In the broad view of cache management, updates should be performed only when a miss occurs and only unique vectors should be stored in the cache. At a more refined level, four specific update strategies have been considered, all of which are borrowed from well-known algorithms for the organization of computer memories [18, 19]: the least recently used (LRU) strategy, the least frequently used strategy (LFU), the first-in/first-out strategy (FIFO), and the working set model (WSM). The first three of these are self explanatory. With the working set model, no explicit cache replacement is done and the cache memory becomes simply a list of distinct addresses that have been used during the time interval $[n - W(n) + 1, n]$, where $W(n)$ is the window size which, in general, may vary with time. The cache size is time varying and depends on the number of unique addresses in the given time interval. The method used to index the image blocks can also affect the hit probability significantly when the cache size, L, is small.

The LRU update strategy can encode image data with approximately 30% fewer bits than the LFU strategy and 15% fewer bits that the FIFO strategy, but is slightly inferior to the working set model. The method used in the coded results discussed later is based on the working set model [15].

9.3.4 Subband and pyramid coding with VQ

Subband image coding or SBIC is now widely accepted as an effective technique for high quality coding of image frames at various bit rates [20]. It may be viewed as having two basic component parts:

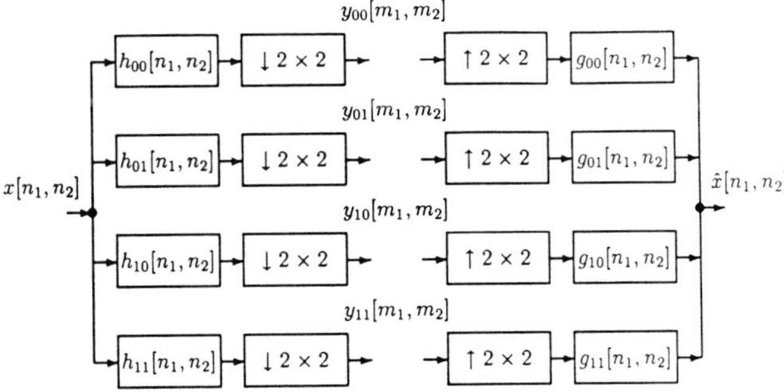

Figure 9.4: Block diagram of a subband image coder.

1. The subband analysis/synthesis subsystem, which is composed of filter banks. The analysis section splits the signal into subband images each of which is decimated to its effective Nyquist rate. The synthesis system is a bank of interpolators that merges the subband images into a reconstructed version of the input. This is illustrated in the block diagram shown in Figure 9.4.

2. The coding/decoding subsystem, in which quantization is performed and bits are assigned to the quantization levels. Coders in this section, which operate on the analysis section outputs, may employ some form of PCM, differential coding, or vector quantization and often use variable length entropy coding techniques to reduce the average bit rate [21, 22]. The decoder converts the bit representation back into amplitude levels.

Both components must be carefully designed because poor performance in either subsystem degrades the overall system quality.

Subband analysis/synthesis systems have been studied extensively and, in the context of coding image sequence frames, should be designed to have several performance related properties. First, to avoid an increase in the amount of data, the subband images should be maximally decimated so that the total number of pixels to be coded will not be larger than the number of pixels in the original image frame. Decimation alone will not achieve this goal because the convolution involved in filtering results in an expansion of the image and is dependent on the spatial extent of the filter impulse response. Several methods have been proposed for handling this data expansion problem: circular convolution, symmetric extension [23], and, more recently, generalizations of these approaches [24, 25]. Second, the decimators and interpolators introduce aliasing and imaging that can lead to distortion in the reconstructed signal. In addition, the filtering operations can also contribute spectral magnitude and phase distortions. These difficulties are typically addressed by using quadrature mirror filters (QMFs) and exact reconstruction filter banks [23]. Finally, implementation complexity should be considered and may be addressed in many ways. Most often rectangularly separable implementations have been used for this reason, although efficient non-rectangularly separable analysis/synthesis fil-

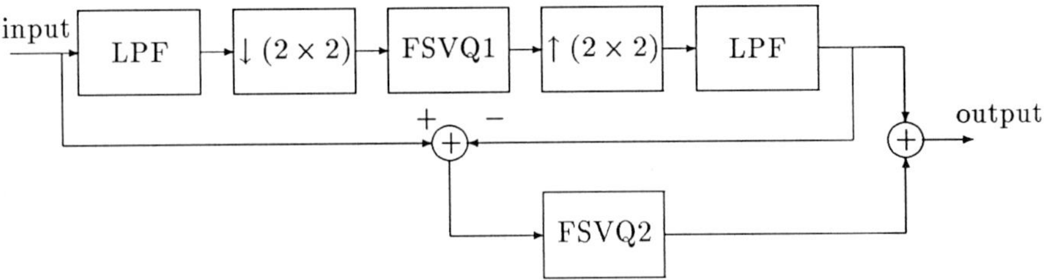

Figure 9.5: Two-stage coder with FSVQ.

ter banks are also known [26]. Among the most efficient systems are those based on recursive allpass polyphase filters [23]. These recursive filters result in perfectly reconstructing analysis/synthesis and achieve tremendous savings in terms of computational complexity. In comparison to FIR quadrature mirror filters with comparable magnitude response characteristics, these recursive filters reduce complexity by an order of magnitude.

Pyramid coding is a related form of decomposition. Here the image is successively filtered and downsampled. At each stage, the decimated signal of the previous stage is interpolated back to the rate of the current stage and the difference signal is computed and coded. A fundamental difference between this and subband image coding is that the number of samples needed to represent the image in a pyramid decomposition is always greater than the number of input pixels because the decomposition is inherently redundant. The rationale for using this data expansive decomposition for coding is that the residual difference components can be coded very efficiently—enough to offset the increased redundancy.

Such a system is illustrated in Figure 9.5 [27] which shows a two-stage system that employs an FSVQ. In this system, the original $N \times M$ image frame is lowpass filtered and downsampled by a factor of two in each direction. The downsampled $N/2 \times M/2$ image is then coded using the FSVQ system described previously, which is denoted by FSVQ1 in Figure 9.5. The quantized image is then interpolated back to its original size. The $N \times M$ residual image, formed by taking the difference between the interpolated FSVQ coded image and the original, is encoded using another finite state vector quantizer, FSVQ2. The residual image contains most of the high frequency information that was lost by downsampling. Hence, the inclusion of a residual component tends to lessen the lowpass appearance that is common with many coding procedures. Bits are allocated between the two quantizers to optimize subjective appearance. FSVQ1 operates at a higher bit rate (but on the small decimated subband image) to preserve the important low frequency information that accounts for most of the signal energy. Higher dimensional vectors are used for FSVQ2 to code the residual. This approach is effective because the larger vectors in FSVQ2 tend to be better suited to exploiting the correlation in the residual.

An important advantage of subband and pyramid encoding approaches is that they allow for bits to be preferentially allocated among the various bands or components in the decomposition to exploit the well-known frequency domain structure of the image frame and the noise masking property of the human visual system. Experimental results show that the majority of the bit investment is best devoted to coding the low frequency subband, which in this case is the $N/2 \times M/2$ decimated image. In the specific case of the coder in Figure 9.5 FSVQ1 uses 4×4 vectors and is designed to have 8 edge classes and 4 shade classes. This results in 144 codebooks, the sizes of which vary between 1024 and 16 depending upon the desired bit rate. Some reduction in the number of codebooks can be achieved by merging certain infrequently used state codebooks. If done carefully, the resulting loss in quality is not significant.

The residual image quantizer, FSVQ2 is based on 8×8 vectors. Simulations have been performed varying the number of classes between 26 and 11 and varying the codebook sizes from 64 to 8 [27].

9.3.5 Performance comparisons

Several different VQ approaches were discussed in the last few sections, each of which offers some unique benefits. In this section, we attempt to compare these techniques in terms of their ability to exploit spatial redundancies and discuss how these methods might be used in combination with other schemes.

For a base-line comparison, examine Figure 9.6 which shows an original and coded image frame taken from the "Mary" sequence. The coding was performed using conventional VQ at a bit rate of 0.25 bits/pixel. The codebook was designed using the LBG algorithm with vector blocks of size 4×4. Notice that at this relatively low bit rate, the quality is rather poor. If the same image is now coded at the same rate but using larger vectors, the quality is expected to be better. To illustrate the subjective improvement in quality, examine Figure 9.7 which shows two images, both of which are coded at 0.25 bits/pixel. The one on the left is coded with 8×8 vectors while the one on the right is based on 16×16 vectors. In spite of the constrained nature of the RVQ codebook, the gain obtained by employing larger block sizes outweighs the degradation incurred from the RVQ. When memory, complexity, and performance are all taken into consideration, RVQ block sizes of 8×8 seem to represent a reasonable compromise.

Additional gains can be obtained because RVQ allows for variable rate implementations. This can be achieved by dynamically limiting the number of stages used for encoding or decoding. Thus, once an RVQ codebook is designed for a peak operating rate of R bits/pixel, other rates less than R can be obtained by simply using fewer RVQ stages. The individual blocks within the frame can be coded at the minimum rate necessary to achieve a desired level of quality for that block. The number of RVQ stages used (or equivalently the bit rate for that block) must be sent as side information. However, with vector sizes of 8×8 and higher, the amount of side information is negligible. Figure 9.7b shows a coded image using variable rate RVQ with 16×16 vectors at a rate of 0.25

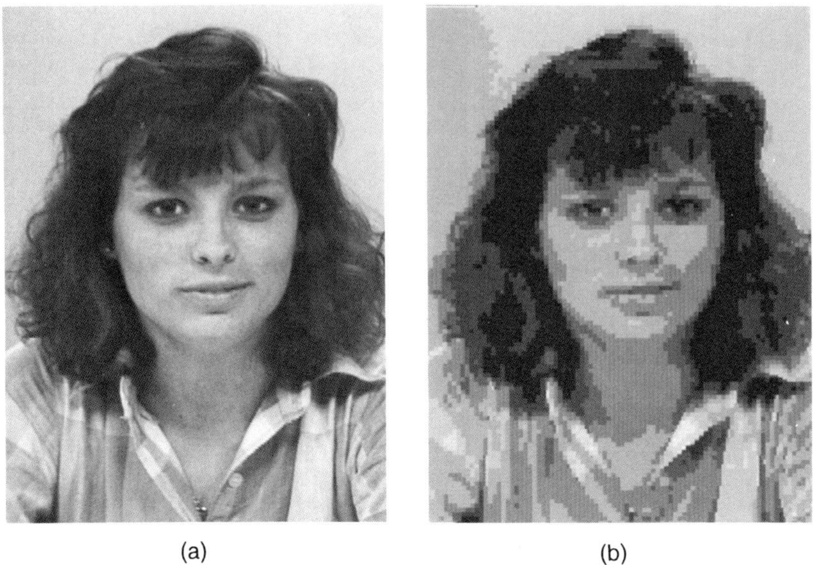

Figure 9.6: a) Original image frame from the test sequence "Mary" of size 352×480. b) The same frame coded at 0.25 bits/pixel using an LBG codebook with 4×4 vector blocks. Peak SNR is 28.88 dB.

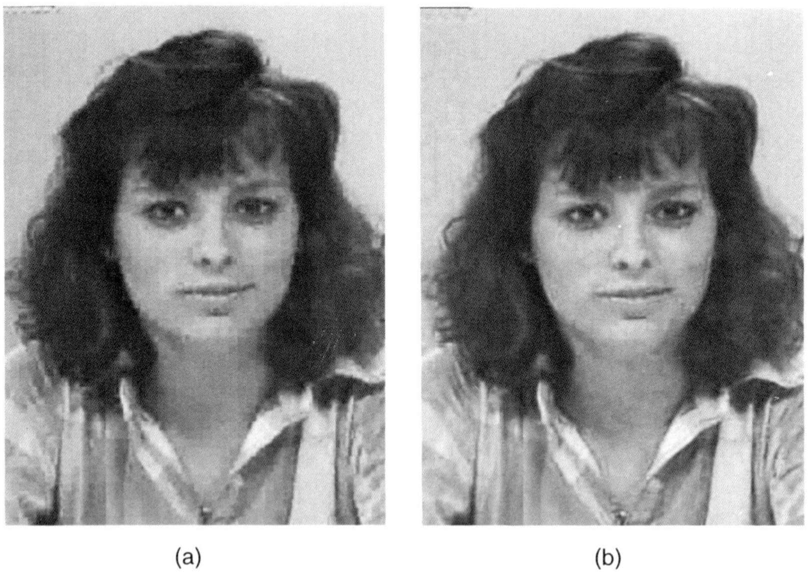

Figure 9.7: a) Image frame coded at 0.25 bits/pixel using a 16-stage RVQ with 8×8 vector blocks. Peak SNR is 33.31 dB. b) Image frame coded at 0.25 bits/pixel using a 64-stage RVQ with 16×16 vector blocks. Peak SNR is 34.48 dB.

(a)

(b)

Figure 9.8: a) Image frame coded at 0.25 bits/pixel using a variable rate RVQ with 16×16 vector blocks. Peak SNR is 35.33 dB. b) Image frame coded at 0.25 bits/pixel using FSVQ with variable rate RVQ and 8×8 vector blocks. Peak SNR is 36.11 dB.

bits/pixel. In comparison with Figure 9.7b notice the improvement due to the variable rate strategy.

Further gains can be achieved by combining variable rate RVQ with FSVQ. This allows the system to utilize information from neighboring blocks to exploit correlation. Examine the two pictures shown in Figure 9.8. Both employ variable rate RVQ. The one on the left uses 16×16 vectors while the one on the right uses smaller, less efficient 8×8 vectors. But, because FSVQ is used in conjunction with smaller vectors, it is able to achieve performance similar to that of the larger vector size case. In fact, the results are even a little better in this case.

It is interesting to compare the variable rate RVQ results with the conventional FSVQ systems reported in the literature based on 4×4 vectors [9, 21]. Although a picture is not provided here for visual comparison, the variable rate finite state RVQ with 8×8 vectors performs noticeably better than conventional FSVQ with 4×4 vectors, both in terms of subjective quality and peak SNR performance.

Cache VQ also results in significant improvement over conventional VQ as illustrated by comparing Figures 9.6 and 9.9. The latter shows the same image frame coded at the same rate of 0.25 bits/pixel with 4×4 vectors. The quality improvement here is not due to the use of larger block sizes or by prediction from adjacent vector blocks. Rather it comes from the adaptive updates of the cache memory. The principle being exploited here shares much in common with that of adaptive PCM in the scalar quantization case. It should be observed that RVQ can easily be used with cache VQ which would allow larger block sizes to

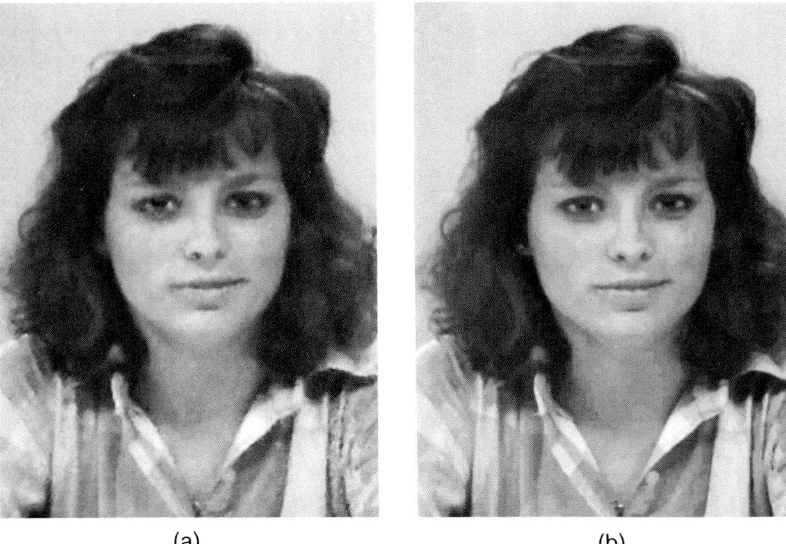

Figure 9.9: a) Image frame coded at 0.25 bits/pixel using cache VQ. Peak SNR is 35.7 dB. b) Image frame coded at 0.25 bits/pixel using a two-stage subband system with FSVQ. Peak SNR is 34.8 dB.

be used.

As a final illustration of how the coding techniques discussed previously can be combined, consider the two-stage coder that employs a high rate 4×4 and low rate 8×8 FSVQ in a pyramid coding scheme. Figure 9.9b shows the image coded with this two-stage system. The improvements here come from two primary sources: first, from exploiting the spatial redundancies due to the use of two FSVQ coders and to the large block sizes in the second stage; and second from exploiting the frequency variant characteristics of natural images. The signal-to-noise ratios for the images with variable rate finite state RVQ, cache VQ, and the two-stage coder are roughly similar, but the images appear subjectively different because the types of errors introduced by the three coders are different.

The ability to exploit spatial redundancies in an image frame is of primary importance in video coding. The VQ ideas discussed thus far are very effective in this regard and can be applied in a number of ways to exploit temporal redundancies.

9.4 Exploiting Spatial and Temporal Redundancies in Video

The amount of temporal or inter-frame redundancy commonly seen in video can be much greater than the spatial redundancies within individual frames, but it can also be highly variable, because there are often periods of little or no motion which boost temporal redundancy dramatically. Consequently the expected gain from inter-frame coding is greater for many image sequences. This does not minimize the importance of intra-frame coding methods. Intra-frame methods are often used to obtain initial frames and reference frames in the video coding system. They can also serve as the basis for coding motion compensated residuals or frame difference signals.

This section will specifically look at exploiting both temporal and spatial redundancy. The starting point is a brief discussion of the problems of motion estimation and motion compensated prediction. Following this, two video coders are introduced and discussed in the context of the VQ principles treated earlier. The first is a system in which FSVQ is used to code motion-compensated residual signals that have been passed through a spatial subband coder. The second coder is one that combines motion compensation with a cache VQ and is able to operate effectively at very low bit rates.

9.4.1 The problem

There are many issues associated with image sequence compression that constrain the design of a coder and strongly depend on the applications to which the coder will be applied. The issues affect all video coders, not only those designed using VQ. One tradeoff is between the cost of the coder and the amount of compression that is desired; another is how that cost should be distributed between the transmitter and receiver. Other questions are: Does the encoding need to be performed in real time? Can a variable rate transmission be tolerated? Will the scene be changing or will attention be focused on a single subject? How much "action" will there be? How long is the sequence? Is there a need for the sequence to have "fast play" or "reverse" capability?

In general, a reduction in the target bit rate often comes at the expense of being able to accommodate these special features. The coders that are described in this section are designed for video conferencing or video telephone applications. This means that the sequences are assumed to be color video sequences of size 352×288 pixels and frame rates between 10 and 30 frames per second. In addition, it is assumed that the amount of motion is modest, and abrupt changes of scene are relatively infrequent. Since video conferencing systems are envisioned for operation over ISDN links, the video compression should be consistent with the 64 kbps ISDN channel rates.

9.4.2 Motion estimation and compensation

Motion prediction is a common technique for reducing the temporal correlation in a video sequence. The coders described in this chapter use a fairly straightforward form of block matching. Block matching takes a block of the current image frame and seeks the block of the same size in the previous frame that is the closest match. Its underlying assumptions are that motion from one frame to the next is translational and that all pixels in the block have the same motion.

The basic block matching geometry is illustrated in Figure 9.10. The current

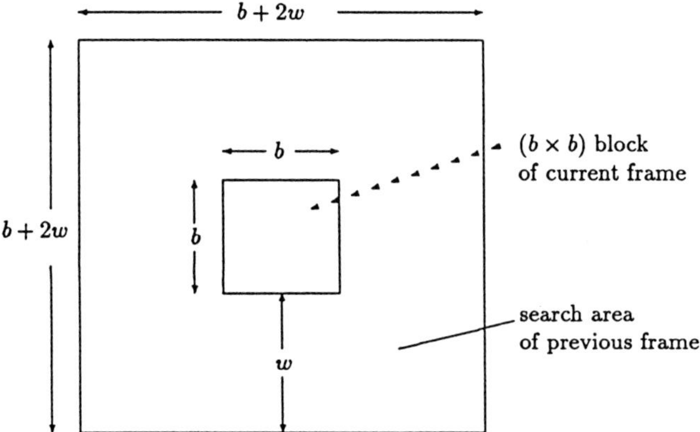

Figure 9.10: Block matching.

frame is uniformly divided into non-overlapping blocks of size $b \times b$. Each block is compared with all $b \times b$ segments of the previous frame that lie within a search window of size $(b + 2w + 1) \times (b + 2w + 1)$, and the best match is found using a matching criterion. Here w is the maximum allowable search shift. The search window is centered on the center of the current block on the previous frame. The location of the best matched block defines the *motion vector* (or *displacement vector*). The procedure for obtaining the motion vector is called *motion estimation.* The motion vectors can be used to form a prediction image for the current frame—commonly called *motion compensation.*

The block matching technique for motion estimation is simple but computationally expensive. For the exhaustive search just mentioned $(2w + 1)^2$ shifts are considered for each block of the input signal and, if the mean squared error is used as a matching criterion, each shift considered requires b^2 multiplications and additions. For this reason, a number of sub-optimal fast search algorithms and alternative matching criteria have been considered. One of these is the three-step search algorithm [28], which is illustrated in Figure 9.11 for $w = 8$. In the first step the 9 shift values labeled s1 are considered (4 corner points, 4 middle points of the sides, and the center sample). In the second step the size of the square is reduced by 1 and the center point of the square is moved to the best matching point of the previous step and the 8 new points are searched. A similar procedure is followed in the third step. The total number of searches is 25, as opposed to 289 required by an exhaustive search. Additional savings have been achieved by replacing the mean squared error matching criterion by a mean absolute difference, which can be implemented without multiplications.

Figure 9.12 demonstrates the effectiveness of motion compensation for several frames of the *salesman* sequence. The three-step search algorithm is seen to perform nearly as well as the exhaustive search, and the motion compensated frame difference is seen to have 2–4 dB less energy than the uncompensated frame difference. The performance varies significantly from frame to frame because of

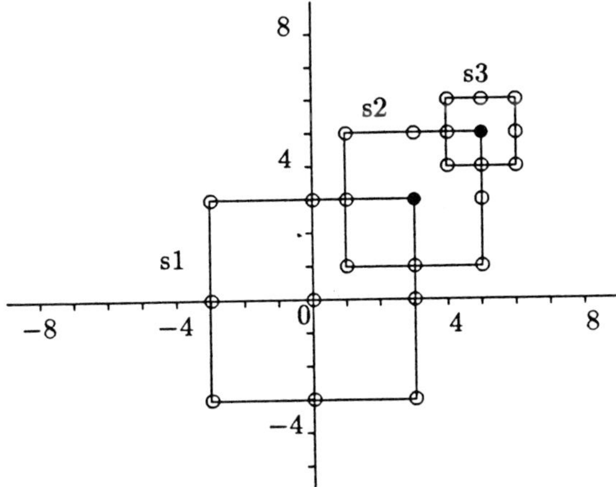

Figure 9.11: The three-step search algorithm.

different amounts of motion in the various frames. The residual energy is due primarily to the inadequacy of the uniform translational motion assumption.

9.4.3 Subband video coding with RVQ

The concepts of subband decomposition, motion compensation and VQ can be integrated to form effective low bit rate video coding systems because they reduce the intrinsic spatial and temporal redundancies in video signals. The general issues underlying this approach are presented next and discussed in the context of a specific system that was investigated at Georgia Tech. The baseline architecture is illustrated in the block diagram shown in Figure 9.13. It employs the well-known motion-compensation-based DPCM structure [29] which also appears in the CCITT H.261 recommendation and the proposed ISO MPEG standard.

This implementation utilizes a three-step search block matching algorithm with 16×16 blocks and a maximum displacement of ± 8 for the motion estimation. Motion information is coded using 2-D Huffman coding [17] in which motion vectors are represented as a combination of horizontal and vertical components. The Huffman coding table is derived from the probability distributions of motion vectors extracted from an ensemble of video sequences. To accommodate the variation in motion information, the bit rate associated with the motion vectors changes accordingly. This adaptivity is achieved by setting a threshold for the distortion in the block matching algorithm. If the average frame difference of a block is larger than that threshold, the motion vector is transmitted. Otherwise, it is set to zero. Motion vectors, the number of which is related to the motion information, are transmitted as side information. On average, they account for about one-eighth of the total bits transmitted.

The input to the analysis filter bank (denoted by "d" in Figure 9.13) is the motion compensated residual signal. Its variance is significantly smaller than the variance of the input samples because much of the temporal redundancy has

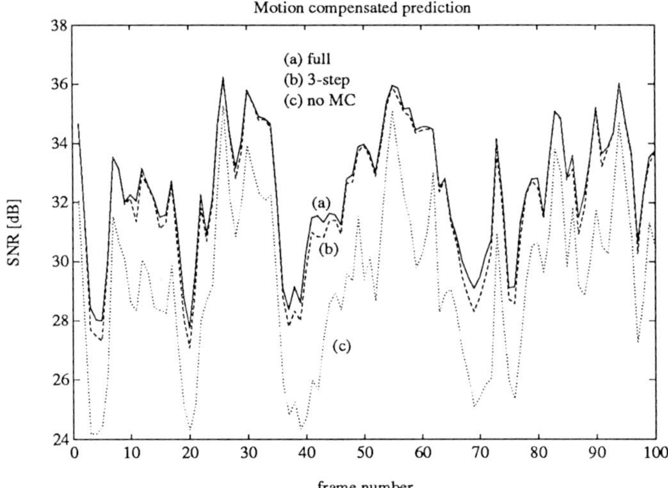

Figure 9.12: Performance of block matching for $b = 16$. a) Exhaustive search. b) Three-step search. c) No motion compensation.

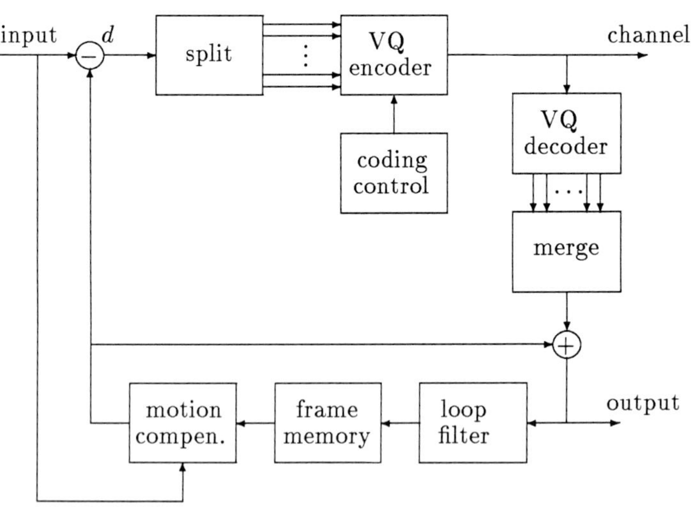

Figure 9.13: A video coding system.

been removed, making it better suited for coding.

The residual signal "d" is then decomposed spatially into 16 equal sized subband images using the efficient allpass polyphase subband decomposition mentioned in Section 9.3.4 [23]. The residual signal energy is typically distributed unevenly among the subbands which allows some to be omitted without a loss of perceptual quality. Due to this property, an energy-based subband selection rule is employed to restrict the transmitted subbands to only those that are perceptually important.

Once the dominant subband images are selected, each is uniformly divided into contiguous (4×4) blocks after which they are VQ encoded following a simple replenishment rule. The replenishment strategy is motivated by the fact that many of the vectors within each subband will contain very little energy. Thus, only vectors that will correct significant errors in the motion compensation prediction, i.e. those with energy above a pre-set threshold, are transmitted as replenishment vectors. These vectors are vector quantized with RVQ. The locations of these vectors, which must also be transmitted, are run-length encoded.

The performance of the system is very good at medium to high rates but the output contains distortions at low rates due primarily to quantization errors introduced in the VQ of the replenishment vectors. The reconstructed image frames may have a *blocky* appearance in the areas specified for replenishment. This distortion may also introduce noise accumulation and cause erroneous motion vector estimates in subsequent frames. Hence several issues must be considered in order to minimize the effects of these errors at low rates.

The first is the inclusion of a simple 3×3 smoothing filter, which is labeled *loop filter* in Figure 9.13. This helps to smooth the blocking artifacts introduced by the motion compensation. The idea of using a loop filter is common practice for this reason and is a recommendation of CCITT study group XV [30].

Most important for good performance, however, is a bit allocation that allows for good compromises among the spatial and temporal components to be coded. In cases where there is considerable motion, more motion vector information is transmitted and residual replenishment vectors are directed to spatial regions where the motion is occurring. When there is little motion, the amount of motion vector information is small. The remaining bits are used in the replenishment mode to update background vectors and correct propagation errors.

The key is the flexible control of the bit distribution among these components. Motion vector information is controlled easily by the motion estimation threshold. A novel feature of this system is that dynamic changes in the allocation to the replenishment vectors can be accommodated because RVQ is being employed. Although the subband/VQ system discussed here was restricted to 2-stage RVQ (for the replenishment vectors) in simulation, many potentially more powerful alternatives are evident. These include employing FS-RVQ, VR-RVQ and FS-VR-RVQ to quantize the replenishment vectors.

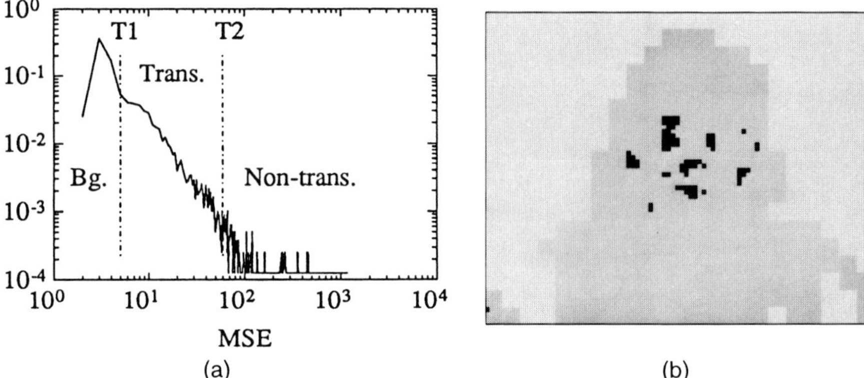

Figure 9.14: a) Normalized mean squared error plotted against the probability of occurrence for several blocks of differing sizes of the Miss America sequence. b) One frame from that sequence showing blocks whose energy is below T_1 (light), blocks whose energy is above T_2 (black), and blocks with intermediate error energy (gray).

9.4.4 Hierarchical video coding with cache VQ

This section describes a second video coder [31] that is appropriate for low rate video conferencing applications. It combines some features of a quad-tree coder with conditional replenishment, motion estimation, and cache VQ. The overall performance of the coder is impressive and both the encoding and decoding complexity are modest, which suggest that this coder might be an inexpensive candidate for hardware implementation.

The plot in Figure 9.14 provides useful insight into the operation of the coder. The video sequence frames were partitioned into square blocks and the displacement vector was found for each block that produced the best fit with the previous frame. The error between the block and its displaced estimate was computed and a histogram of the errors was plotted. The figure on the right shows a single frame of the Miss America sequence. Those blocks whose displacement error is less than T_1 are shown as lightly shaded. Those blocks whose error lies between T_1 and T_2 are shown as a darker gray, and those blocks whose errors are greater than T_2 are highlighted in black.

It is clear from that figure that those blocks with very small motion compensated difference errors occur mainly in the background, where there is no motion. The blocks with the largest errors are concentrated in the eyes, the mouth, and in areas of uncovered background where the motion is not translational. These blocks do not occur often, but they are extremely important; in fact, for video conferencing these are probably the most important features. The gray blocks represent areas where the assumption of uniform linear motion is reasonable. For the most part they are adjacent to one another and the motion estimates of each block are correlated with those of their neighbors. The coder calculates a displacement estimate for each block, measures the normalized displacement

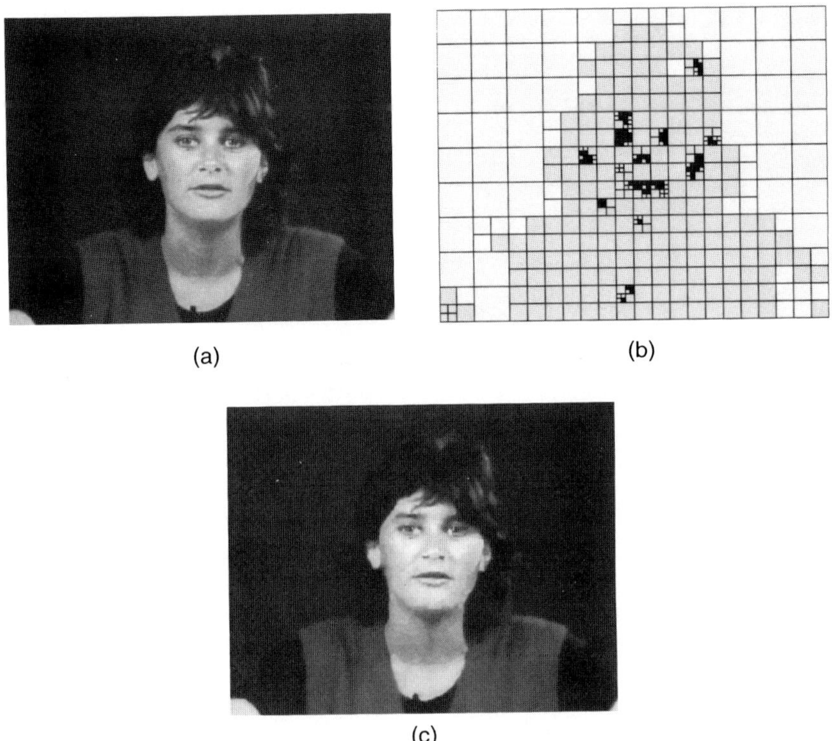

Figure 9.15: a) A single frame (#23) from the Miss America sequence. b) Illustration of the relative distribution of the six levels of the coder for this frame. c) Coded frame using 2000 bits. PSNR=37.1 dB

error and uses this error as a feature to divide the image blocks into these three classes. Each of the classes is coded differently. The background blocks are coded using replenishment; the translational motion blocks are coded using a motion coder; and the remaining perceptually important blocks are coded using a fairly high rate vector quantizer.

The coder is implemented as a hierarchy of several coders; for the example in this chapter six coders were used. Proceeding down the hierarchy the coders have progressively less distortion and require progressively more bits. When a block is to be coded the first coder is tried. If it will code the block satisfactorily, it is used and no other coders are tried; otherwise, the next coder is tried, etc. The various background and motion coders differ in the sizes of the blocks that are considered, in a manner similar to a quad-tree coder. Figure 9.15 shows the relative distribution of the six levels of the coder for a frame of the Miss America sequence. To calculate the average bit rate for the coder, let r_i be the average bit rate for the coder at level i and let h_i denote the conditional probability that a hit occurs at level i given that it did not occur at level $i-1$. Then if r_{ei}

denotes the effective rate at level i for encoding all of the blocks at higher levels, we see

$$\begin{aligned} r_{e6} &= r_6 \\ r_{e5} &= r_{o5} + h_5 r_5 + (1 - h_5) r_{e6} \\ r_{e4} &= r_{o4} + h_4 r_4 + (1 - h_4) r_{e5} \\ r_{e3} &= r_{o3} + h_3 r_3 + (1 - h_3) r_{e4} \\ r_{e2} &= r_{o2} + h_2 r_2 + (1 - h_2) r_{e3} \\ r_{e1} &= r_{o1} + h_1 r_1 + (1 - h_1) r_{e2}. \end{aligned} \quad (9.4)$$

Note that $r_1 = r_2 = 0$ because these represent the background coder and no bits are needed to decode the background blocks. The quantities r_{oi} represent the overhead bits that must be transmitted to inform the receiver which coder is being used. The number of bits required to encode the information at the various layers depends upon how the coding is done. For this implementation cache VQ, regular VQ, and entropy coding were used, although other methods, such as RVQ, could be incorporated to effect further computational savings.

The method used to calculate successive frames described above cannot be used for the first frame. For the first frame one of the fixed frame methods described in Section 9.2 must be used and the bits required to transmit it must be amortized over the entire sequence, along with any reference frames that are sent periodically.

9.4.5 Performance comparison

Both the subband coding and hierarchical VQ systems discussed here perform well and have unique advantages. They were designed assuming slightly different input size conditions and target bit rates. Nonetheless, some useful insight can be gained by a direct comparison. The performance of the subband and hierarchical VQ coders are shown in Figures 9.16. The plot shows the PSNR produced by each coder over 30 frames taken from the Miss America standard test sequence. Some care must be taken in interpreting this plot because the two examples shown were coded at different bit rates, and they were trained on different sequences. For the case shown, the spatial extent of the sequence was 256 $\times$ 256 and the frame rate was 10 frames per second. Training data for the subband VQ was obtained from the first twenty frames of the sequence and the performance was measured on the following thirty frames. Over the thirty frames the average PSNR was slightly greater than 38.1 dB. This method of training biases the performance of the coder slightly and some degradation would be expected if an unrelated sequence were used for VQ training. However, the coder still performs well on a variety of other image sequences using the Miss America trained codebooks.

The hierarchical VQ system codes at a significantly lower average bit rate. The rate for the examples shown is 25.2 kbps, although more recent adjustments to the coder have reduced this rate to 20.1 kbps while improving the quality

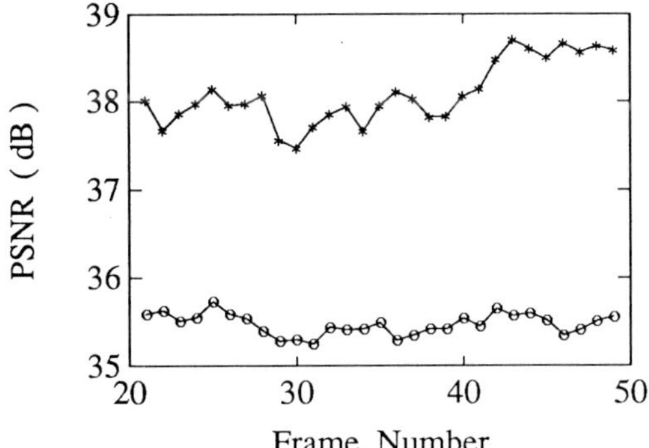

Figure 9.16: Plot of the PSNR of the two video coders versus frame number for a segment of the Miss America sequence. The sequence size was 256 × 256 operating at 10 frames per second. The top curve corresponds to the subband VQ coder operating at 64 kbps; the lower curve corresponds to the hierarchical coder operating at 25.2 kbps.

slightly. The actual sequence coded was 352 × 288 of which the center 256 × 256 samples are displayed. Its average PSNR was 35.5 dB (later improved to 37.72 kbps) over 80 frames. The coder was trained using a portfolio of head and shoulder still images, none of which consisted of frames from the sequence being encoded. PSNR data can be useful in gauging the relative performance of the coders. It is evident from Figure 9.16 that the performance of the two coders is not uniform from frame to frame, because the amount of motion varies. It should be noted that while the performance varies from frame to frame, the local maxima and minima of the PSNR curves for the two systems do not occur at the same places. This is due to the inherent differences in the methods being employed. Thus comparisons should be made with these facts in mind. We can get a rough subjective measure of quality by examining selected frames in the image sequence, one that represents a high quality image (i.e. only a small amount of motion), one that represents a low quality image (i.e. a large amount of motion), and one that represents an average quality level for each of the coders. Figure 9.17 provides such a comparison. Nine pictures are shown in this picture. The rows correspond to the image frames #25, #48, and #31 which represent high SNR, typical SNR and low SNR cases. The images in each row (from left to right) correspond to the original frame, the subband VQ coder, and the hierarchical VQ coder, respectively. Thus by examining the columns you can see the quality associated with the two coders in comparison to each other and to the original.

The quality of the coded sequences is good in both cases, but the nature of the distortion that appears is different. As expected, the quality for the subband system (which operated at the 64 kbps rate) is better that the hierarchical VQ system (which operated at the 25.2 kbps rate). But the hierarchical system has a couple of attractive features of great practical importance. Its complexity is

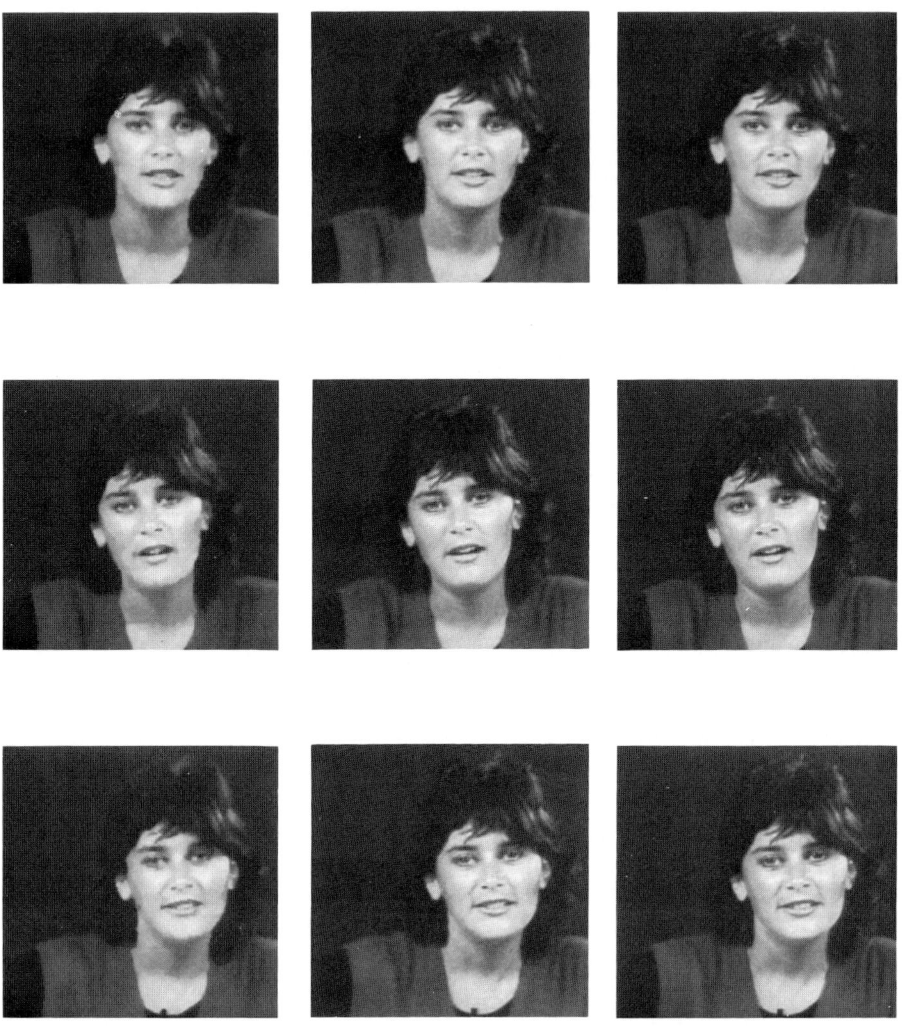

Figure 9.17: Comparison of three different frames of the Miss America sequence. The three rows represent frames #25, #48, and #31. The first column shows the original image frames. The second column shows those frames coded with subband VQ. The third shows frames coded with hierarchical VQ.

considerably less than that of the subband VQ coder and it is suitable for real-time implementation. In addition, the operational bit rate can be changed easily without a redesign of the codebooks, which makes it suitable for many practical applications. This task is slightly more difficult for the subband VQ system.

9.5 Summary and Discussion

Unquestionably, VQ is an attractive technique for signal coding offering many advantages over scalar quantization. Its application to video coding is a broad and interesting problem area with much remaining to explore and understand. The spectrum of VQ methods that has been reported in the literature addresses a broad range of tradeoffs among the important design objectives: quality of performance, encoder/decoder numerical complexity, bit rate, vector size, and required memory for codebook storage. In this chapter, RVQ and its variates (finite state RVQ, variable rate RVQ, etc.) were shown to provide a tradeoff among these objectives that was particularly attractive for video coding applications. In particular, it allows for the use of large dimensional vectors thereby providing that additional latitude to improve performance while imposing only a modest cost in computational complexity and memory.

The finite-state and large block RVQ methods were shown to achieve dramatic improvement over classified VQ with 4×4 vectors due to the way in which they exploit spatial correlation. It was further shown that adaptivity could be incorporated into the VQ coding by using ideas borrowed from cache memory principles in the computer literature. The specific intent here is to indicate ways in which these developments can be applied to the video sequence coding problem. None of the systems described have been fully engineered and optimized and thus the ultimate limits on their performance are an area of continuing investigation. Moreover, the two video coding systems developed in this chapter are mere examples of possible ways in which video coding performance gains can be achieved using VQ principles.

The subband VQ system, for example, illustrates one way in which subband coding, motion compensation and vector quantization can be effectively combined. A key feature in the success of this system is the dynamic allocation of bits among the subband images. This variable rate coding of the subband images is made possible by the inherent structure of RVQ. Similarly, the hierarchical VQ system illustrates a way in which quad-tree principles, conditional replenishment, motion estimation, cache VQ, and VQ can be effectively combined.

Much of the success of the hierarchical VQ system is owed to the cache memory ideas employed which effectively make the VQ adaptive. Not much imagination is needed to conceive of other combinations and permutations of the principles discussed in this chapter, some of which may lead to quality improvements. Many aspects of this work remain topics of continuing investigation.

Acknowledgment

This research was supported in part by the Joint Services Electronics Program under contract number DAAC-03-90-C-0004 and by the National Science Foundation under contract number MIP-9116113.

References

[1] R. Gray, "Vector quantization," *ASSP Magazine,* pp. 4–29, April 1984.

[2] A. Gersho and R. Gray, "Vector quantization and signal compression," Kluwer Academic Publishers, 1992.

[3] W. H. Equitz, "New vector quantization clustering algorithm," *IEEE Trans. Acoustics, Speech, Signal Processing,* vol. ASSP-37, pp. 1568–1575, October 1989.

[4] K. K. Truong and R. M. Mersereau, "Structural image codebooks and the self-organizing feature map algorithm," *Proc. 1990 IEEE Int. Conf. Acoustics, Speech, Signal Processing,* pp. 2789–2792.

[5] N. Nasrabadi and R. Feng, "Vector quantization of images based upon Kohonen self-organizing feature map," *Proc. 1988 IEEE Int. Conf. Neural Networks,* pp. 101–108.

[6] K. K. Truong, "Multilayer Kohonen image codebooks with a logarithmic search complexity," *Proc. 1991 IEEE Int. Conf. Acoustics, Speech, and Signal Processing,* pp. 2789-2792.

[7] B. H. Juang and A. H. Gray, "Multiple stage vector quantization for speech coding," *Proc. 1982 IEEE Int. Conf. Acoustics, Speech, Signal Processing,* pp. 597–600.

[8] J. Foster, R. M. Gray and M. O. Dunham, "Finite-State Vector Quantization for Waveform Coding," *IEEE Trans. Info. Th.,* vol. IT-31, pp. 348–359, May 1985.

[9] R. Aravind and A. Gersho, "Low-rate image coding with finite-state vector quantization," *Proc. 1986 IEEE Int. Conf. Acoust., Speech, Signal Processing,* pp. 137–140.

[10] J. Makhoul, S. Roucos, and H. Gish, "Vector quantization in speech coding," *Proc. IEEE,* vol. 73, pp. 1551–1581, November 1985.

[11] F. Kossentini, M. Smith, and C. Barnes, "Image coding with variable rate RVQ," *Proc. 1992 IEEE Int. Conf. Acoustics, Speech Signal Proc.,* pp. 369–372.

[12] C. F. Barnes, *Residual Quantizers*, PhD thesis, Brigham Young University, Provo, Utah, Dec. 1989.

[13] F. Kossentini, M. Smith, and C. Barnes, "Large block RVQ with multipath searching," Proc. 1992 IEEE Int. Symp. Circ. Syst., pp. 300–303.

[14] F. Kossentini, M. Smith, and C. Barnes, "Finite-state residual vector quantization" submitted to *J. Vis. Commun. Im. Rep.*, 1992.

[15] K. K. Truong and R. M. Mersereau, "Variable rate vector quantization for images based on principles of cache memory design,"submitted to *IEEE Trans. Image Processing,* 1992.

[16] M. Liou, "Overview of the p×64 kbit/s video coding standard," *Comm. ACM,* vol. 34, pp. 59–63, April 1991.

[17] D. A. Huffman, "A method for the construction of minimum redundancy codes," *Proc. IEEE*, vol. 40, pp. 1098–1101, 1952.

[18] E. G. Coffman and P. J. Denning, *Operating Systems Theory,* Prentice-Hall, 1973.

[19] H. S. Stone, *High-Performance Computer Architecture,* Addison-Wesley, 1987.

[20] J. W. Woods and S. D. O'Neil, "Subband coding of images," *IEEE Trans. Acoustics, Speech, Signal Processing,* vol. ASSP–34, pp. 1278–1288, October 1986.

[21] C. S. Kim, J. Bruder, M. J. T. Smith, and R. M. Mersereau, "Subband coding of color images using finite state vector quantization," *Proc. 1988 IEEE Int. Conf. Acoustics, Speech, Signal Processing,* pp. 753–756.

[22] P. H. Westerink, *Subband Coding of Images,* Ph.D. Thesis, Technische Universiteit Delft, 1989.

[23] M. Smith and S. Eddins, "Analysis/synthesis systems for subband image coding," *Trans. Acoustics, Speech, Signal Processing,* vol. ASSP-38, pp. 1446–1457, August 1990.

[24] R. de Quieroz, "Subband processing of finite length signals without border distortion," *Proc. 1992 Int. Conf. Acoustics, Speech, Signal Processing,* vol. 4, pp. 613–616.

[25] S. A. Martucci, "Signal extension and noncausal filtering for subband coding of images," *SPIE Conf. Vis. Commun. Image Proc. '91: Visual Communication,* vol. 1605, pp. 137–148.

[26] R. Ansari, "Advanced television coding using exact reconstruction filter banks," in *Subband Image Coding,* (J. Woods, editor), Kluwer Academic Publishers, 1991.

[27] C. S. Kim, M. J. T. Smith, and R. M. Mersereau, "Two-stage multirate coding of color images," *Signal Processing: Image Communications,* vol. 3, pp. 79–89, February 1991.

[28] T. Koga et al., "Motion compensation interframe coding for video conferencing," *Proc. Nat. Telecommun. Conf.,* pp. G5.3.1–5, 1981.

[29] N. S. Jayant and P. Noll, *Digital Coding of Waveforms*, Englewood Cliffs, NJ:Prentice-Hall, Inc., 1984.

[30] CCITT SGXV Recommendation Ref. Model 8, "Coding for visual telephony, " Document 525, June 1989.

[31] K. K. Truong and R. M. Mersereau, "Low rate video coding using hierarchical vector quantization," submitted to *IEEE Trans. Circ. Syst. Vid. Tech.,* 1992.

10

Model-Based Image Sequence Coding

M. Buck and N. Diehl

Daimler-Benz Research Center, Ulm, Germany

10.1 Introduction

Conventional schemes for low bitrate coding of image sequences such as the video coding standard for audiovisual services at low data rates (CCITT Rec. H.261, [8]) are developed from statistical communication theory. Thus, they utilize statistical dependencies both within each picture (intra-frame mode) and between successive pictures (inter-frame mode). This leads to a block based hybrid coding structure with a frame to frame DPCM-loop and a cosine transform of the prediction error. In order to reduce the prediction error due to motion within the coding loop motion compensation techniques are used. Thus, moving regions can be predicted as well.

These algorithms work quite well for typical videophone scenes, if there is not too heavy motion. However, they have some disadvantages. First, they are not explicitely adapted to the real contents of the images. Second, the underlying block structure does not fit to object borders and leads to block artefacts. Finally, the motion estimation cannot cope with complex motions as only pure translation is considered.

A significant improvement of the picture quality with respect to a given bitrate can be obtained if images are segmented and described according to their real contents rather then being divided into static rectangular blocks. To this end, the scene is split up into separate objects. These objects are modelled and their three-dimensional motion is estimated.

This leads to a so-called model based or analysis-synthesis (coding) system. On the coder side the scene is analysed and specific model parameters and motion parameters are extracted. These are transmitted to the decoder where the scene is synthesized and displayed again.

In Section 10.2 the idea of model based coding is discussed in detail and compared to coding schemes based on image statistics. Section 10.3 describes different approaches of image description and object modelling. The most important models are the completely three-dimensional models. Beside a classification of these models in Subsection 10.3.2 several examples are given in Subsection 10.3.3. In Subsections 10.3.4 and 10.3.5. an approach implicitely describing surface shape and motion of objects by a set of mapping parameters is discussed. These parameters characterize the relation of an object region between two sucessive frames assuming that the object can be modelled by a parameterized (e.g. quadratic surface). Section 10.4. discusses motion estimation in the context of model based coding.

10.2 From Image Statistics To Image Understanding

When talking about image sequence coding, we mainly have in mind the task of data compression for digital image sequences. The importance of this task can be judged when comparing the data rates of digital image sources with the transmission capacity of available digital communication networks:

Digital studio format (CCIR Rec. 601):	216 000 Kbit/s
ISDN channel	64 Kbit/s

How can the factor of about 3000 between the two be covered? Digital image sequences contain, in terms of information theory, not only pure information, but also redundancy, which of course does not need be transmitted. In other words, such components of image data which are already known on the receiver side or which can be derived from other parts of the data stream, can be removed.

The classical approach to remove redundancy from digital images takes the image data to be a two-dimensional signal. Examining the statistics of this signal, correlations between adjacent areas in the two-dimensional spatial domain are representing redundancy. Spatial redundancy appears where pixels have similar luminance or chrominance values as their adjacent or surrounding pixels. Such pixel values are spatially correlated to each other. This corresponds to a concentration of the signal energy in the low frequency part of the signal spectrum. Transformations from the original (spatial) domain to the frequency domain, using for example a cosine transform, make this dominance in the low frequency spectral components explicit, which allows to decorrelate the data.

In the case of continuous image sequences, redundancy can be detected also in the temporal domain. Redundancy in the temporal domain appears where image regions remain unchanged or at least similar in successive images of a sequence. This leads to the concept of inter-frame prediction, which means that each image is predicted by the previous one(s), and only the prediction error (the difference between original and prediction image) has to be transmitted (frame-to-frame DPCM). The prediction error is small, of course, if there are only small changes between successive frames, which corresponds to a large degree of reduncancy.

In 1990, a video coding standard for audiovisual services at low data rates based on such methods was issued (CCITT Rec. H.261, [8]), which decomposes each image into a set of square blocks of 16x16 or 8x8 pixels each, which are then processed rather independently. This leads to a quite regular structure, which is particularly suitable for parallel processing. Due to this block decomposition, such methods are also called block based coding schemes.

Block based coding schemes are basing on pure signal statistics, and do not use any knowledge about the real contents of the scene to be coded. However, real scenes generally consist of real objects, a fact which leads to statistical dependencies in image regions that belong to one object. Such dependencies correspond to redundancy. Redundancy always refers to data, which do not actually carry information, because these data are already known, or can be derived from other parts of the data stream.

A well known application of this idea is the introduction of motion compensation into the frame-to-frame DPCM loop. If we assume that parts of an image get displaced from one image to another, a displacement vector can be estimated, which describes where we can find the content of a given region of the current image in the previous image(s). In this way the inter-frame prediction can be extended to displaced regions. This method is already contained in the CCITT Rec. H.261.

As a matter of fact, we do know more about the behavior of real objects. Generally, real objects do not only perform translatory motion, but also perform three-dimensional rotations, which the mentioned displacement compensation cannot cope with. With rotations, parts of an object appear which were invisible before, and others, which have been visible so far, may get occluded. So, in the two-dimensional image sequence, new image contents may appear even if the present objects have not changed their shape or color at all, but only their orientation. If we understand the real three-dimensional scene which is shown in the two-dimensional image sequence, changes of the image contents due to three-dimensional motion can be described in terms of objects, and not only in terms of pixels or square blocks. In other words, the basic principle holds again: If we know that there is a three-dimensional object in the scene and if the geometry and color of such an object have been determined, it is sufficient to describe these properties once. The repeated coding of shape and color details would be redundant, at least as long as they remain unchanged.

Here the principle of model based coding comes in: On the transmitter side, a three-dimensional model (shape + color) has to be constructed and adapted to the object in the corresponding input image sequence and transmitted once. During the scene, only dynamic model parameters such as three-dimensional motion parameters or illumination parameters have to be transmitted, and if necessary some data describing object deformation have to be additionally transmitted in the case of non-rigid objects. By using this information and the knowledge about the model the scene can be completely reconstructed on the receiver side.

In terms of information theory, the information which is transmitted to the receiver (transinformation) is reduced if some of the image contents are already

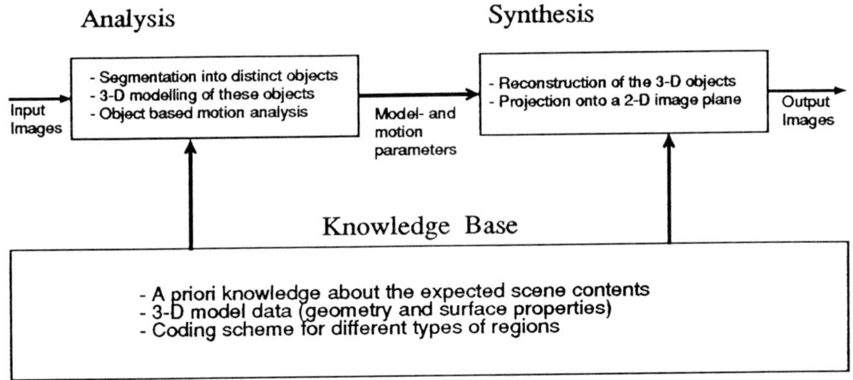

Fig. 10.1: A model based analysis-synthesis coding scheme

known, for example if there is a wellknown object in the scene. If such already known image contents were transmitted explicitly, the information would contain a large amount of redundancy.

This structure of an image sequence transmission system is also known as 'Analysis-Synthesis Coding'. Principally, this is a generalisation of the idea of model based coding but in most cases both model based coding and analysis-synthesis coding, are used as synonyms.

The structure of a model based analysis-synthesis transmission system is depicted in Figure 10.1. On the transmitter side, image analysis methods are applied to the incoming image sequence to extract specific and compact data like object description parameters, motion parameters etc. These need not necessarily be restricted to deterministic three-dimensional model parameters. Another example would be the parameters which control so-called 'Iterative Function Systems', which are used to realize a fractal modelling and coding system [29].

The analysed parameters are then transmitted to the receiver side, where they are used to control the reverse process, i.e. the reconstruction of realistic images. For any kind of parameters, there must be some general knowledge about the underlying image description concept, in terms of what kind of model is used in the case of deterministic models, or what kind of image creation process in the case of stochastic models.

This sounds logical and easy to do. However, the most difficult task, which has to be fulfilled, is the construction of the three-dimensional models from two-dimensional image data. The two-dimensional image data is a projection of the three-dimensional world onto the image plane, and this process has to be reversed during the modelling procedure, depending on the the available perspectives.

Fig. 10.2: Ambiguity in image interpretation [4]

The human visual system performs this task extremely well. Nevertheless, there are situations where a scene may be interpreted incorrectly, because what is perceived eyes sometimes can be interpreted in several ways. Figure 10.2 shows such ambiguous image contents, taken from [4], which can be interpreted as a young or an old lady.

However, in most cases the human visual system finds the correct interpretation, because it not only relies on the visual impression which it perceives, but it also has some expectations of what should be seen, according to knowledge about the whole situation. If we know that a given object is convex, the concave interpretation is not taken into account.

Similarly in computer vision, image understanding relies on both, on the image data from the input image sequence and on a priori knowledge about what is expected to be seen. A priori knowledge may refer to the whole scene or to specific properties of the contained objects. Depending on the particular application the amount of a priori knowledge, which is available to contribute to image understanding and object modelling, can vary from quite general constraints (e.g. with rigid objects) to very specific ones (e.g. with traffic scenes or video phone situations). These questions in the context of image modelling are discussed in detail in Section 10.3.

10.2.1 Differences between model based coding and block based coding

What are the major differences between the two approaches?
The motivation for the model based approach is, of course, to have a coding scheme with a good relation between image quality and compression rate. Model based coding schemes are under consideration for a wide range of applications, from very low bitrates up to highest bitrates for HDTV transmission. However, it seems that the major advantages of model based coding methods lie primarily in the field of very low bitrate applications, like mobile communication networks which offer 8 kBit/s channels.

Another difference is that model based coding schemes tend to produce a highly varying bit rate. At the beginning of a transmission all the model data have to be transmitted, which produces a data burst. During the scene, mainly some model parameters and motion parameters have to be transmitted constituting very little amounts of data. From time to time a model update might be necessary if, for example, new objects have entered the scene, which again causes data bursts. In reality, modelling failures can occur. These have to be corrected and cause data bursts as well. So, such methods are particularly suited to be used in connection with packet oriented communication networks, or ATM networks (ATM = Asynchronous transfer mode) with variable bitrates. Future digital broadband networks are planned to work on that basis.

To give an estimation of the bitrates which can be expected to occur in a model based coding system, the different transmitted types of data are listed in table 10.1 which also includes estimations of the corresponding amounts of data. The given figures are based on a scene with some rigid objects performing three-dimensional movements in front of still background.

The corresponding transmitted data for a block based coding scheme, according to CCITT rec. H.261 are given in table 10.2.

Table 10.1: Transmission data of a model based coding scheme

Type of data	transmission frequency	bit estimation
Description of the model	Once (at the first appearance of an object)	100.000 Bit
Motion Parameters	for each image	80 bit/frame per object

Table 10.2: Transmission data of a block based coding scheme

Type of data	transmission frequency	bit estimation
block addresses, motion vectors, spectral components	for each image	64 Kbit/s

Another important difference concerns the scope of application for which the two types of coding schemes are suitable. One major advantage of the block based methods is, of course, their universality, as they are able to process image sequences of any contents. So this type of coding is suited for situations where it is difficult to apply a priori knowledge because the scene contents are not known, or because the objects in the scene are too complex to be modelled, e.g. trees or a crowd of people.

On the other hand, if the scene and the contained objects can be restricted in a way which allows to describe them adequately with three-dimensional models, the coding efficiency can be increased significantly by applying model based coding schemes (see tables 10.1 and 10.2). This advantage is, however, is qualified by the risk of model specific artefacts, if an object in the scene does not perfectly correspond to the properties of the model used.

10.2.2 Artefacts

Using conventional coding schemes such as H.261 artefacts mainly result from the underlying block structure used both for the displacement estimation and the cosine transform of the prediction error. Normally the block boundaries and the boundaries of the objects in the scene do not coincide, because the blocks are fixed and not adapted to the image contents. This can lead to disturbing artefacts. If ,for instance, the boundary between two differently moving objects is in the middle of one block, the motion estimation will be unreliable and the prediction image will be of low quality. Furthermore, if there is a large region with homogeneous motion, this region might be split up into different blocks which may not be processed identically. Situations such as rotations or zoom cannot be described correctly if, as in conventional coding schemes, only pure translation is considered. This may lead to prediction images of poor quality.

By using model based coding schemes we will have no blocking artefacts. However, so called model artefacts may occur. Model artefacts result from the fact that the model does not sufficiently fit to the real object to be modelled, or that the model is not adjusted correctly. This may occur if the model does not have enough degrees of freedom to describe the real object. For instance, when modelling a person's head, in a given model the size of the nose might be fixed in relation to the size of the head. Thus different forms and sizes of noses could not be described correctly.

Other problems occur when a rigid object model is used but the real object is not rigid at all. Thus, when a rigid head model is used, the changing facial expression may not be described correctly when a person begins to laugh or when a person speaks. Here we have to find a compromise between very accurate and detailed models with a lot of parameters to be continously updated and very general models, which cannot sufficiently cope with the objects to be modelled.

Further problems occur when the adaption of the parameters does not work properly so that the model does not exactly correspond to the real object. This problem may be found when estimating the three-dimensional motion parameters. Motion estimation may result in slightly wrong motion parameters. For instance, small rotations may be confused with small displacements. Thus the model is moved to a slightly wrong position compared to the real object. Although the overall subjective impression of the synthesized image is good this image does not exactly correspond to the original image. The question has to be answered whether a good subjective impression is sufficient or if the synthesized image has also to be quite identical to the original image.

To assess the image quality of model based coding schemes, new criteria have to be introduced. As conventional coding schemes are mainly based on signal statistics the mean square error between original and prediction image or the signal to noise ratio are suitable, however not perfect, fidelity criteria. For model based schemes these criteria alone are not sufficient. Instead, an additional subjective assessment is necessary. Here we have to assess how "natural" a synthesized scene looks. However, it is difficult to find general subjective fidelity criteria.

10.3 Image Modelling

To describe and code the objects in an image sequence using a model based analysis-synthesis coding scheme according to Figure 10.1, first an adequate modelling of the image contents has to be performed. Adequate means that a lot of factors have to be taken into consoderation to select the modelling procedure which gives the best results under given circumstances. Such considerations are for example:

- Which kind of model representations corresponds best to the kind of objects in the scene?
- How much a priori knowledge about the scene contents is available?
- Is there a real-time constraint given for the modelling procedure?
- Is there a separate modelling phase with controllable conditions?

In the following sections such questions are raised and discussed. To illustrate these rather theoretical aspects, examples of some modelling procedures will be given, considering dedicated three-dimensional models as well as a more general modelling approach based on an implicit description of surface shape and three-dimensional motion of the objects.

Beside the modelling the segmentation of a scene into several objects and the estimation of the three-dimensional motion of these objects is an important task. Often a hierarchical scene description is used. These points are dealt with in Sections 10.3.4. and 10.4.

10.3.1 Straight forward vs. analysis-by-synthesis modelling

As we could see in the previous chapter, the basic idea of model based image coding is to reconstruct the three-dimensional scene which leads to the two-dimensional image sequence, and to use a combination of a priori knowledge and an analysis of the incoming image data to succeed reconstructing the scene.

Here are two completely different types of approaches to achieve this, namely straight forward methods and analysis-by-synthesis methods.

Please note that analysis-by-synthesis modelling has to be distinguished strictly from analysis-synthesis coding. The first one refers to modelling methods that are performed exclusively on the transmitter side. The second one refers to a complete coding and transmission system, comprising image analysis on the transmitter and image synthesis on the receiver side (see Fig. 10.1).

The straight forward approach uses analysis data and a priori knowledge to construct directly the three-dimensional models for the objects in the scene The resulting model based scene description is supposed to correspond sufficiently to the original input sequence. The straight forward approach can, however, fail if for example the input sequence does not meet the a priori expectations, or if the image analysis algorithms do not work properly. To cope with such situations failure detection is necessary, which in case of a modelling failure activates a fallback mode, for example one of the block based coding schemes, which do not employ knowledge about the scenes and therefore work with all types of image contents.

The reliability of the straight forward modelling approach depends on quite precise expectations about the type of objects to be encountered, and depends directly on the reliability and robustness of the image analysis algorithms which are used to construct the three-dimensional model. If these premises are fulfilled, a fast and efficient modelling can be achieved. On the other hand, if the modelling premises are inadequate or the analysis methods are not working properly, the modelling procedure can fail completely because there is no feedback provided for in order to allow corrections.

Here the idea of analysis-by-synthesis modelling comes in. Per definition, methods that are based on this idea consist of an iterative loop in which the model is successively approximated to the original object. To achieve this, in addition to the analysis procedures which yield the model parameters, a synthesis path is inserted, which recovers two-dimensional perspectives of the model, corresponding to the two-dimensional input images. By comparing these two kinds of two-dimensional images, the original ones and the synthesized ones, a measure for the deviation between the three-dimensional model and the real object can be derived, which can be used as an 'error signal' to modify the model to get a better approximation. This principle is illustrated in Figure 10.3.

Having this possibility of iterative model approximation increases, of course, the flexibility and scope of applications significantly, however, with the disadvantage of an increased computational effort. The increased effort can be seen

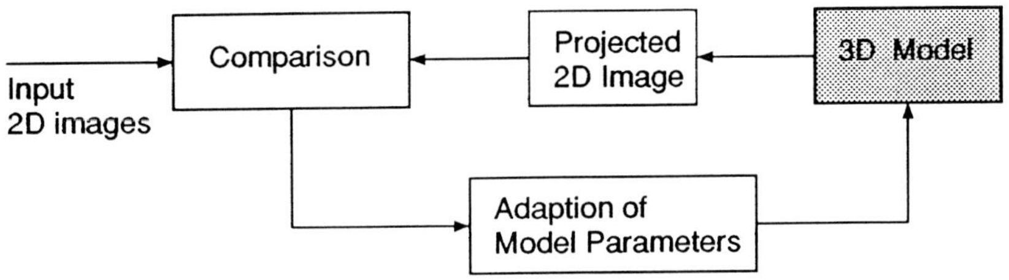

Fig. 10.3: The Analysis-by-synthesis modelling strategy

from the fact that during the modelling phase several iterations are necessary. Furthermore, the iterative approximations in a multi-parameter space have difficulties in reliably finding the absolute and not only a local minimum of the error function. Global optimization methods using simulated annealing or neural nets are quite reliable to find the absolute optimum, but with extremely high computational effort.

10.3.2 Three-dimensional modelling

Explicit vs. parametric object descriptions

As we could see in the introductory sections, the crucial characteristics of three-dimensional models in the context of image sequence coding are:

- They describe the three-dimensional process of image creation, which after the projection onto the two-dimensional image plane leads to a two-dimensional image sequence.
- They deal with the relevant parameters which are necessary to globally control the changing appearances of moving objects in a scene.

According to this distinction, existing model types can be grouped into those which particularly underline the three-dimensional shape aspect by locally describing the particular three-dimensional shape (explicit shape descriptions), and others which concentrate on the global control of the model (parametric shape description). These two aspects do not necessarily imply a contradiction, however explicit shape descriptions tend to need more data to control them, whereas parametric ones tend to neglect minor details in the shape representation. Another aspect is that explicit shape descriptions are quite general, whereas the advantage of parametric shape descriptions in terms of easy control is accompanied in most cases by some degree of dedication to particular kinds of objects.

Surface representations vs. volume representations

Another possible classification of the models is the distinction between surface and volume description. This classification is independent of the explicit vs. parametric object description. When explicitly describing the three-dimensional shape of an object, there are two approaches which theoretically are completely equivalent. These two are the surface description and the volume description. Obviously, given the surface of an object, all the inner points represent the corressponding volume description, and vice versa. In the practical context, however, there are preferences for one of the two, depending on the chosen modelling procedure, depnding on the necessary amount of data, and depending on the reconstruction. As for image sequence coding, which in most cases imposes a strong limitation of the amount of data, the surface models are clearly dominating. This is easy to understand, as in most cases the modelled objects are opaque, which means that only the object surfaces are of interest, which reduces the data effort. Moreover, most objects are in most parts locally smooth, which allows to apply data saving interpolation techniques.

a) Volume models. A very simple volume representation is the voxel representation ('voxel' corresponds to 'volume element', in analogy to 'pixel' = 'picture element') where each element in a three- dimensional bit array corresponds to a cubic volume element, set to '1' inside the object and '0' outside. This representation can be hierachically structured by grouping the object voxels to bigger cubes which are homogeneously filled with '1' or '0', called Octree structure, in analogy to the two-dimensional Quadtree structure. A detailed description of Octrees is given in [13].
Another volume representation, motivated by CAD methods, is CSG (Constructive Solid Geometry), where complex object shapes are represented by simple geometric primitives (cube, sphere, cylinder etc.), which are combined using boolean set operations (intersection, union, complement). This approach is of advantage when modelling objects with mainly geometric shapes [13].

b) Surface models. Existing surface models can be classified, according to the distinction explicit / parametric models. Most popular, thanks to their flexibility and adaptability, are wire frame models, which belong to the explicit models. Here the surface is approximated by planar polygonal patches, in most cases of triangular shape (polygon mesh, triangular mesh). Thus, the surface can be represented by the set of points defining the vertices of these triangles.

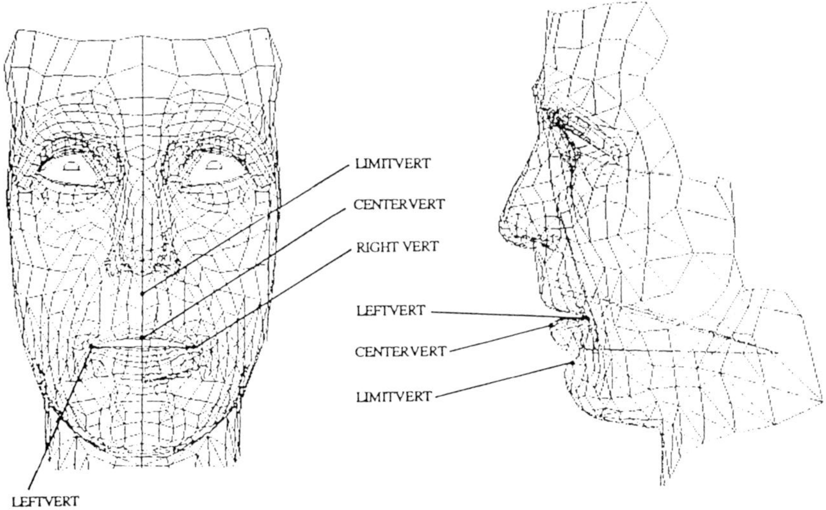

Fig. 10.4: Polygon mesh of a human face with control points for lip movements [34]

As the patch sizes may vary locally, depending on the surface complexity, they are capable of describing any object with a given demand of precision (see Figure 10.4).

The lack of smoothness of such polygon meshes can be overcome by using the set of vertices as control points for smooth surfaces like 2nd order Bezier Curves, or NURBS [13], with a similar amount of data to describe the shape. One important feature of polygon meshes, as already mentioned, is their flexibility and generality. This property however may become disadvantageous, if an object shape implies hidden or implicit constraints, which then remain unused. Such constraints, containing useful information as well, could be global geometric parameters, like the ones defining the surface of a sphere (center, radius), or could be restrictions in the variety of possible object shapes, under the assumption that only a certain set of objects may occur. To cope with such specific situations, parametric surface models are suited. Basically, the property of being parametric does not automatically imply a particular data structure. A variety of such representations can be imagined, corresponding to the variety of possible specific objects to be modelled. Simple examples are geometric objects like spheres, more specific ones are spherical harmonics to model chemical molecules [24], or potential surfaces to model smooth range data [26]. Finally, there are completely dedicated models, such as models describing human faces and facial expressions [34].

Three-dimensional shape modelling techniques

The advantage of three-dimensional models in the context of image sequence coding is that they are capable of reflecting the three-dimensional image creation process. This allows to synthesize a two-dimensional image sequence by projecting the three-dimensional objects onto the two-dimensional image plane, considering three-dimensional object motion as well. However, image sequence coding starts from two-dimensional input sequences, and the three-dimensional models have to be constructed and/or adapted using exactly this two-dimensional information, possibly together with some a priori knowledge when applying dedicated models.

Depending on the application for which image sequence coding is intended, on the kind of objects, and on the kind of model, a variety of modelling techniques can be used. We try to make some important aspects transparent in the following paragraphs.

a) Online modelling vs. offline modelling. The actual coding application influences the possible modelling techniques. For real time applications like videophone systems, the modelling and all other processing must immediately be done as the two-dimensional sequence is available from the image source. This implies, that the modelling can only be based on information that has been revealed so far during the available part of the sequence ('the past'). If the latest image contains a new perspective of an object, this can be used to modify the model accordingly. Object parts which are not visible during the scene can't be modelled - and need not be modelled: As the model is used to reproduce at the receiver side a close copy of the original input sequence, these hidden parts can be neglected. Another difficulty occurs if the image source can't be controlled very well, which also is true for the video phone: the lighting conditions might be bad, the user's position or motion in front of the camera might be inappropriate etc.

Things are different for offline applications, like image sequence storage on CD ROMs, or the production of synthetic but natural looking films. The modelling and coding procedure can be controlled very much better if necessary, input sequences can be modified and repeated, and if the very objects that appear in the scene are known in advance, they can be modelled separately in highly controlled situations, and even active sensors like laser scanners might be used.

b) Automatic vs. interactive modelling. This distinction is in a way connected with the online / offline issue. In offline applications, the modelling procedure can be designed as to extract all the necessary shape information in a well defined process, under perfect control of all the circumstances like illumination, motion etc. This procedure might consist, for example, of putting an object onto a turning platform and taking a number of views from all necessary perspectives. These can be used then to extract the three-dimensional shape using, for instance, e.g. the 'depth from disparity' method (explained below). There are however a lot of other such straight forward methods, like depth from focusing, shape from shading, contour clipping, structured lighting etc. These and other methods are described in [19].

In applications that need online modelling, automatic modelling procedures have to be used. They can be straight forward ones, which in one deterministic procedure determine the relevant model paramters, or iterative methods, typically represented by analysis-by-synthesis modelling approaches (see above). With respect to the time constraints imposed by realtime applications, iterative methods suffer from the need to complete in a fixed time interval. This might be a severe problem, as iterative methods normally depend on an error criterium and not on a given number of iterations.

Starting from some initial model, derived from a priori knowledge or other available initial data, an initial synthesized image of the modelled object is compared with the original input image of the object. If there are significant differences, a strategy must be applied to modify the model parameters as to obtain an improved approximation of the model's three-dimensional shape with respect to the shape of the original object, or more precisely spoken, to obtain an improved approximation of the object's two-dimensional projection(s) with respect to the corresponding two-dimensional projection(s) of the original object. It can be necessary to repeat this approximation step several times until a sufficiently small discrepancy between model and original has been reached (see Figure 10.3).

c) 'Low level' modelling vs. 'high level' modelling. This distinction refers to the measure of available a priori knowledge about the scene to be coded, and about the assumptions and constraints that are available when modelling the contained objects. 'Low level' in this context means, that the mere three-dimensional shape of an object is extracted, without using any semantic attributes, and with only little or no knowledge at all about the possible properties of the occurring objects. 'High level', on the other hand, refers to the fact that semantic attributes are attached to given objects ('human head', 'car', 'oxygen molecule'), and in most cases implies that more or less dedicated models are used. Dedicated models ('face model', 'car model') take advantage from the fact that a great amount of possible object shapes can be excluded beforehand, and the modelling can be restricted to adapt a reduced set of specific model parameters. This advantage, however, is accompanied by the risk of being unable to cope with unexpected objects, which do not meet the a priori assumptions.

10.3.3 Examples of modelling methods

Shape from disparity (straight forward, low level)

The object to be modelled is viewed from two cameras. For each object point which can be identified in both images, a depth value can be calculated for this point, considering the relevant geometric camera parameters as well as the positions of that point in each of the images. If this can be done for a sufficient number of points, and from different perspectives, the three-dimensional shape of the object can be obtained. However, the so called correspondence problem has to be solved for each point. This means that the positions of one object point in both images have to be determined correctly in both images, which is not trivial, due to occlusions, reflections or insufficient texture. This method is quite general, and if the correspondence problem is solved, it is possible to determine the three-dimensional shape of objects without any a priori knowledge. Object regions which can not be modelled (for example because they do not have texture) can be interpolated between modelled regions. If the resulting model is too rough, it can be used as initial step to a further model adaption, as described in the following paragraph.

Iterative wire frame adaption (analysis by synthesis, low level)

This modelling approach is maybe one of the most important ones in the field of image sequence coding, because it combines several advantageous properties.

- general (it works for objects of any shape)
- adaptive (it adapts automatically to changing object shapes)
- memory efficient
- fast (there are efficient methods to render two-dimensional images from three-dimensional models)

As already indicated by the naming, this method adapts an already existing initial wire frame model. Depending on the availability of knowledge about the present object, various methods are convenient to obtain such an initial model. Of course, the more specific the knowledge is, the closer the initial model will be to the final adapted model. In [3], the construction and adaption of a wire frame model of a human face is described. According to the restricted variety of possible objects, the modelling procedure starts with a general (average) face model, which already contains details like cheeks and a nose.
With less specific a priori knowledge, a more general initial model is suitable. For the case of an assumed head-and-shoulder scene, [22] proposes to start with the object's silhouette, (which can be obtained, for example, by motion analysis or stereoscopic information). This silhouette is then rotated, which leads to an

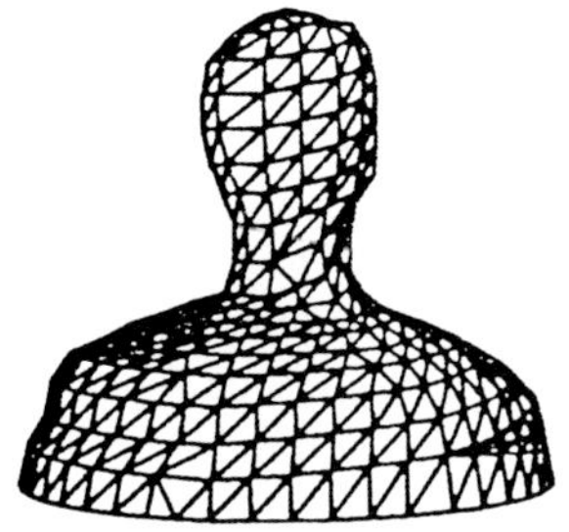

Fig. 10.5: Initial wire frame model from a silhouette (from [22])

approximate three-dimensional shape (volume model) of a head-and-shoulder object. Finally, a triangle mesh is spread onto it (Fig. 10.5).
Without any information about the object to be modelled, and if the coding application requires real time shape analysis methods, similar approaches such as the latter one may be used. However, without any assumptions about the object shape such methods may easily fail (an object with a cubic shape can not be described adequately by a rotational volume). Without realtime requirements, shape analysis methods like the ones already mentioned (three-dimensional shape from contour clipping, from stereo disparity etc) can be applied.

An important aspect of wire frame models is the choice of the polygons' sizes and their distribution over the object's surface. In fact, this question implies a trade-off between the level of detail which can be achieved (with the extreme case of infinitely small polygons), and the data efficiency of the model description. Moreover, different parts of the same object may require different polygon resolutions, depending on the local shape. In the case of a face model, the eye region should be modelled with a finer resolution than the forehead.
Another aspect is the positioning of the polygons. This is of particular importance for geometrically shaped objects with flat surface parts and sharp edges. In such cases, the vertices of the polygon mesh should ideally be positioned at the edges, to preserve the geometrical shape characteristic.
Both aspects should be taken into consideration when the initial model is initial model is chosen (if any is possible), and particularly during the consecutive model adaption.

The iterative model adaption consists mainly of modifications of the positions of the vertices (control points) of the polygon mesh, and of modifications of the mesh structure, like the introduction of new control points or the deletion of existing ones, as far as necessary. The basis for such modifications is the availability of additional information about shape details, as revealed by new perspectives of the object due to motion of the object relative to the camera. For flexible objects, even without global motion, new shape information may come up due to deformation of the object [22].

For a given view, each of the control points of the polygon mesh can be moved along the line of perspective with no (parallel projection) or small (central projection) changes of the object's appearance. If in a successive image the object has rotated relative to the camera, sliding a control point along its former line of perspective leads to a translation of that point in the current image (disparity), which can be used to determine the correct position - theoretically. Practically there are several difficulties.

- The angle of rotation should not be too small, otherwise the disparity will be small as well, and the depth calculation becomes instable.
- The mentioned disparity effect due to object rotation can easily be mixed up with translatory motion of the object. To be able to separate the two effects exactly, the motion of the object relative to the camera (translation and rotation) should be estimated as precisely as possible. This however is difficult, as the shape of the object is not exactly known.
- The correspondence problem has to be solved, i.e. in successive images the locations have to be identified which correspond to the same point of the object's surface. In regions without sufficient texture, or when there are shadow effects, this can be a very hard task.
- Iterative shape adaption not only allows to adapt and refine initial shape models, but also to update the variable shape of flexible objects. In this case, another complication may occur: local motion due to deformation superimposes the global object motion. To adapt the shape model according to the object deformation, the global motion must be separated from the deformation effects - again a difficult task if the shape is not perfectly known.

Adaption of a human face model (straight forward, high level)

The kind of methods described in the following use a priori knowledge to model particular objects with specific properties. Consequently, the structure of such models can be adapted to the particular structure of the real object, and can reflect its specific properties. Several authors use specialized wire-frame models [3, 14, 33], others dedicated models [34, 5].
In the general case, objects can have static properties which do not change along a scene, and dynamic ones which do change, so the modelling procedure can be divided into these steps:

1. Model definition (once for a given type of objects)
   Determination which object features are relevant, and construction of a specific model template which reflects these features.
2. Model initiation (once at the beginning of the image sequence)
   Identification of the selected (static and/or dynamic) features of the real

objects in the input images, and extraction of a descriptive set of parameters; these are then applied to adapt the model template.

3. Model update (during the image sequence)
Determination of the current values of the dynamic feature parameters, and updating of the model. Also motion properties have to be coped with at this stage (see chapter 10.3)

In the case of modelling a human face, relevant features could be:

- the general three-dimensional geometry of human heads
- attributes like beards or glasses
- the locations of eyes, eyebrows, nose, mouth
- a set of face muscles which perform changes in the facial expression
- the color of skin, hair etc.
- the subjectively increased importance of subregions like eyes or mouth

According to this, a model template should be used which can describe the typical shape properties, i.e. smooth surfaces (forehead, cheeks) as well as detailed elements (e.g. the eyes). Subjectively important regions can be modelled with more detail than less important or smooth ones. A given set of face muscles can be reflected by defining specific degrees of freedom for shape deformations.
Wire frame models meet most of these demands. However, a high degree of smoothness requires a detailed wireframe even for regions of low detail like the cheeks, which implies a great amount of data to descibe the model. The disadvantage of such highly detailed models is twofold: all the data must be extracted correctly from the two-dimensional input scene to adapt the model, which might be difficult and time consuming; secondly, the data compression will be less efficient when a lot of details have to be transmitted. An approach to a smooth face model is described in [6]. This model composes the three-dimensional shape of a set of horizontal slices, which are outlined by an elliptic function of the form

$$z(x,y) = z_{max}(y) \cdot \left(1 - \frac{x^2}{A^2}\right)^{\mu(y)} \qquad (10.1)$$

Some of the slices are complemented by a parabolic shape for the nose, or circular shapes for the eyes, respectively (see Fig. 10.6). The advantage of such a model is not only its smoothness, but also the compactness of the shape parameters it is controlled by.

The particular geometry and the locations of eyes, nose etc. have to be extracted from the two-dimensional input sequence to determine A, μ(y) and z_{max}(y). Features like eyes or mouth can be identified by a combination of local

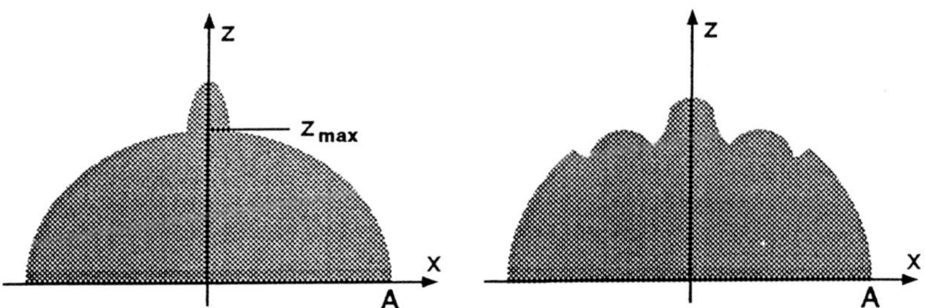

Fig. 10.6: Typical cross sections from a human face

operators and of a priori geometric relations. More precisely, the first step of this face modelling procedure is the determination of the face contour in a frontal view. From this face contour, a first three-dimensional shape without features like eyes or nose is derived by adapting a general face shape model to the given outline geometry. At that stage, the parameters of the slice functions (Eq. 10.1) are adapted. The second step is to determine the locations of eyes, nose, and mouth, using a combination of local properties, like luminance gradients or luminance/color classification, together with a priori knowledge about the relative positions and distances between these features. At these locations the model is corrected by adding shape details for eyes, nose etc, according to the sizes measured in the image.

10.3.4 Implicit models

The complete three-dimensional modelling of a scene can be very difficult if the explicit three-dimensional structure of the contained objects is not accessible. However, for some applications of model based image sequence coding a simpler modelling may be sufficient, which does not describe the object's three-dimensional shape explicitly, but only the image transformation due to three-dimensional object motion. In the work described in the following, the object's surface shape and motion is taken into account implicitely [12]. As we are primarily interested in describing scenes with moving objects, those parts of an image are defined as single objects which can be consistently characterized by a uniform motion. Successive images should be segmented with respect to inter-frame or temporal information. In addition, the boundaries of these regions should coincide with the boundaries of the real objects. To this end, information extracted from the single images (intra-frame or spatial information) such as contour, color, or texture information should be used.

By using the implicit modelling, not the complete three-dimensional structure and motion of the objects are considered but only their projection onto the camera plane. This means that only a description that reflects the connection between two successive images in the camera plane can be obtained for each object. This information is required in order to predict the image contents of

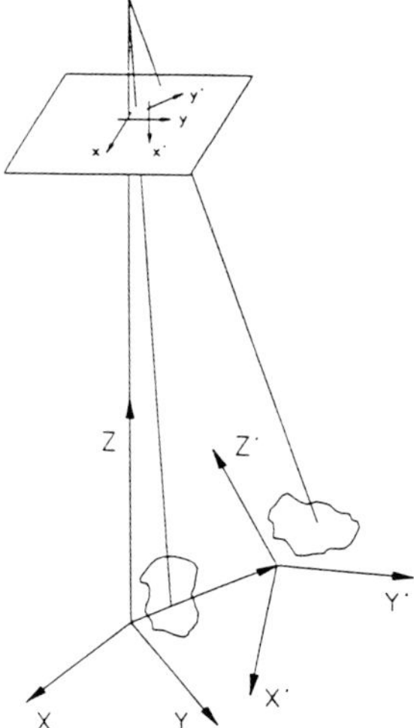

Fig. 10.7: Moving object in the three-dimensional space and central projection onto the image plane.

the second image $I_{n+1}(\mathbf{x})$ of the two successive images with respect to the first image $I_n(\mathbf{x})$. This connection is characterized by a transform $\mathbf{h}(\mathbf{x}, \mathbf{T})$ with a specific mapping vector $\mathbf{T}$ for each object. The parameters of this vector implicitely describe the three-dimensional motion (rotation, tilt, zoom and pan), the object surface and the camera model used.

To get a better understanding of the implicit modelling, a simplified signal model will be discussed. A camera looks at a single moving rigid object in the three-dimensional space in front of a homogeneous background. No occlusion effects should occur.

$$\begin{aligned} I_n(\mathbf{x}) &= S(\mathbf{x}) + N_n(\mathbf{x}) \\ I_{n+1}(\mathbf{x}) &= S(\mathbf{x}') + N_{n+1}(\mathbf{x}) \end{aligned} \tag{10.2}$$

are two successive grey-scale images of this moving object within a video scene. $S(\mathbf{x})$ and $S(\mathbf{x}')$ are the projections of the moving object onto the image plane. $N_n(\mathbf{x})$ and $N_{n+1}(\mathbf{x})$ are additive noise terms. The situation is illustrated in Figure 10.7.

Generally, the connection between two images $S(\mathbf{x})$ and $S(\mathbf{x}')$, and thus the connection between the coordinate systems $\{\mathbf{x}\}$ and $\{\mathbf{x}'\}$ in the image plane is given by a transform $\mathbf{x}' = \mathbf{h}(\mathbf{x}, \mathbf{T})$ with the mapping vector $\mathbf{T}$. The elements of $\mathbf{T}$, the mapping parameters, describe the connection between the two images if the structure of the transform $\mathbf{h}(\mathbf{x}, \mathbf{T})$ is known. They determine how a region of $I_n(\mathbf{x})$ is mapped to the corresponding region of $I_{n+1}(\mathbf{x})$.
The transform $\mathbf{h}(\mathbf{x}, \mathbf{T})$ mainly depends on three factors:

- the three-dimensional motion of the object, which is described by a three-dimensional rotation matrix $\mathbf{R}$ and a three-dimensional translation vector $\mathbf{d}$:

$$\mathbf{X}' = \mathbf{R}\,\mathbf{X} + \mathbf{d} \tag{10.3}$$

or more precisely

$$\begin{pmatrix} X' \\ Y' \\ Z' \end{pmatrix} = \begin{pmatrix} r_{11} & r_{12} & r_{13} \\ r_{21} & r_{22} & r_{23} \\ r_{31} & r_{32} & r_{33} \end{pmatrix} \begin{pmatrix} X \\ Y \\ Z \end{pmatrix} + \begin{pmatrix} d_1 \\ d_2 \\ d_3 \end{pmatrix} \tag{10.4}$$

$\mathbf{X} = (X, Y, Z)^T$ and $\mathbf{X}' = (X', Y', Z')^T$ are the three-dimensional coordinates.

- the mathematical model which describes the projection from the three-dimensional space onto the camera plane. This can be accomplished by central or parallel projection.

- the approximation of the object surface, for instance as a planar or parabolic surface.

Three examples will illustrate the combination of these factors to yield different types of implicit object models:

- The transform

$$x' = \frac{(1 + a_1)\,x + a_2\,y + a_3}{a_7\,x + a_8\,y + 1}, \qquad y' = \frac{a_4\,x + (1 + a_5)\,y + a_6}{a_7\,x + a_8\,y + 1} \tag{10.5}$$

describes the connection of the coordinates $\{\mathbf{x} = (x, y)^T\}$ and $\{\mathbf{x}' = (x', y')^T\}$ in the camera plane if central projection is utilized. The surface of the moving object is assumed to be planar ($Z = aX + bY + c$). The parameters $\mathbf{T} = (a_1, a_2, a_3, a_4, a_5, a_6, a_7, a_8)^T$ ('pure parameters' [36]) implicitely contain the motion parameters $(\mathbf{R}, \mathbf{d})$ and the surface information (a, b, c).

- For parallel projection and again a rigid planar patch, the coordinate transform is the affine transform

$$x' = (1 + c_1)\,x + c_2\,y + c_3\,, \qquad y' = c_4\,x + (1 + c_5)\,y + c_6 \tag{10.6}$$

with $\mathbf{T} = (c_1, c_2, c_3, c_4, c_5, c_6)^T$. This approximation of equation 10.5 assumes that the distance from the object to the camera is large compared to the object size.

- The quadratic transform

$$\begin{aligned} x' &= a_1\,x^2 + a_2\,y^2 + a_3\,xy + (1 + a_4)\,x + a_5\,y + a_6 \\ y' &= b_1\,x^2 + b_2\,y^2 + b_3\,xy + b_4\,x + (1 + b_5)\,y + b_6 \end{aligned} \tag{10.7}$$

can be used for several situations. The vector $\mathbf{T} = (a_1, ..., a_6, b_1, ..., b_6)^T$ has 12 components.

The quadratic transform describes the case of parabolic surfaces and parallel projection for the image formation exactly. The surface is characterized by

$$Z = a_{11} X^2 + a_{12} XY + a_{22} Y^2 + a_{13} X + a_{23} Y + a_{33} \tag{10.8}$$

and the equations

$$x = m X , \quad y = m Y \quad \text{and} \quad x' = m' X' , \quad y' = m' Y' \tag{10.9}$$

describe the parallel projection. The constants m and m' are determined by the focal length of the camera and the distance of the object. From Equations 10.4, 10.8, and 10.9 we obtain

$$\begin{aligned} x' &= \tfrac{m'}{m}\{ \; r_{11} x + r_{12} y + m d_1 + \\ &\quad r_{13} (a_{11} x^2 + a_{12} xy + a_{22} y^2 + m a_{13} x + m a_{23} y + m^2 a_{33})\} \\ y' &= \tfrac{m'}{m}\{ \; r_{21} x + r_{22} y + m d_2 + \\ &\quad r_{23} (a_{11} x^2 + a_{12} xy + a_{22} y^2 + m a_{13} x + m a_{23} y + m^2 a_{33})\} \end{aligned} \tag{10.10}$$

If terms with the same exponent are grouped, we finally achieve the quadratic transform of equation 10.7. Thus, $\mathbf{T}$ implicitely includes both motion information and information about the object surface.

Beside the exact description for parabolic surfaces assuming parallel projection, the quadratic transform can be used as an approximation for other situations. If Equation 10.5 is expanded in a Taylor series, the quadratic transform is a good approximation. The same is true if the objects have a quadratic surface and the image formation is described by central projection. Also, changes in the shape of non-rigid objects can be described with no additional effort if these changes can be modelled by a quadratic transform $\mathbf{h}(\mathbf{x}, \mathbf{T})$. Effects due to global zoom or pan of the camera are also included in this description since the mapping parameters implicitly contain this information.

Although the assumption of a parabolic surface is quite general and many objects can be described this way, there are some drawbacks. Objects with sharp edges such as cubes cannot be modelled directly. Each of the visible sides of the cube has to be parameterized by its own transform $\mathbf{h}(\mathbf{x}, \mathbf{T})$. Thus, the cube is split up into several objects with planar surfaces although its real motion could be described by a single rotation matrix and a single translation vector.

Nevertheless, the quadratic transform can describe many situations using twelve parameters. This is an important aspect if an object-oriented coding scheme is to be developed. To keep the side information low, only a few parameters should describe the motion of a large image region. The more flexible the transform is, the larger the region that can be described by one parameter set will be. If a quite general model is used, a high coding efficiency can be expected. The quadratic transform can be efficiently used for image segmentation and object-oriented description.

When using implicit modelling it is not completely possible to extract the real three-dimensional motion and the explicit surface structure using only the mapping parameters of the transform $\mathbf{h}(\mathbf{x}, \mathbf{T})$. This problem will not be discussed in this chapter because for image coding, as discussed before, the geometrical connection between successive images is sufficient. Since the temporal and spatial prediction of two successive images in the camera plane is of main interest, the exact three-dimensional description is not needed.

A restriction is constituted by the fact that the implicit object description is not capable of handling occlusion effects. It is only valid for those parts of objects that can be actually seen in the two images $I_n(\mathbf{x})$ and $I_{n+1}(\mathbf{x})$. Parts which are uncovered cannot be described without further information. They always represent new image contents. Of course the same is true for image contents that cannot be predicted at all, such as the opening or closing eyes or mouths.

The implicit object modelling has the advantage that little has to be known about the objects in the scene. Especially no explicit three-dimensional models of the objects are necessary. But the disadvantage is that those objects that differ too much from the assumed (e.g. parabolic) surface shape cannot be modelled accurately. For example, in a human face, the nose may occlude part of the cheek, when the head is turning to the side. The implicit model cannot deal with this occlusion effect. Instead, a quite disturbing distortion of the face would be the result. Such distortions of a face are of course not acceptable to a videophone user, therefore the implicit modelling is of no use for faces. However, it serves well for simpler objects such as the chest or the arms of a person, where it is quite effective.

10.3.5 Implicit object modelling and scene segmentation

A great advantage of the implicit modelling is the ability to combine object modelling and segmentation easily. The scene is divided into several differently moving objects und subobjects. Each object or subobject is characterized by a specific mapping vector $\mathbf{T}$, which implicitely describes its motion and surface shape. Thus, segmentation and motion estimation are treated as a combined problem.

In contrast to the simplified signal model in the preceding chapter, there are generally several moving objects in front of a structured background. Furthermore, occlusion effects occur and the objects are not necessarily rigid.

For the segmentation the term *object* has to be defined more precisely, in accordance with the described implicit modelling concept:

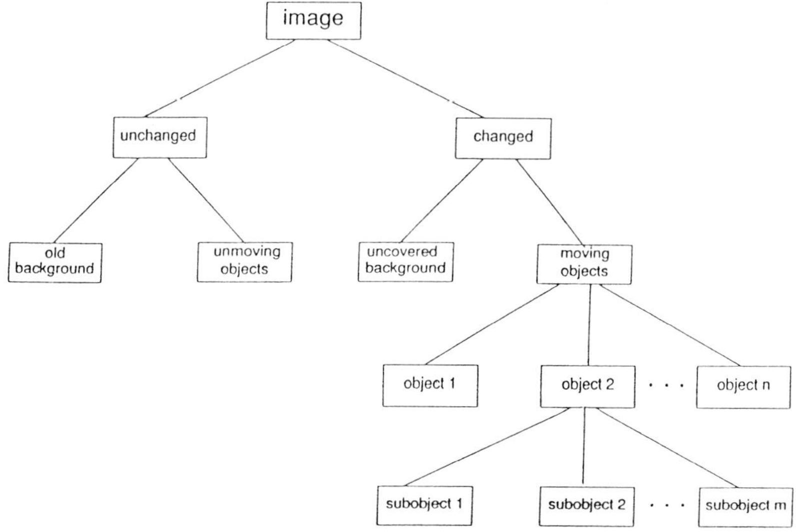

Fig. 10.8: Scheme for a hierarchical scene description.

Fig. 10.9: Image $I_n(\mathbf{x})$ from the 'Salesman' scene.

Fig. 10.10: Segmentation mask of image $I_n(\mathbf{x})$ with the objects of the first and second hierarchical level.

- Each connected region of the image is defined as a single object if for this region the geometrical connection between two successive images can be described (within some limits) by a transform $\mathbf{h}(\mathbf{x},\mathbf{T})$ with a special vector $\mathbf{T}$. In addition, this segmentation based on inter-frame information must be consistent with a segmentation based on single images (intra-frame information).

With this definition a hierarchical algorithm can be formulated, which corresponds to the analysis by synthesis modelling principle (see Fig. 10.3), and has the following features. Using two successive images $I_n(\mathbf{x})$ and $I_{n+1}(\mathbf{x})$ of a video scene a segmentation $\mathcal{S}_n$ of image $I_n(\mathbf{x})$ into several objects and subobjects is obtained. Furthermore, each object or subobject is characterized by a specific vector $\mathbf{T}$. With these vectors a prediction image $I_{np}(\mathbf{x},\widehat{\mathbf{T}})$ and a predicted segmentation $\mathcal{S}_{np}$ can be computed. $I_{np}(\mathbf{x},\widehat{\mathbf{T}})$ is relatively similar to $I_{n+1}(\mathbf{x})$ and $\mathcal{S}_{np}$ is similar to the segmentation $\mathcal{S}_{n+1}$ of $I_{n+1}(\mathbf{x})$. The algorithm is characterized by the following points:

- In a first step the regions which have changed between $I_n(\mathbf{x})$ and $I_{n+1}(\mathbf{x})$ are determined. All isolated regions of the resulting segmentation are defined as objects of hierarchical level one.
- For each of these objects a parameter vector $\mathbf{T}$ of a transform $\mathbf{h}(\mathbf{x},\mathbf{T})$ which connects the two images is estimated.
- Then the consistency of these mapping parameters is checked and those regions of each object are removed where this vector $\mathbf{T}$ is not valid. These regions are defined as objects of the second hierarchical level.

- For the objects of level two and the remaining parts of level one, the parameter vectors **T** are estimated.
- This procedure may be repeated recursively to obtain more hierarchical levels.

In addition, the segmentation is refined by the intra-frame information of $I_n(\mathbf{x})$. Contour and texture information [21], which is extracted from the single images, is combined with the segmentation based on the mapping parameters. Thus, the boundaries of the segmented objects become identical to those of the real objects of the scene. With this improved segmentation also the mapping parameter estimation can be improved furthermore.

As a result of the segmentation process we get a hierarchical object oriented description of the scene. Figure 10.8 shows a scheme for such a hierarchical description. Each object is characterized by its motion, surface, contour, and texture. An example of the segmentation is given in Figures 10.9 and 10.10 using the 'Salesman' scene.

10.4 Motion Estimation

Beside the modelling of the objects the estimation of the object motion is an important point in model based image sequence coding.

The general problem of motion estimation has become of great interest in several areas of image processing [16, 17, 27, 28] and cannot be discussed in full detail within this chapter. So far, most algorithms for three-dimensional motion estimation [2, 18] utilize special features such as corresponding points, or they are based on the displacement vector field (optical flow). In the first approach [23, 31, 36, 37, 38] special features such as corresponding points and lines are extracted from each image. Then,inter-frame correspondence is established and the observed displacement of corresponding features are utilized to compute the motion parameters of objects in the scene. The main problem of feature based methods is extracting and establishing feature correspondencies. In general this is complicated by occlusion which causes features to be hidden and hidden features to reappear. Furthermore, these methods are quite sensitive to noise in flat regions, because small errors in the correspondencies can lead to relatively large errors of the motion parameters.

The second approach [1, 15, 20, 30] is based on computing the optical flow. The optical flow is then used in conjuction with additional constraints to compute the actual three-dimensional relative velocities between objects in the scene and the camera. Computing the optical flow is quite sensitive to noise since first and second partial derivatives of image brightness values have to be evaluated. In addition, errors in the flow field will arise at object boundaries because these methods are based on certain continuity assumptions of the optical flow. Furthermore, computing the optical flow is a quite time-consuming task. Another way to compute the displacement vector field is based on hierarchical block matching [35].

However, in the context of model based image sequence coding, another approach is very successful. Ordinary grey-scale images are used and brightness constancy is assumed, except global illumination changes. Then, basically, successive images are matched to get the motion parameters. In three-dimensional motion a direct matching in the high-dimensional parameter space would be very time consuming. Instead, efficient iterative parameter estimation techniques are used [9, 10, 11], which are similar to the ones used at the implicit modelling approach in the last section.

According to equation 10.2 the connection between two successive images $I_n(\mathbf{x})$ and $I_{n+1}(\mathbf{x})$ can be descibed by

$$I_n(\mathbf{x}) = S(\mathbf{x}) \qquad I_{n+1}(\mathbf{x}) = S(\mathbf{x}') \tag{10.11}$$

neglecting the noise terms. $S(\mathbf{x})$ and $S(\mathbf{x}')$ are the projections of the moving object onto the image plane.

For the implicit modelling the transform between the coordinate systems $\{\mathbf{x}\}$ and $\{\mathbf{x}'\}$ in the image planes of the images $I_{n+1}(\mathbf{x})$ and $I_n(\mathbf{x})$ was given by the explicitely known transform

$$\mathbf{x}' = \mathbf{h}(\mathbf{x}, \mathbf{T}) \tag{10.12}$$

according to Section 10.3.4. To describe the motion of the object the mapping vector $\mathbf{T}$ of the transform $\mathbf{h}(\mathbf{x}, \mathbf{T})$ has to be estimated.

For the explictly modelled objects the situation is more complicated. Here the connection is given by

$$\mathbf{x}' = \mathbf{f}(\mathbf{x}, \mathbf{R}, \mathbf{d}) \tag{10.13}$$

The transform $\mathbf{f}(\mathbf{x}, \mathbf{R}, \mathbf{d})$ cannot be given explicitely as it depends on the three-dimensional object shape (which is supposed to be static), the three-dimensional motion characterized by the rotation matrix $\mathbf{R}$ and the three-dimensional translation vector $\mathbf{d}$, and the projection of the model onto the image plane. The three-dimensional motion parameters $\mathbf{R}$ and $\mathbf{d}$ have to be estimated.

Nevertheless, for both descriptions the method of parameter estimation is very similar. In the case of $\mathbf{h}(\mathbf{x}, \mathbf{T})$ a prediction image $I_{np}(\mathbf{x}, \widehat{\mathbf{T}})$ with the estimate $\widehat{\mathbf{T}}$ of the true vector $\mathbf{T}$ is introduced. This prediction image results from $I_n(\mathbf{x})$ through the same transformation $\mathbf{h}(\mathbf{x}, \mathbf{T})$ as $I_{n+1}(\mathbf{x})$ but now with the model parameter vector $\widehat{\mathbf{T}}$. The vector $\widehat{\mathbf{T}}$ is changed iteratively until the prediction image $I_{np}(\mathbf{x}, \widehat{\mathbf{T}})$ is as close as possible to $I_{n+1}(\mathbf{x})$ i.e. until an error criterion describing the difference between $I_{np}(\mathbf{x}, \widehat{\mathbf{T}})$ and $I_{n+1}(\mathbf{x})$ is a minimum. Thus, it is possible to predict $I_{n+1}(\mathbf{x})$ exactly from $I_n(\mathbf{x})$ if the correct transform $\mathbf{h}(\mathbf{x}, \mathbf{T})$ is chosen and $\widehat{\mathbf{T}}$ is estimated correctly.

In the case of $\mathbf{f}(\mathbf{x}, \mathbf{R}, \mathbf{d})$ the estimation of the motion parameters $\mathbf{R}, \mathbf{d}$ works quite similar. So at first the three-dimensional model is moved with a rotation

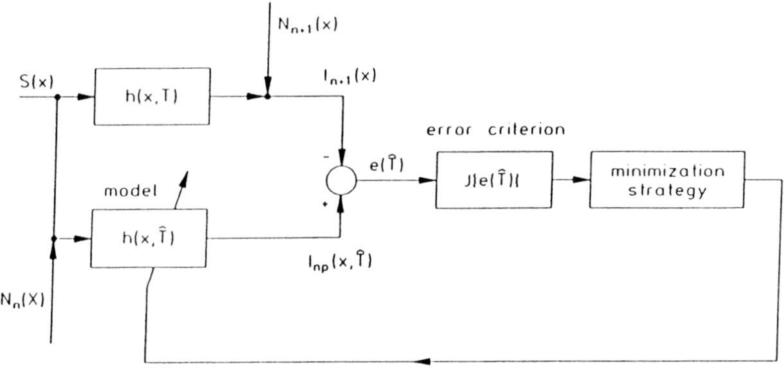

Fig. 10.11: Structure of the parameter estimation scheme.

vector $\mathbf{R}$ and a translation vector $\mathbf{d}$ to a new three-demensional position. Then it is projected onto the image plane resulting in the prediction Image $I_{np}(\mathbf{x}, \widehat{\mathbf{R}}, \widehat{\mathbf{d}})$, which is compared with $I_{n+1}(\mathbf{x})$. If there is a significant difference between the prediction image and the original, then the motion parameters are modified (according to a kind of Newton algorithm) iteratively until the difference is a minimum.

In both cases the mean square error between $I_{n+1}(\mathbf{x})$ and $I_{np}(\mathbf{x}, \widehat{\mathbf{T}})$ or $I_{np}(\mathbf{x}, \widehat{\mathbf{R}}, \widehat{\mathbf{d}})$ is used as the error criterion to measure the quality of the estimation. To find the optimal estimate, which leads to a minimum of the error criterion, very powerful algorithms are necessary to adapt the parameters systematically. It is not possible to test each combination of motion values in a multidimensional space if there are real time requirements. Therefore iterative algorithms with little computational effort are used for minimizing the estimation error. A very efficient, so-called modified Newton algorithm combined with quasi-Newton-methods is described in [10, 12]. This iterative algorithm combines several good performance features such as a large region of stability and a high, bandwidth-adaptive convergence rate with a relatively low amount of calculations within each iteration. Furthermore, the algorithm is relatively insensitive to additive noise because whole regions of the images are utilized. Similar algorithms for planar objects can be found in [20, 32].

10.5 Summary

The described methods of model based image sequence coding are a consequent application of the principle of redundancy reduction, extending the block coding schemes to three-dimensional descriptions of the image sequence contents. A variety of model types and modelling procedures have been presented, which are suited for particular coding conditions and image contents. Therefore, problems will occur if the image contents to be coded are not compatible with the chosen model or the expected scene. This is especially true for dedicated high-level models, which are less flexible than more general models. Consequently, model based coding schemes should be used preferably in applications with well defined scene and possible image contents.

For situations where modelling procedures fail, in practical systems a fall back mode has to be provided for to take over, for example one of the conventional coding schemes. To avoid such modelling failures, and to improve critical items such as the stability of the modelling procedures or the probability of artefacts, the described methods of model based image sequence coding are still being developed further.

Acknowledgement

Parts of this work have been supported by the European Community within the RACE 1018 (HIVITS) and RACE 2052 (MONALISA) projects.

References

[1] G. Adiv, "Determining three-dimensional motion and structure from optical flow generated by several moving objects", *IEEE Trans. Pattern Analysis and Machine Intelligence*, Vol. 7, No. 4, 1985, pp. 384–401.

[2] J. K. Aggarwal, N. Nandhakumar, "On the computation of motion from sequences of images — a review", *Proc. IEEE* , Vol. 76, No. 8, 1988, pp. 917–935.

[3] K.Aizawa, H.Harashima, T.Saito, *Model Based Analysis Synthesis Image Coding (MBASIC) System for a Person's Face*, Signal Processing: Image Communication Vol.1, No. 2 (1989)

[4] F.Bartlett, *Denken und Begreifen*, Kiepenheuer & Witsch 1951

[5] M.Buck, N.Diehl, *Segmentation and Modelling of Head-and-shoulder Scenes*, 3rd Int. Workshop on 64 kBit/s Coding of Moving Video, Rotterdam, 1990

[6] M.Buck, L.Groten, N.Diehl, *Construction and Adaption of a 3D Face Model from Colour Image Sequences*, Daimler-Benz Technical Report, FAU 027/91, 1991

[7] H. Busch, *Automatic Modelling of Rigid 3D Objects Using an Analysis by Synthesis System*, SPIE Symposium on Visual Communications and Image Processing, Philadelphia, USA, 1989

[8] *CCITT Recommendation H.261, VIDEO CODEC FOR AUDIOVISUAL SERVICES AT p * 64 Kbit/s*, CCITT doc. COM XV-R 37-E, 1990

[9] N. Diehl, "Methoden zur allgemeinen Bewegungsschätzung in Bildfolgen", Ph.D. Thesis, TU Hamburg-Harburg, Fortschritt-Bericht VDI, Vol. 10, No. 92, 1988.

[10] N. Diehl, H. Burkhardt, "Motion estimation in image sequences", in: K. Linkwitz, U. Hangleiter, ed., *High Precision Navigation*, Springer, 1989, pp. 297–312.

[11] N. Diehl, "Motion estimation including surface models", *Proc. SPIE: Information Processing — Applications of Digital Image Processing*, San Diego, 1989.

[12] N. Diehl, "Object-oriented motion estimation and segmentation in image sequences". *Signal Processing: Image Communication*, Vol.3, No. 12 (1991.

[13] J.D.Foley, A.v.Dam, S.K.Feiner, J.F.Hughes, *Computer Graphics, Principles and Practice*, Addison-Wesley, 1990

[14] T.Fukuhara, K.Asai, T.Murakami, *Model-Based Image Coding Using Stereoscopic Images and Hierarchical Structuring of New 3-D Wireframe Model*, VISICOM Picture Coding Symposium, Tokyo, Japan, 1991

[15] B. K. P. Horn, B. G. Schunck, "Determining optical flow", *Artificial Intelligence*, Vol. 17, 1981, pp. 185–203.

[16] T. S. Huang, ed., *Image Sequence Analysis*, Springer, 1981.

[17] T. S. Huang, ed., *Image Sequence Processing and Dynamic Scene Analysis*, Springer, 1983.

[18] T. S. Huang, "Motion analyis", in: S. C. Shapiro et al., eds., *Encyclopedia of Artificial Intelligence*, John Wiley, 1987, Vol. 1, pp. 620–632.

[19] R.A. Jarvis, *A Perspective on Range Finding Techniques for Computer Vision* IEEE Trans. PAMI-5 No.2, March 1983

[20] K.-I. Kanatani, "Structure and motion from optical flow under orthographic Projection", *Computer Vision, Graphics, and Image Processing*, Vol. 35, 1986, pp. 181–199.

[21] M. Kunt, A. Ikomonopoulos, M. Kocher, "Second generation image coding techniques", *Proc. IEEE*, Vol. 73, No. 4, 1985, pp. 549–574.

[22] C.-E. Liedtke, H. Busch, R. Koch *Automatic Modelling of 3D Moving Objects from a TV Image Sequence* SPIE/SPSE Symposium on Electronic Imaging, Santa Clara, USA, 1990

[23] H. C. Longuet-Higgins, "A computer algorithm for reconstructing a scene from projections. *Nature*, Vol. 293, 1981, pp. 133–135.

[24] N.L.Max, E.D.Getzoff, *Spherical Harmonic Molecular Surfaces* IEEE Computer Graphics & Applications, pp.42-50, July 1988

[25] F. May, "Codierung von Bildfolgen mit objektbezogener Bewegungskompensation und Signifikanzklassifikation", Ph.D. Thesis, Univ. Karlsruhe, 1984.

[26] Sh. Muraki, *Volumetric shape Description of Range Data using "Blobby Model"*, Computer Graphics, Vol. 25, No. 4, July 91

[27] H. G. Musmann, P. Pirsch, H. J. Grallert, "Advances in picture coding", *Proc. IEEE*, Vol. 73, No. 4, 1985, pp. 523–548.

[28] H. H. Nagel, "Image sequences — ten (octal) years — from phenomenology towards a theoretical foundation", *Proc. 8th Intern. Conference on Pattern Recognition*, 1986, pp. 1099–1102.

[29] H.-O. Peitgen, D. Saupe (eds.), *The Science of Fractal Images*, Springer, 1988

[30] K. Prazdny, "Egomotion and relative depth map from optical flow", *Biological Cybernetics*, Vol. 36, 1980, pp. 87–102.

[31] J. W. Roach, J. K. Aggarwal, "Determining the movements of objects from a sequence of images", *IEEE Trans. Pattern Analysis and Machine Intelligence*, Vol. 2, No. 6, 1980, pp. 554–562.

[32] P. Spoer, "Schätzung der 3-dimensionalen Bewegungsvorgänge starrer, ebener Objekte in digitalen Fernsehbildfolgen mit Hilfe von Bewegungsparametern", Ph.D. Thesis, Univ. Hannover, 1987.

[33] I.So, O.Nakamura, T.Minami, *A Study on a Model-Based Coding System Based on Isodensity Maps of Facial Images*, VISICOM Picture Coding Symposium, Tokyo, Japan, 1991

[34] N.M.Thalmann, D.Thalmann, *Synthetic Actors in Computer-Generated 3D Films*, Springer, 1990

[35] R. Thoma, M. Bierling, "Motion compensating interpolation considering covered and uncovered background", *Image communication*, Vol. 1, No. 2, 1989, pp. 191–212.

[36] R. Y. Tsai, T. S. Huang, "Estimating three-dimensional motion parameters of a rigid planar patch", *IEEE Trans. Acoustics, Speech, and Signal Processing*, Vol. 29, No. 6, 1981, pp. 1147–1152.

[37] R. Y. Tsai, T. S. Huang, "Estimating three-dimensional motion parameters of a rigid planar patch, II: Singular value decomposition", *IEEE Trans. Acoustics, Speech, and Signal Processing*, Vol. 30, No. 4, 1982, pp. 525–534.

[38] R. Y. Tsai, T. S. Huang, "Uniqueness and estimation of three-dimensional motion parameters of rigid objects with curved surfaces", *IEEE Trans. Pattern Analysis and Machine Intelligence*, Vol. 6, No. 1, 1984, pp. 13–27.

11

Human Facial Motion Analysis and Synthesis with Application to Model-Based Coding

K. Aizawa

University of Tokyo, Japan and University of Illinois, U.S.A

C. S. Choi

Myong Ji University, Korea

H. Harashima

University of Tokyo, Japan

T. S. Huang

University of Illinois, U.S.A

11.1 Introduction

Model-based coding is a new framework for image compression [1, 11, 15, 33]. In contrast to the conventional waveform coding schemes which encode and reproduce waveforms of image signals, model-based coding makes use of 3-D properties of the objects in the scene, and analyzes and transmits parameters for the objects. Because the encoder transmits only the required parameters, extremely low bit rate image transmission can be potentially realized.

There have been two approaches to model-based image coding. One approach uses general models such as planar patches and smooth surfaces to represent the 3-D properties of the objects (e.g.[29, 16, 27, 9]). The other approach utilizes detailed parameterized 3-D object models such as a parameterized facial model (e.g.[2, 13, 19, 24, 30, 37, 7, 26]). The former approach has been discussed in the context of advanced motion compensation. The latter approach is clearly different from the conventional approaches, since it uses an explicit 3-D model, and it encodes images with computer-vision-oriented techniques and it reproduces images with computer-graphics-oriented techniques. It is expected to have far broader applications than conventional waveform coding techniques. Hereafter we shall refer to the latter approach as model-based coding or 3-D model-based coding. In the following sections, we focus on facial image processing issues related to the model-based coding.

3-D model-based coding requires detailed parameterized object models. To obtain such a detailed model from general scenes is extremely difficult. However, when the object to be coded is restricted to specific classes, special knowledge obtained from a 3-D model of the object can be used in the coding system.

From this point of view, Aizawa, Harashima [4, 3, 2] and Welsh[38, 37] have proposed model-based coding schemes, independently, which utilize parameterized 3-D models of a person's face. Prior to their proposals, *semantic coding* had been proposed by Forchheimer[12], but the image of his original proposal was too synthetic for image communication, though its concept was very similar to model-based coding.

Model-based analysis synthesis image coding for a person's face

The diagram of the system proposed by Aizawa et al.[2] is shown in Fig. 11.1. This model-based coding system consists of three main components, viz. a 3-D facial model, an encoder and a decoder.

The encoder separates the object from the background, estimates the motion of the person's face, analyzes the facial expressions, and then transmits the necessary analysis parameters. The encoder will add new depth information and initially unseen portions of the object into the model by updating and correcting it if required. The decoder synthesizes and generates the output images by using a 3-D facial model and the received analysis parameters. A very low-rate transmission can be achieved by the use of this model-based image coding scheme because the encoder sends only the analysis parameters.

Analysis and synthesis of human facial images plays very important role for this model-based coding system. In this chapter, we will overview synthesis and analysis methods of human facial motions from the point of view of model-based coding, and we will describe our contributions to face modeling, synthesis[2] and analysis[7].

11.2 Face Modeling and Synthesis: Overview

Face modeling

Modeling an object is the most important part in model based coding because both analysis and synthesis methods strongly depend on the model they use.

There have been previous studies, in the computer graphics field, that handle human face animation. For example, Parke developed a parameterized model of a human face with geometrical details like eyes, mouth, etc [32]. However, such work has produced results which lack surface details and reality because only wire frame models and shading techniques were used to reconstruct a human

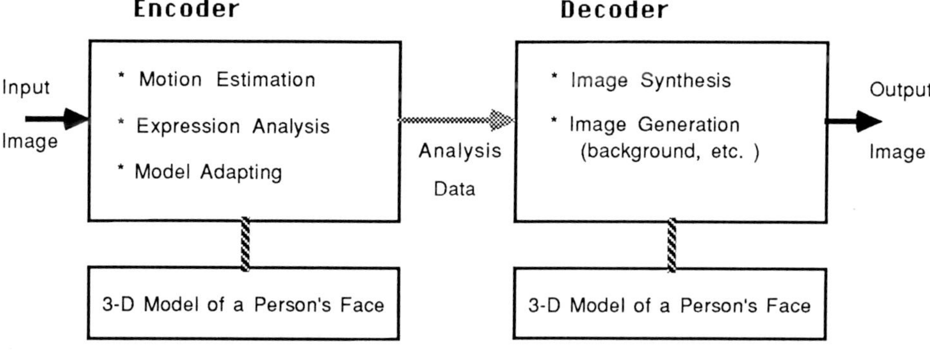

Figure 11.1: General description of a model-based analysis synthesis image coding system.

face. For image communication purposes, a person's face must be modeled in sufficient detail.

In order to develop an accurate model, an original intensity facial image is used through texture mapping technique[4, 2]. A 3-D wire frame generic face model which approximately represents a human face is utilized. The modeling process is as follows: The 3-D wire frame generic face model is scaled and adjusted so that it can fit a frontal face image of the object person. Next, the original face image is texture mapped on the adjusted wire frame model. Even this modeling process is hard to be fully automated. An interactive system for building a facial model has been developed at the University of Tokyo, which is not only for model-based coding but also for experimental studies of facial expressions by psychologists[41]. This system is also being used and improved at the University of of Illinois at Urbana-Champaign.

Once a 3-D facial model has been obtained it can be easily manipulated, e.g. rotated in any direction. The rotated images still appear natural, although the depth information of the created 3-D facial model is only a rough estimate.

Most of model-based coding systems for a person's face so far have taken a similar approach, viz. using a 3-D wire frame model and texture-mapping original images onto the model. Additional information such as side views of a face[5], continuous aspect view of a face[42] and range data [13] can be used for increasing the accuracy of the initial 3-D facial model. Typically, the wire frame model is composed of a surface triangularized by 100 - 500 triangle elements. The larger the number of triangle elements is, the better the quality of the synthesis image is, although the complexity of modeling and analysis will grow. Recently, use of 3-D range data has been attempted to model a person's face[36].

The data were obtained by using a special 3-D scanner which can acquire both range and color data with sufficiently high density.

As for the human body, a stick figure model has been proposed for describing and coding the human body motion[21].

Synthesis of facial movements

Texture mapping of original facial images onto a 3-D wire frame model gives rise to natural looking images. In addition, in order to synthesize naturally animating images, synthesis of facial movements(expressions) plays a very important role. When the 3-D wire frame model is available, facial movements can be reproduced in a number of ways. There is hierarchy of levels for the parameterization for synthesizing facial actions.

(1) Texture Level Reproduction: Reproduce facial movements by updating the texture; for example, clip-and-paste methods[2], template methods[37], model-based/waveform combined approaches[31] etc.

(2) Node Control Level: Synthesize facial movements by controlling the nodes of the wire frame model of the face; interpolation of both intensities and node positions of 3-D model templates[6].

(3) Shape Control Level: Synthesize facial movements by controlling shape of the wire frame model of the face; for example, shape parameterization [2, 7, 24] by making use of Facial Action Coding System(FACS)[10], heuristic control of shapes of facial components such as eyes and mouth[18] etc.

(4) Muscle Control Level: Parameterize and build muscle models in the wire frame model and synthesize facial actions by controlling the muscles[34, 35].

(5) Abstract Level: Control the above parameters according to more abstract parameters such as description of emotions.

In computer graphics, so far, the majority have adopted the shape control method[32]. In their work, they seems to attempt to come up with their own descriptions, though they start with some general description such as FACS. A muscle model approach has been applied[34]. Recently, a facial model which has models of three skin layers and muscles has been investigated[35].

In model-based coding proposals, the shape control method and the texture reproduction method have been commonly utilized. For example, one proposal of the texture reproduction approach uses templates for facial components(eyes, mouth), which are stored in advance, and images are updated using these templates[37]. There are a variety of shape control approaches, from heuristic control to more refined control based on facial anatomy. Another intersting approach is an interpolation method in terms of the facial shape as well as facial texture, in which a number of stored templates of 3-D facial model are interpolated into a different facial expressions[6].

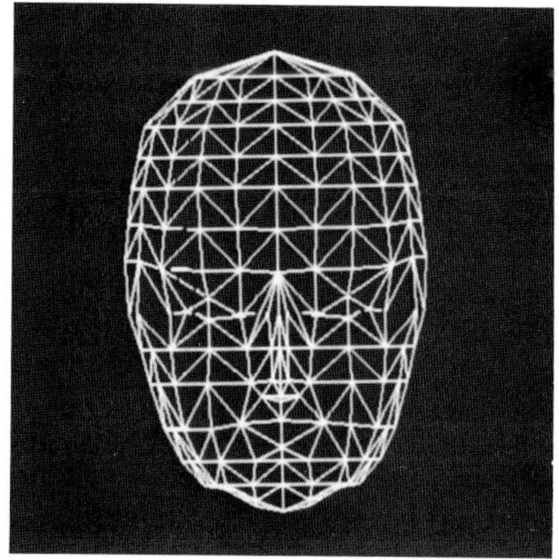
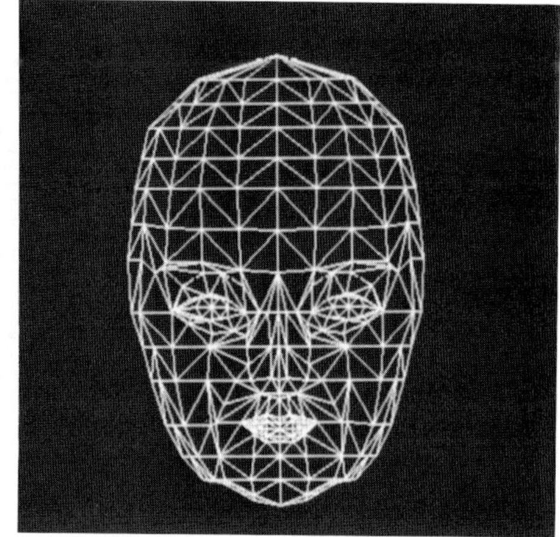

Figure 11.2: Three dimensional wire frame generic face model

In the work of Aizawa et al.[2], two different methods for synthesizing facial expressions were addressed. One technique used a *clip-and-paste method* which belong to the texture reproduction level and the other a *facial structure deformation method* which belong to the shape control method. In the next section, this work is described in detail.

11.3 Facial Modeling and Synthesis by using a Generic Face Model

11.3.1 Modeling a person's face

For image communication purposes a person's face must be modeled in sufficient detail so that the reconstructed facial image looks like the person's face. Thus in order to develop an accurate model, we utilized an original facial image and a texture mapping technique. We assumed that we are able to use only the first frame of the image sequence and that the correct 3-D information of the object is not available. We have a rough generic model and a more detailed generic model as shown in Fig. 11.2: The rough one is used for a clip-and-paste synthesis method which does not require the facial component models for eyes, mouth etc., and the detailed one is used for a structure deformation synthesis method.

A 3-D model of a person's face was constructed as follows:

(1) A 3-D wire frame *generic face model*, which approximately represents a human face, was first constructed. The wire frame model was composed of about 500 triangles (See Fig. 11.2).

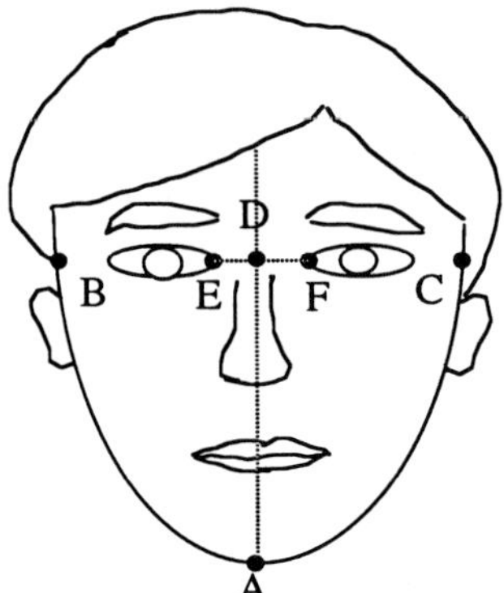

Figure 11.3: The four feature points (A,B,C,D) that are used to roughly fit the wire frame generic face model to the full-face image. D is a points which equally divides line EF.

(2) The generic face model was then 3-D affine-transformed to harmonize its four feature point positions (see Fig. 11.3) with the corresponding feature points located on a full-face 2-D image that does not have depth information. The depth of the four feature points on the full-face 2-D image (Z_{face}) are then estimated using the general face model as follows:

$$Z_{face} = Z_{model} \frac{AD_{face}}{AD_{model}} \tag{11.1}$$

where AD_{face} is length from A to D in full-face 2-D image, AD_{model} is length from A to D on the 3-D wire frame generic face model, and Z_{model} is depth of each feature point on the 3-D wire frame generic face model.

This process roughly scales and warps the 3-D wire frame generic model to fit the full-face 2-D image.

(3) The points on the lower face outline of the affine-transformed wire frame model are moved so that they are located on that of the full-face image (Fig. 11.4). The other points not on the lower face outline are also moved towards the direction of the wire frame center axis in proportion to the translations of the points on the lower face outline (Fig. 11.5).

(4) For the detailed model, the facial component models of eye brows, eyes, lips and nose need to be located so as to match those of the frontal image. Four control points are defined on each component (see. Fig. 11.13).

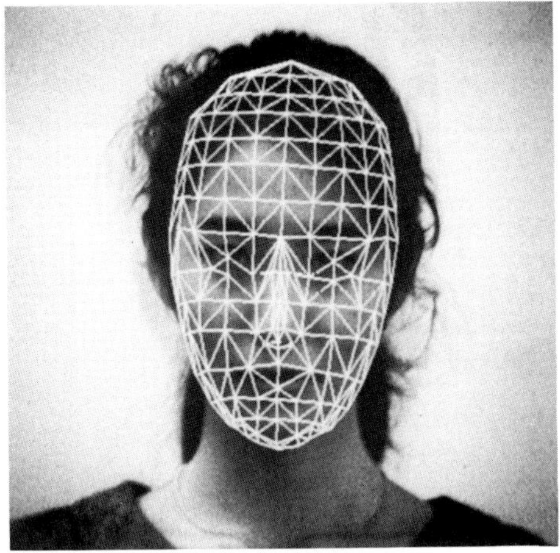

Figure 11.4: A full-face image and the adjusted wire frame model.

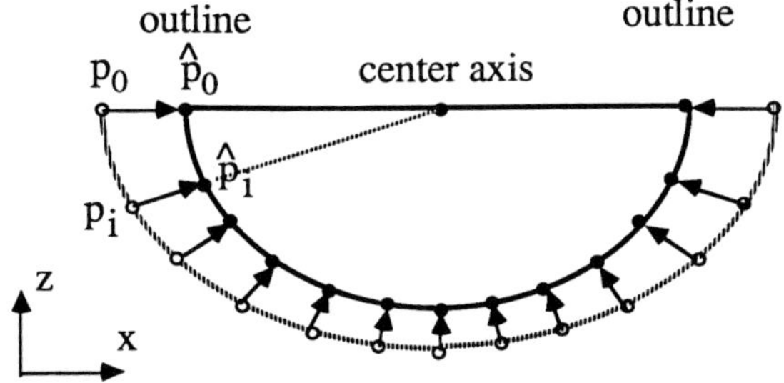

Figure 11.5: This figure shows a horizontal slice of the head. Adjustment of the points excluding the lower face outline points: Point P_0 is adjusted onto the lower face outline. The other points P_i are moved towards the direction of the wire frame center axis, such that $\hat{P}_i = P_i(1 - |\hat{P}_0 - P_0|/|P_0|)$.

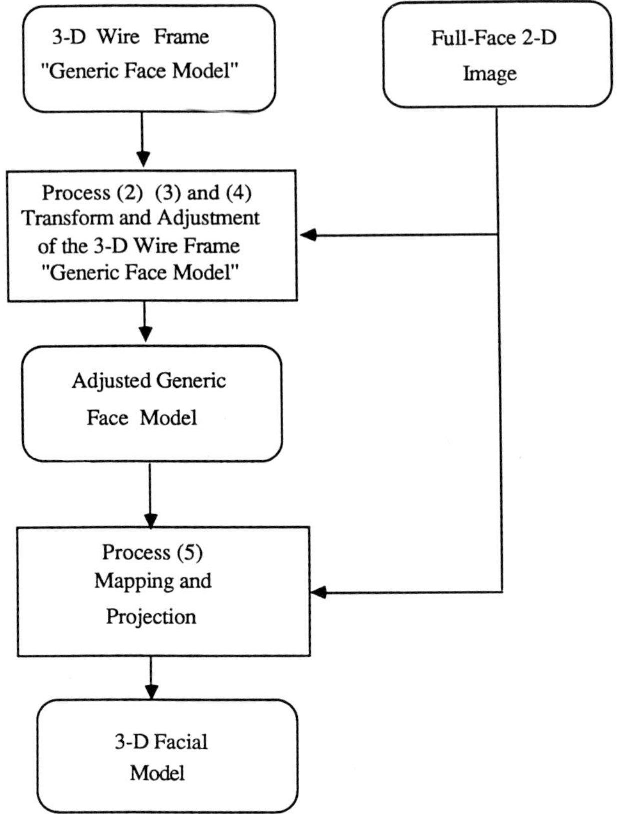

Figure 11.6: A diagram illustrating the construction of 3-D facial model.

They are located on corresponding positions of those of the frontal image. The positions of the other points which belong to the components are interpolated from the positions of the control points.

(5) The full-face 2-D image is then projected and mapped onto the *adjusted generic face model* and a *3-D facial model* is created which consists of a set of points which have 3-D coordinate values and intensities.

Figure 11.6 shows the flow diagram of this facial modeling method.

Once the 3-D facial model has been obtained it can be easily rotated in any direction. Synthesized images which have been rotated are shown in the frames of Fig. 11.7. Although the depth information on the created 3-D facial model is only a rough estimate, the synthesized images still appear natural.

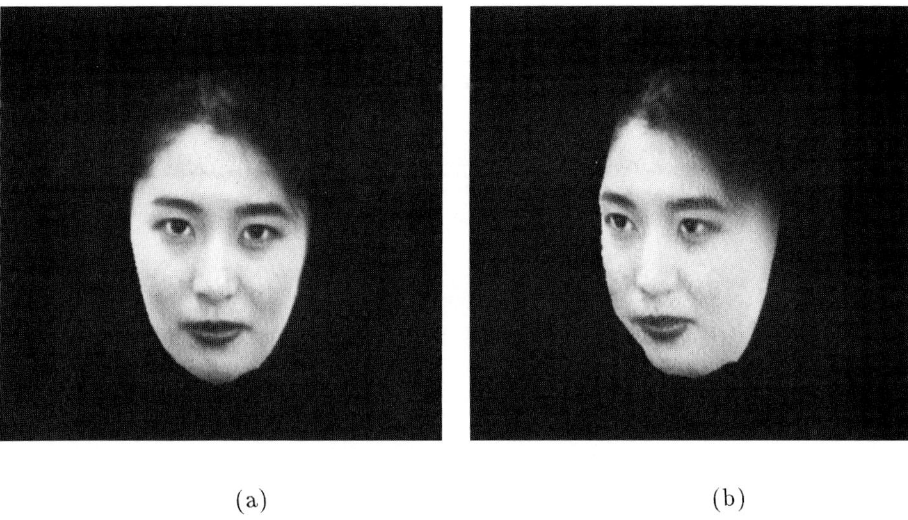

(a) (b)

Figure 11.7: Synthesized image by rotating 3-D model (a) Frontal view frame and (b) Side view frame.

11.3.2 Facial expression synthesis

The representation of facial expression plays a very important role in the model-based analysis synthesis coding system. Two different methods for synthesizing facial expressions will now be described. One technique used a *clip-and-paste method* and the other a *facial structure deformation method.* As discussed earlier, the former belongs to the texture reproduction level representation, and the latter the shape control level representation.

The clip-and-paste method is performed by extracting specific expressive regions from an input image, transmitting them, and then placing them in the corresponding regions of the synthesized image. This method is suitable for the representation of fine facial movements such as eye expressions.

The facial structure deformation method deforms the structure of the 3-D wire frame generic face model to simulate facial expressions. This method controls the locations of the vertices of the wire frame model according to *deformation rules* that are based on psychological and anatomical knowledge.

Clip and paste method

The clip-and-paste method is performed by the following procedure: Regions that must be clipped are identified on the adjusted generic face model which moves in unison with the person's head.

(1) Triangular elements of the adjusted general face model that correspond to specific expressive regions are first defined.

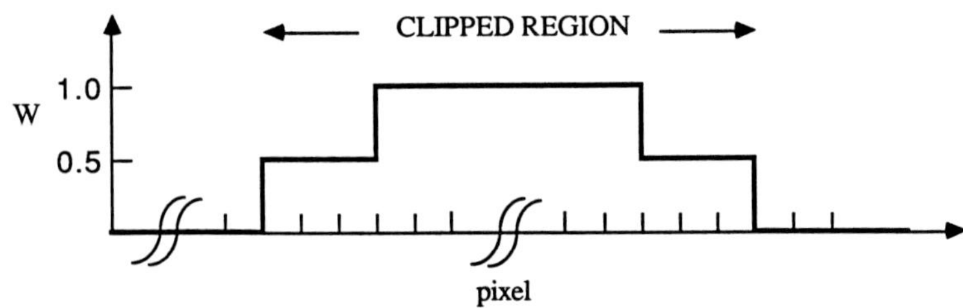

Figure 11.8: A sectional view of the weighting function $w(x, y)$. At $w = 1$, the output is the clipped signal. At $w = 0$, the output is the image synthesized from 3-D facial model.

(2) Three dimensional motion parameters of the person's head are then estimated using successive image frames. To simplify motion estimation these 3-D head motion parameters are calculated using two dimensional translations of the "white points" that were initially marked on the person's face (See Fig. 11.11).

(3) The 3-D adjusted generic face model is then translated and rotated in unison with the person's head using the previously estimated parameters.

(4) The expressive regions are then clipped from the encoder's input image using the pre-defined triangular elements.

(5) A difference signal between the clipped image and the image synthesized from the 3-D facial model is then transmitted.

(6) The difference signal is finally put into the corresponding regions of the image synthesized from the 3-D facial model.

The difference signal $d(x, y)$ is defined as follows:

$$d(x, y) = w(x, y)[f(x, y) - g(x, y)] \tag{11.2}$$

where $f(x, y)$ is the clipped signal, $g(x, y)$ is the image synthesized from 3-D facial model, and $w(x, y)$ is a weighting function used to smooth the boundaries around the clip-and-paste regions. Figure 11.8 illustrates a sectional view of $w(x, y)$.

Figure 11.9 shows a block diagram of this clip-and-paste procedure.

Figure 11.10 shows an input image and the regions to be clipped. Figure 11.11 shows the input image sequence and the synthesized images.

The 3-D movement of the head is analyzed by a least squares method using six *white points* which are initially marked on the subject's face. The depth information of these points is first roughly estimated using the 3-D facial model

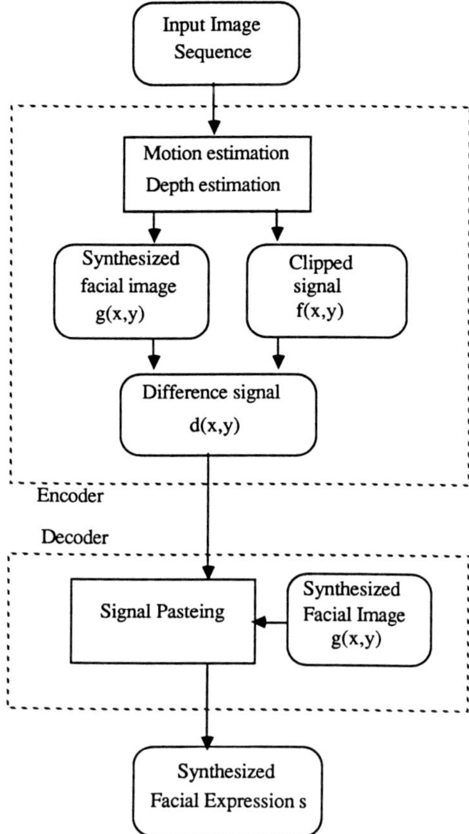

Figure 11.9: A diagram illustrating the clip and paste method.

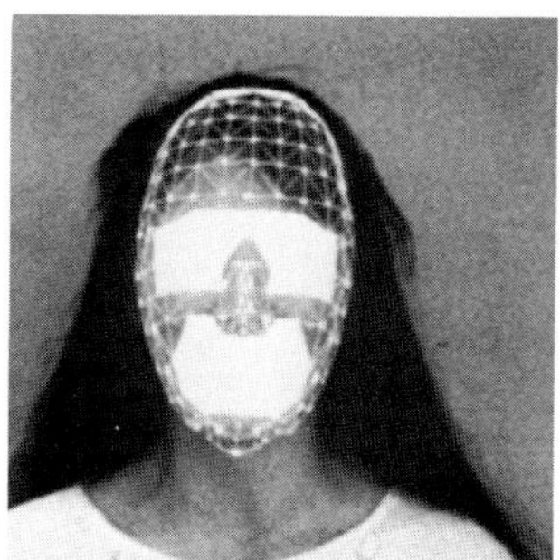

Figure 11.10: Eye and mouth area as clipped regions

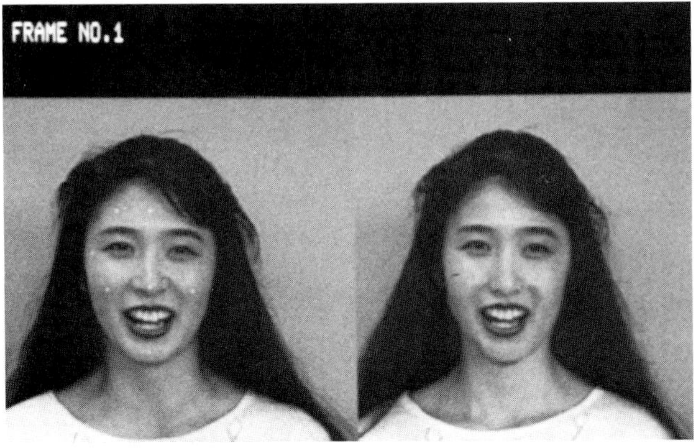

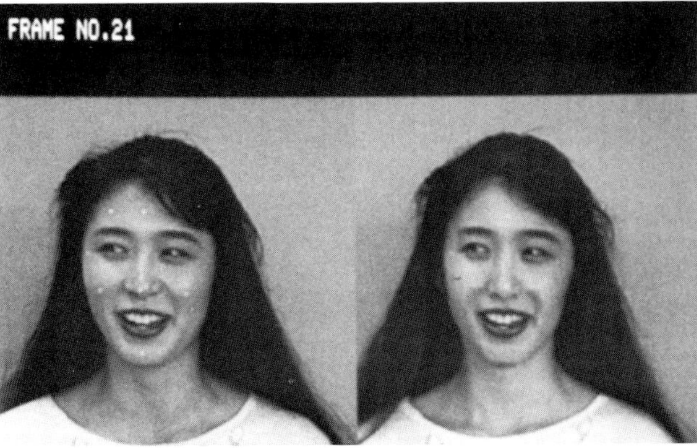

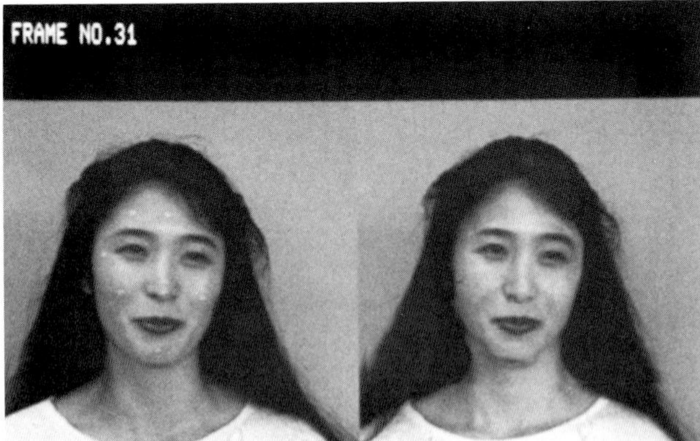

Figure 11.11: Three frames showing images synthesized by the clip-and-paste method. The input image are on the left side and the reproduced images are on the right side.

and then successively updated in conjunction with the estimation of the motion parameters[2].

The hair and shoulders are modeled by a 2-D plane and they are translated two dimensionally. When a sparsely-mapped part of the face appears, e.g. the side of the face, or a newly visible part of the face appears, additional texture data should be transmitted.

Facial structure deformation method

This method simulates facial expressions by deforming the 3-D facial model, which can be done in in many ways. Here we adopt Facial Action Coding System (FACS)[10] to describe facial actions. However, FACS itself was originally used in psychological studies and does not utilize numeric representation.

FACS describes a set of the minimal basic actions (called Action Units or AUs) performable on a human face e.g. inner brow raise, outer brow raise, lip tighten, etc. There are 44 possible AUs that are designed to be closely connected with the anatomy of the human face. Facial actions can be decomposed into AU combinations when FACS is used. If the AUs are properly parameterized on the 3-D facial model using *deformation rules*, any facial action can be synthesized. These deformation rules require the location of several control point vertices for the facial components such as the eyes, mouth, jaw, etc. Currently, 34 AUs out of 44 are parameterized in our system.

A finely detailed wire frame model (Fig. 11.2) that contains a mouth and eyes is used so as to flexibly synthesize expressions. Figure 11.12 shows an example of a deformed wire frame model. In order to use this finely detailed model, the mouth and eye shapes as well as the outline of the wire frame generic face model must be adjusted before the full-face image can be mapped onto the adjusted generic face model.

The wire frame model is deformed as follows:

(1) Control points at the vertices of each facial component are set at locations determined by the AUs deformation rule. Figure 11.13 shows these control points.

(2) Secondary control points are then located based on the AUs and the originally selected control points.

(3) The locations of all the other points are then interpolated using either a line or a quadratic curve which connects the original control points and the secondary control points.

Figure 11.14 shows an example of a lip shape deformation. The original control points are represented by black dots and the secondary control points are marked by white circles.

(4) The mapped data is then transformed at every triangular element according to the deformed wire frame model.

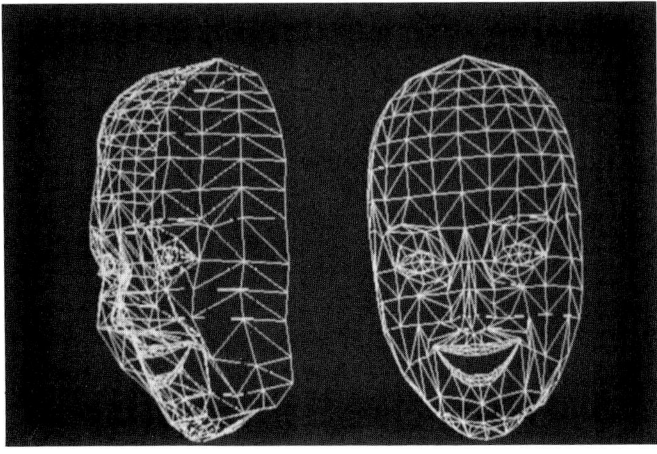

Figure 11.12: An example of a deformed wire frame model.

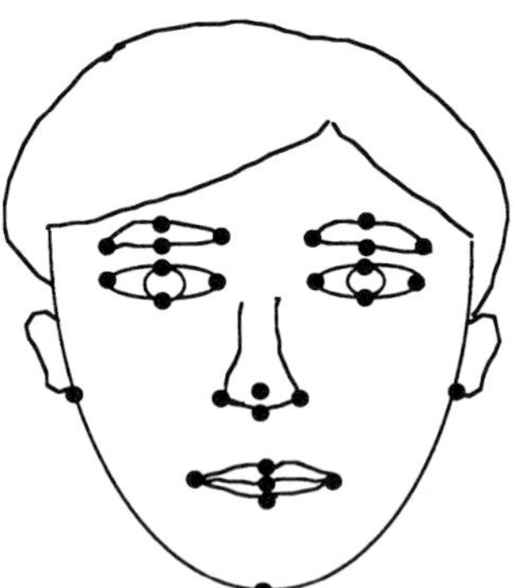

Figure 11.13: The original control points which are used to get expressions by deforming the 3-D facial model

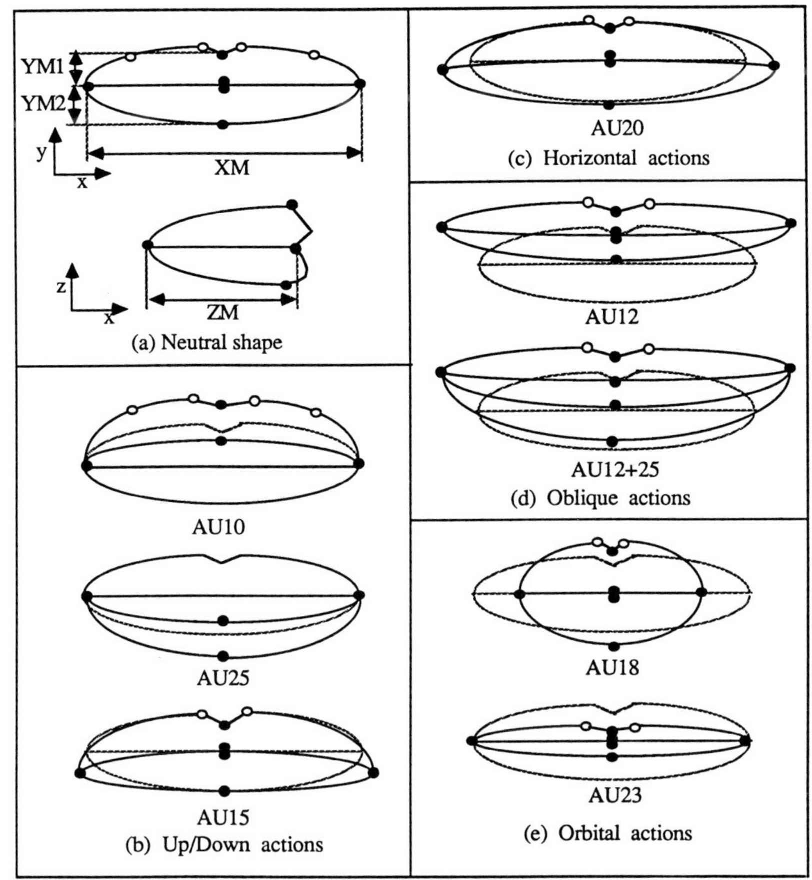

Figure 11.14: Deformation of a lip shape. The control points are marked by black dots and secondary control points by white circles. The dashed line shows the neutral shape and the solid line shows the deformed shape. The values YM1,YM2,XM, and ZM are the reference values that are used to determine the amount of movements. AU1: Upper lip raise, AU12: Lip corner pull, AU15: Lip corner depression, AU18: Lip pucker, AU20: Lip stretch, AU23: Lip tighten, AU25: Lip apart.

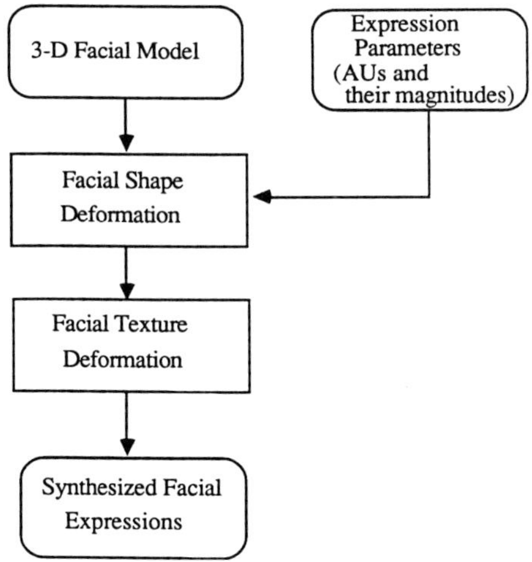

Figure 11.15: A diagram illustrated the facial structure deformation method.

Figure 11.15 summarizes the *facial structure deformation method.* The initial model and resulting synthesized images are shown in Fig. 11.16.

Expected bit-rates

Table 11.1 shows the expected bit transmission rates for both methods when a frame rate of approximately 10 frames/sec is used.

Table 11.1: Expected transmission rates

Synthesis method	Information to be transmitted	Expected transmission rates
Clip-and-paste method	3-D motion parameters and clipped image signals	1K - 10K bits/s
Facial structure deformation method	3-D motion parameters and expression parameters	less than 1K bits/s

The clip-and-paste method needs both motion parameters and clipped image signals. The frame size of the image in Fig. 11.11 is 256x256, and approximately 5% of these are clipped and transmitted. The clipped signals are expected to be compressed to much less than 1 bit/pixel when correlation with successive frames is taken advantage of, i.e. the bit rates of this method is expected to drop below 10 Kbits/s.

When the facial structure deformation method is used, the bit transmission

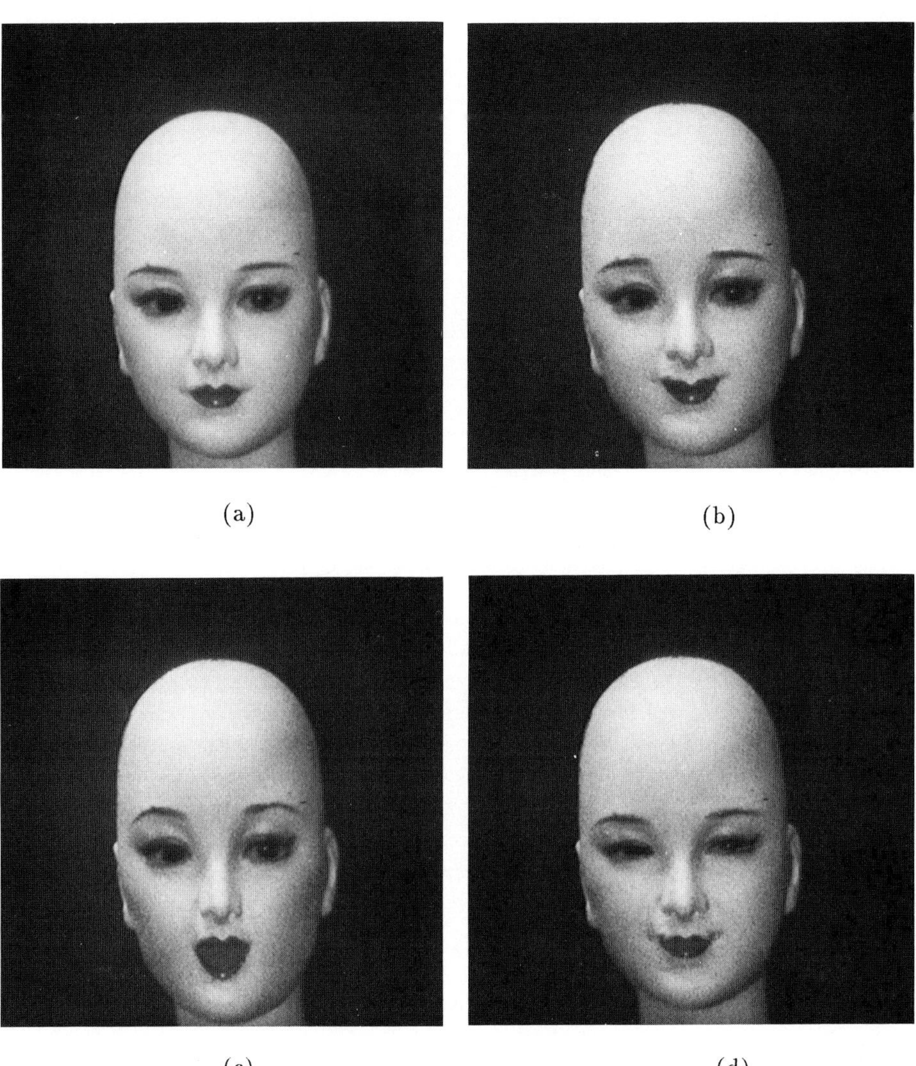

(a) (b)

(c) (d)

Figure 11.16: Images synthesized by the facial structure deformation method. (a) Initial facial image, (b) Synthesized image with AU1(Inner brow raise)+AU13(Cheek puff) (c) Synthesized image with AU1+AU2(Outer brow raise)+AU13+AU26(Jaw Drop) (d) Synthesized image with AU6(Cheek raise)+AU23(Lip tighten)+AU25(Lip part)

Figure 11.17: *Mona Lisa* synthesized with expressions

rate is expected to be under 1Kbits/s because the necessary image information contains only motion and expression parameters.

New image contents such as an uncovered background or newly appearing side part of the face will have to be transmitted as additional information, although this extra data will not be necessary if the model initially contains this information. It is also possible to supply this additional information during the image transmission. This will gradually reduce the required amount of transmission data and result in reduced transmission rates. In the particular case of uncovered background images, supplemental data will not be required if all the background data is initially provided. One of the updating method is discussed in the final section.

Application of the synthesis system

Model-based coding has much wider range of applications than the conventional waveform coding techniques. For example, one can synthesize an animated image sequence from a script, using any 3-D face model.

Figure 11.17 illustrates this. When a 3-D facial model is made from the *Mona Lisa*, and the facial structure deformation technique is applied, her movements can be visualized and her face can become more expressive. If the decoder also contains the *Mona Lisa* model, then Mona Lisa's movements can be created by the person's expressions which are generated at the encoder.

Other interesting applications have been carried out. For example, a speech-driven facial animation system[28] is developed, in which the the face model animates according to the parameters extracted from speech. *Virtual space conference*[39] is an interesting idea, in which various 3-D databases will be used together in model-based coding system.

11.4 Facial Motion Analysis: Overview

Analysis problems are much more difficult than synthesis problems. There appears to be no system, at present, which can work automatically both in modeling objects and in analyzing the sequence, even for simple head-and-shoulder images.

The analysis methods strongly depend on what is assumed as the model and what is synthesized as output images. Besides, analysis problems include several different aspects such as segmentation of objects, estimation of global motion and estimation of local motion. In the case of facial images, global motion and local motion corresponds to head motion and facial expressions, respectively. Results of motion analysis have been reported under some restrictive situations such as that the model is made in advance and the initial position of the face is known. One feature of those facial motion analysis methods for model-based coding is that they attempt to use the information of the 3D model when analyzing the image. Approaches to analyzing facial motion images includes;

- feature points extraction by thresholding[18]
- motion estimation and tracking of marks plotted on the moving object[2]
- feature point extraction and tracking by using active contour model[17]
- estimation of facial expressions (Action Units) by using feature point displacements[8]
- direct estimation of head motion and facial expressions (Action Units) using spatiotemporal gradient[7]
- global motion estimation of the head using displacement vector field obtained by the block matching technique[23]
- facial muscle movement estimation based on optical flow[25]
- estimation of facial muscle movement by using the feature lines extracted by active contour model[35]

In the next section, we describe our work of a direct estimation of facial motion by using spatiotemporal gradient method.

11.5 Facial Motion Analysis under Constraint of the Action Units

In this section, a facial motion analysis method proposed by Choi et al. [7], is described in detail. In this scheme, the facial motion is separated into global

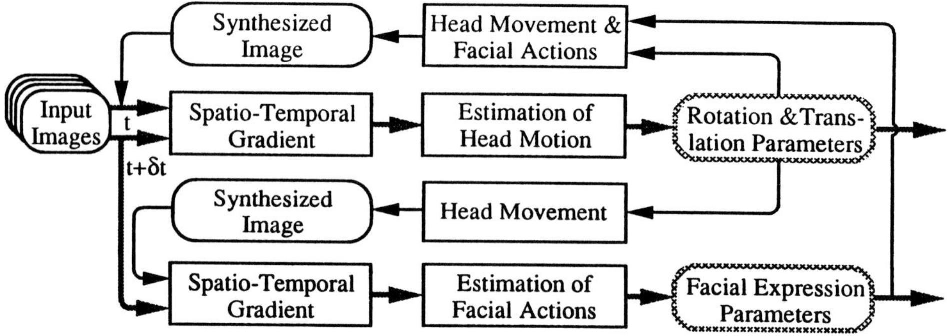

Figure 11.18: A diagram illustrating the procedure of direct estimation of head motion and facial expressions.

motion parameters and local motion parameters; the global motion parameters are the head motion parameters, and the local motion parameters are the facial actions caused by the facial expressions. Those head motion parameters and facial expression parameters are hierarchically estimated, where the head motion is treated as a rigid motion and the facial expressions are modeled as the Action Units described in the previous section.

As illustrated in Fig. 11.18, analysis of the facial image sequence consists of three steps: Head motion estimation, compensation of head motion, facial expression estimation. Those step are iterated for improving the precision. It is assumed that the the initial location of the head is known.

The motion of a nonrigid body, such as a human face, which has small deviation from a rigid body is formulated as;

$$V \quad = \quad \Omega \times (P + \delta P) + U \tag{11.3}$$

where $V = (v_x, v_y, v_z)^T, \Omega = (\omega_x, \omega_y, \omega_z)^T$ and $U = (u_x, u_y, u_z)^T$ represent velocity of a point on the object, angular velocity and translational velocity, respectively. $P = (x, y, z)^T$ is the 3-D coordinate of the point, and δP represents a small amount of deviation caused by the nonrigid movements. In the case of face motion, $V_h = \Omega \times P + U$ represents the head motion and $V_f = \Omega \times \delta P$ does the facial actions.

11.5.1 Estimation of head motion parameters

Following the usual optical flow approach, we use the following constraint between the velocity and the spatiotemporal gradient of the intensity between two

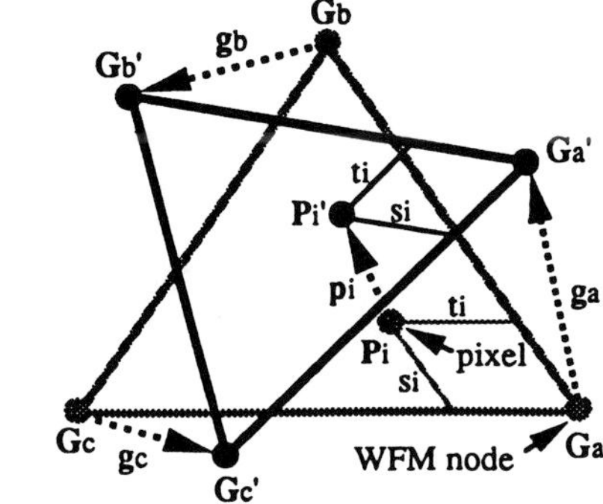

Figure 11.19: Linear approximation of the pixel points from nodes of the wire frame model.

consecutive frames, $I(x, y, t)$ and $I(x, y, t + \delta t)$:

$$I_x v_x + I_y v_y + I_t \quad = \quad 0 \tag{11.4}$$

where I_x, I_y and I_t are the partial derivatives of $I(x, y, t)$ with respect to x, y and t respectively.

Using the 3-D shape of the face as constraints, eq. 11.4 leads to direct estimation of the head motion(Ω and U) [40]. Assume that the image is an orthographic projection. Then the x and y components of V comprise the optical flow. The head motion is estimated as follows: First, the head motion is roughly estimated by ignoring the facial actions V_f in eq. 11.3. Next, the pixels affected by the facial actions are removed and the head motion is estimated again to improve accuracy. The detail is described below.

By substituting V_h into V in eq. 11.3, eq. 11.4 leads to

$$(-zI_y, zI_x, xI_y - yI_x, I_x, I_y)(\omega_x, \omega_y, \omega_z, u_x, u_y)^T \quad = \quad -I_t. \tag{11.5}$$

The z-values (depth) of points other than the nodes of the wire frame model are obtained by linearly interpolating the z-values of the nodes. Let G_a, G_b and G_c be the 3-D locations of the nodes of a triangle. Let P_i be the 3-D location of the point to be interpolated. Since the point P_i is assumed to be on the surface of the triangle, the relationship among G_a, G_b, G_c and P_i is (See Fig 11.19):

$$P_i \quad = \quad s_i(G_b - G_a) + t_i(G_c - G_a) + G_a. \tag{11.6}$$

The coefficients s_i and t_i are uniquely determined by the x and y components of G_a, G_b, G_c and P_i. Then the z-value of P_i can be obtained by using s_i, t_i and z-values of G_a, G_b, G_c.

Applying eq. 11.5 to all points(pixels) in the face area gives

$$HX = Y \tag{11.7}$$

where i-th row of H ($l \times 5$ matrix) is $(-zI_{y_i}, zI_{x_i}, xI_{y_i} - yI_{x_i}, I_{x_i}, I_{y_i})$, X is $(\omega_x, \omega_y, \omega_z, u_x, u_y)^T$, and Y is $-(I_{t_1}, I_{t_2}, ..., I_{t_l})$. l is the total number of pixels to be processed. The least mean square method estimates the unknown vector X to minimize error $\|HX - Y\|^2$. Thus, X is given by

$$X = [H^T H]^{-1} H^T Y \tag{11.8}$$

This X is the first rough estimation of the head motion because the facial action V_f is ignored. To improve the estimation, the points affected by facial actions have to be removed. Thus, eq. 11.5 is evaluated for each points by using the first rough estimation of X. If the squared difference of the two sides of eq. 11.5 is larger than the mean error $\|HX - Y\|^2/l$, the point is regarded as being influenced by facial actions. After removing those points affected by facial actions, the head motion is calculated again using the same procedure.

The above method can estimate the head motion parameters directly without any point correspondence. It seems robust because it exploits all pixels in the face area. For estimation of the facial expressions, we synthesize a reference image $I_s(x, y, t)$ by moving the facial model using the estimated head motion parameters. Then, the differences between $I_s(x, y, t)$ and $I(x, y, t + \delta t)$ is used for facial action estimation.

11.5.2 Estimation of facial expression parameters

The local facial actions cause the differences between the image $I(x, y, t + \delta t)$ and the image $I_s(x, y, t)$ which is synthesized by using the 3-D facial model and the head motion parameters. In order to analyze the facial actions, we make use of the parameterization by the Action Units(AUs), discussed in the previous section.

Similar to the constraint eq. 11.4, we can write

$$I_{sx} v_{fx} + I_{sy} v_{fy} + I_{st} = 0. \tag{11.9}$$

An AU deformation rule can be interpreted as a set of displacement vectors of nodes of the wire frame model. Conversely, assume that the displacements of the nodes can be represented as the weighted sum of those of AUs. Then, the relationship between the node displacements and the AUs is

$$g = (A_1, A_2, ..., A_m)\, a \tag{11.10}$$

$$= \tilde{A} a \tag{11.11}$$

where A_i is the set of displacements which act on the nodes according to i-th AU of its unit intensity, and $a = (a_1, a2, ..., a_m)$ are intensities of AUs. $g = (g_{x1}, g_{y1}, g_{z1}, ..., g_{zn})^T$ is a vector which is supposed to act on the nodes in the input image. The n and m are the numbers of the nodes and AUs, respectively.

Consider the relationship between the displacements of the pixels and those of the nodes. As illustrated in Fig. 11.19, suppose the triangle G_a, G_b and G_c is deformed to the triangle G'_a, G'_b and G'_c, and assume that s_i and t_i are not changed by the deformation. Let g_a, g_b and g_c be the 3-D displacement vectors of the nodes and δP_i be the 3-D displacement vector of the point P_i in the triangle. The relationship between them is

$$\delta P_i = s_i(g_b - g_a) + t_i(g_c - g_a) + g_a. \tag{11.12}$$

Applying this relationship for all triangles of the wire frame model leads to

$$\delta P = Sg \tag{11.13}$$

where $\delta P = (\delta P_1^T, \delta P_2^T, ..., \delta P_l^T)^T$ and the matrix S $(3l \times 3n)$ describes the relationship between the movements of the points and those of nodes. Combining eq. 11.11 and eq. 11.13, eq. 11.13 can be written as

$$\begin{aligned} \delta P &= S\tilde{A}a & (11.14) \\ &= \tilde{A}_s a. & (11.15) \end{aligned}$$

Since δP is the displacement of the points between two frames and this causes local facial actions V_f,

$$V_f = \Omega \times \delta P. \tag{11.16}$$

Combining eq. 11.16 and eq. 11.15 results in linear equations in a, the form of which is similar to 11.7. Consequently, a can be calculated by using least square estimation in the same way as eq. 11.8. Thus, our Action Unit parameterization enables us to directly estimate the facial actions.

11.5.3 Improving precision of estimation by iteration

Since we use first order approximations in equations such as eq. 11.3, eq. 11.4 and eq. 11.9, larger displacements between the frames will lead to poor estimation. Thus, an iterative estimation method is employed to reduce the errors. The procedure is as follows: First, the head motion parameters and facial action parameters are estimated by using the frame #1 and the frame #2, and an image is synthesized by the facial model driven by those analysis parameters. Displacements between this synthesized image and the frame #2 will be smaller than between the two original frames. A second estimation is performed between the synthesized image and the frame #2. This iteration continues until the error becomes smaller than some threshold.

11.5.4 Experimental results

Two experiments were carried out to verify the algorithm. The first experiment is for quantitative evaluation. In this case, image sequences are synthesized by using the facial model and utilized for the experiment; they are

(1) head translation up to 10 pixels with facial expressions of AU1, AU2, AU25 of intensities 0.1.

(2) head rotation up to 10 degree with facial expressions of AU1, AU2, AU25 of intensities 0.1.

(3) facial actions AU1, AU2 and AU25 of intensities from 0.1 to 1.0, without global head motion.

Fig. 11.20 shows the RMSE for nodes of the wire frame model. In each case, the RMSE is below 0.2 within six iterations. It shows that the iterative estimation works well.

The second experiment uses the test sequence of CCITT standard image sequence *Clare*. Fig. 11.21 shows the original images of the first frame and the eighth frame in the sequence. Fig. 11.22 shows the image synthesized with the analysis parameters for the eighth frame. We can observe that Fig. 11.22 is a fair reproduction of the corresponding original frame.

11.5.5 Updating of 3-D facial model

The 3-D facial model constructed by using the texture and the approximated depth of only the first frame might be insufficient. This section discusses a method to update both the texture and the depth information by using more frames in the image sequence.

11.5.6 Update of texture information

The texture of the first frame might be insufficient in synthesizing images after any deformations of the face model. The texture information should be updated in analyzing image sequence in order to improve the reproduction quality. We shall describe two schemes for updating the texture information[7].

Texture update: method I

The first method updates the texture on the basis of the estimated facial action parameters. Suppose the lip corners are pulled upward (AU12), the nasolabial furrows become deep and clear, which cannot be synthesized by the initial facial model with the neutral face. Let $I(x, y, 0)$ and $I(x, y, t_f)$ be the first neutral face image and the image with the furrow texture, respectively. Let $I_m(x, y, t_f)$ be the facial texture after normalization, in which the input facial texture $I(x, y, t_f)$ is warped so that the facial shape becomes the same as the initial face model at $t = 0$. Then, the furrow texture $I_f(x, y, t_f)$ in the region F_j, where the furrow appears, is defined as

$$I_f(x, y, t_f) = W(x, y)\,[I_m(x, y, t_f) - I(x, y, 0)] \quad (x, y) \in F_j \tag{11.17}$$

where $W(x, y)$ is a function which smooths boundaries of the region.

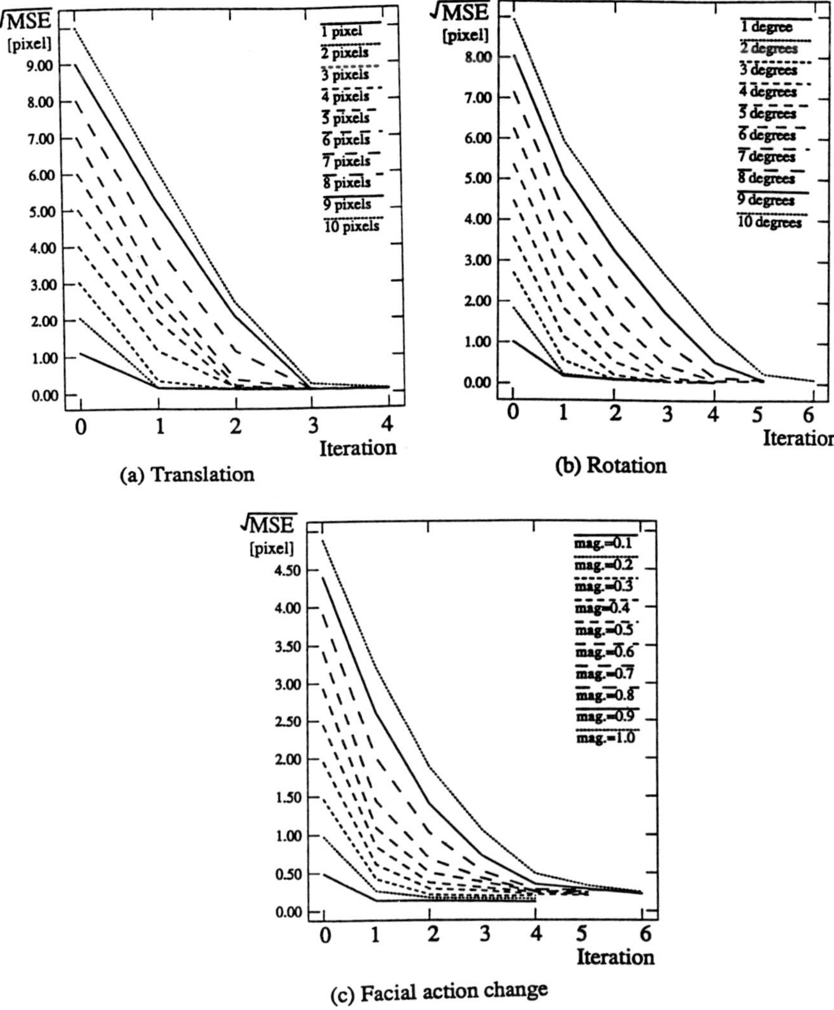

Figure 11.20: Accuracy of the estimation of the head motion and the facial expressions. Root mean square errors of the node positions are shown for three different sequences; (1)head translation up to 10 pixels with facial expressions of AU1, AU2, AU25 of intensities 0.1, (2)head rotation up to 10 degree with facial expressions of AU1, AU2, AU25 of intensities 0.1, (3)facial actions AU1, AU2 and AU25 of intensities from 0.1 to 1.0, without global head motion.

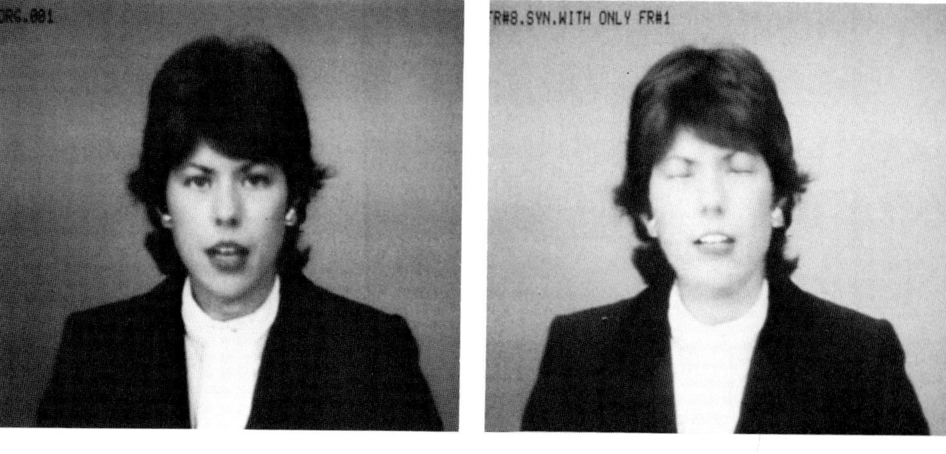

(a) (b)

Figure 11.21: Original images of *Clare*. The left had side (a) is frame #1, and the right hand side (b) is the frame #8.

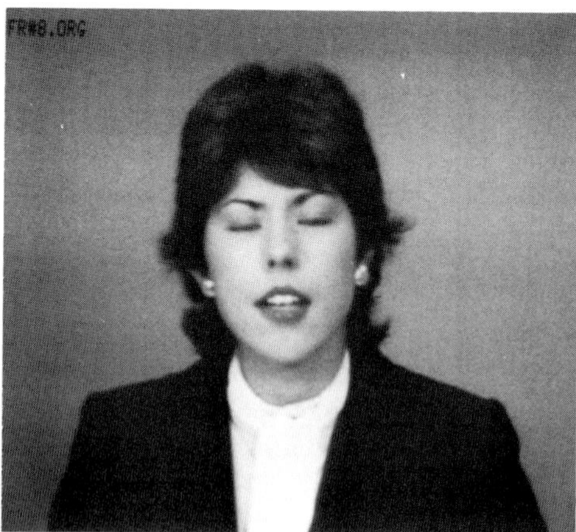

Figure 11.22: Images synthesized with the analysis parameters extracted for frame #8. The texture of the frame #1 is used for the synthesis.

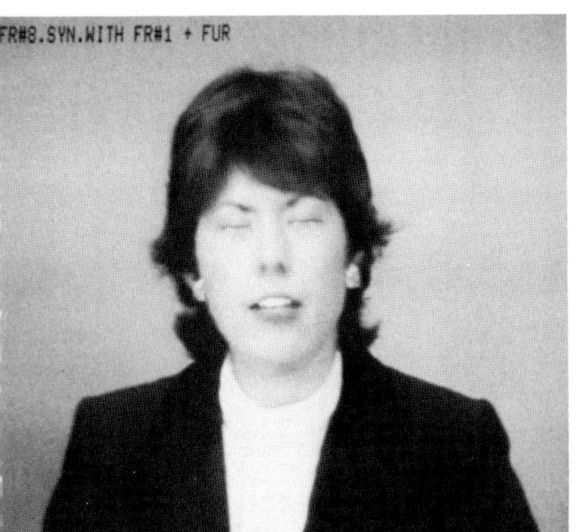

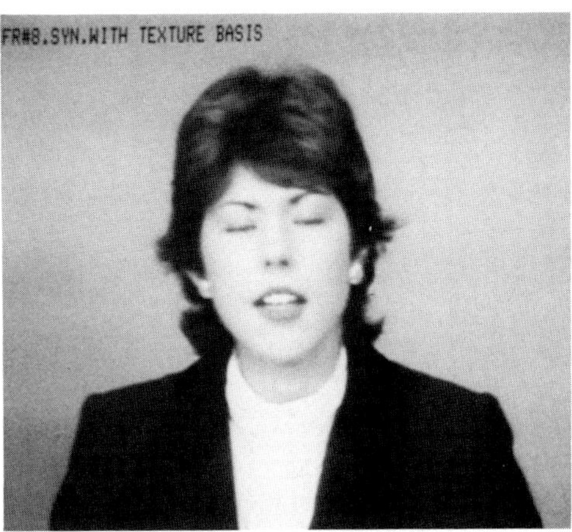

(a) (b)

Figure 11.23: Images synthesized with the analysis parameters extracted for frame #8 of *Clare* by using the texture update methods. (a) image synthesized using texture update method I (b) image synthesized using texture update method II.

In a similar way, the texture of the teeth area is updated when the mouth opens. The textures are not necessary updated every frame. They are updated when the AUs are clearly observed. Those additional textures are used in the synthesis process as follows;

$$I_s(x, y, t_s) \quad = \quad I(x, y, 0) + (a_{is}/a_{if})I_f(x, y, t_f) \tag{11.18}$$

where a_{is} and a_{if} are AU intensities of the synthesized image and the image used for texture updating. Effect of the use of updated texture is clearly observed in Fig. 11.24, in which the areas around nasolabial furrows, glabella, chin, root of the nose, teeth are substantially improved. In Fig. 11.23(a) the upper eye lids are improved.

Texture update: method II

The second method generates *texture basis* from the input images by Gram-Shumdt orthonormalization and describes the textures with the texture basis. The procedure is as follows:

(1) An input image is analyzed and the facial texture is normalized, this means that the input texture is deformed and mapped onto the initial wire frame model. The facial part of the image is defined as vector I_i, components of which are the pixel brightness of the normalized texture.

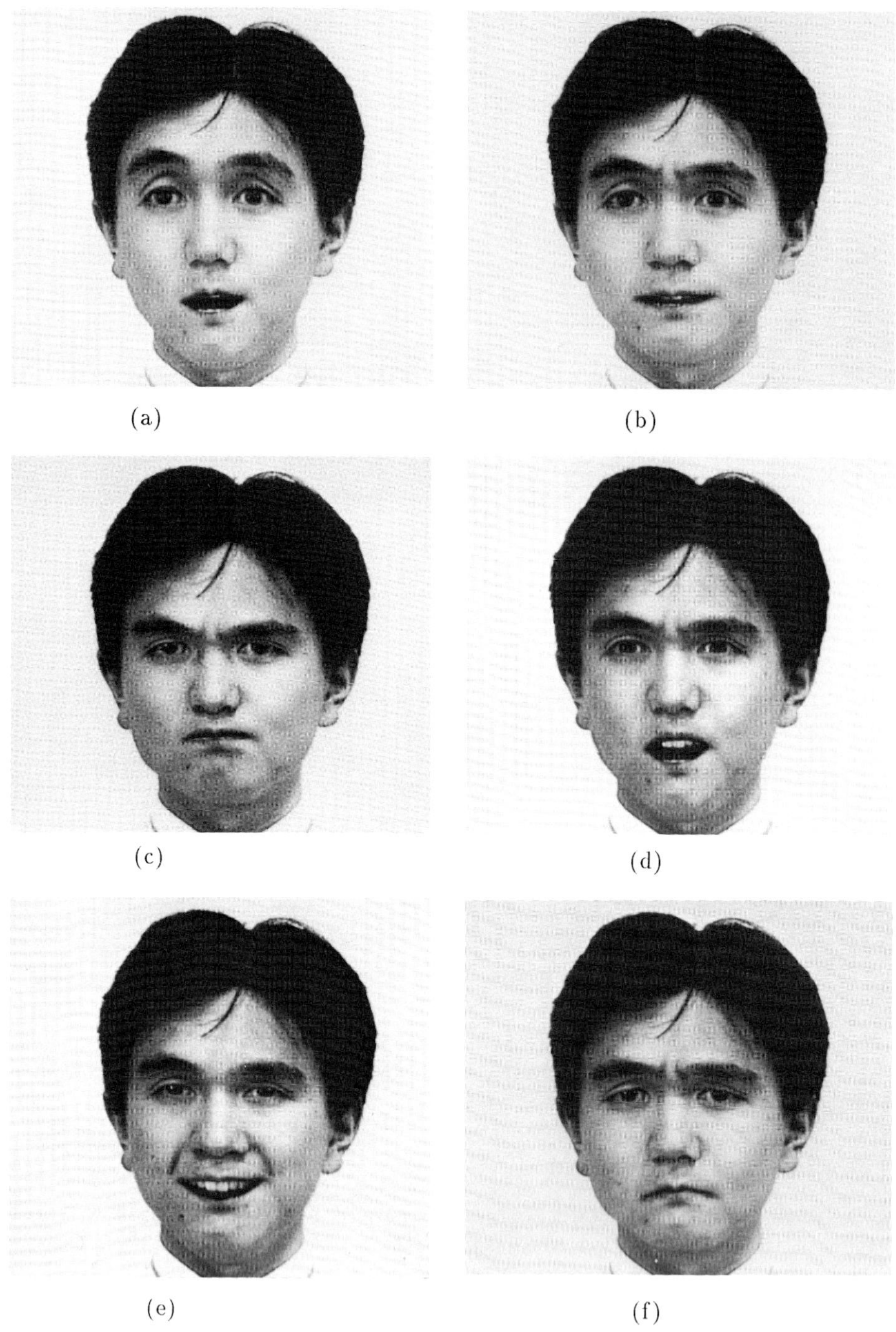

Figure 11.24: Images synthesized with textures updated by method I (a)AU1+2+5+26 (surprise) (b)AU1+2+4+5+20 (fear) (c)AU4+9+17 (disgust) (d)AU4+5+7+10+26 (anger) (e)AU6+12+26 (happiness) (f)AU1+4+15 (sadness)

(2) The first texture basis is $B_1 = I_1/||I_1||$.

(3) The normalized second image frame I_2 is represented with B_1. Then, the remaining vector is $R_2 = I_2 - (B_1, I_2)B_1$. R_2 is orthogonal to B_1. If $||R_2|| > Threshold$, $B_2 = R_2/||R_2||$ is adopted as the second texture basis. If $||R_2|| < Threshold$, $||R_2||$ is neglected. Continuing this process produces a set of orthogonal vectors which represents the facial image sequence efficiently.

Parameters which represents an image texture are inner products of the image and the texture basis. The texture is reproduced with the texture basis and those inner product values, that is, the texture at jth frame is $\sum_{i=1}^{q}(B_i, I_j)B_i$. q is the number of the texture bases. This method can reproduce the details of the original images as shown in Fig. 11.23(b), though it may require a large storage for the texture basis. Dividing the facial part into several areas may be necessary for saving space for the texture basis. The texture information should be transmitted continuously when coding sequences, but it may not be necessary to send additional information after a while.

11.5.7 Update of depth information

So far, the motion parameters are estimated under the assumption that the depth of each node is known. Conversely, the depth can be also estimated by making use of the estimated motion parameters.

Let $P' = P + \delta P$ and $P' = (x', y', z')^T$. Substituting eq. 11.3 into eq. 11.4 leads to

$$(I_x\omega_y - I_y\omega_x)\, z' \quad = \quad -(x'I_y\omega_z - y'I_x\omega_z + I_x u_x + I_y u_y + I_t). \qquad (11.19)$$

From eq. 11.19, depth (z') is obtained directly. To improve the precision and the robustness, applying eq. 11.19 to all points gives

$$H_z Z' \quad = \quad Y_z \qquad (11.20)$$

where $Z' = (z'_1, z'_2, .., z'_l)^T$ and components of H_z and Y_z are the left and right hand side of eq. 11.19. Relationship between z-vales of the nodes and the points in triangles is given by eq. 11.6. Thus,

$$Z' \quad = \quad S_z Z'_g \qquad (11.21)$$

where $Z'_g = (z_{g1}, z_{g2}, .., z_{gn})^T$ is a vector representing depth of the nodes, and S_z is a matrix ($l \times n$) determined by using eq. 11.6. Substituting eq. 11.21 into eq. 11.20 leads to a form similar to eq. 11.7 so that Z'_g can be calculated by least square method. Thus, we can do the motion estimation and depth correction alternately until the results converge.

References

[1] K. Aizawa and T.S. Huang, Model-Based Image Coding , Paper Summaries of Dynamic Scene Understanding, 1991 IJCAI Workshop, Sydney, 1991.

[2] K. Aizawa, H. Harashima, and T. Saito, Model-Based Analysis Synthesis Image Coding for a Person's Face , Image Communication, vol. 1, no. 2, pp. 139-152, 1989.

[3] K. Aizawa, H. Harashima, and T. Saito, Model-Based Synthetic Image Coding System , Picture Coding Symposium '87, Stockholm, 1987.

[4] K.Aizawa,T.Saito and H.Harashima, Construction of a 3-dimensional personal face model for knowledge-based image data compression , National Conference Record of IEICEJ, Musashino, Japan, Sep. 3-6 1986, 542, p.I-221 (in Japanese)

[5] T.Akimoto and Y.Suenaga, 3D Facial Model Creation Using Generic Model and Front and Side Views of Face, IEICE Trans. Inf. & Sys. Vol.75-D, No.2, pp.191-197 (1992)

[6] C.S.Choi, T.Okazaki, H.Harahsima and T.Takebe, A System of Analyzing and Synthesizing Facial Images, IEEE International Symposium of Circuit and Systems, pp.2665-2668, Singapore, 1991

[7] C.S. Choi and T. Takebe, Analysis and Synthesis of Facial Expressions in Knowledge-Based Coding System for Facial Image Sequence , ICASSP91, Toronto, 1991.

[8] C.S. Choi, K. Aizawa, H. Harashima, and T. Takebe, Analysis and Synthesis of Facial Expressions in Model-Based Image Coding , Picture Coding Symposium '90, Boston, 1990.

[9] N. Diehl, "Object-Oriented Motion Estimation and Segmentation in Image Sequences," Signal Processing: Image Communication, vol. 3, pp. 23-56, 1991.

[10] P.Ekman and W.V.Friesen Facial action coding system, Consulting psychologists press 1977

[11] R. Forchheimer and T. Kronander, Image Coding-From Waveforms to Animation , IEEE Trans. on ASSP, vol. 37, no. 12, pp. 2008-2023, December, 1989.

[12] R. Forchheimer, O. Fahlander, and T. Kronander, Low Bit-Rate Coding Through Animation , Proc. PCS (Picture Coding Symposium) '83, Davis, CA, pp. 113-114, March, 1983.

[13] T. Fukuhara, K. Asai, and T. Murakami, Model-Based Image Coding Using Stereoscopic Images and Hierarchical Structuring of New 3-D Wire-Frame Model , Picture Coding Symposium '91, Tokyo, 1991.

[14] T. Fukuhara, K. Asai, and T. Murakami, Hierarchical Division of 3-D Wire-Frame Model and Vector Quantization in a Model-Based Coding of Facial Images , Picture Coding Symposium '90, Boston, 1990.

[15] H.Harashima, K.Aizawa and T.Saito, Model-based analysis synthesis coding of videophone images - conception and basic study of intelligent image coding, IEICE Trans. Vol. E72-5, pp.452-459 (1989)

[16] M. Hötter and J. Ostermann, Analysis Synthesis Coding Based on Planar Rigid Moving Objects , Int. Workshop on 64 Kbps Coding of Moving Video, Hannover, 1988.

[17] T.S. Huang, S. Reddy, and K. Aizawa, Human Facial Motion Modeling, Analysis and Synthesis for Video Compression , Proc. Visual Communication and Image Processing SPIE, pp. 234-241, Boston, 1991.

[18] M. Kaneko, A. Koike, and Y. Hatori, Real-Time Analysis and Synthesis of Moving Facial Images Applied to Model-Based Image Coding , Picture Coding Symposium '91, Tokyo, 1991.

[19] M. Kaneko, A. Koike, and Y. Hatori, Coding of Facial Image Sequence Based on a 3-D Model of the Head and Motion Detection. , Journal of Visual Communication and Image Representation, vol. 2, No. 1, pp. 39-54, March, 1991.

[20] M. Kaneko, A. Koike and Y. Hatori, Coding with Knowledge-Based Analysis of Motion Pictures. , Proc. PCS (Picture Coding Symposium), Stockholm, pp. 167-178, June 1987.

[21] T. Kimoto and Y. Yasuda, Hierarchical Representation of the Motion of a Walker and Motion Reconstruction for Model-Based Image Coding , Optical Engineering, vol. 20, no. 7, pp. 888-903, 1991.

[22] R. Koch, Adaptation of a 3D Facial Mask to Human Faces in Videophone Sequences Using Model Base Image Analysis , Picture Coding Symposium '91, Tokyo, 1991.

[23] A.Koike, M.Kaneko and Y.Hatori, Model-based image coding with 3-D motion estimation and shape change detection, Picture Coding Symposium90, Boston, 1990

[24] H. Li, P. Roivainen, and R. Forchheimer, Recursive Estimation of Facial Expression and Movement , ICASSP 92, pp. 593-596, San Francisco, CA, 1992.

[25] K. Mase, An Application of Optical Flow - Extraction of Facial Expressions , Proc. of MVA90, Tokyo, 1990.

[26] T. Minami, I. So, T. Mizuno, and O. Nakamura, Knowledge-Based Coding of Facial Images , Picture Coding Symposium '90, Boston, 1990.

[27] H. Morikawa and H. Harashima, 3-D Structure Extraction Coding of Image Sequences , Journal of Visual Communication and Image Representation, Vol.2, No.4, pp.332-344 (1991)

[28] S. Morishima, K. Aizawa and H. Harashima, An Intelligent Facial Image Coding Driven by Speech and Phoneme , ICASSP89, Glasgow, 1989.

[29] H. G. Musmann, M. Hötter and J. Ostermann, Object-Oriented Analysis-Synthesis Coding of Moving Images , Image Communication, vol. 1, no. 2, pp. 117-138, Oct., 1989.

[30] Y.Nakaya and H.Harashima, Model-based/Waveform Hybrid Coding for Low-Rate Transmission of Facial Images, IEICE Trans. Commun., Vol. E75-B, No.5, pp.377-384 (1992)

[31] Y. Nakaya, K. Aizawa, and H. Harashima, Texture Updating Methods in Model-Based Coding of Facial Images, Picture Coding Symposium '90, Boston, 1990.

[32] F.I. Parke, Parameterized Models for Facial Animation , IEEE Computer Graphics and Applications, vol. 12, pp. 61-68, Nov. 1982.

[33] D. Pearson, Model-Based Image Coding , Proc. GLOBECOM '89, Dallas, 16.1, pp. 554-558, Nov., 1989.

[34] S.M. Platt and N.I. Badler, Animating Facial Expressions, Computer Graphics, vol. 13, pp. 245-242, Aug. 1981.

[35] D. Terzopoulos, and K. Waters, Analysis of Facial Images Using Physical and Anatomical Models , ICCV90, Osaka, 1990.

[36] K. Waters and D. Terzopoulos, Modeling and Animating Faces Using Scanned Data , The Journal of Visualization and Computer Animation, vol. 2, pp. 129-131, 1991.

[37] W.J. Welsh, Model-Based Coding of Videophone Images , Electronics & Communication Engineering Journal, vol. 3, no. 1, pp. 29-36, Feb, 1991.

[38] W.J. Welsh, Model-Based Coding of Moving Images at Very Low Bit Rate , Picture Coding Symposium '87, Stockholm, 1987.

[39] G.Xu, H.Agawa, Y.Nagashima, F.Kishino and Y.Kobayashi: Three-Dimensional Face Modeling for Virtual Space Teleconferencing Systems, IEICE Trans.,Vol.E 73,No.10,1990

[40] M.Yamamoto, P.Boulanger, J.-A Beraldin, M.Rioux, and J.Domey Direct Estimation of Deformable Motion Parameters from Range Image Sequence, 3rd International conference on computer vision, pp.460-464, Osaka, 1991

[41] H.Yamada, H.Chiba, K.Tsuda, and K.Maiya, New approach to the research on the facial expressions: Model-based facial Action Synthesizing Computer (MASC) system, International Journal of Psychology, 1992, vol. 27, p.47.

[42] J.Y. Zheng, Y. Nagashima, and F. Kishino, 3-D Modeling From Continuous Aspect Views , Picture Coding Symposium '91, Tokyo, 1991.

[43] In addition, many papers related to model-based coding have appeared in Picture Coding Symposium87,88,90,91, Picture Coding Symposium of Japan 86,87,88,89,90,91 (in Japanese).

12

Motion Compensated Spatiotemporal Kalman Filter[1]

J. W. Woods and J. Kim

Center for Image Processing Research
Rensselaer Polytechnic Institute, Troy NY, USA

12.1 Introduction

In the recent past there has been much research in spatial-interaction models (i.e. 2-D random field models) [1, 2]. They have been used in image restoration, image coding, and image analysis. In the field of image sequence processing much attention is currently being given to noise reduction and deblurring. Noise can occur in the storage, transmission, or processing of image sequences. The reduction of this noise is very important for improving visual quality and also potentially to the success of subsequent image processing.

In a time sequence of images, each pair of consecutive frames has a high correlation in the temporal direction, while the noise is mostly uncorrelated. When this temporal correlation is used for noise reduction, the noise can be reduced without losing spatial high frequencies, if there is little or no movement between frames. However, when the image changes significantly from one frame to the next, due to object or camera movement, then these moving parts of the image sequence will become blurred. A moving object still has a strong correlation with the previous frame. Yet, this strong correlation is now along the trajectories of the motion. In this case, motion compensation is required for successful spatio-temporal filtering.

Noise reduction by temporal filtering was described in [3, 4, 5]. In these approaches, they used only temporal correlation. However, image sequences have spatial correlation as well as temporal correlation. When the image motion vector flow is low, then the image sequence may be assumed to be stationary random field sequence [6]. For a stationary random field, Wiener filtering is classically used for reducing noise. The Wiener filter is designed based on the image and noise power spectral density [2]. In order to use both temporal correlation and spatial correlation, a 3-D Wiener filter and motion-compensated

[1]Research partially supported by NSF grant No. MIP-9013247.

3-D Wiener filter were introduced in [7].

When considering a Markovian assumption, one can see that such an hypothesis can be applied to the time varying and space varying image field sequence (3-D Markovian field sequence). Under the appropriate 3-D Markovian assumption, the spatio-temporal Kalman filter was described in [6]. In this approach, they separated the temporal part from the spatial part to reduce the computational burden. They performed this simplification under the so-called separability assumption, which is known to be not valid in general. In two dimensions, the reduced update Kalman filter (RUKF) and reduced-order model Kalman filter (ROMKF) has been proposed to solve the same computational burden problem[8, 9], but in such a way as to well approximate the optimal 2-D fixed lag smoother.

In this chapter we will first introduce spatio-temporal reduced update Kalman filter (3-D RUKF) which is the natural extension of the spatial 2-D RUKF to 3-D image sequences. In this spatio-temporal Kalman filter, the image sequence is filtered line-by-line and then frame-by-frame based on a scalar 3-D image model. This corresponds to a normal TV raster scan that is progressive, i.e. without interlace.

When image motion vector flow is large, then the stationary assumption can not be valid [6]. To deal with this case, adaptive filtering and motion-compensated filtering have been suggested [10, 5, 11, 7]. When the image sequence is aligned along a motion trajectory via motion-compensation, then the motion compensated (MC) image sequence can be approximated by a temporally stationary random field sequence [12]. We will also introduce a new motion-compensated spatio-temporal Kalman filter, MC-RUKF. In this motion-compensated filter, the 3-D RUKF is applied to the MC image sequence, and iterated once.

This chapter is organized as follows: In Section 12.2, spatio-temporal Kalman filtering is introduced. In Section 12.3, hierachical block matching motion estimation is reviewed. In Section 12.4, spatio-temporal Kalman filtering is extended by the introduction of motion compensation. In Section 12.5, experimental results are presented and discussed. Finally in Section 12.6, we make some conclusions.

12.2 Spatio-temporal Kalman Filter

In a global sense, images can be assumed to be sequences of stationary random fields, or random field sequences, but they have apparently nonstationary local structure. Nevertheless, using the stationary assumption, the 2-D space-invariant Kalman filter has shown quite good results in image deblurring and restoration. Although the space-invariant Kalman filter shows good performance in deblurring, its performance has been known to be limited by ringing artifacts. For noise-only filtering, linear space invariant filtering has not been very successful.

In general we have found that temporal correlation is usually stronger than

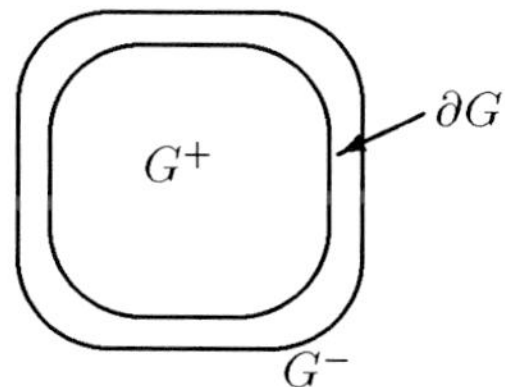

Figure 12.1: Regions used in definition of Markov random field sequence

spatial correlation. Thus, in a time sequence of images, one image frame has great information about the next few frames. In noise reduction filtering, this temporal information can be used to preserve local properties of the image frames.

When the motion vector flow of an image sequence is low, the sequence can be assumed to be temporally stationary Markov field sequence. Since one previous frame has the information from all the previous frames in the nonmoving areas, the hypothesis of the temporal Markov random sequence can be used for slowly moving areas. On the basis of this spatial and temporal stationarity assumption, the spatiotemporal invariant 3-D Kalman filter can then be used on such image sequence.

In the next subsection, we review properties of Markov random field sequences. Then we introduce the observation model for the image sequences of interest here. The 3-D reduced update Kalman filter (3-D RUKF) is then introduced. Various issues related to the 3-D RUKF are then mentioned, including the development of suitable boundary conditions, the Kalman gain, and choice of update region size for the covariance update region. This will then complete this section on the spatial-invariant 3-D Kalman filter.

12.2.1 Image model

Here we present a summary of the properties of discrete Markov random field sequences. It will be shown that the nonsymmetric half-space (NSHS) recursive models generate a special class of Markov random fields, which are an extension of the 2-D nonsymmetric half-plane (NSHP) recursive models presented in [13].

Definition. A discrete Markov random field sequence: Let s be a random field sequence on I^3, the 3-D lattice. Let a band of minimum width p, ∂G("the present") separate I^3 into two regions: G^+("the future") and G^-("the past"). Then s is Markov-p if $\{s|_{G^+}$ given $s|_{\partial G}\}$ is independent of $\{s|_{G^-}\}$ for all ∂G (see Figure 12.1).

For the homogeneous, zero-mean, Gaussian case, this definition can be expressed by the following recursive equation model:

$$s(m,n,t) \quad = \quad \sum_{D_p} c(k,l,\tau)s(m-k,n-l,t-\tau)+u(m,n,t) \qquad (12.1)$$

where
$E\{s(m,n,t)u(k,l,\tau)\} = \sigma_u^2 \delta_{mk}\delta_{nl}\delta_{t\tau}$, and
$D_p = \{k,l,\tau | k^2 + l^2 + \tau^2 \leq p^2 \text{ and } (k,l,\tau) \neq (0,0,0)\}$.

The input term $u(m,n,t)$ is a Gaussian, zero-mean, homogeneous random field sequence with correlation (covariance) function of bounded support,

$$R_u(m,n,t) = \begin{cases} \sigma_u^2, & (m,n,t) = (0,0,0) \\ c_{mnt}\sigma_u^2, & (m,n,t) \in D_p \\ 0, & \text{elsewhere.} \end{cases} \tag{12.2}$$

The $c_{kl\tau}$ are the interpolation coefficients of the minimum mean-square error (mmse) linear interpolation: given $\{s \text{ for } (m,n,t) \neq (0,0,0)\}$

The solution is given as the conditional mean,

$$E\{s(0,0,0) | s \text{ for } (m,n,t) \neq (0,0,0)\} = \sum_{D_p} c_{kl\tau} s(-k,-l,-\tau). \tag{12.3}$$

In this formulation σ_u^2 is the mean-square interpolation error.

For a 3-D recursive filter, we must separate the *past* from *future* of the 3-D random field sequence. One way is to assume that the random field sequence is scanned in line-by-line and frame-by-frame. In this way, we can divide the image sequence into these two parts, i.e. past and future. In the 2-D case, the NSHP (nonsymmetric half-plane) has been widely used. A simple extension of the NSHP concept is NSHS (nonsymmetric half-space). Then we can arrive at a general Markov image sequence model of the form,

$$s(m,n,t) = \sum_{S_c} c(k,l,\tau) s(m-k, n-l, t-\tau) + w(m,n,t) \tag{12.4}$$

where $\mathcal{S}_c = \{k \geq 0, l \geq 0, \tau = 0\} \cup \{k < 0, l > 0, \tau = 0\} \cup \{\tau > 0\}$ as shown in Figure 12.2. and where w(m,n,t) is a white noise field sequence.

In recursive filtering, only a finite subset of the NSHS is updated at each step. This updated region is called the *local state region.* This update region is slightly enlarged from the model support of the right-hand side of (12.1) for an (M1,M2,M3)th order $\oplus$ model of the NSHS variety.

Our observed image model is given by:

$$r(m,n,t) \quad = \sum_{k,l,\tau \in S_h} h(k,l,\tau) s(m-k, n-l, t-\tau) + v(m,n,t) \tag{12.5}$$

The $\mathcal{S}_h$ is the support of the point-spread function (psf) $h(k,l,\tau)$. In $h(k,l,\tau)$, q makes it possible to shift the psf in the temporal direction. The observation noise $v(m,n,t)$ is assumed to be an additive, zero-mean, homogeneous Gaussian field with covariance $Q_v(m,n,t) = \sigma_v^2 \delta(m,n,t)$.

12.2.2 Reduced update Kalman filter (RUKF)

Kalman filtering is well established for the 1-D and 2-D cases [14, 15]. In the 2-D case many methods were introduced to address the computation time problem. The reduced update Kalman filter (RUKF) is widely known as a good

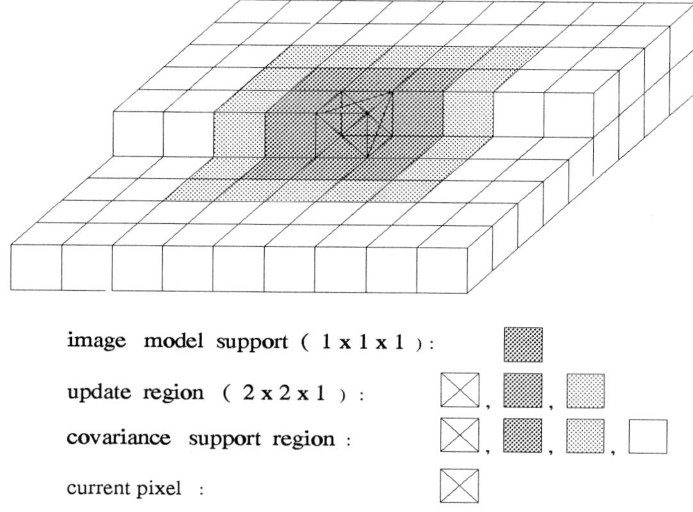

Figure 12.2: 3-D Kalman Support Regions

suboptimal filter. Here we explain our extension of RUKF to three dimensions as required for the spatiotemporal filtering application to image sequences.

The signal model and observation equation are shown in (12.4, 12.5). When the original image is modeled with an $(M \times M \times M)$th order model, the local state vector is given by:

$$\begin{aligned} s(m,n,t) = [&s(m,n,t), \quad \ldots \quad, s(m-M+1,n,t); \\ &s(m+M+1,n-1,t), \ldots, s(m-M+1,n-1,t); \\ &\ldots : s(m+M+1,n-M,t), \ldots, s(m-M+1,n-M,t); \\ &s(m+M+1,n+M,t-1), \ldots, s(m-M+1,n+M,t-1); \\ &\ldots : s(m+M+1,n-M,t-1), \ldots, s(m-M+1,n-M,t-1)] \end{aligned} \tag{12.6}$$

The local state region, $\mathcal{R}_\oplus$, which is to be updated recursively, is shown in Figure 12.2. As mentioned above, we assume that the received image sequence $r(m,n,t)$ is scanned in line-by-line and frame-by-frame. Then the appropriate Kalman filter equations can be derived as follows:

$$\begin{aligned} \hat{s}_b(m,n,t) &= \sum_{k,l,\tau \in S_c} c_{kl\tau} \hat{s}_a(m-k,n-l,t-\tau) \\ \hat{s}_a(x,y,z) &= \hat{s}_b(x,y,z) + \kappa(m-x,n-y,t-z) \\ &\quad \cdot \left[r(m,n,t) - \sum_{k,l,\tau \in S_h} h_{kl\tau} \hat{s}_b(m-k,n-l,t-\tau)\right] \\ &\quad (x,y,z) \in \mathcal{R}_\oplus^{(m,n,t)} \end{aligned} \tag{12.7}$$

$$
\begin{aligned}
R_b(m,n,t;k,l,\tau) &= \sum_{o,p,q\in S_c} c_{opq} R_a(m-o,n-p,t-q;k,l,\tau) \\
&\quad (k,l,\tau) \in \mathcal{T}_{\oplus}^{(m,n,t)} \\
R_b(m,n,t;m,n,t) &= \sum_{k,l,\tau\in S_c} c_{kl\tau} R_a(m,n,t;m-k,n-l,t-\tau) + \sigma_w^2 \\
\kappa(x,y,z) &= \frac{\sum_{k,l,\tau\in S_h} h_{kl\tau} R_b(m-k,n-l,t-\tau;m-x,n-y,t-z)}{\sum_{k,l,\tau}\sum_{o,p,q} h_{kl\tau} h_{opq} R_b(m-k,n-l,t-\tau;m-o,n-p,t-q) + \sigma_v^2} \\
&\quad (x,y,z) \in \mathcal{R}_{\oplus}^{(m,n,t)} \\
R_a(x,y,z;k,l,\tau) &= R_b(x,y,z;k,l,\tau) - \kappa(m-x,n-y,t-z) \\
&\quad \cdot \sum_{o,p,q} h_{opq} R_b(m-o,n-p,t-q;k,l,\tau) \\
&\quad (x,y,z) \in \mathcal{R}_{\oplus}^{(m,n,t)}, \qquad (k,l,\tau) \in \mathcal{T}_{\oplus}^{(m,n,t)}
\end{aligned}
\tag{12.8}
$$

The error covariance support region $\mathcal{T}_{\oplus}$ is shown in Figure 12.2. This region is selected somewhat larger than the update region $\mathcal{R}_{\oplus}^{(m,n,t)}$, to allow for spatial buildup of the approximate error covariance values.

In the above equations, as in one dimension, we see that only the first two linear equations involve the observed data. The last four equations are nonlinear and do not involve the data directly. Since we assume stationary random field sequences, we can expect a steady state RUKF to develop as we move into the data, i.e. away from the initial conditions and boundaries. In fact, one can then breakup these six equations into two parts: the first two are the linear filtering parts and will be approximated below with the steady-state gain. The other four non-linear equations can be calculated off-line to determine the steady-state gain, assuming for the moment that one exists. Effectively, we run these non-linear equations on a fictitious and small image sequence, which is just big enough to determine a close approximation to the steady-state gain. In the experiments below, we find that a 20x20x20 sized region is quite sufficient for the statistics encountered in image sequence filtering. This is shown in Figures 12.3a, b for both spatial and temporal directions. The saw-tooth presentation codes results, before and after updating.

12.2.3 Boundary conditions

All recursive filters have the problem of choosing appropriate initial conditions [14]. In two and three dimensions, this initial condition problem also acquires a boundary value aspect. The effect of the boundary values is decreasing as the filter moves away from the boundaries, if the filter is stable. In RUKF, a stable image model should be chosen to control this boundary effect, and restrict it to a transient response effect. This boundary value problem was described in [8]. There two methods were advanced: random boundary conditions and fixed boundary conditions. An alternative option is the use of zero derivative boundary values [16], borrowed from the partial differential equations literature.

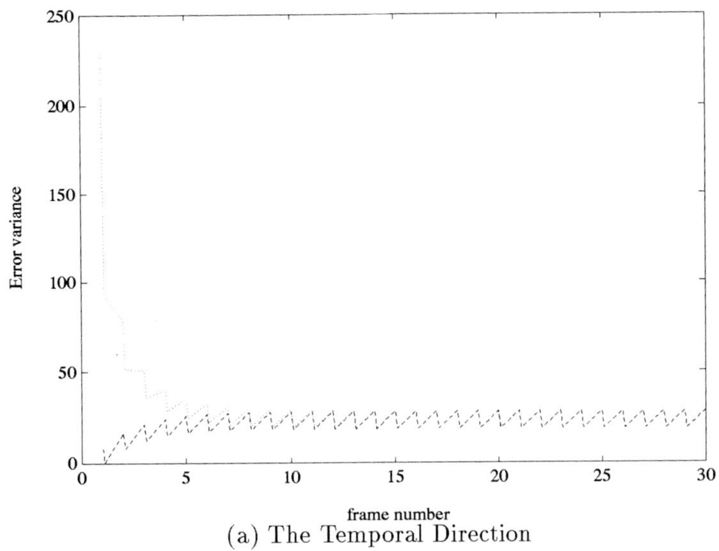

(a) The Temporal Direction

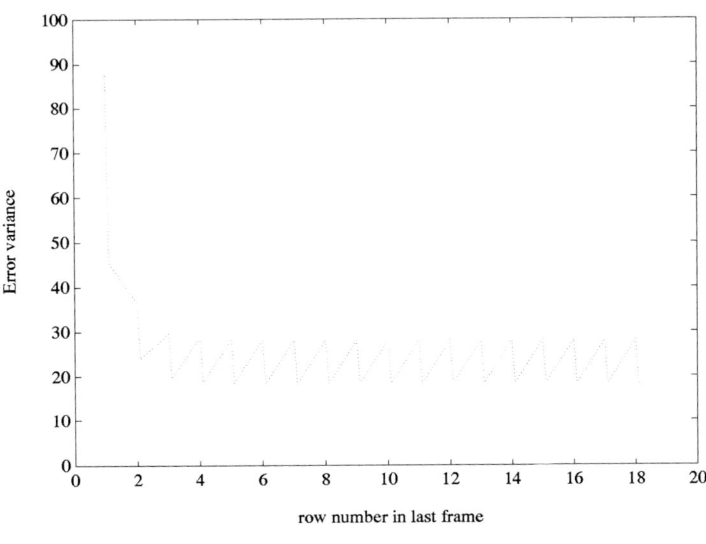

(b) The Spatial Direction

Figure 12.3: Error Variance Convergence

Another simple method is to shift the starting point and ending point inside of each frame and inside the image sequence in time, effectively not filtering right up to the boundaries. This is the so-called *noisy boundary* condition. This boundary condition, which was suggested for the 2-D RUKF, can be easily extended to the 3-D RUKF.

In calculating the Kalman gain, since we intend to make the training image size as small as possible, the boundary condition is very important. We used the noisy boundary condition for calculating the steady-state Kalman gain, since it gave us a good result.

12.2.4 Boundary condition for Kalman gain

For this boundary condition, M columns and M rows are appended to the sides of each frame, where M is the same value used to describe the local state support size. In the calculation step, the update region is selected as M pixels wider in each dimension than the coefficient support of the local state region. The boundary is assigned a precalculated initial error covariance which is determined by the random boundary condition. This error covariance is decreasing as it propagates inside the fictitious training image. In the temporal direction, M frames are used to specify this random boundary condition.

12.2.5 Kalman gain

A Kalman gain is separately calculated on a fictitious image sequence, as mentioned above. In the 3-D RUKF, we update only the local update region instead of the enormous global state.

Remember that the model support regions has $M \times M \times M$ order. Then the support of the local state is approximately $(2M+1) \times (2M+1) \times (M+1)$, as can be seen with reference to Figure 12.2. Since a Kalman gain is calculated from the error covariance of the local state, the error covariance of the local state should be updated at each pixel. As the error covariance processing progresses, the local state slides column by column and row by row. At the next pixel of the same row, the new local state has a new column, and at the next pixel of the same column, the new local state has a new row. Hence, the reduced update region of the error covariance should be wider than the Kalman gain region in both the vertical and horizontal directions. This wider area is called the covariance support region, and is also shown in Figure 12.2. The error covariance between all pixels in the local state and the covariance region is updated at each observation. The total number of error covariance values which are updated at each pixel is about $((4M+1) \times (4M+1) \times (M+1)) \times ((2M+1) \times (2M+1) \times (M+1))$.

So far, we have explained the updating of one frame at a time. In the spatio-temporal Kalman filter, each frame is updated one after another. Since the temporal support size is M, we need error covariances for the past M frames. After one frame is processed, these error covariances are shifted temporally, and the most recent $(M + 1)$ frames error covariances are saved. Hence the total number of error covariance values to be retained in high speed storage is

$((4M+1)\times(4M+1)\times(M+1))\times((N+M)\times(N+2M)\times(M+1))$, or about $16M^4N^2$ since $N \gg M$, where N is the fictitious image size. For $M = 1$, as used here, we get about 16K words of RAM needed to design the steady-state 3-D Kalman gain.

12.3 Hierarchical Block Matching Motion Estimation

When motion vector flow is large, spatiotemporal invariant filtering can cause blurring of moving objects. To handle this problem, the invariant filter must be applied along an estimated motion trajectory. There are two different popular kinds of motion estimation: block matching methods and differential methods [17]. Only block matching methods will be considered here, although differential methods may be equally valid.

When we estimate a displacement vector at the location (x, y, t), we match a block which includes the point (x, y, t) to blocks within a search area in the previous frame. The criterion can be mean-square error, mean-absolute error, or correlation [17]. In the block matching method, the block size and search area are critical for the performance of motion estimation.

A small block size can detect small moving objects, but the resulting displacement estimate is very sensitive to noise. A large block size is less sensitive to noise, but cannot detect the motion of small objects. A large search area can detect fast motion, but also requires a large computation time, and is also sensitive to noise. The best matching , if not well matched, block is often good enough for a coding algorithm, however in our filtering algorithm, when the estimated motion is not the true physical motion, visible artifacts can be created.

12.3.1 Hierarchical method

Hierarchical motion estimation has been proposed and discussed to handle these problems [18, 19]. The basic idea of such algorithms is that one should initially estimate a coarse motion vector at low resolution, preferably in a subband of the image which has only low spatial frequencies(subbands). Then one should refine this motion vector by increasingly introducing high frequencies.

The motion vector is also subsampled in subbands the low frequency subband image. Therefore this method can estimate fast motion with a small block size and a small search-area size. Since the estimated motion in the low subband is applied to predicting motion at the next resolution step, these hierarchical algorithms are less sensitive to noise.

To get a higher level(lower resolution) image in the pyramid, the lower level image is filtered with a lowpass filter and is subsampled. So 'high level' means low resolution, and vice versa. The hierarchical motion vector estimate is thus conducted as follows. First, a motion vector is estimated at the highest pyramid level. This estimated motion vector is then propagated to the next lower level in the pyramid. The search area of the next level is centered around the propagated motion vector. A refined motion vector is a vector sum of the new estimated one and the propagated one. This procedure is repeated at each resolution level.

12.3.2 Change detection

Hierachical motion estimation gives us a much smoother motion vector field, but it can result in erroneous motion vectors on the outside boundaries of moving objects. When the SNR is very low, the estimated motion vectors can also be erroneous in a relatively flat region of an image frame. The resulting motion estimation error can cause visible distortion in motion-compensated image filtering. However, this effect can be alleviated by detecting changed areas [19]. A given threshold value is used to detect the changed areas. When the uncompensated frame difference is less than a given threshold value, then the estimated motion vector is assumed to be erroneous. In the case of a low SNR, the measure of frame difference is very sensitive to noise. The frame difference is filtered to decrease the effect of noise on the change detection. A box filter is used for this purpose. The size of this box filter is important. A small sized box filter detects only the boundary area of a moving object, while a large sized box filter detects an unnecessarily large area. In our experiments reported below, a 7×7 box filter was used, and the detection criterion was local variance. Since a moving object cannot be an isolated pixel, all isolated pixels should be removed.

12.3.3 Unpredictable region detection

In a region where an estimated motion vector is poor, or an uncovered scene appears, the correlation between the current frame and the previous one is low that the application of the homogeneous 3-D AR model causes some distortion. Let us call this area *unpredictable*. A 3×3 box filter detects of the unpredictable region . The detection criterion is local variance. These results are quite sensitive to threshold value. A well chosen threshold value decreases the effect of the unpredictable area. When an unpredictable region is detected, then a 2-D or spatial-only RUKF is applied to the supposed uncovered scene. We used the displaced frame difference to detect the unpredictable scene.

12.4 Motion Compensated 3-D RUKF

When motion vector flow is large, the spatiotemporal invariant filter has a limitation on its performance. When a fast moving object is aligned along the trajectory of its motion, we can then apply the spatiotemporal invariant filter to the motion compensated image sequence. Since this motion-compensated image sequence has a strong temporal correlation, its image sequence model will have small prediction error. This means that observation noise can be much reduced while retaining high spatial frequencies, with such motion compensated filtering.

The overall block diagram of our motion compensated spatiotemporal Kalman filter, or MC RUKF for short, is shown in Figure 12.4. The motion compensated

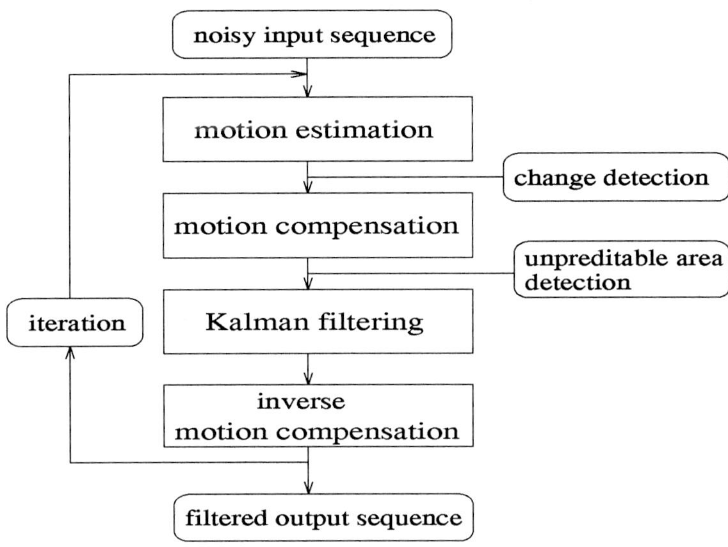

Figure 12.4: The Overall MC RUKF Block Diagram

spatiotemporal filter consists of three major parts: motion estimator, motion compensator, and reduced update Kalman filters.

12.4.1 Motion estimation in spatiotemporal Kalman filter

While filtering an image sequence, two different previous frames can be used for motion estimation; first, one could use the smoothed frame $E\{s(t-1)|r(t), r(t-1), ...\}$, or second, use the original noisy frame $r(t-1)$. The smoothed frame has reduced noise but also has some degree of distortion due to the uncovered scene, a possible motion estimation error, and the smoothing error. The noisy frame has no such signal distortions but does have a strong noise component. Both choices have advantages and disadvantages. The best one to choose may also depend on the kind of motion estimator used. In our experiment we have generally used the best smoothed previous frame currently available.

In our hierarchical block matching, the number of hierarchical stages is four, with resolution indicated as LLLLLL, LLLL, LL, and full resolution. The maximum search distance is two at all pyramid levels. In the experiments, as the block size increases, the global SNR improvement due to motion compensated filtering increases. The SNR improvement was almost saturated when the block size is 13×13. However, we also found that a large block size may cause distortion to the boundary of fast moving objects, thus a 9×9 block size was chosen to provide a relatively good local and global improvement. The resulting SNR improvement is shown in Figure 12.5 as a function of block size.

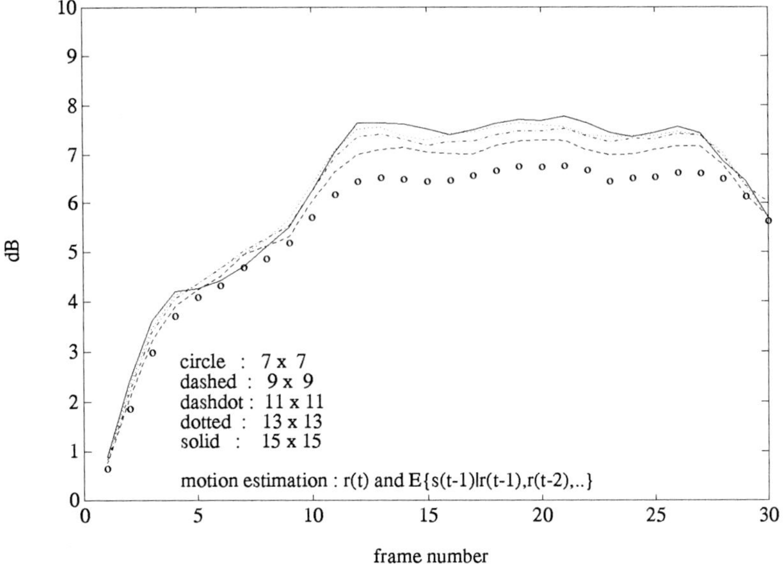

Figure 12.5: The SNR Improvement of Different Block Size

12.4.2 Motion compensation

Following the motion estimation, we align along the estimated motion trajectory the set of local frames which will be filtered by the 3-D RUKF. The current frame is the original noisy image, $r(t)$, and the previous frame is the spatially smoothed previous frame image $E\{s(t-1)|r(t-1), r(t-2), ...\}$. For this local alignment, two methods have been used. The first method is to displace the previous frame with respect to the current frame. The second method is to displace the current frame, keeping the previous frame unshifted. In our experiments, we generally displaced the previous frame. Displacing the previous frame with respect to the current frame is shown in Figure 12.6.

Motion estimation chooses the motion vector which best matches local areas of two frames corrupted by noise. Therefore the motion estimation is not one to one, and the estimated motion may not be the true motion. Due to this estimation error, some pixels of the previous frame cannot be located in the motion compensated frames, and other pixels of the previous frame may be located several times on the compensated frames. Let us call the missed pixels the 'gap area', and the latter ones the 'overlapped area'. These gap areas and overlapped areas typically occur near the boundary of a moving object due to the moderate block size used for motion vector estimation. In the gap areas, a 2-D (spatial) filter or low temporal correlation model (in a multi-model case) can be applied. In the overlapped areas, one location has greater temporal correlation than the others. This region can be chosen in the inverse motion compensation

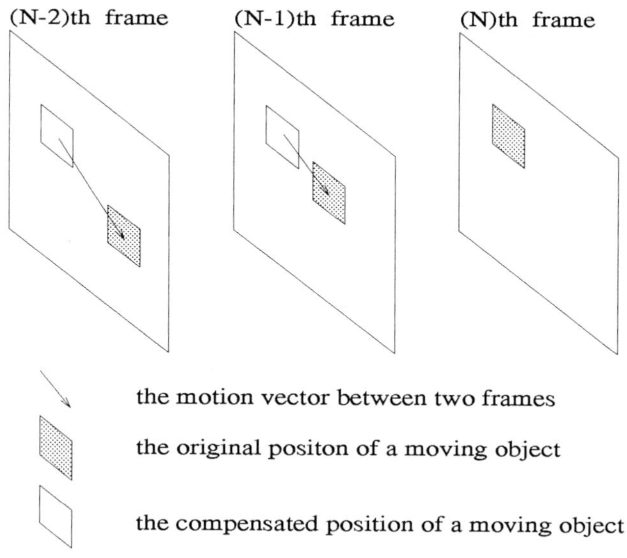

Figure 12.6: Motion Compensation

step. The simple way is to take an average. In our experiments, in order to make the problem simple, we did not treat these gap and overlapped areas specially.

The local correlation between two MC frames depends on the accuracy of the motion estimation. When the motion estimation is very accurate, the correlation is very high, and when the motion estimation is not accurate, the correlation can be low. Since there is no matching area in the occlusion areas, either covered or uncovered, the correlation between two MC frames is very little there.

Since the motion estimation error can be assumed a stationary Gaussian random variable in a global sense by the central limit theorem, the temporal correlation of the compensated frames can be assumed globally stationary. Hence the motion compensated image sequence can be assumed to be a stationary random field sequence in a global sense. Therefore we applied the steady state Kalman filter to the motion compensated image sequence. The results of this motion compensated filter will be discussed in the experimental section below.

Even though the motion compensated image sequence can be globally stationary, a local failure of motion estimation can cause visible artifacts, because we assume a strong temporal correlation in the compensated frames, and because of the effect of finite non-unity block size.

12.4.3 Motion compensated reduced update Kalman filtering

The spatiotemporal invariant RUKF is here applied to the motion compensated image frames. The image model is chosen an $1 \times 1 \times 1$-order AR model for simplicity. This model support was illustrated in Figure 12.2. This 3-D AR model was obtained from the residual image of the original sequence for our simulation below. The number of updated frames can be two or three. As for the update region support, a temporal order of 2 is only slightly better than an order of 1. This is thought due to the error in the estimated motion vectors and the fact that motion is only estimated with pixel accuracy. Thus we only update 1 previous frame in the experiment below. The update region was also shown in Figure 12.2.

12.4.4 Iterative method

When the input SNR is very low, the motion estimation is not reliable. To handle this problem, each image frame can be pre-filtered with a spatial filter. However this pre-filtered image looses some high frequencies, so that motion estimation with these pre-filtered image frames has a limitation on its performance. We indicate an improved approach next, via filter iteration.

In our iterative method for filtering noisy image sequences, two smoothed frames are saved for the iterative method. These smoothed frames retain spatial high frequencies and have reduced noise, so that these frames can now be used for motion estimation with a smaller size block. A motion vector field is estimated from these two frames, i.e. $E\{s(t-1)|r(t), r(t-1), ...\}$ and$E\{s(t)|r(t+1), r(t), ...\}$. The maximum search areas of the pyramid levels are 2 for all stages. The block size is 5 for all stages. With this estimated motion vector, the MC RUKF is applied to the original noisy sequence again. The motion vector estimated in the first iteration is less reliable than the motion vector estimated in the second iteration. The error variance of 3-D AR model used in the first iteration should be more than the error variance estimated from the orginal sequence [20]. In our current work we used an empirical increment in the variance. This iterated method gives an improved visual result, and the SNR improvement is shown in Figure 12.7.

12.5 Experimental Results

Our experimental image sequence is the *salesman* sequence. In this image sequence we used only the even numbered frames, because there is a significant difference between odd and even numbered frames. Due to a distortion present on the odd numbered frames, only the even numbered sequence can be used in our study on motion in image linear filtering. The resulting frame rate is then 15 per second. For our experiment, we shifted the mean values of the salesman sequence to obtain zero mean as required for the filtering. We used this modified sequence as an original noiseless sequence.

A noisy image was obtained by adding a white Gaussian noise to this image sequence. The resulting SNR was 10 dB. The size of image model support was $1 \times 1 \times 1$-order, so that is M is one. The gain support region is $2 \times 2 \times 1$.

The 3-D or 2-D AR models obtained from the original (modified) image

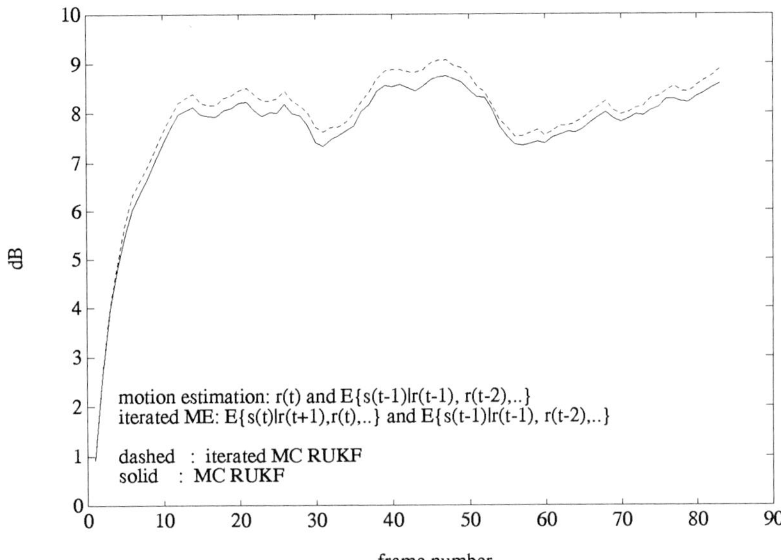

Figure 12.7: The SNR improvement of motion compensated RUKF: the first motion estimation and the iterated motion estimation

sequence are used. These models could be obtained from the noisy sequence or a prototype image sequence. Many image model identification algorithms have been proposed in [21, 13] for this purpose. One of these algorithms, after extension to 3-D, could probably have be used to obtain the required sequence models. While the result would not have been as good as the model identification on the noise-free original image, still it can be expected that it would be quite close.

12.5.1 Spatiotemporal invariant RUKF

When calculating the Kalman gain separately, the initial value (σ^2) of the error covariance matrix R should be greater than the model error and the noise variances. The resulting spatiotemporal Kalman filter converges to steady-state at about the 12th frame. In the salesman sequence, the sequence consists of nonmoving background, a large size moving body, and a small size moving box and hands. The speeds of the moving objects are different and are changing at different rates. When these speeds are slow, the filtered image yields good performance, compared to the results of just spatial filtering. This result is shown in Figure 12.9c. However, one global image model cannot satisfy the temporal correlation of fast moving objects and nonmoving objects simultaneously. Therefore, the noise in the nonmoving background cannot be reduced enough, and fast moving objects are blurred. However, spatiotemporal invariant RUKF

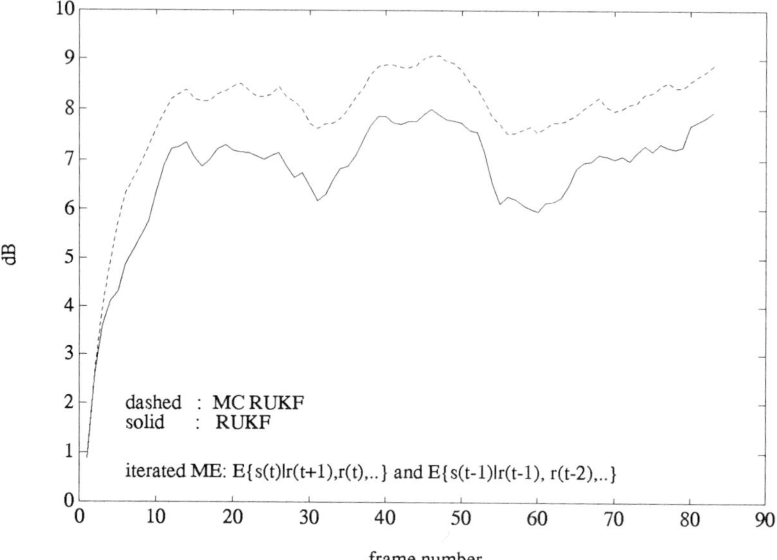

Figure 12.8: The SNR improvement of spatio-temporal invariant RUKF and MC RUKF

globally yields better result, compared to the result of just spatial filtering. Even though the input SNR is 10 dB, the filtered image sequence retains sharpness. The global SNR improvement at steady state, frame 20, is about 7 dB. The graph of SNR improvement versus frame is shown in Figure 12.8.

12.5.2 Motion-compensated spatiotemporal RUKF

In this section, we consider the iterated motion estimation algorithm. In the motion compensated spatiotemporal Kalman filter case, the noise in the non-moving background was decreased significantly, and the blurring of moving objects, which occurs in the spatiotemporal invariant RUKF, was reduced. These MC RUKF results are shown in Figure 12.9d - 12.11d. Since the input SNR is very low, only 10 dB, the motion estimation is not quite accurate. In our experiments, the motion was estimated only to pixel accuracy, and this has put a limitation on the SNR improvement. The global SNR improvement at steady state, frame 20, is about 8 dB. The graph of SNR improvement versus frame number is shown in Figure 12.8.

There are some problems when dealing with a real motion picture. Since a fast moving object is blurred in the original sequence, it is difficult to get a good motion estimate. The displacement of the fast moving 'hand' between frame 29 and 30 (the original frame numbers are 56 and 58) is greater than 20 pixels. In this fast motion case, the motion estimation is not reliable. This unreliable motion estimation yields blurring in the filtered image sequence. However a fast moving object (such as the 'left hand' in frame 58) is so severely blurred in the original that the blurring caused by filtering can be allowed in such a region. An example of acceptable blurring is shown in Figure 12.11d. Another problem, when dealing with a real motion picture, is that the pixel intensity of moving objects (such as the 'moving box') changes because of lighting angles. Yet another example of a problem is illustrated when the salesman's mouth moves. The mouth moves very fast and changes its shape, which means that uncovered and covered areas occur. Therefore, getting a good motion estimation in this region is quite difficult.

When an estimated motion vector is not reliable, the motion prediction from the previous frame could be relied upon less. One way to address the problem caused by motion estimation error is with a multiple-model algorithm which can be an extension of multi-model algorithm of the 2-D case proposed in [22, 23]. Another possible approach is to simply modulate the 3-D model prediction error with the local block matching error. This modulation would be somewhat analogous to the inhomogeneous Gaussian image models proposed in [24]. Such methods are presently under study.

12.6 Conclusion

We have introduced a spatiotemporal reduced update Kalman filter (3-D RUKF) and motion compensated extension (MC RUKF). These filters can be applied to random field sequences (i.e. time sequences of images). In an image sequence with motion vector flow, the MC RUKF has incorporated a hierarchical motion estimation technique. When the motion estimation fails, it causes some artifacts which depend on the characteristics of the image sequences. To deal with this motion estimation failure and newly exposed areas, a multi-model or inhomogeneous algorithm can be suggested. We are presently studying such an approach.

Acknowledgment

The authors would like to thank T. Naveen for providing the motion estimation program.

(a) the original image sequence: frame number = 13

(b) the noisy image sequence: frame number = 13

continued ...

(c) the image sequence filtered with RUKF : frame number = 13

(d) the image sequence filtered with MC-RUKF : frame number = 13

Figure 12.9: The filtering results when motion vector flow is slow.

(a) the original image sequence: frame number = 21

(b) the noisy image sequence: frame number = 21

continued ...

(c) the image sequence filtered with RUKF : frame number = 21

(d) the image sequence filtered with MC-RUKF : frame number = 21

Figure 12.10: The filtering results when motion vector flow is medium.

(a) the original image sequence: frame number = 30

(b) the noisy image sequence: frame number = 30

continued ...

(c) the image sequence filtered with RUKF : frame number = 30

(d) the image sequence filtered with MC-RUKF : frame number = 30

Figure 12.11: The filtering results when motion vector flow is fast.

References

[1] J. S. Lim, *Two-Dimensional Signal and Image Processing.* Englewood Cliffs, NJ: Prentice-Hall, 1990.

[2] A. K. Jain, "Advances in mathematical models for image processing," *Proceedings of the IEEE*, vol. 69, pp. 502–528, 1981.

[3] T. S. Huang and Y. P. Hsu, "Image sequence enhancement," in *Image Sequence Analysis*, pp. 289–309, Berlin, Germany: Spinger-Verlag, 1981.

[4] T. J. Dennis, "Nonlinear temporal filter for television picture noise reduction," *IEE Proc. Pt, G*, vol. 127, 1980.

[5] E. Dubois and M. S. Sabri, "Noise reduction in image sequences using motion-compensated temporal filtering," *IEEE Trans. Commun.*, vol. 32, pp. 826–831, 1984.

[6] D. Cano and M. Benard, "3D Kalman filtering of image sequences," in *Image Sequence Processing and Dynamic Scene Analysis*, pp. 563–579, Berlin Germany: Springer-Verlag, 1983.

[7] M. K. Ozkan, I. Sezan, A. T. Erdem, and A. M. Tekalp, "Motion compensated multiframe Wiener restoration of blurred and noisy image sequences," in *Proc. IEEE Int. Conf. Acoust., Speech, Signal Processing*, San Francisco, Califonia, Mar. 1992.

[8] J. W. Woods and V. K. Ingle, "Kalman filtering in two dimensions: Futher results," *IEEE Trans. Acoust., Speech, Signal Process.*, vol. 29, pp. 188–97, 1981.

[9] D. L. Angwin and H. Kaufman, "Image restoration using reduced order models," *Signal Processing*, vol. 16, pp. 21–28, 1988.

[10] A. K. Katsaggelos, R. P. Kleihorst, S. N. Efstratiadis, and R. L. Lagendijk, "Adaptive image sequence noise filtering methods," *Proc. SPIE Conf. Visual Communications and Image Process.*, vol. 1606, pp. 716–727, Nov. 1991.

[11] D. M. Martinez, *Model-Based Motion Estimation and its Application to Restoration and Interpolation.* MIT, EECS Dept.: Technical Report No. 530, 1987.

[12] T. Kronander, *Some Aspects of Perception Based Image Coding.* PhD thesis, Linkoping University, Linkoping, Sweden, 1989.

[13] J. W. Woods, "Two-Dimensional Kalman Filtering," in *Two-Dimensional Digital Signal Processing*, pp. 154–205, Berlin Germany: Springer-Verlag, 1983.

[14] A. Gelb, *Applied Optimal Estimation.* Cambridge, MA: MIT Press, 1974.

[15] D. L. Angwin, *Adaptive Image Restoration using Reduced Order Model Based Kalman Filters.* PhD thesis, Rensselaer Polytechnic Institute, Troy, NY, 1989.

[16] A. M. Tekalp, *Identification and Restoration of Noisy and Blurred Images.* PhD thesis, Rensselaer Polytechnic Institute, Troy, NY, 1984.

[17] H. G. Musmann, P. Pirsch, and H. J. Grallert, "Advances in picture coding," *Proceedings of the IEEE*, vol. 73, pp. 523–548, Apr. 1985.

[18] F. Glazer, G. Reynolds, and P. Anandan, "Scene matching by hierarchical correlation," in *Proceedings of the IEEE Computer vision and pattern recognition*, Washington DC., June 1983.

[19] M. Bierling and R. Thoma, "Motion compensating fields interpolation using a hierarchically structured motion estimator," *Signal Processing*, pp. 387–404, 1986.

[20] J. N. Driessen, *Motion Estimation for Digital Video.* PhD thesis, Delft University of Tech., Delft, Netherlands, 1992.

[21] R. L. Lagendijk, M. Tekalp, and J. Biemond, "Maximum likelihood image and blur identification: A unifying approach," *Journal of Optical Engineering*, May 1990.

[22] J. W. Woods, S. Dravida, and R. Mediavilla, "Image estimation using double stochatic Gaussian random field models," *IEEE Trans. Pattern Analysis and Machine Intelligence*, vol. PAMI-9, pp. 245–253, Mar. 1987.

[23] F. C. Jeng and J. W. Woods, "Simulated annealing in compound Gaussian random fields," *IEEE Trans. Inform. Theory*, vol. 36, pp. 94–107, Jan. 1990.

[24] F. C. Jeng and J. W. Woods, "Inhomogeneous Gaussian image models for estimation and restoration," *IEEE Trans. Acoust., Speech, Signal Process.*, vol. 36, pp. 1305–1312, 1988.

13

Multiframe Wiener Restoration of Image Sequences

M. K. Ozkan*, M. I. Sezan, A. T. Erdem, and A. M. Tekalp†

Eastman Kodak Company, Rochester NY, USA

**Thomson Consumer Electronics, Indianapolis IN, USA*

†University of Rochester, Rochester NY, USA

13.1 Introduction

Imagine an image sequence whose frames are both blurred and noisy. Three frames of such a sequence is shown in Fig. 13.1 where each frame suffers from focus blur as well as additive white Gaussian noise. Due to blur and noise contamination, the amount of information that a human observer or a machine can extract from this sequence is rather limited. It is therefore desirable to restore this image sequence. By that we mean the estimation of the original sequence from its blurred and noisy rendition. (The original sequence is shown in Fig. 13.2)

In this chapter, we address the problem of restoring image sequences that suffer from spatial blurring, such as motion and/or focus blur, as well as noise contamination due to quantization and/or the characteristics of the imaging sensor. Motion blur is common in image sequences recorded by high-speed cameras that monitor high-speed events in various industrial and scientific applications, surveillance cameras, and cameras mounted on moving vehicles such as an aircraft. Focus blur may be encountered in image sequences acquired by autofocus cameras due to the inertia of the autofocusing system. (The time interval between consequtive frames is in general smaller than the time needed by the autofocusing system to correct the focus.) Restoration of degraded image sequences increases their overall visual quality, and it is of particular importance in freeze frame applications such as hardcopy/softcopy display and automated analysis of single frames.

Figure 13.1: Three frames of an image sequence suffering from focus blur as well as noise contamination.

At this point, one may raise the following question: Why not utilize the vast variety of restoration algorithms that have been developed for single images and perform restoration on a frame-by-frame basis, treating each frame as a single image. This is indeed a valid and a straightforward approach. However, it is not an optimal one. The single-frame approach does not make use of the information contained in other frames in restoring a particular frame. In the particular case of Wiener restoration, which is of interest in this chapter, this amounts to disregarding the existing temporal correlation among the image frames. An optimal approach that does take into account the interframe correlations is the multiframe Wiener restoration where the problem of simultaneously restoring multiple frames is addressed. Wiener restoration of multiple images has been first discussed by Galatsanos and Chin [1] in the context of restoration of multispectral images. Recently, Srinavas *et al.* [2] applied the same multispectral Wiener filter to the restoration of misregistered radar imagery where the multiple images are treated as image data at different spectral bands. The Galatsanos-Chin filter [1] requires the recursive inversion of an $NM^2 \times NM^2$ block matrix with diagonal subblocks, where M^2 is the number of total pixels in a single frame and N is the number of frames that are simultaneously restored.

In the following, we first discuss a general computationally efficient multiframe Wiener restoration algorithm—the cross-correlated multiframe (CCMF) Wiener filter—which is based on the same filter expression used by Galatsanos and Chin for multispectral filtering. This efficient implementation requires the inversion of M^2 matrices of size only $N \times N$, which can be performed in parallel. Indeed, when N is sufficiently small (e.g., $N = 3$), inversions can be performed analytically. Extensive experiments that we have performed [3] show that the quality of the cross-correlated multiframe restoration is closely related to the accuracy of the estimates of the cross power and power spectra of the ideal image frames. In this chapter, we summarize the results of these experiments and provide strategies for determining the spectral estimates such that the cross-correlated multiframe approach becomes a worthwhile alternative to the single-frame approach.

In certain special cases, the CCMF approach lends itself to a closed-form solution that does not involve any matrix inversion. We discuss two such cases in which the noise is assumed to be temporally wide-sense stationary. One of these cases occurs when power and cross power spectra are estimated using the periodogram method. A more interesting special case is obtained when each frame is assumed to be a globally shifted version of the previous frame. This assumption is usually valid in cases where the interframe motion is due to either a shift of the camera with respect to a stationary scene (e.g., a camera attached to an aircraft), or a global shift of a scene with respect to a stationary camera (e.g., a camera overlooking objects on a conveyor belt). In this case, we first propose estimating the relative shift (misregistration) between the frames using a motion estimation algorithm recently developed by Fogel [4]. The Fogel algorithm has been shown to be both accurate and well-behaved in the presence of noise unlike other approaches [5]. Using the estimated motion information, which implicitly accounts for the interframe correlations, we then derive a closed-form expression for the multiframe Wiener filter. This filter, called the motion-compensated multiframe (MCMF) Wiener filter, thus requires neither the explicit estimation of cross correlations among the frames, nor any matrix inversion.

In what follows, we first formulate the multiframe Wiener restoration problem. A general efficient solution to the multiframe Wiener restoration problem, namely the CCMF approach, is discussed in Section 13.3. Special cases that lend themselves to closed-form solutions, including the MCMF Wiener filter, are discussed in Section 13.4. In Section 13.5, we provide the major results of an extensive experimental study on the sensitivity of the CCMF and the MCMF approaches to the spectra and interframe motion estimates, respectively, under varying blur, noise and interframe motion. We also compare the performance of these two approaches in restoring an image sequence with global interframe shifts both with each other and with the single-frame approach. Summary and conclusions are furnished in Section 13.6.

13.2 Problem Formulation

Let us consider restoration of N frames of an image sequence which are $M \times M$ pixels each. Suppose that each frame is degraded by linear shift-invariant spatial blur[1] and additive noise, where the blur point spread function (PSF) and noise statistics may change from frame to frame. Thus, the degraded frames, $g_i(m,n)$, $i = 1, \cdots, N$, can be expressed as

$$g_i(m,n) = \sum_{k,\ell=0,\cdots,M-1} h_i(m-k, n-\ell) f_i(k,\ell) + v_i(m,n), \quad i = 1, \cdots, N, \tag{13.1}$$

where $f_i(m,n)$, $v_i(m,n)$, $m,n = 0,1,\cdots,M-1$, represent, respectively, the original image frame and observation noise samples, and $h_i(m,n)$ represents the blurring PSF for the ith frame, where $1 \leq i \leq N$. Let $\mathbf{g}_i$, $\mathbf{f}_i$, and $\mathbf{v}_i$ denote the $M^2 \times 1$ vectors obtained by lexicographically ordering $g_i(m,n)$, $f_i(m,n)$, $v_i(m,n)$, respectively, into $M^2 \times 1$ vectors, and $\mathbf{D}_i$ denote the $M^2 \times M^2$ matrix of blur PSF coefficients for the ith frame [6]. Then, the observation model (13.1) can be expressed as

$$\boldsymbol{g} = \boldsymbol{\mathcal{D}} \boldsymbol{f} + \boldsymbol{v}, \tag{13.2}$$

where

$$\boldsymbol{g} \doteq \begin{bmatrix} \mathbf{g}_1 \\ \vdots \\ \mathbf{g}_N \end{bmatrix}, \quad \boldsymbol{f} \doteq \begin{bmatrix} \mathbf{f}_1 \\ \vdots \\ \mathbf{f}_N \end{bmatrix}, \quad \boldsymbol{v} \doteq \begin{bmatrix} \mathbf{v}_1 \\ \vdots \\ \mathbf{v}_N \end{bmatrix}, \text{and } \boldsymbol{\mathcal{D}} \doteq \begin{bmatrix} \mathbf{D}_1 & \cdots & 0 \\ \vdots & \ddots & \vdots \\ 0 & \cdots & \mathbf{D}_N \end{bmatrix}. \tag{13.3}$$

If we assume that the image and noise sequences are uncorrelated, i.e., $\mathcal{E}\{\mathbf{v}_i \mathbf{f}_j^T\} = \mathbf{0}$, $i,j = 1,2,\cdots,N$, (where $\mathcal{E}$ denotes the statistical expectation

[1] Here, without loss of generality, we assume that there is no blurring in the temporal direction. It should be noted that filters for noise suppression only can be obtained by setting the blur PSF to an impulse in the resulting filter expressions given in this and the following sections.

operation, and $\mathbf{0}$ denotes the $M^2 \times M^2$ matrix with all zero elements), the linear minimum mean square error (LMMSE) estimate, also known as the Wiener estimate, $\hat{\boldsymbol{f}}$ of $\boldsymbol{f}$ given $\boldsymbol{g}$ is [1]

$$\hat{\boldsymbol{f}} = \mathcal{R}_f \mathcal{D}^T (\mathcal{D} \mathcal{R}_f \mathcal{D}^T + \mathcal{R}_v)^{-1} \boldsymbol{g}, \tag{13.4}$$

where

$$\hat{\boldsymbol{f}} \doteq \begin{bmatrix} \hat{\mathbf{f}}_1 \\ \vdots \\ \hat{\mathbf{f}}_N \end{bmatrix}, \quad \mathcal{R}_f \doteq \begin{bmatrix} \mathbf{R}_{f;11} & \cdots & \mathbf{R}_{f;1N} \\ \vdots & \ddots & \vdots \\ \mathbf{R}_{f;N1} & \cdots & \mathbf{R}_{f;NN} \end{bmatrix}, \quad \mathcal{R}_v \doteq \begin{bmatrix} \mathbf{R}_{v;11} & \cdots & \mathbf{R}_{v;1N} \\ \vdots & \ddots & \vdots \\ \mathbf{R}_{v;N1} & \cdots & \mathbf{R}_{v;NN} \end{bmatrix}, \tag{13.5}$$

and $\mathbf{R}_{f;ij} \doteq \mathcal{E}\{\mathbf{f}_i \mathbf{f}_j^T\}$ and $\mathbf{R}_{v;ij} \doteq \mathcal{E}\{\mathbf{v}_i \mathbf{v}_j^T\}$, $i, j = 1, 2, .., N$, denote the cross-correlation matrices between the ith and j^{th} frames of the ideal image sequence and those of the noise sequence, respectively. In the following, we assume that the image and noise frames are individually wide-sense stationary in the spatial coordinates, which implies that the submatrices $\mathbf{R}_{f;ij}$ and $\mathbf{R}_{v;ij}$, $i, j = 1, 2, .., N$, in (13.5) are block Toeplitz [6]. However, we do not assume that the ideal image and noise frames are wide-sense stationary in the temporal coordinate. Hence the matrices $\mathcal{R}_f$ and $\mathcal{R}_v$ are not, in general, block Toeplitz. We also assume that the noise frames are jointly uncorrelated, i.e., $\mathbf{R}_{v;ij} = \mathbf{0}$, $i \neq j$, $i, j = 1, 2, \cdots, N$; and let $\mathbf{R}_{v;i} \doteq \mathbf{R}_{v;ii}$, $i, j = 1, 2, \cdots, N$, for notational simplicity. Note that the multiframe formulation reduces to independent single-frame restoration if the image frames are also jointly uncorrelated, i.e., if $\mathbf{R}_{f;ij} = \mathbf{0}$ for $i \neq j$, $i, j = 1, 2, \cdots, N$.

The direct computation of (13.4) requires the inversion of the $M^2N \times M^2N$ matrix $(\mathcal{D} \mathcal{R}_f \mathcal{D}^T + \mathcal{R}_v)$. For a typical sequence of 512×512 images, the size of this matrix is $(512)^2 N \times (512)^2 N$, and therefore its inversion is not practical. Fortunately, efficient computation of (13.4) is possible since the submatrices $\mathbf{D}_i$, $\mathbf{R}_{f;ij}$, and $\mathbf{R}_{v;ij}$ are block Toeplitz and can be diagonalized individually, under the block circulant approximation[2], using 2-D discrete Fourier transformation (DFT) [1]. In the following, we show, in detail, the diagonalization of the individual submatrices, and then determine a Fourier-domain expression for the Wiener estimate (13.4), which requires the inversion of a block matrix with diagonal subblocks. We note that because the matrices $\mathcal{D}$, $\mathcal{R}_f$ and $\mathcal{R}_v$ are not block Toeplitz in general, they cannot be diagonalized via 3-D DFT.

We let $\mathbf{W}$ denote the matrix that diagonalizes an $M^2 \times M^2$ block circulant matrix through a similarity transformation [6]. That is, if $\mathbf{C}$ is an $M^2 \times M^2$ block circulant matrix, then $\mathbf{W}^{-1}\mathbf{C}\mathbf{W}$ is a diagonal matrix whose elements can be computed using 2-D DFT, and also, if $\mathbf{b}$ is a vector of lexicographically-ordered samples of an image frame, then $\mathbf{W}^{-1}\mathbf{b}$ corresponds to the lexicographically-ordered samples of the 2-D DFT of the image frame. Then, as in [1], we let $\mathcal{W}$

[2]The (block) circulant approximation for (block) Toeplitz matrices has been investigated in [7] where it is shown that a Toeplitz matrix approaches to a circulant one, in the Euclidean norm sense, as the size of the matrix increases.

denote the $M^2N \times M^2N$ transformation matrix, defined as

$$\mathcal{W} \doteq \begin{bmatrix} \mathbf{W} & \cdots & 0 \\ \vdots & \ddots & \vdots \\ 0 & \cdots & \mathbf{W} \end{bmatrix}. \tag{13.6}$$

The operator $\mathcal{W}$ when applied to $\boldsymbol{f}$ stacks the 2-D DFTs of the individual frames (i.e., it is not the 3-D DFT operator). We premultiply both sides of (13.4) with $\mathcal{W}^{-1}$ to obtain

$$\mathcal{W}^{-1}\hat{\boldsymbol{f}} = (\mathcal{W}^{-1}\mathcal{R}_f\mathcal{W})(\mathcal{W}^{-1}\mathcal{D}^T\mathcal{W})[\mathcal{W}^{-1}(\mathcal{D}\mathcal{R}_f\mathcal{D}^T + \mathcal{R}_v)\mathcal{W}]^{-1}\mathcal{W}^{-1}\boldsymbol{g}, \tag{13.7}$$

which can be expressed in a compact form as

$$\hat{\mathcal{F}} = \mathcal{P}_f\mathcal{H}^*\mathcal{Q}^{-1}\mathcal{G}, \tag{13.8}$$

where the quantities $\hat{\mathcal{F}}$, $\mathcal{P}_f$, $\mathcal{H}^*$, $\mathcal{Q}$ and $\mathcal{G}$ are defined as follows:

$$\hat{\mathcal{F}} \doteq \mathcal{W}^{-1}\hat{\boldsymbol{f}} = \begin{bmatrix} \hat{\mathbf{F}}_1 \\ \vdots \\ \hat{\mathbf{F}}_N \end{bmatrix}, \quad \text{where,} \quad \hat{\mathbf{F}}_i \doteq \mathbf{W}^{-1}\hat{\mathbf{f}}_i = \begin{bmatrix} \hat{F}_{i,1} \\ \vdots \\ \hat{F}_{i,M^2} \end{bmatrix}, \quad i = 1, \cdots, N, \tag{13.9}$$

denote the lexicographical ordering of the 2-D DFT of the LMMSE estimate of the ith frame;

$$\mathcal{P}_f \doteq \mathcal{W}^{-1}\mathcal{R}_f\mathcal{W} = \begin{bmatrix} \mathbf{P}_{f;11} & \cdots & \mathbf{P}_{f;1N} \\ \vdots & \ddots & \vdots \\ \mathbf{P}_{f;N1} & \cdots & \mathbf{P}_{f;NN} \end{bmatrix}, \tag{13.10}$$

where the block $\mathbf{P}_{f;ij}$ denotes the diagonalized cross-correlation matrix of the actual (non-degraded) i^{th} and the jth frames (note that $\mathcal{P}_f$ itself is not diagonal), and

$$\mathbf{P}_{f;ij} \doteq \mathbf{W}^{-1}\mathbf{R}_{f;ij}\mathbf{W} = \begin{bmatrix} P_{f;ij,1} & \cdots & 0 \\ \vdots & \ddots & \vdots \\ 0 & \cdots & P_{f;ij,M^2} \end{bmatrix}, \quad i,j = 1, \cdots, N; \tag{13.11}$$

$$\mathcal{H}^* \doteq \mathcal{W}^{-1}\mathcal{D}^T\mathcal{W} = \begin{bmatrix} \mathbf{H}_1^* & \cdots & 0 \\ \vdots & \ddots & \vdots \\ 0 & \cdots & \mathbf{H}_N^* \end{bmatrix}, \tag{13.12}$$

where

$$\mathbf{H}_i^* \doteq \mathbf{W}^{-1}\mathbf{D}_i^T\mathbf{W} = \begin{bmatrix} H_{i,1}^* & \cdots & 0 \\ \vdots & \ddots & \vdots \\ 0 & \cdots & H_{i,M^2}^* \end{bmatrix}, \quad i = 1, \cdots, N, \tag{13.13}$$

denotes the diagonalized blur matrix (* denotes the Hermitian operation, i.e., matrix transposition and complex-conjugation);

$$\mathcal{Q} \doteq \mathcal{W}^{-1}(\mathcal{D}\mathcal{R}_f\mathcal{D}^T + \mathcal{R}_v)\mathcal{W} = \begin{bmatrix} \mathbf{Q}_{11} & \cdots & \mathbf{Q}_{1N} \\ \vdots & \ddots & \vdots \\ \mathbf{Q}_{N1} & \cdots & \mathbf{Q}_{NN} \end{bmatrix}, \qquad (13.14)$$

where

$$\mathbf{Q}_{ij} \doteq \mathbf{W}^{-1}(\mathbf{D}_i\mathbf{R}_{f;ij}\mathbf{D}_j^T + \delta_{ij}\mathbf{R}_{v;ii})\mathbf{W} = \begin{bmatrix} Q_{ij,1} & \cdots & 0 \\ \vdots & \ddots & \vdots \\ 0 & \cdots & Q_{ij,M^2} \end{bmatrix}, \qquad (13.15)$$

(δ_{ij}, $i, j = 1, \cdots, N$, denotes the Kronecker delta function, i.e., $\delta_{ij} = 1$ if $i = j$, and $\delta_{ij} = 0$ if $i \neq j$), and letting

$$\mathbf{P}_{v;i} \doteq \mathbf{W}^{-1}\mathbf{R}_{v;i}\mathbf{W} = \begin{bmatrix} P_{v;i,1} & \cdots & 0 \\ \vdots & \ddots & \vdots \\ 0 & \cdots & P_{v;i,M^2} \end{bmatrix}, \quad i, j = 1, \cdots, N, \qquad (13.16)$$

we have

$$Q_{ij,k} = H_{i,k}P_{f;ij,k}H_{j,k}^* + \delta_{ij}P_{v;i,k}, \quad i, j = 1, \cdots, N, \quad k = 1, \cdots, M^2; \qquad (13.17)$$

and finally,

$$\mathcal{G} \doteq \mathcal{W}^{-1}\boldsymbol{g} = \begin{bmatrix} \mathbf{G}_1 \\ \vdots \\ \mathbf{G}_N \end{bmatrix}, \quad \text{where} \quad \mathbf{G}_i \doteq \mathbf{W}^{-1}\mathbf{g}_i = \begin{bmatrix} G_{i,1} \\ \vdots \\ G_{i,M^2} \end{bmatrix}, \quad i = 1, \cdots, N, \qquad (13.18)$$

is the lexicographical ordering of the 2-D DFT of the degraded version of the ith frame.

Thus, the computation of (13.8) requires the inversion of the block matrix $\mathcal{Q}$, given by (13.14), whose blocks are diagonal matrices. It is essential to use a well-conditioned and computationally efficient algorithm for the inversion of the matrix $\mathcal{Q}$. It can be shown that the inverse of a block matrix with diagonal blocks is also a block matrix with diagonal blocks [1]. Thus, the blocks of $\mathcal{Q}^{-1}$ are also diagonal. Therefore, once $\mathcal{Q}^{-1}$ is evaluated, the computation of $\hat{\mathcal{F}}$ given by (13.8) becomes straightforward since it involves multiplication of block matrices with diagonal blocks. Galatsanos and Chin [1] use a recursive method for evaluating the inverse $\mathcal{Q}^{-1}$. Their method involves successive partitioning of the matrix $\mathcal{Q}$ and recursively computing the inverse of the partitions. In the following, we discuss an alternative approach for the computation of the Wiener solution (13.8), which requires the inversion of $N \times N$ matrices only.[3] This approach lends itself to efficient implementation in parallel processors where each matrix inversion can be implemented by a separate processor.

[3] A similar approach, in the context of multispectral image restoration, recently appeared in [8] and was brought to our attention after our work was completed and this particular approach was presented in part in [9].

13.3 An Efficient Multiframe Wiener Filter: General Solution

In this section we show that the inverse of the $NM^2 \times NM^2$ matrix $\mathcal{Q}$ that appears in the Wiener solution (13.8) can be determined by inverting M^2 matrices of size $N \times N$ only. This result is based on the following lemma.

Lemma 1: Let $\mathcal{A}$ be an $NM^2 \times NM^2$ nonsingular diagonal-blocks matrix given as

$$\mathcal{A} = \begin{bmatrix} \mathbf{A}_{11} & \cdots & \mathbf{A}_{1N} \\ \vdots & \ddots & \vdots \\ \mathbf{A}_{N1} & \cdots & \mathbf{A}_{NN} \end{bmatrix},$$

where the blocks $\mathbf{A}_{ij}$ $i,j = 1,\cdots,N$, are $M^2 \times M^2$ diagonal matrices denoted as

$$\mathbf{A}_{ij} = \begin{bmatrix} A_{ij,1} & \cdots & 0 \\ \vdots & \ddots & \vdots \\ 0 & \cdots & A_{ij,M^2} \end{bmatrix}.$$

Let $\mathcal{B}$ denote the inverse of $\mathcal{A}$ given as

$$\mathcal{A}^{-1} \equiv \mathcal{B} = \begin{bmatrix} \mathbf{B}_{11} & \cdots & \mathbf{B}_{1N} \\ \vdots & \ddots & \vdots \\ \mathbf{B}_{N1} & \cdots & \mathbf{B}_{NN} \end{bmatrix},$$

where $\mathbf{B}_{ij}$ $i,j = 1,\cdots,N$, are $M^2 \times M^2$ diagonal matrices denoted as

$$\mathbf{B}_{ij} = \begin{bmatrix} B_{ij,1} & \cdots & 0 \\ \vdots & \ddots & 0 \\ 0 & \cdots & B_{ij,M^2} \end{bmatrix}.$$

Define the following $N \times N$ matrices

$$\tilde{\mathbf{A}}_k \doteq \begin{bmatrix} A_{11,k} & \cdots & A_{1N,k} \\ \vdots & \ddots & \vdots \\ A_{N1,k} & \cdots & A_{NN,k} \end{bmatrix}, \; k = 1,\cdots,M^2,$$

and

$$\tilde{\mathbf{B}}_k \doteq \begin{bmatrix} B_{11,k} & \cdots & B_{1N,k} \\ \vdots & \ddots & \vdots \\ B_{N1,k} & \cdots & B_{NN,k} \end{bmatrix}, \; k = 1,\cdots,M^2,$$

where $A_{ij,k}$, and $B_{ij,k}$ $i,j = 1,\cdots,N$, $k = 1,\cdots,M^2$, are the kth diagonal elements of $\mathbf{A}_{ij}$ and $\mathbf{B}_{ij}$, respectively. The following equation then holds

$$\tilde{\mathbf{B}}_k = \tilde{\mathbf{A}}_k^{-1}, \qquad k = 1,\cdots,M^2. \tag{13.19}$$

Therefore, the elements of the inverse matrix $\mathcal{B}$ can be obtained by inverting matrices, $\tilde{\mathbf{A}}_k$, $k = 1, \cdots, M^2$, each of which is $N \times N$. $\diamond$

The proof of Lemma 1 is given in [3].

In order to compute the inverse of $\mathcal{Q}$ using Lemma 1, we let

$$\mathcal{Z} = \begin{bmatrix} \mathbf{Z}_{11} & \cdots & \mathbf{Z}_{1N} \\ \vdots & \ddots & \vdots \\ \mathbf{Z}_{N1} & \cdots & \mathbf{Z}_{NN} \end{bmatrix} = \begin{bmatrix} \mathbf{Q}_{11} & \cdots & \mathbf{Q}_{1N} \\ \vdots & \ddots & \vdots \\ \mathbf{Q}_{N1} & \cdots & \mathbf{Q}_{NN} \end{bmatrix}^{-1}, \tag{13.20}$$

where $\mathbf{Z}_{ij}$, $i, j = 1, \cdots, N$, are $M^2 \times M^2$ diagonal matrices given as

$$\mathbf{Z}_{ij} \doteq \begin{bmatrix} Z_{ij,1} & \cdots & 0 \\ \vdots & \ddots & \vdots \\ 0 & \cdots & Z_{ij,M^2} \end{bmatrix}. \tag{13.21}$$

Invoking Lemma 1, the elements $Z_{ij,k}$, $i, j = 1, \cdots, N$, $k = 1, \cdots, M^2$, of $\mathcal{Z}$ can be computed from

$$\tilde{\mathbf{Z}}_k = \tilde{\mathbf{Q}}_k^{-1}, \qquad k = 1, \cdots, M^2, \tag{13.22}$$

where $\tilde{\mathbf{Z}}_k$ and $\tilde{\mathbf{Q}}_k$ are defined as

$$\tilde{\mathbf{Z}}_k \doteq \begin{bmatrix} Z_{11,k} & \cdots & Z_{1N,k} \\ \vdots & \ddots & \vdots \\ Z_{N1,k} & \cdots & Z_{NN,k} \end{bmatrix}, \quad k = 1, \cdots, M^2, \tag{13.23}$$

and

$$\tilde{\mathbf{Q}}_k \doteq \begin{bmatrix} Q_{11,k} & \cdots & Q_{1N,k} \\ \vdots & \ddots & \vdots \\ Q_{N1,k} & \cdots & Q_{NN,k} \end{bmatrix}, \quad k = 1, \cdots, M^2, \tag{13.24}$$

where $Q_{ij,k}$ $i, j = 1, \cdots, N$, $k = 1, \cdots, M^2$, are as defined in (13.17). Thus the inverse of the $NM^2 \times NM^2$ matrix $\mathcal{Q}$ can be computed by inverting the $N \times N$ matrices $\tilde{\mathbf{Q}}_k$, $k = 1, \cdots, M^2$, only. Note that each submatrix $\tilde{\mathbf{Q}}_k$ is associated with a spatial frequency k. The M^2 matrix inversions can be carried out in parallel, achieving significant gains in computational speed.

Once $\mathcal{Q}^{-1}$ is computed, the samples of the 2-D DFTs of the frame estimates, $\hat{F}_{i,k}$, $i = 1, \cdots, N$, $k = 1, \cdots, M^2$, can be obtained from (13.8) as

$$\boxed{\hat{F}_{i,k} = \sum_{p=1}^{N} P_{f;ip,k} H^*_{p,k} \sum_{q=1}^{N} Z_{pq,k} G_{q,k}, \quad k = 1, \cdots, M^2, \ i = 1, \cdots, N,} \tag{13.25}$$

where $P_{f;ip,k}$, $H^*_{p,k}$, $G_{q,k}$ and $Z_{pq,k}$ are as defined in (13.11), (13.13), (13.18), and (13.21) respectively. Finally, the frame estimates, $\hat{\mathbf{f}}_i$, $i = 1, \cdots, N$, are obtained, from (13.9), as

$$\hat{\mathbf{f}}_i = \mathbf{W}\hat{\mathbf{F}}_i, \quad i = 1, \cdots, N. \tag{13.26}$$

We refer to this efficient multiframe restoration filter defined by (13.25) and (13.26) as the cross-correlated multiframe (CCMF) Wiener filter.

13.4 Closed Form Solution: Special Cases

In this section, we discuss some special cases of interest where the inverse of the matrices $\tilde{\mathbf{Q}}_k$, $k = 1, \cdots, M^2$, can be expressed analytically, and hence a closed-form solution to the CCMF Wiener filter can be obtained. We first introduce a sufficient condition for the existence of a closed-form solution, and then elaborate on two specific cases where a closed-form solution can be found.[4]

A sufficient condition for the existence of a closed-form solution is that the matrix $\tilde{\mathbf{Q}}_k$ can be expressed as the sum of a vector outer product and a scaled identity matrix for every k. This is because such matrices can be analytically inverted using Lemma 2 (Sherman-Morrison formula, [10]) stated below. In the following, we first state this sufficient condition in terms of the cross power spectra matrix $\mathcal{P}_f$, defined by (13.10), and the second-order statistics of the noise. Let us define the matrices $\tilde{\mathbf{P}}_{f;k}$, $k = 1, \cdots, M^2$, formed by arranging the kth diagonal elements of the subblocks of the cross power spectra matrix $\mathcal{P}_f$ into $N \times N$ matrices as

$$\tilde{\mathbf{P}}_{f;k} \doteq \begin{bmatrix} P_{f;11,k} & \cdots & P_{f;1N,k} \\ \vdots & \ddots & \vdots \\ P_{f;N1,k} & \cdots & P_{f;NN,k} \end{bmatrix}, \quad k = 1, \cdots, M^2,$$

where $P_{f;ij,k}$ denotes the value of the cross power spectrum of the ith and the jth frames at frequency k and is defined as in (13.11). Suppose now that $P_{f;ij,k}$ can be expressed in the form of

$$P_{f;ij,k} \equiv S_{f;i,k} S^*_{f;j,k}, \quad i,j = 1, \cdots, N, \; k = 1, \cdots, M^2, \tag{13.27}$$

and that the power spectra of the noise frames are identical for every frame, i.e.,

$$P_{v;i,k} \equiv P_{v;k}, \quad i = 1, \cdots, N, \; k = 1, \cdots, M^2. \tag{13.28}$$

Then, substituting (13.27) and (13.28) in (13.17) we obtain

$$Q_{ij,k} = H_{i,k} S_{f;i,k} S^*_{f;j,k} H^*_{j,k} + P_{v;k}\delta_{ij}, \quad i,j = 1, \cdots, N, \quad k = 1, \cdots, M^2. \tag{13.29}$$

Equation (13.29) expresses $\tilde{\mathbf{Q}}_k$ as a vector outer product plus a scaled identity matrix for any k, given (13.27) and (13.28). Thus, (13.27) and (13.28) are sufficient conditions for the existence of a closed-form solution. Next, we state Lemma 2 and then derive the closed-form expression for the CCMF Wiener filter using (13.29).

Lemma 2: (Sherman-Morrison formula, [10]) Let $\mathbf{A}$ be a matrix which can be expressed as the sum of a vector outer product and a scaled identity matrix, i.e.,

$$\mathbf{A} \doteq \mathbf{x}\mathbf{x}^* + \alpha\mathbf{I}.$$

Then, the inverse of $\mathbf{A}$ is given by

$$\mathbf{A}^{-1} = \frac{1}{\alpha\Lambda}(\Lambda\mathbf{I} - \mathbf{x}\mathbf{x}^*), \quad \text{where,} \quad \Lambda \doteq \mathbf{x}^*\mathbf{x} + \alpha. \qquad \diamond$$

---

[4]We specifically assume here that the blur is in the spatial domain only.

Using Lemma 2, we can analytically determine the elements of $\tilde{\mathbf{Z}}_k$, defined by (13.23), as

$$Z_{ij,k} = \frac{\left(\sum_{\ell=1}^{N} |S_{f;\ell,k} H_{\ell,k}|^2 + P_{v;k}\right) \delta_{ij} - H_{i,k} S_{f;i,k} S^*_{f;j,k} H^*_{j,k}}{P_{v;k}\left(\sum_{\ell=1}^{N} |H_{\ell,k} S_{f;\ell,k}|^2 + P_{v;k}\right)},$$

$$i,j = 1,\cdots,N, \;\; k = 1,\cdots,M^2. \quad (13.30)$$

Substituting (13.27) and (13.30) in (13.25) and simplifying the result, we obtain the closed-form solution for the CCMF Wiener filter, under the assumptions stated by (13.27) and (13.28), as

$$\boxed{\hat{F}_{i,k} = \frac{S_{f;i,k} \sum_{q=1}^{N} S^*_{f;q,k} H^*_{q,k} G_{q,k}}{\sum_{\ell=1}^{N} |S_{f;\ell,k} H_{\ell,k}|^2 + P_{v;k}}, \quad k = 1,\cdots,M^2.} \quad (13.31)$$

In the remainder of this section, we discuss two special cases where the matrices $\tilde{\mathbf{P}}_{f;k}$, $k = 1,\cdots,M^2$, lend themselves to an outer product representation. The first case is when the cross power spectra is computed using the periodogram method. The second case is when the interframe motion is a global displacement, which results in the motion-compensated multiframe (MCMF) Wiener filter.

13.4.1 Periodogram-based spectral estimation

If the cross power spectrum matrices are estimated using the periodogram method, we in fact have $P_{f;ij,k}$ in the form of (13.27), where $\{S_{f;i,k}\}_{k=1}^{M^2}$ denotes the 2-D DFT of the ith frame of a prototype image sequence. Thus, assuming that (13.28) also holds, we can use (13.31) to obtain a closed-form CCMF Wiener solution for the case of peridogram-based spectral estimation.

13.4.2 Global interframe displacement

A more interesting special case is when the consecutive frames are globally shifted versions of each other, i.e.,

$$f_i(x,y) = f_1(x + d_{x,i}, y + d_{y,i}), \quad i = 2,\cdots,N, \quad (13.32)$$

where x and y denote the continuous spatial coordinates, and $[d_{x,i}\ d_{y,i}]^T$ represents the real-valued displacement vector of the ith frame with respect to an arbitrary but predetermined reference frame.[5]

In this section, we derive another closed-form solution, called the motion-compensated multiframe (MCMF) Wiener filter under the assumptions stated by

[5] Here, the first frame is taken as the reference frame, without loss of generality.

(13.32) and (13.28). In the MCMF Wiener filter, the interframe correlations are implicitly utilized through the interframe motion information that may either be available *a priori*, or estimated via a motion estimation (registration) algorithm, such as the phase correlation method [11] or the Fogel algorithm [4]. Thus, the MCMF filter does not require the estimation of cross-correlations (or cross power spectra) between the frames, nor does it require any matrix inversion.

In order to express (13.32) in the discrete-spatial domain, we define the quantities p_i, q_i, ϵ_i, and ζ_i, $i = 2, \cdots, N$, such that

$$d_{x,i} = p_i + \epsilon_i, \quad \text{and} \quad d_{y,i} = q_i + \zeta_i, \tag{13.33}$$

where p_i and q_i are integers, and $0 \leq \epsilon_i, \zeta_i < 1$. We then propose the following discrete-space model for (13.32) that accounts for non-integer-valued spatial displacements between the frames:

$$f_i(m,n) = f_1(m+p_i, n+q_i) ** \phi(m,n;\epsilon_i,\zeta_i), \quad i = 2, \cdots, N, \tag{13.34}$$

where $f_i(m,n)$ represents the sampled version of $f_i(x,y)$, $i = 1, \cdots, N$, (we assume, without loss of generality, that the size of the sampling interval is unity in both dimensions), $**$ denotes the 2-D circular convolution, and $\phi(m,n;\epsilon_i,\zeta_i)$ models the effect of non-integer displacements. Note that if the components of the displacement vector $[d_{x,i}\ d_{y,i}]^T$ are integer-valued, i.e., $\epsilon_i, \zeta_i = 0$, we have $\phi(m,n;\epsilon_i,\zeta_i) = 1$, for $i = 2, \cdots, N$. Otherwise, there are two possible approaches for defining the function $\phi(m,n;\epsilon_i,\zeta_i)$. One of the approaches is motivated by the fact that a frame that is displaced by a non-integer amount from the reference frame can be obtained through an interpolation of the reference frame, where the interpolation can be expressed as a convolution of the reference frame with an appropriate kernel $\phi(m,n;\epsilon_i,\zeta_i)$, as in (13.34). In [3], we have derived the expression for $\phi(m,n;\epsilon_i,\zeta_i)$ for the inverse-distance weighted interpolation suggested by Shepard [12]. This derivation is not discussed here. The other approach is a Fourier-domain approach and is described below.

In the discrete Fourier transform domain, (13.34) becomes

$$F_i(k,\ell) = F_1(k,\ell)\, e^{j2\pi k p_i/M} e^{j2\pi \ell q_i/M}\, \Phi(k,\ell;\epsilon_i,\zeta_i), \quad k,\ell = 0, \cdots, M-1, \tag{13.35}$$

where $F_i(k,\ell)$, $F_1(k,\ell)$, and $\Phi(k,\ell;\epsilon_i,\zeta_i)$ represent the 2-D DFTs of $f_i(m,n)$, $f_1(m,n)$, and $\phi(m,n;\epsilon_i,\zeta_i)$, respectively. This approach is motivated by the fact that an integer-valued shift of a signal in the discrete-space domain corresponds to the addition of a linear phase in the discrete Fourier domain as in (13.35). Based on this property, we also model any non-integer-valued spatial displacement by the addition of an appropriate linear phase in the discrete Fourier domain. Thus, in this approach, $\phi(m,n;\epsilon_i,\zeta_i)$ is defined such that its Fourier transform $\Phi(k,\ell;\epsilon_i,\zeta_i)$ is given by

$$\Phi(k,\ell;\epsilon_i,\zeta_i) \doteq \begin{cases} e^{j2\pi k\epsilon_i/M} e^{j2\pi \ell\zeta_i/M} & \text{if} \quad 0 \leq k,\ell < M/2 \\ e^{j2\pi k\epsilon_i/M} e^{j2\pi(\ell-M)\zeta_i/M} & \text{if} \quad 0 \leq k < M/2 \leq \ell < M \\ e^{j2\pi(k-M)\epsilon_i/M} e^{j2\pi \ell\zeta_i/M} & \text{if} \quad 0 \leq \ell < M/2 \leq k < M \\ e^{j2\pi(k-M)\epsilon_i/M} e^{j2\pi(\ell-M)\zeta_i/M} & \text{if} \quad M/2 \leq k,\ell < M \end{cases}. \tag{13.36}$$

It is important to note that the linear Fourier phase implied by the definition of $\Phi(k,\ell;\epsilon_i,\zeta_i)$ in (13.36) is chosen to be an odd function of k and ℓ so that $\phi(m,n;\epsilon_i,\zeta_i)$ corresponds to a real function. Here, M is assumed to be even without loss of generality.

We now proceed with the development of the MCMF method. We assume, for the purpose of mathematical tractability, that the interframe shift is cyclic and therefore it can be represented by a circulant matrix operating on the image vector. In practice, this assumption is valid for all pixels except for the ones belonging to a boundary region whose size is directly proportional to the amount of the shift. Then, we can express (13.34) in the vector-matrix form as

$$\mathbf{f}_i = \boldsymbol{\Phi}_i \mathbf{J}_i \mathbf{f}_1, \qquad i = 1, \cdots, N, \tag{13.37}$$

where $\mathbf{J}_i$, $i = 2, \cdots, N$, are cyclic shift matrices corresponding to integer-valued spatial displacements (p_i, q_i) and are formed by 0's and 1's placed in appropriate locations. The matrices $\boldsymbol{\Phi}_i$, $i = 2, \cdots, N$, are circulant convolution matrices formed by the coefficients of $\phi_i(m,n;\epsilon_i,\zeta_i)$ which are determined by using one of the two approaches outlined above, and $\boldsymbol{\Phi}_1 \equiv \mathbf{J}_1 \equiv \mathbf{I}$, the $M^2 \times M^2$ identity matrix.

Defining

$$\mathbf{R}_f \doteq \mathbf{R}_{f;11} \tag{13.38}$$

for simplicity of notation, using (13.37), and the definition $\mathbf{R}_{f;ij} \doteq \mathcal{E}\{\mathbf{f}_i \mathbf{f}_j^T\}$, the correlation matrix for the ith and the jth frames can be expressed as

$$\mathbf{R}_{f;ij} = \boldsymbol{\Phi}_i \mathbf{J}_i \mathbf{R}_f \mathbf{J}_j^T \boldsymbol{\Phi}_j^T.$$

Therefore, from (13.11), we have

$$\mathbf{P}_{f;ij} = \boldsymbol{\Psi}_i \mathbf{C}_i \mathbf{P}_f \mathbf{C}_j^* \boldsymbol{\Psi}_j^* \tag{13.39}$$

where

$$\boldsymbol{\Psi}_i \doteq \mathbf{W}^{-1} \boldsymbol{\Phi}_i \mathbf{W} = \begin{bmatrix} \Psi_{i,1} & \cdots & 0 \\ \vdots & \ddots & \vdots \\ 0 & \cdots & \Psi_{i,M^2} \end{bmatrix}, \quad i = 1, \cdots, N, \tag{13.40}$$

and

$$\mathbf{C}_i \doteq \mathbf{W}^{-1} \mathbf{J}_i \mathbf{W} = \begin{bmatrix} C_{i,1} & \cdots & 0 \\ \vdots & \ddots & \vdots \\ 0 & \cdots & C_{i,M^2} \end{bmatrix}, \quad i = 1, \cdots, N, \tag{13.41}$$

are the diagonalized shift matrices. Note that, from (13.35), and due to the fact that $\mathbf{J}_i$ and $\boldsymbol{\Phi}_i$ are circulant, we have

$$\Psi_{i,kM+\ell+1} = \Phi(k,\ell;\epsilon_i,\zeta_i), \quad k,\ell = 0, \cdots, M-1. \tag{13.42}$$

and

$$C_{i,kM+\ell+1} = e^{j2\pi k p_i/M} e^{j2\pi \ell q_i/M}, \quad k,\ell = 0, \cdots, M-1. \tag{13.43}$$

Hence, for this special case, we have the cross power spectrum samples expressed as

$$P_{f;ij,k} = \Psi_{i,k} C_{i,k} P_{f;k} C^*_{j,k} \Psi^*_{j,k}, \quad i = 1,\cdots,N, \ k = 0,\cdots,M-1, \tag{13.44}$$

which are in fact in the form of (13.27) with $S_{f;i,k}$ given by

$$S_{f;i,k} = \Psi_{i,k} C_{i,k} P^{1/2}_{f;k}, \quad i = 1,\cdots,N, \ k = 0,\cdots,M-1. \tag{13.45}$$

Then, in order to find the MCMF Wiener solution, we substitute (13.45) in (13.31) and obtain

$$\hat{F}_{i,k} \quad = \quad \frac{\Psi_{i,k} C_{i,k} \sum_{q=1}^{N} C^*_{q,k} \Psi^*_{q,k} H^*_{q,k} G_{q,k}}{\sum_{\ell=1}^{N} |H_{\ell,k} \Psi_{\ell,k}|^2 + \dfrac{P_{v;k}}{P_{f;k}}}. \tag{13.46}$$

where we used that $|C_{j,k}| \equiv 1$, $k = 1,\cdots,M^2$, $i = 1,\cdots,N$, from (13.43).

For the reference frame, i.e., frame 1 in our case, we have $\Psi_{1,k} = C_{1,k} = 1$, for all k, and therefore, the MCMF Wiener solution given by (13.46) becomes

$$\boxed{\hat{F}_{1,k} = \frac{\sum_{q=1}^{N} \Psi^*_{q,k} C^*_{q,k} H^*_{q,k} G_{q,k}}{\sum_{\ell=1}^{N} |H_{\ell,k} \Psi_{\ell,k}|^2 + \dfrac{P_{v;k}}{P_{f;k}}}, \qquad k = 1,\cdots,M^2.} \tag{13.47}$$

The MCMF estimate of the reference frame, $\hat{\mathbf{f}}_1$, is then computed using (13.26).

In principle, it is possible to determine the estimates of the rest of the frames by shifting the reference frame estimate appropriately:

$$\hat{\mathbf{f}}_i = \boldsymbol{\Phi}_i \mathbf{S}_i \hat{\mathbf{f}}_1, \qquad i = 2,\cdots,N. \tag{13.48}$$

In practice, however, this may not result in a satisfactory solution due to the fact that the actual shift is not entirely cyclic and the modeling of non-integer shifts is inexact. We therefore propose MCMF filtering of the image sequence one frame at a time (this is shown in Fig. 13.12).

13.5 Experimental Results

We performed experiments with the proposed CCMF and the MCMF Wiener filters in order to evaluate their performance in general, and also compare them with the single-frame Wiener filter under various different conditions. In our comparisons, we investigated the effects of the following factors: (i) amount of blur and noise, (ii) methods for estimation of power and cross power spectra, (iii) accuracy of interframe motion estimation, and (iv) number of frames used in multiframe restoration. In the following, we first give a brief description of our experiments and then discuss the major results. For a more detailed discussion of these results, the reader is referred to [3].

13.5.1 Test sequences

We experimented with three different image sequences synthesized from real-life still images. Each frame was 256×256 and 8 bits/pixel. In the first sequence (Seq1), shown in the lefthandside of Fig. 13.2, the disk containing Lena's face was moved from frame-to-frame. In addition, the background was also moved horizontally (2 pixels/frame) to simulate camera panning. In comparison, the second sequence (Seq2), shown in the right hand side of Fig. 13.2, had only global interframe motion. The three frames of Seq2 were created by considering three 256×256 overlapping sections cut from a large aerial image. The first two sections were displaced from each other by $[-4 \quad 2]^T$, and the third section was displaced from the second by $[6 \quad -3]^T$. The third sequence, Seq3 (Fig. 13.3), was created similar to Seq2 using the same aerial image. Seq3 had seven frames created by sections whose displacements were $[-10 \quad -2]^T$, $[-8 \quad -3]^T$, $[-8 \quad -4]^T$, $[-6 \quad -6]^T$, $[-4 \quad -7]^T$, and $[-4 \quad -8]^T$ between the second and first, third and second, fourth and third, fifth and fourth, sixth and fifth, and seven and sixth frames, respectively. Note that CCMF is applicable to all of these three sequences whereas MCMF is applicable to only Seq2 and Seq3 where the interframe motion is a global shift. Since the derivation of the MCMF Wiener estimate assumes cyclic global shift for the purpose of mathematical tractability (Section 13.4.2), the interframe shift in Seq2 and Seq3 is implicitly assumed to be cyclic in the case of MCMF restoration. As we have noted previously, this assumption is invalid only within a certain boundary region whose size is directly proportional to the amount of interframe shift.

13.5.2 Blur and noise degradation

In our experiments, we considered simulated focus blur. The blur PSF is modeled by a uniform distribution with circular support whose diameter determines the degree of defocus [13]. We blurred the image sequences in two different ways: (i) by using identical blur diameters for each frame, and (ii) by using different blur diameters for each frame. We experimented with varying degrees of blur in both of these cases. In the case of experiments using Seq1 and Seq2, we used the following four sets of blur diameters (in pixels) $\{8,8,8\}$, $\{7.5, 8.0, 8.5\}$, $\{4,4,4\}$, and $\{3.5, 4.0, 4.5\}$. In the case of Seq3, we used the set $\{9.0, 8.9, 8.7, 8.4, 8.0, 7.5, 6.9\}$. (In each set, the first number refers to the diameter for the first frame, and so on.) We simulated the noise degradation by adding white Gaussian noise to the blurred image frames. We used signal-to-noise ratios (SNR) of 40 dB, 30 dB, and 20 dB.

In practice, if the blur PSF is not readily available, it can be estimated by applying one of the well-known techniques such as the cepstrum and maximum likelihood methods to each frame in the sequence [14]. Further, under white stationarity assumption, the variance of the noise process at each frame can be estimated from the sample variance computed over a uniform image region.

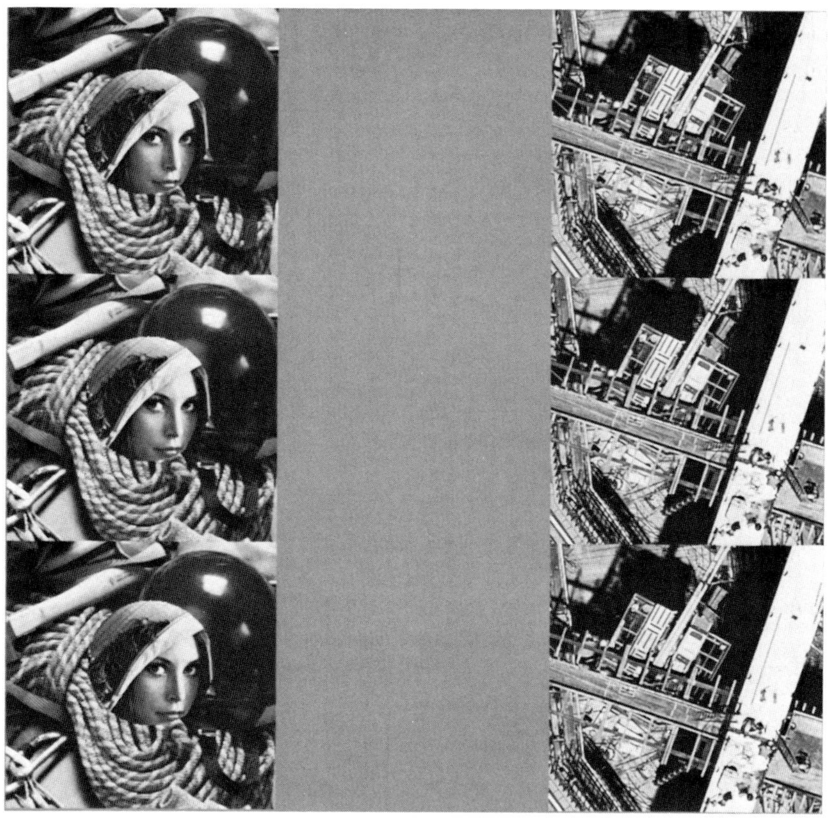

Figure 13.2: The test sequences Seq1 (left) and Seq2 (right). From top to bottom: Frames 1, 2 and 3. In Seq1, the disk containing Lena's face was moved from frame to frame in addition to the background which was also moved horizontally to simulate camera panning. In Seq2, the interframe motion is a global shift only.

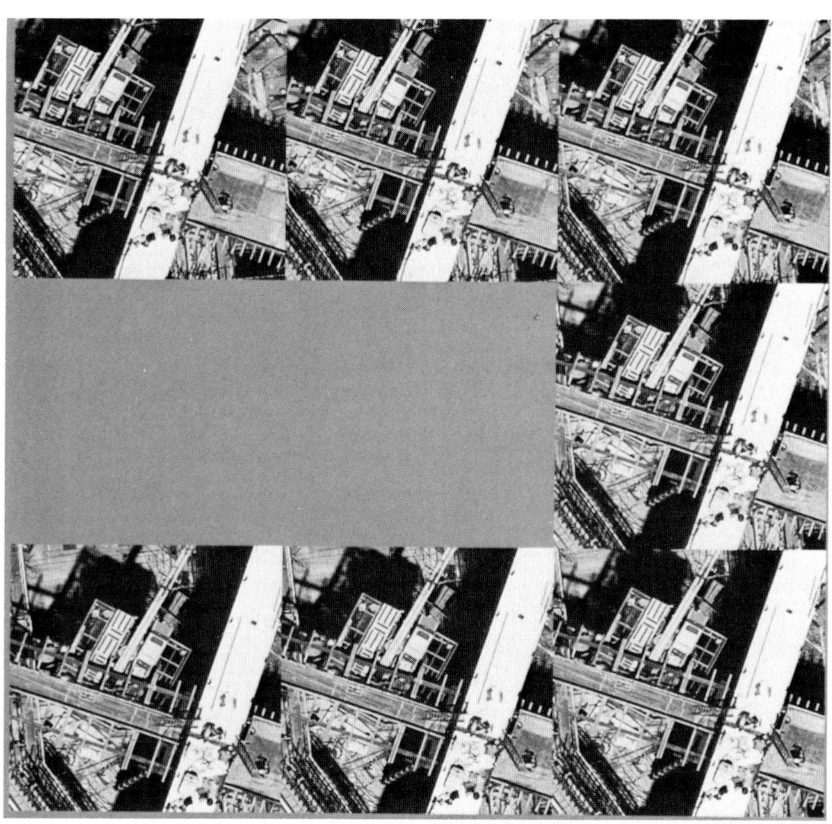

Figure 13.3: The test sequence Seq3. Clockwise from upper left: Frames 1, 2, 3, 4, 5 , 6 and 7. The interframe motion between all frames is a global shift as in Seq2.

13.5.3 Power and cross power spectra estimation

The MCMF Wiener filter requires the estimation of the power spectrum of the actual frame that is chosen as the reference (Section 13.4.2). Similarly, the single-frame Wiener filter requires the estimation of the power spectrum of the actual frame. The CCMF filter, on the other hand, requires the estimation of cross power spectra between the actual frames as well as their power spectra (Section 13.3). There are two issues of concern in power spectrum and cross power spectrum estimation: (1) selection of a prototype image or an image sequence whose statistical characteristics resemble that of the actual image or image sequence, and (2) selection of a particular spectral estimation method. We used the following prototypes in estimating the power spectrum in the case of single-frame and MCMF Wiener restoration, and in estimating the power and cross power spectra in the case of CCMF restoration: (i) actual image (sequence); (ii) observed image (sequence); (iii) a sharp image (sequence) other than the actual image (sequence); and (iv) image sequence obtained by individually restoring the frames of the degraded sequence using a single-frame Wiener filter. The use of the actual image data as the prototype was aimed at determining the best possible performance of the algorithms. Indeed, the actual data is not available in practice.

For power spectrum estimation, in the case of MCMF and single-frame Wiener restoration, we used the periodogram, the Welch, and the autoregressive (AR) model-based methods. In the Welch method, we used 9 overlapping segments, each with 128×128 pixels, with an overlap of 64 pixels. The AR-model based power spectrum used a 1×1 nonsymmetric half plane (NSHP) AR model whose parameters were estimated from the prototype image. In the case of CCMF, we used the direct extensions of the periodogram and the Welch methods to power and cross power spectra estimation, as well as a method based on a 3-D multiframe AR modeling of the actual image sequence [15],[16]. As we have discussed in Section 13.4.1, CCMF Wiener filtering admits an analytic closed-form solution in the case of the periodogram method. The 3-D AR model is a frame-to-frame coupled model with 1×1 NSHP spatial support at each frame. The model parameters are estimated from the prototype image sequence.

13.5.4 Interframe motion estimation

In the case of MCMF Wiener filtering, the interframe shift vector between the reference and another frame is estimated as follows. First, the Fogel algorithm [4] is used to determine an estimate of the displacement vector field between the two frames at every p^{th} ($p \geq 1$) pixel of the reference frame. The shift vector is then estimated by the average of the displacement vectors. The average is computed over all the displacement vectors except for the ones associated with pixels within a certain boundary region. This is because the displacement estimates are inaccurate within a certain boundary region due to occlusion. The size of the boundary region depends on the amount of interframe motion. In our experiments, we used $p = 4$. We subsampled the diplacement vector field at those pixels where the estimates were available, e.g., a 64×64 field was obtained in the case of a 256×256 image frame, and then averaged the values of the displacement vectors in a 54×54 area. In other words, we discarded a boundary region that was within 5 pixels from each side of the 64×64 field.

The Fogel algorithm, in general, provides an accurate estimate of the interframe motion and is robust in the presence of noise [5]. In any event, we investigated the sensitivity of the MCMF restoration to possible errors in the estimates of the interframe shift vectors. To this end, we implemented the MCMF filter using a set of errors, ranging from 0.0 to 1.0 pixel, in the displacement vector estimates.

13.5.5 Evaluation of results

Restoration results were evaluated in terms of visual subjective quality as well as the dB-improvement achieved by the restoration filter. For the ith frame the dB-improvement, μ_i, is defined by

$$\mu_i = 10 \log \frac{||\mathbf{f}_i - \mathbf{g}_i||^2}{||\mathbf{f}_i - \hat{\mathbf{f}}_i||^2} \tag{13.49}$$

where $|| \cdot ||$ denotes the usual norm in M^2 dimensional Euclidean space.

13.5.6 Results of CCMF Wiener restoration

First, we performed experiments to evaluate the sensitivity of the performance of CCMF filter to power and cross power spectra estimation, and to determine the spectral estimation strategy that would result in the best practical performance. We considered both Seq1 and Seq2 that were blurred in four different ways and contaminated by noise at three different SNR levels as explained above.

Three prototypes were used for single frame restoration: the sharp original, second frame of Seq1 or Seq2 (depending on which sequence is being restored; second frame of Seq1(2) is used in the restoration of Seq2(1)), (henceforth the sharp prototype) and the observed image. The sequence formed by the single-frame restoration of the image frames, using the AR model of the second frame of Seq1 (or Seq2) for power spectrum estimation, is used as a prototype sequence for the CCMF restoration. The reason for choosing the AR model-based power spectrum estimation in this case was that the single-frame restoration was shown to result in good quality restorations when the power spectra was estimated on the basis of an AR model of a sharp prototype, and further, the performance was robust in case of different choices for the sharp prototype [17]. Our results suggests the following:

(i) In the ideal case where the prototype is the actual image sequence and the power and cross power spectra are estimated using periodogram method, the CCMF filter outperforms the single-frame filter in all cases. However, in practical situations where the actual image data is unavailable, the periodogram method, as well as the others, may result in CCMF restorations that are worse than their single-frame counterparts.

Table 13.1: The dB-improvements obtained for the CCMF and the single-frame restorations of the central frames of Seq1 and Seq2.

		Seq1		Seq2	
SNR	Blur Set	CCMF	Single-Frame	CCMF	Single-Frame
40dB	{7.5 8.0 8.5}	7.1	5.5	7.5	6.3
40dB	{8.0 8.0 8.0}	6.6	5.5	7.4	6.3
40dB	{3.5 4.0 4.5}	7.2	5.8	8.0	6.4
40dB	{4.0 4.0 4.0}	6.8	5.8	7.9	6.4
30dB	{7.5 8.0 8.5}	4.3	3.2	4.1	3.4
30dB	{8.0 8.0 8.0}	4.1	3.2	4.0	3.4
30dB	{3.5 4.0 4.5}	3.4	2.0	3.5	3.0
30dB	{4.0 4.0 4.0}	3.2	2.0	3.3	3.0
20dB	{7.5 8.0 8.5}	2.4	1.7	1.8	1.6
20dB	{8.0 8.0 8.0}	2.3	1.7	1.8	1.6
20dB	{3.5 4.0 4.5}	1.8	1.0	1.8	1.4
20dB	{4.0 4.0 4.0}	1.7	1.0	1.8	1.4

(ii) In practice, the best choice for the prototype image sequence for CCMF filtering is that generated by individually restoring the image frames using the single-frame filter, i.e., the single-frame restored sequence. (Recall that these single-frame restorations are performed using power spectra estimated from the AR model of the sharp prototype.) In this case, the CCMF filter outperforms the single-frame filter.

(iii) The performance of the CCMF filter using the single-frame restored sequence as a prototype is considerably better than that of the single frame restoration when the power and cross power spectra are estimated using either the Welch (for Seq2) or the 3-D AR-method (for Seq1).

The dB-improvements for the CCMF and the single-frame filters are compared in the third and fourth columns in Table 13.1. The CCMF filtering results refer to 3-D multiframe AR-model based power and cross power spectra estimation in the case of Seq1, and to the Welch method for Seq2. These results are for restoration of the central frames of respective sequences. Similar results were obtained for the first and the third frames. One can observe from Table 13.1 that the benefit of using the CCMF approach rather than the single-frame filtering is larger when the blur varies from frame-to-frame. On the other hand, the difference in performance between the CCMF and the single-frame Wiener filtering decreases with decreasing SNR.

Examples of the improved visual quality obtained by the CCMF restoration are shown in Figs. 13.4 and 13.5, at SNR=40 dB and 20 dB, respectively, for the blur set $\{7.5, 8, 8.5\}$. The upper left image shows the degraded second frame of Seq2. Its single-frame Wiener restoration is shown in the bottom left, and the bottom right image is its CCMF Wiener restoration. The CCMF restorations are in general sharper than the single-frame restorations. It is interesting to note in Fig. 13.5 that the visibility of the "E"-like structures on the rectangular area located in a perpendicular direction to the grayish bridge-like structure in the lower half of the image is resolved much better in the case of CCMF restoration at SNR = 20 dB.

The last set of experiments were aimed at investigating the dependence of the CCMF restoration on the number of frames. We used Seq3 and adopted the best strategy obtained in the case of Seq2 above for power and cross power spectra estimation. The dB-improvement obtained in the case of the fourth (i.e., the central) frame of Seq3 versus the number of frames at different SNR levels is plotted in Fig. 13.6. (Number of frames = n means that the central frame and its $n-1$ neighboring frames are considered.) From Fig. 13.6, we observe that, in general, the performance improves considerably as the number of frames goes up to three. For a larger number of frames, however, the increase in performance levels off.

13.5.7 Results of MCMF Wiener restoration

The first set of experiments were aimed at evaluating the sensitivity of MCMF restoration to power spectrum estimation and determining the best practical strategy for power spectrum estimation. In this case, Seq2 was blurred using the four different sets of blurs and contaminated by noise at the three different SNR levels. The central frame was taken as the reference frame. The interframe shift vectors were estimated using the Fogel algorithm. As expected, the best results were obtained in the ideal case where the original frame was used as the prototype and the periodogram method was used for power spectrum estimation. For practical cases, we made the following observation:

In all cases, the best practical MCMF restoration is obtained when the power spectrum of the reference frame (i.e., the central frame) is estimated on the basis of an AR model of a sharp prototype image. (In fact, using a sharp image as the prototype is consistently better in case of all three of the power spectrum estimation methods.) In our experiments, we have found that the difference between the dB improvements obtained by using the well-known Cameraman image versus the second frame of Seq1 as the sharp prototype is insignificant, hence the robustness to the choice of a sharp prototype.

In the second set of experiments, our goal was to compare the MCMF and the single-frame Wiener restorations of the central frame of the degraded Seq2 sequence. We also investigated the amount of error the MCMF method allows in the estimated interframe shift vector, beyond the error obtained by the Fogel algorithm, while resulting in better dB improvements than the single-frame approach. In this set of experiments, we used the best practical strategy for power spectrum estimation method, namely the AR-model based method where the model parameters are estimated from the Cameraman image.

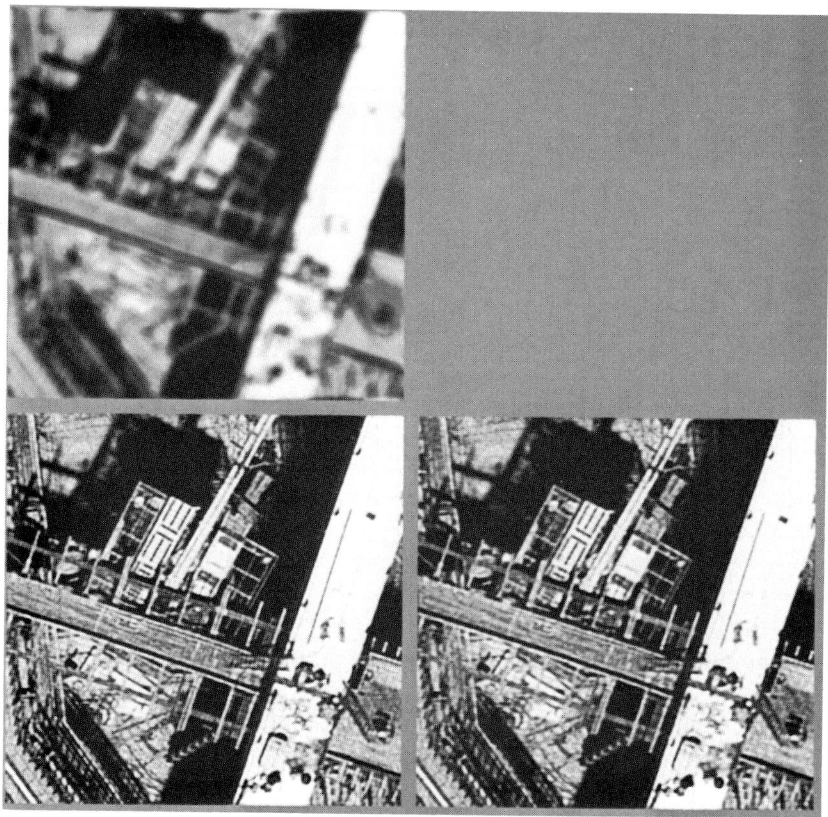

Figure 13.4: Restoration results shown for the second frame of Seq2 at SNR=40dB and in the case of blur diameter set $\{7.5, 8, 8.5\}$. Upper left: The degraded image. Lower left: The CCMF Wiener restoration. Lower right: The single-frame Wiener restoration.

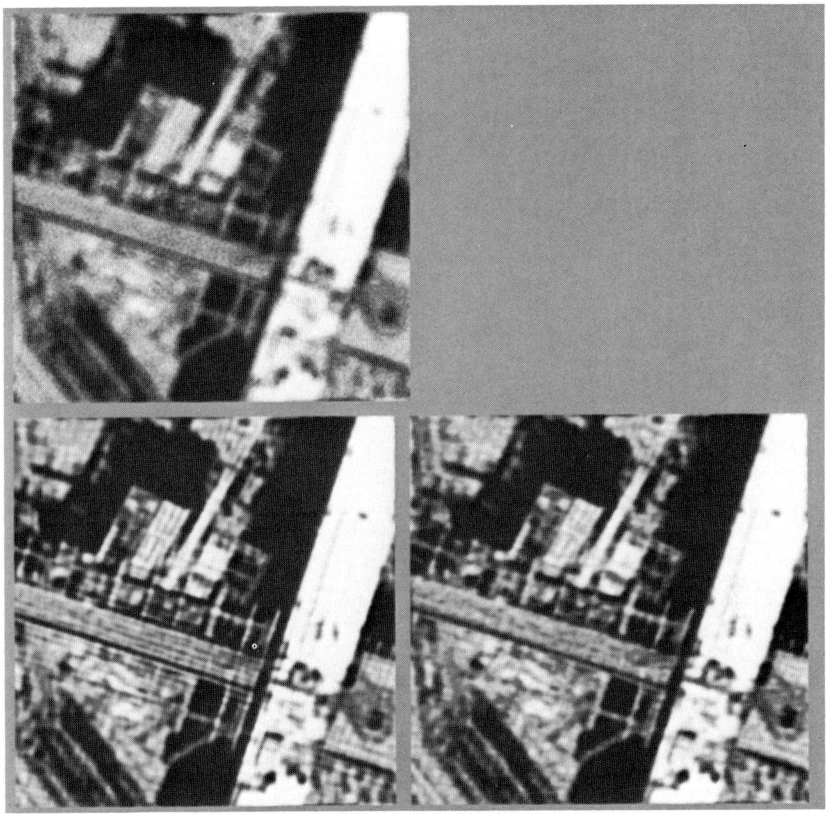

Figure 13.5: Restoration results shown for the second frame of Seq2 at SNR=20dB and in the case of blur diameter set $\{7.5, 8, 8.5\}$. Upper left: The degraded image. Lower left: The CCMF Wiener restoration. Lower right: The single-frame Wiener restoration.

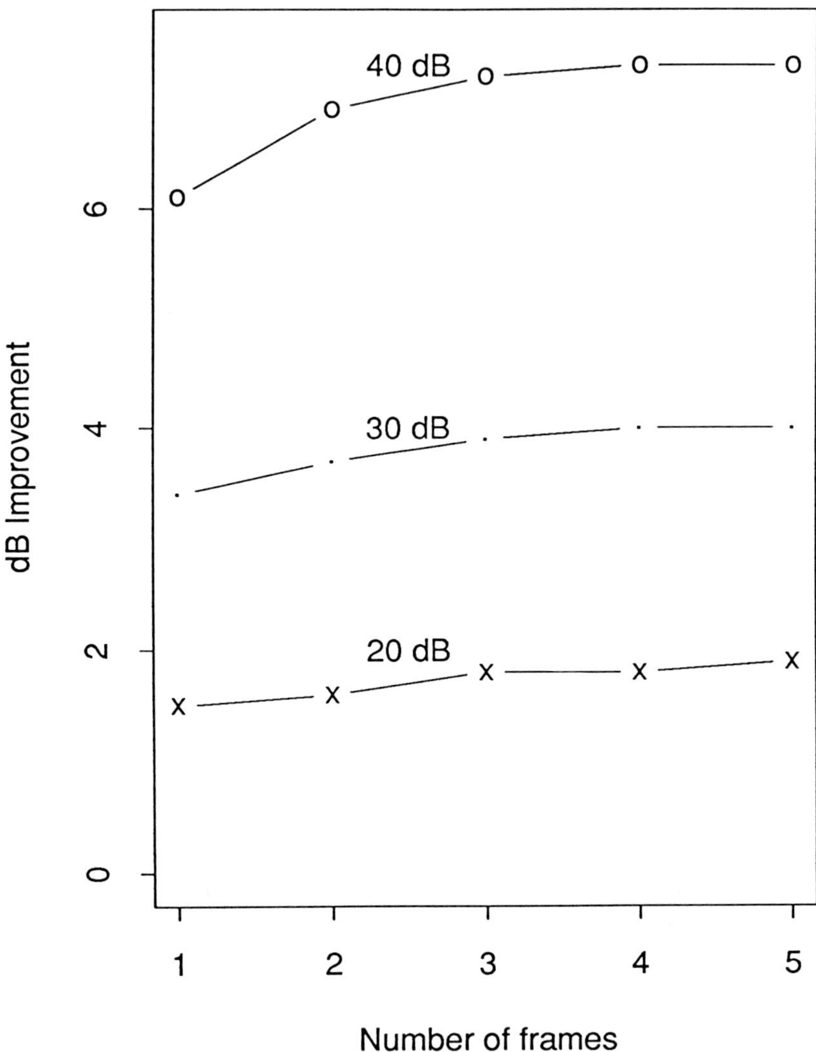

Figure 13.6: Dependence of the CCMF filtering on the number of frames: The dB improvement obtained in the case of the CCMF restoration of the fourth (i.e., the central) frame of Seq3 versus the number frames, at different SNR levels.

Table 13.2: The dB-improvements obtained for the MCMF and the single-frame restoration of the central frames of Seq2. (Interframe shift vectors are estimated using the Fogel algorithm.)

		MCMF		Single-Frame
SNR	Blur Set	Fourier Dom.	Spt. Interp.	
40dB	{7.5 8.0 8.5}	9.8	9.8	6.3
40dB	{8.0 8.0 8.0}	8.1	8.1	6.3
40dB	{3.5 4.0 4.5}	13.5	13.7	6.4
40dB	{4.0 4.0 4.0}	8.7	8.7	6.4
30dB	{7.5 8.0 8.5}	5.4	5.3	3.6
30dB	{8.0 8.0 8.0}	4.8	4.8	3.6
30dB	{3.5 4.0 4.5}	6.1	6.1	3.1
30dB	{4.0 4.0 4.0}	4.3	4.3	3.1
20dB	{7.5 8.0 8.5}	2.7	2.7	1.7
20dB	{8.0 8.0 8.0}	2.5	2.5	1.7
20dB	{3.5 4.0 4.5}	2.8	2.8	1.5
20dB	{4.0 4.0 4.0}	2.3	2.3	1.5

Fig. 13.7 illustrates the dB-improvements obtained by the MCMF method using the exact interframe shift vectors as well as the shift vectors with errors in the case of blur sets $\{7.5, 8.0, 8.5\}$ and $\{8, 8, 8\}$. The same amount of error was introduced to the shift vectors associated with each frame with respect to the reference frame. We experimented with errors of +0.10, +0.25, +0.50, +0.75, and +1.00 pixels in both components of the shift vectors. It should be noted that the absolute values of the actual estimation errors obtained by the Fogel algorithm was below 0.05 pixels for both components of the shift vectors in all cases. Results of using both the Fourier-domain (Section 13.4.2) and spatial interpolation [3] methods for modeling non-integer interframe shifts are shown in Fig. 13.7 . The dB-improvement obtained by the single-frame restoration is indicated by the horizontal line segment $S - S$. The dB-improvements obtained by the MCMF filter using the Fogel algorithm are compared to that obtained by the single-frame filter in Table 13.2 for all sets of blurs and SNR values and for both frequency domain and interpolation methods for modeling non-integer shifts.

From Fig. 13.7 and Table 13.2, we observe the following:
(i) The MCMF Wiener filter using the Fogel algorithm for motion compensation outperforms the single-frame Wiener filter in all cases.
(ii) The benefit of using the MCMF filter is larger when the blur diameters vary from frame-to-frame.
(iii) For each blur diameter set, the performance difference between the MCMF and the single-frame filters decreases with decreasing SNR.

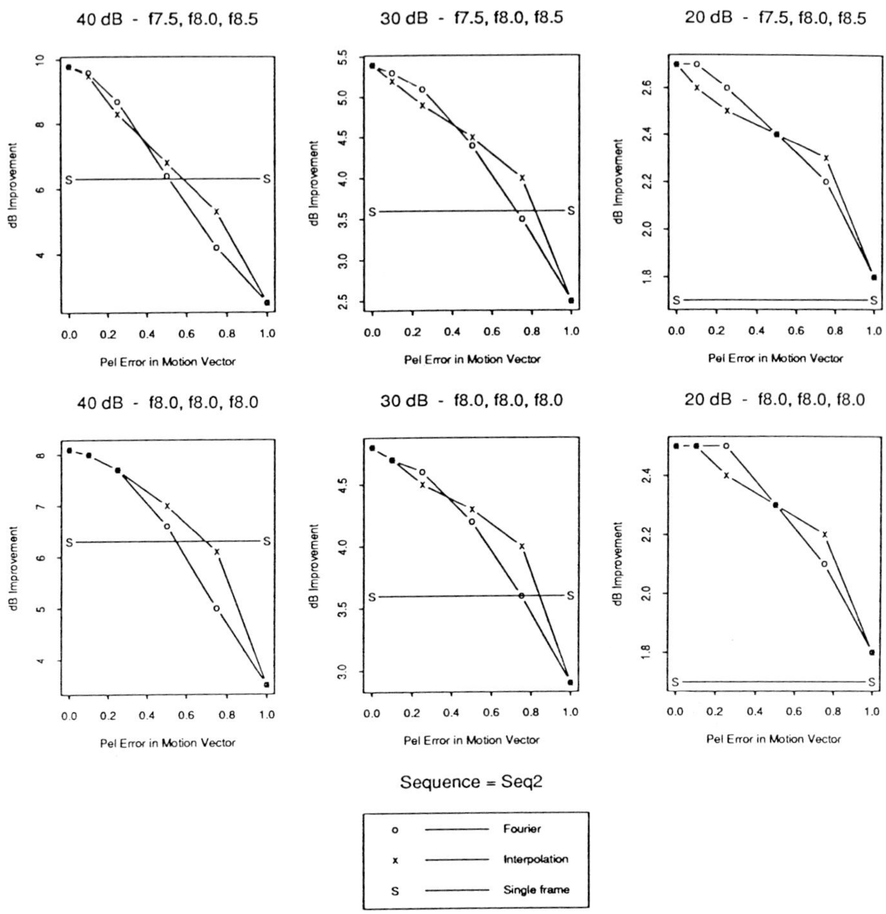

Figure 13.7: Illustration of the dependence of the MCMF restoration to the accuracy of interframe shift estimation. The dB improvements obtained by the MCMF method using the exact interframe shift vectors as well as the shift vectors with errors are shown. First Row: Blur diameter set $\{7.5, 8, 8.5\}$; Second Row: Blur diameter set $\{8, 8, 8\}$.

(iv) The MCMF filter does allow some errors in the estimates of the interframe shift vector in producing better quality restoration results than the single-frame filter. The MCMF filter becomes more tolerant to errors in interframe motion estimation in the case of larger blurs and noise.

Examples of MCMF restoration are shown in Figs. 13.8 and 13.9, where the degraded (upper left), the MCMF-restored using shift estimates obtained by the Fogel algorithm (lower left), and the single-frame restored (lower right) versions of the reference frame of Seq2 are shown for the blur set $\{7.5, 8.0, 8.5\}$ at SNR=40 dB and SNR=20 dB, respectively. At SNR=40 dB (Fig. 13.8), observe the increased visibility of the steps of the ladder, and the small white rectangular detail over the grayish bridge-like structure transversing the lower half of the image. At SNR=20 dB (Fig. 13.9), the "E"-like structures on the rectangular area located in a perpendicular direction to the grayish bridge-like structure can be resolved only in the case of MCMF restoration.

The third set of experiments were carried out to determine how the number of frames affects the performance of the MCMF filter under various amounts of observation noise, and how its performance compares to that of the single-frame filter as the errors in the estimates of the motion vectors are increased. We experimented with the Seq3 sequence blurred using the set of blur diameters $\{9.0, 8.9, 8.7, 8.4, 8.0, 7.5, 6.9\}$. We applied the MCMF filter using three frames (Frames 3, 4 and 5), five frames (Frames 2, 3, 4, 5 and 6), and seven frames (Frames 1, 2, 3, 4, 5, 6 and 7) to restore the reference frame (Frame 4). The results are shown in Fig. 13.10 at theree different SNR levels. The results obtained using the estimates determined by the Fogel algorithm is summarized in Table 13.3. We observe that the increase in the amount of dB improvement due to the use of MCMF filter rather than the single-frame filter becomes larger as the frame number increases, provided that the motion vector estimates are sufficiently accurate (as is the case with the Fogel's algorithm). In general, however, the tolerance of the MCMF filter to errors in interframe shift vector estimates decreases as the frame number increases.

13.6 Summary and Conclusions

In this chapter, we have discussed two approaches, the CCMF and the MCMF Wiener filters, for efficient multiframe restoration of spatially blurred and noisy image sequences. The CCMF approach is a general one that applies to any sequence and to any arbitrary type of interframe motion. The MCMF approach is a special case of the CCMF approach and is applicable to image sequences where the interframe motion is a global shift and the second-order statistics of the noise process remains the same from one frame to another. We have investigated the conditions under which these two multiframe approaches provide significantly improved restorations compared to that obtained by the single-frame approach.

In theory, when the interframe displacement is a global shift and is cyclic, and the second-order noise statistics are identical for all frames, then the CCMF and the MCMF approaches are equivalent. In practice, however, these two ap-

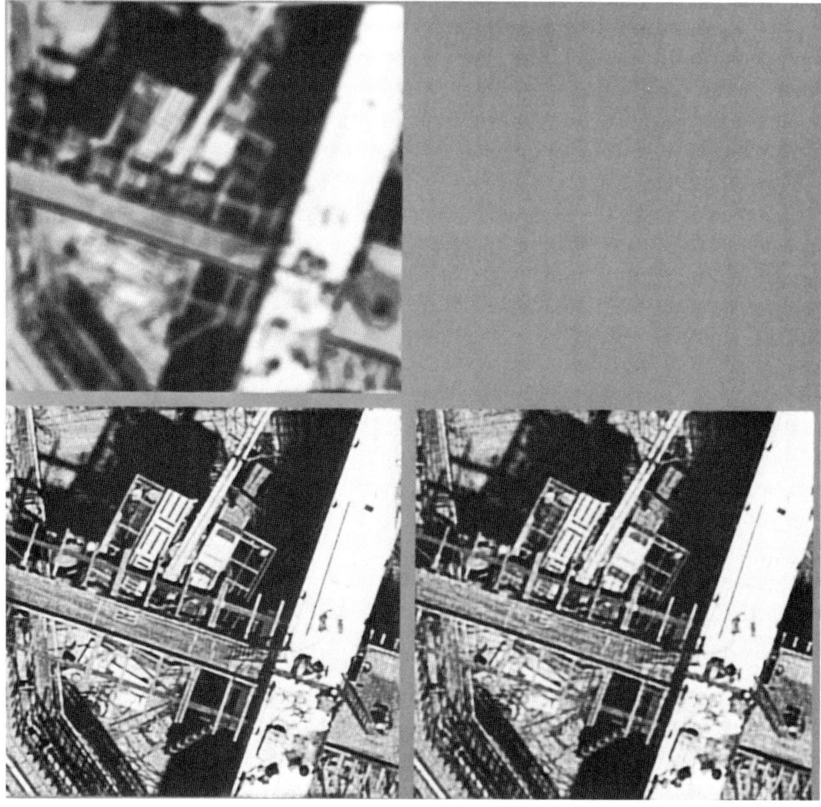

Figure 13.8: Restoration results shown for the second (reference) frame of Seq2 in the case of blur diameter set $\{7.5, 8, 8.5\}$ and at SNR=40dB. Upper left: The degraded image. Lower left: The MCMF restoration. Lower right: The single-frame restoration.

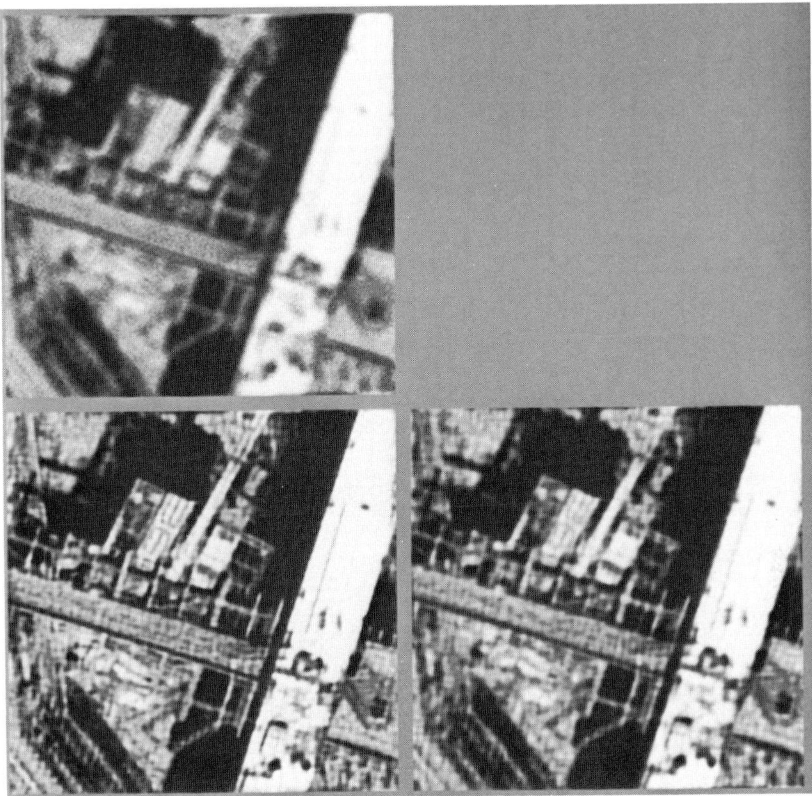

Figure 13.9: Restoration results shown for the second (reference) frame of Seq2 in the case of blur diameter set $\{7.5, 8, 8.5\}$ and at SNR=20dB. Upper left: The degraded image. Lower left: The MCMF restoration. Lower right: The single-frame restoration.

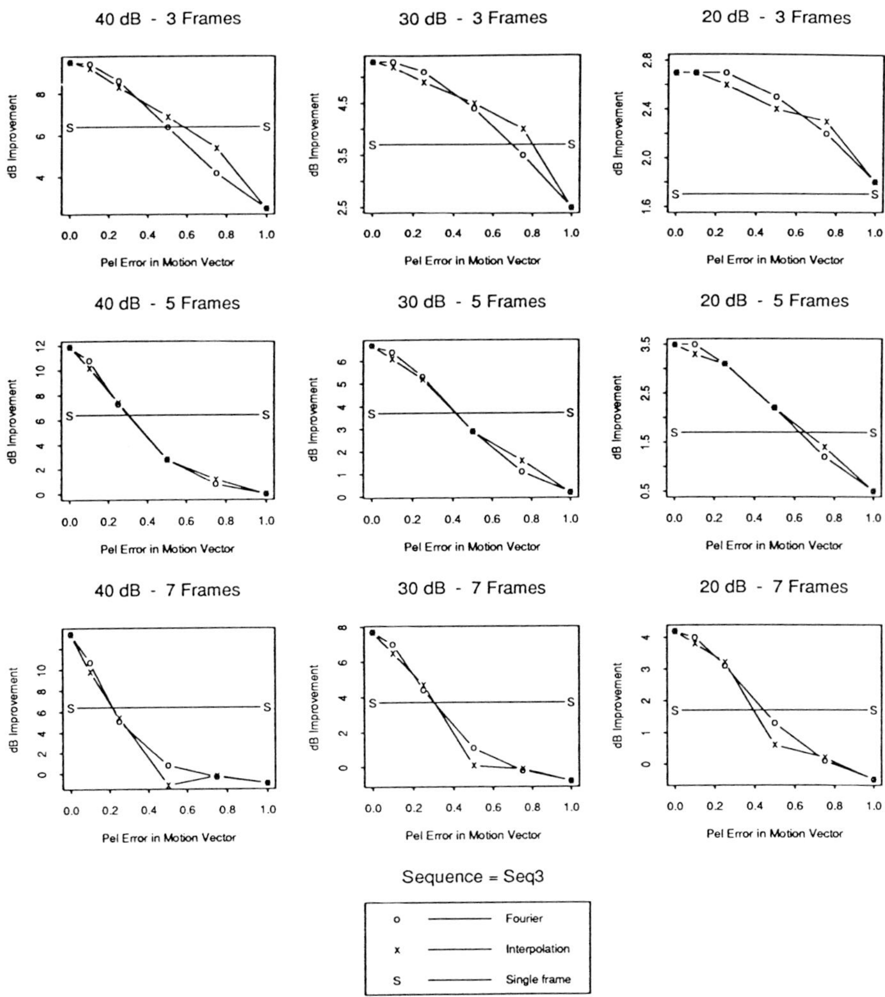

Figure 13.10: Dependence of the performance of the MCMF filter on the number of frames for various amounts of errors in the estimates of motion vectors. The results obtained for the MCMF restoration of the fourth (i.e., the central) frame of Seq3 are shown at different SNR levels.

Table 13.3: Single-frame versus MCMF restoration of the central frame of Seq3 using different number of frames.

SNR	No. of Frames	MCMF Fourier Dom.	MCMF Spt. Interp.	Single-Frame
40dB	3	9.5	9.5	6.4
40dB	5	11.8	11.8	6.4
40dB	7	13.2	12.9	6.4
30dB	3	5.3	5.3	3.7
30dB	5	6.6	6.6	3.7
30dB	7	7.7	7.6	3.7
20dB	3	2.7	2.7	1.7
20dB	5	3.5	3.5	1.7
20dB	7	4.2	4.2	1.7

proaches almost always produce different results due to the following reasons. First, a global shift is hardly ever cyclic in practice. Secondly, the power spectrum and the interframe motion information used by the MCMF, and the power and cross power spectra estimates used by the CCMF are not exact. In fact, we have observed in the previous section that the MCMF approach may outperform the CCMF approach in restoring sequences containing global interframe shifts, resulting in significant performance improvement over the single-frame approach. Therefore, in cases where the interframe displacement is a global shift whose amount is either readily available, or can be estimated accurately using a motion estimation technique such as the Fogel algorithm and a possible increase in computational requirements due to motion estimation can be afforded, the MCMF approach should be preferred over the CCMF approach.

Our results suggest that the best choice for a prototype sequence for power and cross power spectra estimation in the case of CCMF filtering is the one formed by single-frame restored versions of the degraded frames. In fact, the CCMF restoration outperforms the single-frame restoration only when this is the case. We have also observed from that the best method for power and cross power spectra estimation may be image dependent. The 3-D AR model-based power and cross power spectra estimation method, that proved to be satisfactory in CCMF restoration of Seq1, resulted in unsatisfactory restorations in the case of Seq2. The AR model was not adequate in accounting for the high spatial and temporal frequency content of Seq2. This was apparent from the high plant noise (or modeling error) variance associated with the model parameters. The Welch method, on the other hand, resulted in CCMF restorations that are superior to the single-frame restorations in the case of both Seq1 and Seq2. In general, therefore, using the single-frame restored sequence as a prototype and estimating the power and cross power spectra via the Welch method is a good strategy for implementing the CCMF filter.

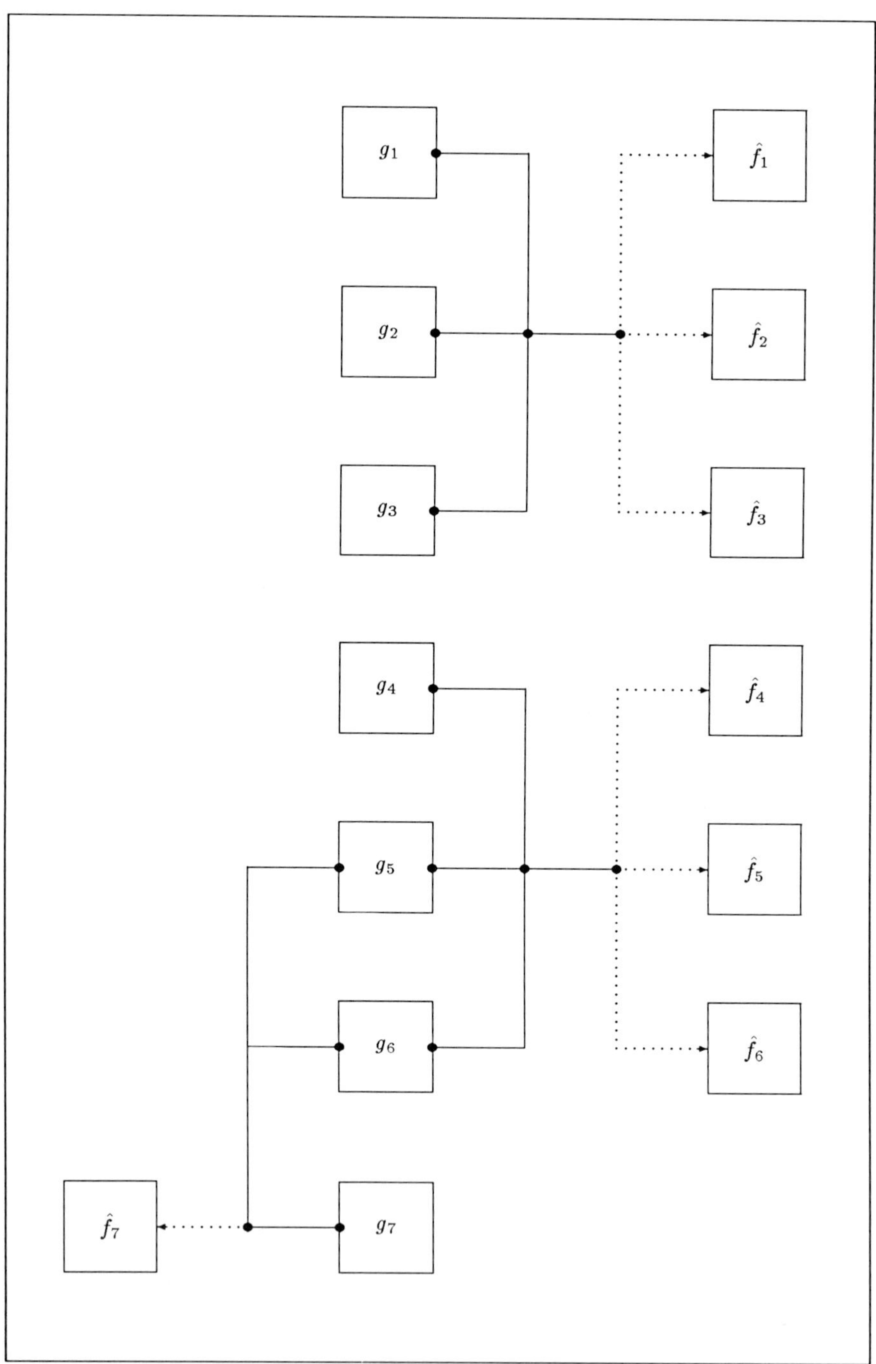

Figure 13.11: An implementation of CCMF for $L = 7$ and $N = 3$. Dotted lines point to the restored frames.

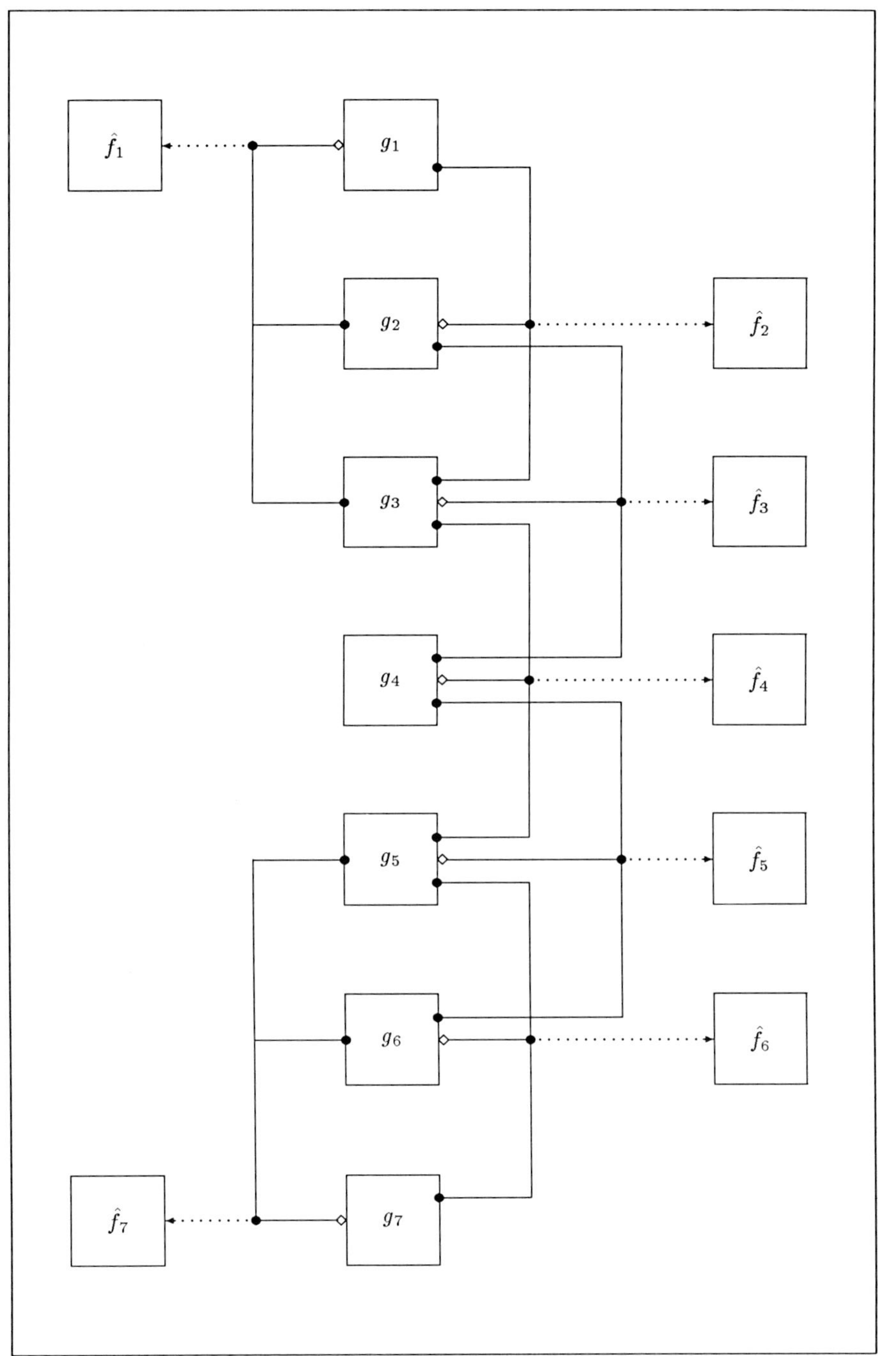

Figure 13.12: An implementation of MCMF for $L = 7$ and $N = 3$. Diamonds ($\diamond$) indicate the reference frames used for a group of three frames. Dotted lines point to the restored frames.

The results that we have obtained with the MCMF filter indicate that the best practical choice for the prototype used in estimating the power spectrum of the reference frame is an undegraded image whose statistical properties are reasonably close to that of the original. In our experiments, the difference between the restoration results obtained by using two different sharp prototypes has been insignificant. The AR-model based power spectrum estimation has provided the best results. The MCMF filter using the interframe shift vectors estimated by Fogel's motion estimation algorithm has outperformed the single-frame filter in all cases.

The performance of both multiframe methods is improved with an increase in the number of frames used. The amount of improvement, however, levels off after a certain number of frames is exceeded (on the order of 5 to 7 frames). In the case of CCMF, this is because the interframe correlations generally decrease with increasing number of frames due to the changes in the scene content over time. In the case of MCMF, however, with increasing number of frames the displacement estimation may become less accurate due to increased amount of occlusions taking place in time. In terms of computational requirements, the size of the matrix to be inverted in the case of CCMF increases with the number of frames. In the case of MCMF, on the other hand, the number of times the motion estimation algorithm is used (per restoration of a reference frame) increases with increasing number of frames, resulting in an increased number of computations. Therefore, a good strategy for multiframe restoration of a sequence of L frames is filtering N frames at a time, where $L > N$ and $N = 3$, or 5. A possible implementation of this strategy is illustrated in Figs. 13.11 and 13.12 for the CCMF and the MCMF filters, respectively, for $N = 3$ and $L = 7$. In Fig. 13.12, the MCMF filter is applied to N frames at a time, each time obtaining the restoration of the reference frame only. In other words, restoration of a certain frame is obtained only when it acts as a reference frame of a group. This is done in order to avoid the shifting of the restored frames as described in Section 13.4.2.

The improvements obtained in the restoration quality due to multiframe filtering are greater when the blur varies from frame-to-frame. This is because different blurs result in the loss of different information. For instance, the frequency components lost due to focus blurs at different amounts of defocus are different due to the different locations of the frequency-domain zero-crossings of the corresponding PSFs. Therefore, certain information that is missing in one frame may be contained in another frame when the frames are blurred differently, hence the advantage of multiframe Wiener filtering.

References

[1] N. P. Galatsanos and R. T. Chin, "Digital restoration of multichannel images," *IEEE Trans. Acoust., Speech, Signal Process.*, vol. 37, pp. 415–421, March 1989.

[2] C. Srinivas and M. D. Srinath, "A stochastic model-based approach for simultaneous restoration of multiple misregistered images," *SPIE*, vol. 1360, pp. 1416–1427, 1990.

[3] M. K. Ozkan, A. T. Erdem, M. I. Sezan, and A. M. Tekalp, "Efficient multiframe Wiener restoration of noisy and blurred image sequences ," *IEEE Trans. Image Process.*, October 1992.

[4] S. V. Fogel, "Estimation of velocity vector field from time-varying image sequences," *Comput. Vision Graphics Image Process.: Image Understanding*, vol. 53, pp. 253–287, May 1991.

[5] M. I. Sezan, M. K. Ozkan, and S. V. Fogel, "Temporally adaptive filtering of noisy image sequences using a robust motion estimation algorithm," in *IEEE Int. Conf. Acoust., Speech, Signal Process.*, (Toronto, Canada), pp. 2429–2432, May 14-17, 1991.

[6] B. R. Hunt, "The application of constrained least squares estimation to image restoration by digital computer," *IEEE Trans. Computers*, vol. 22, pp. 805–812, 1973.

[7] R. M. Gray, "On the asymptotic eigenvalue distribution of toeplitz matrices," *IEEE Trans. Inf. Theory*, vol. IT-18, pp. 725–730, November 1972.

[8] N. P. Galatsanos, A. K. Katsaggelos, R. T. Chin, and A. D. Hillery, "Least squares restoration of multichannel images," *IEEE Trans. Signal Process.*, vol. 39, pp. 2222–2236, October 1991.

[9] M. K. Ozkan, M. I. Sezan, A. T. Erdem, and A. M. Tekalp, "Image sequence restoration using multiframe Wiener filtering," in *Proc. Seventh Workshop on Multidimensional Signal Processing*, (Lake Placid, NY), 23–25 September 1991.

[10] G. H. Golub and C. F. V. Loan, *Matrix Computations*. Baltimore, MD: Johns-Hopkins, 1985.

[11] C. D. Kuglin and D. C. Hines, "The phase correlation allignment method," in *Proc. Int. Conf. on Cybernetics and Society*, (San Fransisco, CA), pp. 163–165, 1975.

[12] D. Shepard, "A two dimensional interpolation function for irregularly spaced data," in *Proc. 23rd Nat. Conf. ACM*, pp. 517–524, 1968.

[13] M. I. Sezan, G. Pavlovic, A. M. Tekalp, and A. T. Erdem, "On modeling the focus blur in image restoration," in *IEEE Int. Conf. Acoust., Speech, Signal Process.*, (Toronto, Canada), pp. 2485–2488, May 14-17, 1991.

[14] M. I. Sezan and A. M. Tekalp, "Survey of recent developments in digital image restoration," *Opt. Eng.*, vol. 29, pp. 393–404, May 1990.

[15] C. W. Therrien and H. T. El-Shaer, "Multichannel 2-D AR spectrum estimation," *IEEE Trans. Acoust., Speech, Signal Process.*, vol. 37, pp. 1798–1800, November 1989.

[16] C. W. Therrien and H. T. El-Shaer, "A direct algorithm for computing 2-D AR power spectrum estimates," *IEEE Trans. Acoust., Speech, Signal Process.*, vol. 37, pp. 1795–1798, November 1989.

[17] H. J. Trussell, M. I. Sezan, and D. Tran, "Sensitivity of color LMMSE restoration of images to the spectral estimate," *IEEE Trans. Acoust., Speech, Signal Process.*, vol. 39, pp. 248–252, 1991.

14
3-D Median Structures for Image Sequence Filtering and Coding

T. Viero and Y. Neuvo

Tampere University of Technology, Finland

14.1 Introduction

Motion detection or estimation and compensation algorithms are used in all areas of image sequence processing, e.g., in image sequence filtering, coding, interpolation, and analysis. In order to give good results, most image sequence processing algorithms require accurate motion information to be available. In this chapter, we take a different approach. We will develop algorithms that perform well under varying motion conditions. Especially, we will show how 3-D median structures can be designed to perform well in both stationary and moving parts of an image sequence without any motion detection or compensation.

One-dimensional temporal processing of image sequences gives optimal results in stationary (nonmoving) regions of image sequences. Temporal filters reduce noise without impairing the spatial resolution. In Differential Pulse Code Modulation (DPCM) systems, temporal predictors give low prediction errors in stationary areas. In moving regions, spatial algorithms give better results if motion information is not available.

Adaptive methods can be used to obtain good results both in stationary and moving regions. In predictive image sequence coding, adaptive intra-interframe predictors have been applied [47, 50, 51, 57]. In these predictors, either a previous frame or intraframe prediction is selected depending on surrounding signal changes. Adaptive algorithms can be used also in noise reduction [11, 15, 16, 48, 67]. Adaptive, temporal 1-D recursive filters are used in television receivers [48]. The filters attenuate noise in stationary regions. In order to prevent distortion of moving objects the filtering action is inhibited in moving regions.

Motion-compensated image sequence processing methods are, in general, more efficient than adaptive methods. In these systems, a motion estimation algorithm provides displacement (motion) vectors for each pixel or a block of pixels in a frame-to-frame basis. Actual processing is then performed utilizing the displacement vector information. In motion-compensated prediction, the current pixel is predicted from previously reconstructed pixels along the motion trajectory. In practice, the current frame is predicted from the previous frame utilizing displacement vectors [17]. In noise removal, 1-D or spatiotemporal filtering is performed along motion trajectories.

The performance of motion-compensated algorithms is highly dependent on the accuracy of motion estimation. Motion estimation methods give usually good results when motion in an image sequence is slow. Their performance deteriorates if motion is complicated or when fast moving objects exceed the effective region of motion estimation. Also noise decreases the performance of motion estimation algorithms. It should be noted that in the case of noninteger displacements of highly detailed moving objects motion estimation should be done at subpixel accuracy. Otherwise, motion-compensated operations may produce poor results. Motion-compensated filters and predictors have been proposed, e.g., in [11, 16, 17, 32, 37, 47, 50, 51, 67].

Filters and predictors designed for image sequence processing may combine motion-compensation and adaptation procedures. Adaptive algorithms are commonly used to change the operation of a filter or predictor when motion estimation produces inaccurate results.

Three-dimensional median structures represent a different approach to image sequence processing. They can be designed to give good results without any motion compensation or adaptation procedure. In other words, a 3-D median filter or predictor can be designed to effectively utilize the spatiotemporal correlation present in image sequences. The median operation[1] is insensitive to outliers and it retains sudden changes, like edges, in signals. These two properties are very desirable both for filters and predictors. Consider, e.g., a median filter in noise reduction. Moving objects, scene changes, and scene cuts create temporal edges in the 3-D signal space, in addition, spatial edges are common in all scenes. Since median filters retain edges they can be designed to give good performance in moving and stationary regions of image sequences. In constant signal plus noise situations noise attenuation capability of a median filter is at its best. Near rapid changes, like edges and ramps, noise attenuation is traded to a good response to the change. This inherent adaptation property is very desirable in image sequence processing applications.

For image sequence processing systems where motion information is available the median operation can be designed to utilize the information. If motion estimation fails, median-based algorithms can still produce acceptable or even good results. This is due to the robustness of the median operation.

In this chapter, we address two image sequence processing areas: noise reduction and DPCM coding. Two kinds of filters for noise reduction are presented: weighted median filters and idempotent weighted median filters. Weighted median filters are designed for effective noise removal while the idempotent filters are suitable for preprocessing. They can remove impulses from image sequences

[1] We use term "median operation" when we refer to both median filters and median predictors.

while having almost no effect in the noise free parts of the sequence. Median predictors are presented for DPCM coders. Median predictors combine linear subpredictors and median operations in order to obtain low prediction errors and robustness against transmission errors. All the 3-D median structures presented in this chapter are simple and easy to implement for real-time applications and simulation purposes.

This chapter is organized as follows. In the following two subsections, basic properties of the median operation and some important extensions of the median are presented. In Section 14.2, the structural design approach to 3-D median operations is discussed after which several 3-D median filtering and prediction algorithms are given. Performance of the 3-D median filters and predictors is discussed in Section 14.3. Finally, conclusions are given.

14.1.1 Properties of the median operation

Development in median and median-related filtering has been especially fast during the last decade and remarkable improvements have been made in the theoretical side as well as in the applications. Despite the advances in analysis and design methods of median and median-related filters, the design of these filters is still more difficult and complex than the design of linear filters. Overviews of median and median-related filters, their analysis methods, and properties can be found, e.g., in [13, 22, 60]. Here we give a short overview of the basic concepts and properties.

A median filter performs a nonlinear filtering operation, where a window of odd length (size) $N = 2M + 1$ is moved over an input signal one sample at a time. In image sequence filtering, the filter window is usually moved at each frame in a raster scan fashion from left to right at each line and from top to bottom for line advances. To account for filtering at the boundaries of frames, M samples are commonly appended to each edge sample having identical values to it. At each window position the samples inside the filter window are sorted by magnitude and the centermost value, the median, is the filter output. Let there be N samples $X_1, X_2, \ldots, X_N$ inside a window. The median filtering procedure is denoted by

$$Y = \mathrm{MED}[X_1, X_2, \ldots, X_N], \tag{14.1}$$

where Y is the resulting output from the median filter. A characteristic property of a median filter is its capability to remove noise effectively and preserve edges, lines, and other image details.

There exists an important analogy between the sample average and the median. The sample average is the maximum likelihood estimate for the mean of normally distributed noise while the median gives the maximum likelihood estimate in the case of Laplacian distributed noise. The median is also the least absolute deviation error estimate of the center of a distribution [36]. Let $\Phi(\beta)$ be defined by

$$\Phi(\beta) = \sum_{i=1}^{N} \mid X_i - \beta \mid^{\gamma} . \tag{14.2}$$

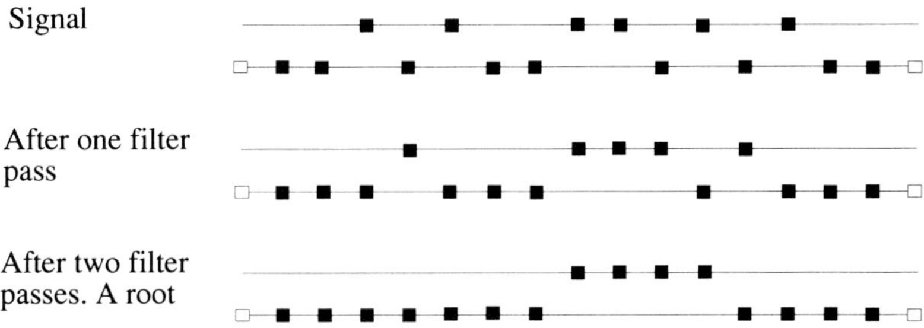

Figure 14.1: Result of repeated median filtering. Length of the filter window is 3.

The value β minimizing $\Phi(\beta)$ for $\gamma = 1$ is the median of $X_1, X_2, \ldots, X_N$. The sample average minimizes $\Phi(\beta)$ when γ is equal to 2. The above results suggest that the performance of a median filter is particularly good for impulse noise and noise with heavy-tailed distributions [36].

Median filters are nonlinear and this fact complicates their mathematical analysis. The superposition property, which makes a very detailed analysis of linear filters possible, does not hold for median filters. Thus, the effects of the filter to the noise and signal cannot be separated like with linear filters. Statistical properties of median (rank-order) filters have been studied, e.g., in [9, 14, 33, 36, 41, 43, 55].

In median filtering, the output is always one of the input samples. This property is the reason for the presence of root signals, signals which are invariant of filtering. Many authors have studied the properties and characteristics of root signals of one-dimensional median filters [6, 19, 24, 56, 71, 74]. It has been shown that a finite length 1-D signal has to be locally monotonic in order to pass through a median filter unchanged. When successively filtering a 1-D signal (i.e., the filtered output is itself again filtered), the original signal will eventually reduce to a root signal [24]. In Figure 14.1 an example of repeated median filtering of a 1-D signal is given. A median filter of length three is used in this example. In the figure, the white squares correspond to the appended edge samples. As can be seen from Figure 14.1, the signal reduces to a root signal after two filtering passes and the root signal contains only constant neighborhoods and edges. For a 1-D median filter of length $N = 2M + 1$ a constant neighborhood is at least $M + 1$ consecutive identically valued samples. An edge is an increasing or decreasing sequence of samples which is immediately preceded and followed by constant neighborhoods. An edge cannot contain any constant neighborhoods. In [74] an upper bound for the convergence of any 1-D signal of length L to a root signal is derived. A signal will converge to a root at most in

$$3 \left\lceil \frac{(L-2)}{2(M+2)} \right\rceil \tag{14.3}$$

passes of a filter of window width $2M + 1$. Root signal analysis of two and three-dimensional signals is much more difficult than the 1-D case. Some results have, however, been reported [44, 54, 71].

Recently, a structural approach to the design and study of median and median-related filters has been introduced [13, 20]. The goal is to determine the type and number of root signals, whether every signal is filtered to a root signal, and which structures are preserved, created, modified, or deleted by median-related filters [22]. A structure is any local or global variation of interest in the magnitude of a signal or image. Theoretical results obtained from the root signal studies are utilized in the structural design approach.

Threshold decomposition is a weak superposition property which is an efficient tool in the theoretical analysis of median and other rank-order based filters [18]. Using threshold decomposition, the properties of median filters can be analyzed in binary domain. In general, the analysis of a median filter is easier in binary domain as there are only two possible signal values.

Threshold decomposition of a P-valued signal $\{X(.)\}$, where all the samples are integer valued and $0 \leq X(.) \leq P-1$, means decomposing it into $P-1$ binary signals $\{x^1(.)\}, \{x^2(.)\}, \ldots, \{x^{P-1}(.)\}$ according to the following rule:

$$x^p(.) = \begin{cases} 1, & \text{if } X(.) \geq p; \\ 0 & \text{otherwise.} \end{cases} \tag{14.4}$$

It is easy to see that the original signal $\{X(.)\}$ is a sum of its thresholded binary signals

$$\{X(.)\} = \sum_{p=1}^{P-1} \{x^p(.)\}. \tag{14.5}$$

Threshold decomposition states that median filtering a P-valued graylevel signal is the same as first decomposing it into a set of binary signals using the algorithm in Equation 14.4 then filtering each binary signal with a median filter and finally, adding up the results of these operations. This procedure is illustrated in Figure 14.2.

In binary domain, median and other rank-order operations can be considered as logical operations. In fact, in binary domain, any median, rank-order, or stack filter [75, 79] is described by a positive Boolean function. Positive Boolean functions contain only AND and OR operations.

14.1.2 Extensions

In this section, three extensions of the median operation are presented. These extensions are the weighted median (WM), the FIR-median hybrid (FMH), and the weighted vector median (WVM) operations.

Weighted median operation

In a WM operation, a nonnegative integer weight $W_i, i = 1, 2, \ldots, N$, is assigned to each input sample and the sum of all weights $S_w = \sum_{i=1}^{N} W_i$ is commonly assumed to be odd. The output of a WM is not uniquely defined if the sum of

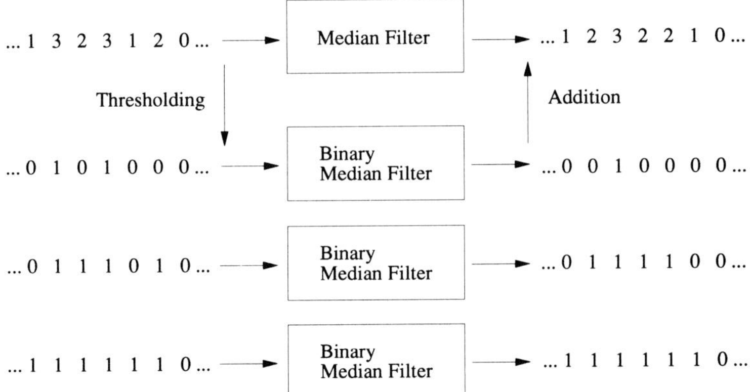

Figure 14.2: Median filtering by threshold decomposition. The signal is a four-valued signal and the filter length $N = 3$.

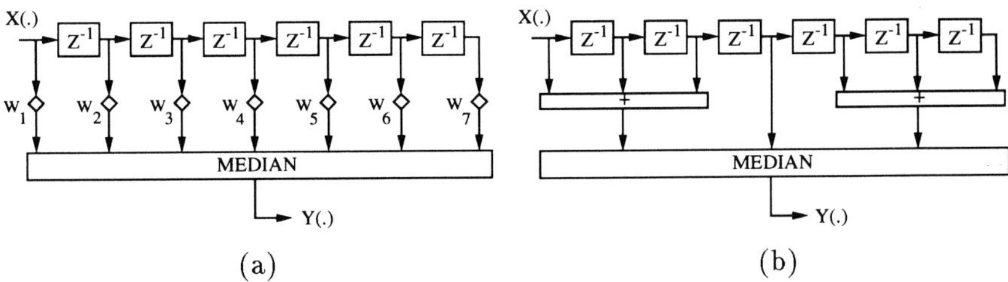

Figure 14.3: One-dimensional median structures. (a) A weighted median filter, (b) a FIR-median hybrid filter.

all weights is even. The output is obtained, e.g., by first duplicating each input sample X_i to the number of the corresponding weight W_i. Then the resulting set of samples is sorted and the median value is selected from this set [36]. Here the duplication operation is expressed as

$$W \Diamond X = \underbrace{X, \ldots, X}_{\text{W times}}. \tag{14.6}$$

The weighted median operation can now be represented as

$$Y = \text{MED}[W_1 \Diamond X_1, W_2 \Diamond X_2, \ldots, W_N \Diamond X_N]. \tag{14.7}$$

An example of a 1-D WM filter of length seven is shown in Figure 14.3(a). The filter structure is similar to the structure of a finite impulse response (FIR) filter. Multiplications have been replaced by duplication operations and averaging by the median operation.

As an example, consider the following WM filter of length five:

$$Y(n) = \text{MED}[1 \Diamond X(n-2), 2 \Diamond X(n-1), 3 \Diamond X(n), 2 \Diamond X(n+1), 1 \Diamond X(n+2)]. \tag{14.8}$$

Now apply the filter of Equation 14.8 to the following sequence

$$X(n): \quad \ldots \quad 1 \quad 5 \quad 8 \quad 11 \quad 2 \quad \ldots \tag{14.9}$$

so that the window is centered on the sample whose value is 8. After duplication and sorting, the samples inside the window are

$$1 \quad 2 \quad 5 \quad 5 \quad 8 \quad 8 \quad 8 \quad 11 \quad 11 \tag{14.10}$$

and the output $Y = 8$, whereas a five-point median filter would have produced the result $Y = 5$.

The weighted median operation can also be defined for positive real-valued weights [63]. In this case, the output is calculated as follows. Starting from either end of the sorted set of samples add up the corresponding weights until the sum $S_w \geq \sum W_i/2$. The weighted median is the sample corresponding to the last weight. If $S_w = \sum W_i/2$ then the WM is not unique and the average of the two samples for which $S_w = \sum W_i/2$ can be defined to be the output value.

The weighted median operation offers more design freedom than the standard median. When processing a multidimensional signal, not only window size and shape can be changed but also the importance of input samples inside the window. Properties of WM operations have been analyzed in [79].

FIR-median hybrid operation

A FIR-median hybrid operation combines the desirable properties of linear operations and the median operation [29, 30, 31]. The window is divided into an odd number of subwindows and a FIR filter is used in each of these subwindows. The output of the FMH operation is the median of the subfilter outputs. Let $H_1(z), H_2(z), \ldots, H_N(z)$ be the transfer functions of the FIR filters. The output of the FMH operation is given by

$$Y = \mathrm{MED}[Y_1, Y_2, \ldots, Y_N], \tag{14.11}$$

where Y_i is the output of the FIR filter defined by $H_i(z)$. A simple FMH structure of length seven is shown in Figure 14.3(b). FMH operations can be designed to give good results in a large variety of applications by changing the FIR subfilters and the overall FMH structure [30, 31, 53].

It has been shown that FMH filters have similar statistical properties and same type of root signals than the standard median filter [30]. There are, however, also differences between these operations. Usually, the number of subwindows in a FMH operation is quite small compared to the number of samples in the subwindows. The number of operations required to find the median is, therefore, significantly reduced. Unlike the standard median operation, the FMH operation generates also new signal values. A FMH operation cannot, in general, reject impulses as efficiently as the median, which operates in the same window.

In an important class of FMH filters averaging filters are used as subfilters [30, 53]. These FMH filters operate differently than median filters near edges.

In the 1-D case, an impulse near an edge may cause edge jitter (a shift of the edge location) in median filter output. There is no edge jitter in the FMH filter output, but the signal level before the edge may change slightly. Root signals of some 2-D FMH filters are analyzed in [53]. By properly designing the FIR subfilters, FMH filters can be tailored to possess a variety of root signals like triangular and sawtooth waveforms [31].

Weighted vector median operation

In many applications, multispectral (multivariate) signals are processed. Examples of these signals are color image sequences, which consist of typically three color components. Median and rank-order operations have been found to be suitable for image processing, so the extension of these operations to multispectral images is of great interest. Different ordering schemes of multispectral data are discussed in [10].

In the restoration and enhancement of multispectral images, utilization of the correlation between different spectral components has been found to give good results [7, 8, 23, 80]. Vector median (VM) operation [7, 8, 25] processes multispectral signals as vector-valued data and, thus uses the correlation between different spectral components.

Definition 14.1 *Let* $\underline{X}_1, \underline{X}_2, \ldots, \underline{X}_N$ *be vectors inside a window and let* $W_1, W_2, \ldots, W_N$ *be their corresponding weights. The weights are restricted to be nonnegative real numbers. The weighted vector median* WVM_{L_1} *is the vector* $\underline{X}_{wvm}$ *such that*

$$\underline{X}_{wvm} \in \{\underline{X}_i; i = 1, 2, \ldots, N\} \tag{14.12}$$

and for all $j = 1, \ldots, N$

$$\sum_{i=1}^{N} W_i \parallel \underline{X}_{wvm} - \underline{X}_i \parallel \leq \sum_{i=1}^{N} W_i \parallel \underline{X}_j - \underline{X}_i \parallel, \tag{14.13}$$

where $\parallel \cdot \parallel$ *stands for* L_1*-norm.*

A more general case of WVM operations is obtained by using the L_p-norm instead of L_1-norm. The selected norm affects the noise reduction and detail preservation capabilities of the WVM operation. WVM operations are simple generalizations of VM operations where $W_i = 1, i = 1, \ldots, N$.

Multispectral signals can be processed componentwise by applying nominally identical filters to the different components separately. Componentwise processing does not utilize the correlation between different spectral components and may result in problems, especially around edges [8, 80]. False colors caused by edge shift are examples of these problems [80]. In Figure 14.4(a), there is a noise corrupted input signal consisting of two color components: red and green. The noise signal was filtered with a componentwise median filter of length three and the result is shown in Figure 14.4(b). As can be seen from this figure, the edge in the red component has moved forward causing the output to be yellow before

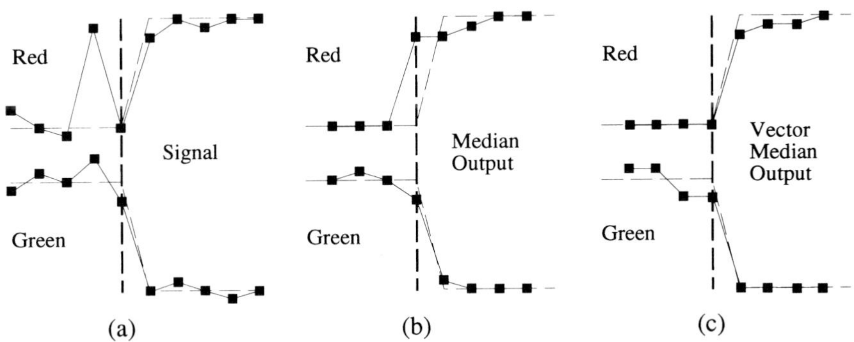

Figure 14.4: Performance of componentwise median and vector-median filters of length 3 near a noisy edge. (a) Noise signal, (b) result of componentwise median filtering, (c) result of vector-median filtering.

changing to red. In Figure 14.4(c), there is the output of a vector median filter of length 3 for the noise signal. As can be seen from this figure, the impulse near the edge is filtered out without causing edge shift. Thus, no false colors appear.

A drawback of VM filtering is the reduced noise attenuation if compared to componentwise filtering. This can be partly overcome by using the extended vector median, which selects the componentwise sample average as the output on smooth parts of the signal [7, 8, 25].

Processing of color images and image sequences is an active area of research. Other "rank-order based" operations for multispectral signal processing have been proposed in [28, 38, 59, 61]. In this chapter, we consider only scalar (B&W) image sequence processing. Color image sequence filtering is studied in [73].

14.2 3-D Median Structures

Three-dimensional median-related operations have been successfully applied to many image sequence processing tasks. In majority of applications, 3-D median operations have been used without any motion compensation or detection. These application areas include, e.g., image sequence enhancement [2, 3, 4, 5, 39, 72, 73], predictive image sequence coding [34, 45, 46, 68, 69], and reconstruction of quincunx coded image sequences [1, 35, 81]. In some applications, median operations have been assisted by motion information. These applications include, e.g., production of high quality still images from high definition television (HDMAC) transmissions [27].

In this section, we present 3-D median structures for image sequence processing, which do not require any motion information. First, guidelines for the determination of good 3-D median operations for image sequence filtering and DPCM coding are given. Then, actual 3-D median structures are presented.

14.2.1 Design of basic 3-D median structures

Let $\{X(.)\}$ be a discrete spatiotemporal graylevel image sequence, such that $\{X(n_1, n_2, n_3); n_1, n_2, n_3 \in \mathbf{Z}\}$, where $\mathbf{Z} = \{..., -1, 0, 1, ...\}$. The horizontal indexing is given by n_1, while n_2 and n_3 refer to vertical and time indexing, respectively. For notational simplicity, when convenient, vector notation $[n_1, n_2, n_3] = \mathbf{n}$ will be used.

Consider the set of samples inside a $(2N_1 + 1) \times (2N_2 + 1) \times (2N_3 + 1)$ window centered at $(\mathbf{n})$. The constants N_1, N_2, N_3 are nonnegative integers and correspond to n_1, n_2, n_3 directions, respectively.

Definition 14.2 *A 3-D WM operation is defined by*

$$Y(n_1, n_2, n_3) = Y(\mathbf{n}) = MED[\mathbf{R}], \tag{14.14}$$

where

$$\begin{aligned} \mathbf{R} = \{ & W_{i_1,i_2,i_3} \Diamond X(n_1 + i_1, n_2 + i_2, n_3 + i_3); \\ & - N_1 \leq i_1 \leq N_1, -N_2 \leq i_2 \leq N_2, -N_3 \leq i_3 \leq N_3\}. \end{aligned} \tag{14.15}$$

An input sample corresponding to a weight having value equal to zero is not taken into account in the calculation of the output. Therefore, window locations inside the $(2N_1+1)\times(2N_2+1)\times(2N_3+1)$ window corresponding to zero weights are not considered to be part of the window. Thus, the actual window can have an arbitrary shape. Properties of a WM operation are controlled by the selection of the window and weights. Large weights are assigned to those samples, which are considered to be important.

Regions in image sequences can be divided into two groups: stationary and moving (changing). The 3-D median operations are designed to give excellent results in stationary regions, which are well perceived by the human visual system. In noise free stationary regions, the filters retain all spatial details independent of their orientation, while the predictors give low (zero) prediction errors. In moving regions, the performance of 3-D median operations is a compromise of the desired overall performance. 3-D median filters may slightly distort moving objects if good noise reduction capability is desired, while median predictors may produce a larger prediction error for the sake of low complexity and fast computation.

In noise free stationary regions, pixel values in consecutive frames are the same, i.e., pixel values along the time axis are equal. A single pixel $X(\mathbf{n})$, which does not change its value in time, is preserved by a 3-D WM operation if the sum of weights inside the window along the time axis (n_3) at the spatial location (n_1, n_2) is equal to or greater than $(\sum W + 1)/2$. More formally:

Property 14.1 *3-D WM operations, which preserve stationary regions, satisfy*

$$\sum_{-N_3 \leq i_3 \leq N_3} W_{0,0,i_3} \geq \Big(\sum_{\substack{-N_1 \leq i_1 \leq N_1 \\ -N_2 \leq i_2 \leq N_2 \\ -N_3 \leq i_3 \leq N_3}} W_{i_1,i_2,i_3} + 1)/2. \tag{14.16}$$

In practice, image sequences contain noise and even in stationary regions pixel values along the time axis are not equal. However, a 3-D median operation, which satisfies Property 14.1, gives good results also in noisy stationary regions.

Consider a 3-D median filter, which satisfies Property 14.1, and the smallest possible stationary region, a temporal one pixel wide line. When noise is "added" to this sequence, pixel values along the line are not equal. We assume that inside the filter window pixel values (samples) lying on the line are adjacent in the ordered set of samples. This is the case when contrast in the image sequence is good and noise variance is sufficiently low. Let $X_1, X_2, \ldots, X_N$ be samples inside the 3-D filter window. When these samples are arranged in ascending order of magnitude, they can be written as

$$X_{(1)}, X_{(2)}, \ldots, X_{(N)}, \tag{14.17}$$

where $X_{(r)}$ is the rth order statistic. Furthermore, let S denote the set of samples inside the filter window that are part of the temporal line. After duplicating and sorting the input samples there are Λ ordered samples $X_{(1)}, \ldots, X_{(\Lambda)}$ of which $L, L \geq (\Lambda+1)/2$, belong to the line (see Property 14.1). Since the samples lying on the line are adjacent in the ordered set of samples, there exist $i, 1 \leq i \leq \Lambda - L + 1$ such that $X_{(i+j)} \in S, j = 0, \ldots, L-1$. For any value of i, one of the samples $X_{(i+j)} \in S, j = 0, \ldots, L-1$, is the centermost sample $X_{((\Lambda+1)/2)}$, i.e., the median. Thus, the line structure is retained, but some filtering is performed to the line.

In moving regions, image information in consecutive frames is different meaning that a spatial rather than temporal operation should be performed in these regions. The selection of spatial samples differs in median filters and predictors. Structural design principles can be applied to the design of median predictors and filters.

Recently, Alp and Neuvo showed [3] that their 3-D median filters reduce to spatial filters in certain moving regions. Ko and Lee showed statistically that spatiotemporal center weighted median filters preserve image structures under motion at the expense of noise suppression [39]. Arce has used test sequences to analyze motion and structure preserving characteristics of some 3-D median filters [5]. All these analysis indicate that properly designed spatiotemporal median filters perform well also in moving regions without motion compensation or detection.

In interlaced scanning, all odd lines of a frame (odd field) are recorded first after which all even lines (even field) are recorded. This procedure introduces time delay between the fields. When scenes containing moving objects are recorded these objects are in different locations in the odd and even fields. When displaying the even and odd fields as a single progressive frame (lines are recorded consecutively from top to bottom) interlacing can be seen as distortion (see Figure 14.5(a)). Filters and predictors should preserve the field structure of interlaced image sequences. Three-dimensional median-based algorithms can also be used in interlaced to progressive conversion [48, 52] where one of the aims is to eliminate the ruggedness of moving objects.

14.2.2 3-D median filters for image sequence filtering

Here we present several 3-D median filters designed to preserve fine temporal and spatial details. From the implementational point of view the filter masks should be as small as possible. Properties of the human visual system can be used in determining the spatial shape of the 3-D median filter window. The human visual system is more sensitive to horizontal and vertical than to diagonal details [65]. Horizontal and vertical details should thus be considered to be more important than diagonal details when designing 3-D median filters for applications like video signal processing and high definition television.

A very simple spatiotemporal filter, which preserves stationary regions, is the L-shaped filter defined by

$$Y_L(\mathbf{n}) = \text{MED}[X(n_1, n_2, n_3), X(n_1 - 1, n_2, n_3), X(n_1, n_2, n_3 - 1)]. \quad (14.18)$$

It is easy to see that this filter preserves also one pixel wide horizontal lines and two pixel wide vertical lines. The window of this filter is not symmetrical. This may cause problems in some applications, since unsymmetrical filters tend to shift edges and other image structures. In Figure 14.5(a), a part of frame 7 of Costgirls image sequence is shown. The Costgirls sequence was filtered with the L-shaped filter and a difference between the original sequence and the filtered sequence was calculated. The same part as in Figure 14.5(a) from this difference sequence after thresholding is shown in Figure 14.5(b). As can be seen from this figure, the L-shaped filter shifts certain spatiotemporal edges. Edge detectors can be constructed by calculating differences between unsymmetrical and symmetrical median filters.

Concepts of shape and weight symmetry have been introduced in the context of median-based idempotent filters [21, 26].

Definition 14.3 *[21, 26] Shape symmetry. A filter window is said to be symmetric w.r.t. the center point of the filter window ($W_{0,0,0}$) if*

$$W_{-i_1,-i_2,-i_3} \neq 0 \Rightarrow W_{i_1,i_2,i_3} \neq 0, \quad (14.19)$$

where $-N_j \leq i_j \leq N_j, j = 1, 2, 3$.

Definition 14.4 *[21, 26] Weight symmetry. A filter window is said to be symmetrically weighted if*

$$W_{-i_1,-i_2,-i_3} = W_{i_1,i_2,i_3}, \quad (14.20)$$

where $-N_j \leq i_j \leq N_j, j = 1, 2, 3$.

As can be seen from these definitions, symmetrically weighted windows are always also symmetrically shaped.

Shape symmetrical, especially weight symmetrical filters preserve edge locations much better than unsymmetrical filters. All the filters presented in the

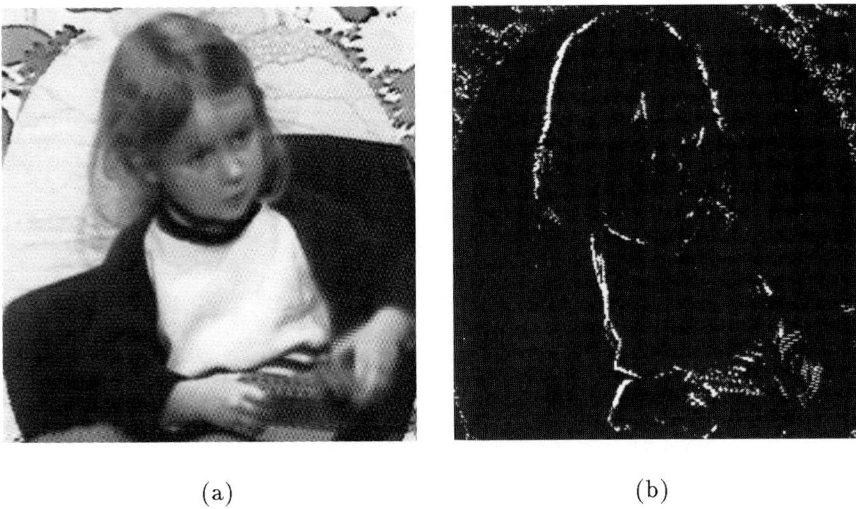

(a) (b)

Figure 14.5: Unsymmetrical median filters shift edges. (a) Part of the Costgirls sequence, (b) corresponding part from the thresholded difference sequence. Difference was calculated between the original sequence and sequence filtered with the L-shaped filter.

remaining part of this subsection are weight symmetrical.[2] The filter structures operate on three image planes. Thus, two frame memories are required to implement these filters.

A five point median filter operating in temporal and horizontal directions is

$$\begin{aligned} Y_{MED5}(\mathbf{n}) = \mathrm{MED}[&X(n_1,n_2,n_3-1), X(n_1-1,n_2,n_3), X(n_1,n_2,n_3), \\ &X(n_1+1,n_2,n_3), X(n_1,n_2,n_3+1)]. \end{aligned} \tag{14.21}$$

In stationary regions, samples from the temporal direction guarantee the preservation of fine spatial details. In moving regions, horizontal fine lines are preserved.

A filter that preserves temporal, horizontal, and vertical fine details is

$$\begin{aligned} Y_{WM7}(\mathbf{n}) = \mathrm{MED}[&X(n_1,n_2,n_3-1), X(n_1,n_2-1,n_3), X(n_1-1,n_2,n_3), \\ &3\diamond X(n_1,n_2,n_3), X(n_1+1,n_2,n_3), X(n_1,n_2+1,n_3), \\ &X(n_1,n_2,n_3+1)]. \end{aligned} \tag{14.22}$$

In Section 14.1.2, we introduced FMH filters. FMH filter extensions of the MED5 and the WM7 filters are given here. Averaging filters are used as subfilters only in the current frame. Thus, also the FMH filters operate on three image planes.

[2] More strict definition of symmetry is used in [40] in the context of center weighted median filters. Our filters are symmetrical also according to this definition.

A FMH filter having similar detail preservation properties as the MED5 filter is

$$\begin{aligned} Y_{FMH5}(\mathbf{n}) = \mathrm{MED}[&X(n_1, n_2, n_3 - 1), ave_1(\mathbf{n}), X(n_1, n_2, n_3), \\ &ave_2(\mathbf{n}), X(n_1, n_2, n_3 + 1)], \end{aligned} \tag{14.23}$$

where linear subfilters ave_1 and ave_2 are

$$\begin{aligned} ave_1(\mathbf{n}) &= (X(n_1 - 1, n_2, n_3) + X(n_1 - 2, n_2, n_3))/2, \\ ave_2(\mathbf{n}) &= (X(n_1 + 1, n_2, n_3) + X(n_1 + 2, n_2, n_3))/2. \end{aligned} \tag{14.24}$$

A FMH extension of the WM7 filter is

$$\begin{aligned} Y_{FMH7}(\mathbf{n}) = \mathrm{MED}[&X(n_1, n_2, n_3 - 1), ave_3(\mathbf{n}), ave_1(\mathbf{n}), 3\diamondsuit X(\mathbf{n}), \\ &ave_2(\mathbf{n}), ave_4(\mathbf{n}), X(n_1, n_2, n_3 + 1)], \end{aligned} \tag{14.25}$$

where ave_1 and ave_2 are defined in Equation 14.24 and ave_3 and ave_4 are given by

$$\begin{aligned} ave_3(\mathbf{n}) &= (X(n_1, n_2 - 1, n_3) + X(n_1, n_2 - 2, n_3))/2, \\ ave_4(\mathbf{n}) &= (X(n_1, n_2 + 1, n_3) + X(n_1, n_2 + 2, n_3))/2. \end{aligned} \tag{14.26}$$

Basic median filter structures can be combined to multistage structures [53]. The Tridirectional Multistage Median (TMM) filter uses the WM7 filter as the first subfilter in the multistage structure. The second subfilter is the WM7d filter defined by

$$\begin{aligned} Y_{WM7d}(\mathbf{n}) = \mathrm{MED}[&X(n_1, n_2, n_3 - 1), X(n_1 - 1, n_2 - 1, n_3), \\ &X(n_1 - 1, n_2 + 1, n_3), 3\diamondsuit X(n_1, n_2, n_3), X(n_1 + 1, n_2 - 1, n_3), \\ &X(n_1 + 1, n_2 + 1, n_3), X(n_1, n_2, n_3 + 1)]. \end{aligned} \tag{14.27}$$

This filter preserves stationary and diagonal details. These two filters guarantee a good performance of the TMM filter in stationary regions. The third subfilter can now be selected to give good performance in moving regions. In order to obtain a good noise attenuation capability, a square median filter given by

$$Y_{MED9}(\mathbf{n}) = \mathrm{MED}[\{X(n_1 + i_1, n_2 + i_2, n_3); -1 \le i_1 \le 1, -1 \le i_2 \le 1\}] \tag{14.28}$$

is used. The outputs of these first-stage subfilters are the inputs of the filter on the second stage. The output of the TMM filter is now

$$Y_{TMM}(\mathbf{n}) = \mathrm{MED}[Y_{WM7}(\mathbf{n}), Y_{WM7d}(\mathbf{n}), Y_{MED9}(\mathbf{n})]. \tag{14.29}$$

This filter is an extension of the multistage median concept presented in [53].

A WM filter, which has approximately the same temporal and spatial detail preservation properties than the TMM filter, can be derived. The window and the weights of the WM11 filter are show in Figure 14.6.

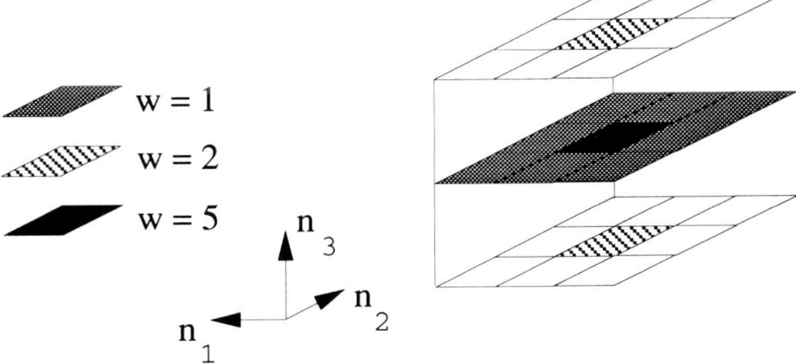

Figure 14.6: Filter window and weights of the WM11 filter.

14.2.3 Idempotent weighted median filters for image sequence filtering

Idempotent filters produce root signals in a single pass, i.e., the filter output is invariant to further filtering with the same filter. Idempotence is a very important concept in morphological filtering [66]. All morphological filters are idempotent. Until recently, the only known median-based idempotent filters have been the 1-D recursive median filters.[3] [56]. An example of a 1-D recursive median filter is

$$Y(n) = \text{MED}[Y(n-2), Y(n-1), X(n), X(n+1), X(n+2)], \qquad (14.30)$$

where $X(n)$ and $Y(n)$ are the values of the input and output, respectively, at position (n). Recursive median and morphological filters could be called "strong" idempotent filters, because they excessively smooth the input signal. Recently, idempotent weighted median filters have been presented [21, 26]. These filters are able to remove impulse noise from signals and yet have almost no effect on the noncorrupted parts of the signal. These filters can be called "soft" idempotent filters. These filters are suitable for image sequence preprocessing tasks. They can be used to remove impulses before other image sequence processing operations, like linear interpolators, whose performance is seriously degraded by impulsive noise.

Idempotent weighted median filters are weighted median filters, where additional constraints have been posed to the filter window shape and to the weights [21, 26]. We will shortly introduce these constraints after which we present some idempotent weighted median filters for image sequence filtering.

At each filter window location, there are $\sum(W_{i_1,i_2,i_3}/W_{i_1,i_2,i_3})$ input samples inside a filter window. The set of indices of these samples is here denoted by

[3] It is easy to verify that two and three-dimensional recursive median filters do not generally reduce a signal to a root in a single pass.

$\Omega_{\mathbf{n}}$. For example, consider the WM7 filter. The index set $\Omega_{\mathbf{n}}$ of this filter is

$$\begin{aligned}\Omega_{\mathbf{n}} = \{&(n_1, n_2, n_3 - 1), (n_1, n_2 - 1, n_3), (n_1 - 1, n_2, n_3), (n_1, n_2, n_3), \\ &(n_1 + 1, n_2, n_3), (n_1, n_2 + 1, n_3), (n_1, n_2, n_3 + 1)\}.\end{aligned} \tag{14.31}$$

Two subclasses of weighted median filters can be defined based on the shape and weights of the filter window.

Definition 14.5 *[21, 26] Class 1. Any weighted median filter whose window is symmetric (Definition 14.3) and the sum of weights* $\sum W_{i_1,i_2,i_3} = 2W_{0,0,0} + 1$, *where* $W_{0,0,0} \geq 1$, *belongs to Class 1.*

Definition 14.6 *[21, 26] Class 2. Any weighted median filter whose window is symmetric,* $\sum W_{i_1,i_2,i_3} = 2W_{0,0,0}+1$, *where* $W_{0,0,0} \geq 1$, *and where for each input sample* $X(\mathbf{n})$ *there exist two input points* $X(\mathbf{o})$ *and* $X(\mathbf{p})$ *such that*

$$\mathbf{o} \in \Omega_{\mathbf{n}}, \mathbf{p} \in \Omega_{\mathbf{n}}, \mathbf{o} \in \Omega_{\mathbf{p}}, \tag{14.32}$$

where $\mathbf{o} \neq \mathbf{p}, \mathbf{o} \neq \mathbf{n}, \mathbf{p} \neq \mathbf{n}$, *belongs to Class 2.*

It is easy to see, that Class 2 filters are also Class 1 filters. It has been proven [21, 26], that any Class 1 recursive and any Class 2 nonrecursive or recursive filter is idempotent. Next we give some examples of temporal, spatial, and spatiotemporal idempotent filters. All these filters are Class 2 nonrecursive filters.

A temporal idempotent weighted median filter is

$$\begin{aligned}Y_{temp_I}(\mathbf{n}) = \mathrm{MED}[&X(n_1, n_2, n_3 - 2), X(n_1, n_2, n_3 - 1), 3\Diamond X(\mathbf{n}), \\ &X(n_1, n_2, n_3 + 1), X(n_1, n_2, n_3 + 2)].\end{aligned} \tag{14.33}$$

A spatial idempotent filter is given by

$$\begin{aligned}Y_{spat_I}(\mathbf{n}) = \mathrm{MED}[&X(n_1 - 1, n_2 - 1, n_3), X(n_1, n_2 - 1, n_3), X(n_1 + 1, n_2 - 1, n_3), \\ &X(n_1 - 1, n_2, n_3), 7\Diamond X(\mathbf{n}), X(n_1 + 1, n_2, n_3), X(n_1 - 1, n_2 + 1, n_3), \\ &X(n_1, n_2 + 1, n_3), X(n_1 + 1, n_2 + 1, n_3)].\end{aligned} \tag{14.34}$$

A spatiotemporal idempotent filter defined in a $3 \times 3 \times 3$ cube would be a logical extension of the temporal and the spatial idempotent filters. Here we, however, define a spatiotemporal idempotent filter, which has a smaller window than a cube. An idempotent filter that preserves stationary regions is defined by

$$\begin{aligned}Y_{spat-temp_I}(\mathbf{n}) = \mathrm{MED}[&X(n_1, n_2, n_3 - 1), X(n_1 - 1, n_2 - 1, n_3), \\ &X(n_1, n_2 - 1, n_3), X(n_1 + 1, n_2 - 1, n_3), X(n_1 - 1, n_2, n_3), \\ &9\Diamond X(\mathbf{n}), X(n_1 + 1, n_2, n_3), X(n_1 - 1, n_2 + 1, n_3), \\ &X(n_1, n_2 + 1, n_3), X(n_1 + 1, n_2 + 1, n_3), X(n_1, n_2, n_3 + 1)].\end{aligned} \tag{14.35}$$

Some of the weighted median filters presented in this chapter can be implemented efficiently by using a property of center weighted median (CWM) filters

[40]. CWM filters are weighted median filters, where all weights inside the filter window have a value equal to one except the center weight, weight of the current input sample. In general, the window of a CWM filter is a square or a cube depending on the signal dimension. Thus, these filters rely on the proper selection of the center weight.

Property 14.2 *[40] The output $Y_{(2M+1,2L+1)}(\mathbf{n})$ of a CWM filter with window size $2M+1$ and center weight $2L+1$ is obtained by*

$$Y_{(2M+1,2L+1)}(\mathbf{n}) = MED[X_{(M+1-L)}, X(\mathbf{n}), X_{(M+1+L)}], \qquad (14.36)$$

where $X_{(r)}$ is the rth smallest one among the $2M+1$ samples within the window.

When the center weight is very large, like in the case of idempotent weighted median filters, utilization of the above property reduces computation time considerably.

14.2.4 3-D median predictors for image sequence coding

In this subsection, several median predictors for DPCM coders are presented. All these spatiotemporal median predictors are designed to perform well without motion compensation or detection. The same design principles used in 3-D median filtering can be applied to predictor design. Three predictors are introduced: a 3-D median (3DM) predictor [34], a 3-D multistage median (3DMM) predictor, and an adaptive weighted median predictor [68]. All these predictors combine linear substructures and median operations, so they can be considered to be FMH predictors. Before introducing the predictors, basic principles of DPCM coders are presented. More detailed studies can be found in [47, 50, 51].

Differential pulse code modulation is a predictive coding scheme which has been studied and applied extensively in various communication systems. The basic structure of a DPCM coder is shown in Figure 14.7. In a causal DPCM coder, prediction $\hat{X}(\mathbf{n})$ of the present sample $X(\mathbf{n})$ is made based upon the previously transmitted and decoded information. The difference between the predicted and the actual value of the sample is quantized, coded, and transmitted as shown in Figure 14.7. After decoding the transmitted code words, the receiver reconstructs the sample by adding the predicted value $\hat{X}$ to the quantized prediction error e'. In order to have the same prediction value at both the transmitter and the receiver, also at the transmitter side the prediction is based on the reconstructed samples. In the DPCM scheme compression is achieved by quantizing the prediction errors more coarsely than the original signal itself. If no transmission errors occur, the only error present in the reconstructed signal is the quantization error. Methods for designing predictors, quantizers, and coders have been investigated to achieve the smallest transmission rate for a desired picture quality.

Median predictors combine linear substructures and the median operation in order to obtain "median" adaptive and robust predictors. The median operation can be considered as a "switch", which selects one of the linear subpredictor

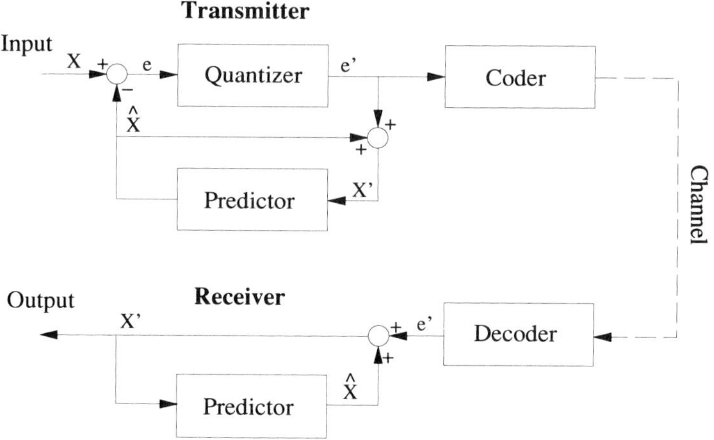

Figure 14.7: DPCM coding scheme. Here X denotes the input sample, e the prediction error, and e' the quantized prediction error. X' is the reconstructed sample and $\hat{X}$ prediction of the present sample.

outputs to be the output of the complete predictor. In multistage median predictors, there are median switches on many stages. A difference between median adaptive and adaptive predictors should be noted. In an adaptive predictor, a switch is controlled by a calculated error criterion (criteria), while in a median adaptive predictor no error criterion is calculated.

The median operation rejects outliers. Due to this property, median predictors are less sensitive to transmission errors than linear predictors [42]. In linear predictors, even a single bit error in the transmission can seriously degrade image quality [51].

A 3-D median predictor [34] is defined by

$$\hat{X}_{3DM}(\mathbf{n}) = \mathrm{MED}[X'(n_1-1,n_2,n_3), X'(n_1,n_2,n_3-1), lsp_1(\mathbf{n}), lsp_2(\mathbf{n}), lsp_3(\mathbf{n})], \tag{14.37}$$

where the linear subpredictors lsp_1, lsp_2, lsp_3 are

$$\begin{aligned} lsp_1(\mathbf{n}) &= X'(n_1, n_2-2, n_3) - X'(n_1, n_2-2, n_3-1) + X'(n_1, n_2, n_3-1), \\ lsp_2(\mathbf{n}) &= X'(n_1-1, n_2, n_3) - X'(n_1-1, n_2, n_3-1) + X'(n_1, n_2, n_3-1), \\ lsp_3(\mathbf{n}) &= X'(n_1-1, n_2, n_3) + X'(n_1, n_2-2, n_3) - X'(n_1-1, n_2-2, n_3). \end{aligned} \tag{14.38}$$

In the above equations, $X'(n_1, n_2-2, n_3)$ is a (reconstructed) pixel directly above the pixel $X'(n_1, n_2, n_3)$ in the current field, while $X'(n_1-1, n_2, n_3)$ is the neighboring pixel on the left hand side. In stationary regions, the lsp_1 and lsp_2 predictors output a value $X'(n_1, n_2, n_3-1)$. Thus, in these regions a previous frame prediction is done. In moving regions, the 3-D predictor tends to behave as a spatial (intrafield) predictor. In particular, a spatial prediction is done if there are horizontal or vertical lines going through the pixel $X'(n_1, n_2, n_3-1)$ in the previous frame. This 3-D FMH predictor is a good example of an effective combination of linear subpredictors and the median operation.

Multistage median predictors (MMP) have been introduced in [45, 46, 69]. A relatively complicated adaptive MMP has been presented in [45], where a switching logic controls whether a prediction of a stationary or a moving region should be done. The switch is controlled according to calculated error criteria. The two different predictions are obtained from MMPs. A MMP, which operates on three different "planes" in the 3-D signal space, was introduced in [46, 69]. We present a new 3-D multistage median (3DMM) predictor, which is constructed from previously presented median predictors [34, 64]. In the predictor, there are three subpredictors on the first stage and a median operation of length three on the second stage. The first subpredictor is the 3DM predictor defined in Equation 14.37. The second subpredictor is a modified version of the first predictor:

$$\hat{X}_{3DM_x}(\mathbf{n}) = \mathrm{MED}[X'(n_1-1, n_2, n_3), X'(n_1, n_2, n_3-1), lsp_4(\mathbf{n}), lsp_5(\mathbf{n}), lsp_6(\mathbf{n})], \tag{14.39}$$

where the linear subpredictors are defined by

$$\begin{aligned} lsp_4(\mathbf{n}) &= X'(n_1-1, n_2-2, n_3) - X'(n_1-1, n_2-2, n_3-1) + \\ &\quad X'(n_1, n_2, n_3-1), \\ lsp_5(\mathbf{n}) &= X'(n_1+1, n_2-2, n_3) - X'(n_1+1, n_2-2, n_3-1) + \\ &\quad X'(n_1, n_2, n_3-1), \\ lsp_6(\mathbf{n}) &= X'(n_1-1, n_2, n_3)/2 + (X'(n_1, n_2-2, n_3) + \\ &\quad X'(n_1+1, n_2-2, n_3))/4. \end{aligned} \tag{14.40}$$

Subpredictors $3DM$ and $3DM_x$ guarantee good performance in stationary regions, so the third subpredictor can be designed to give good performance in moving regions. We select the third subpredictor to be one of the spatial median predictors presented in [64]. This predictor is defined by

$$\hat{X}_{2DM}(\mathbf{n}) = \mathrm{MED}[2\Diamond X'(n_1-1, n_2, n_3), X'(n_1, n_2-2, n_3), lsp_3(\mathbf{n}), lsp_6(\mathbf{n})]. \tag{14.41}$$

The linear subpredictors lsp_3 and lsp_6 are defined in Equations 14.38 and 14.40, respectively. The 3DMM predictor is

$$\hat{X}_{3DMM}(\mathbf{n}) = \mathrm{MED}[\hat{X}_{3DM}(\mathbf{n}), \hat{X}_{3DM_x}(\mathbf{n}), \hat{X}_{2DM}(\mathbf{n})]. \tag{14.42}$$

The last predictor introduced in this section is an adaptive 3-D weighted median (A3DWM) predictor [68]. The weights of this predictor are changed so that the mean absolute prediction error is minimized. The A3DWM predictor is defined by

$$\begin{aligned} \hat{X}_{A3DWM}(\mathbf{n}) = \mathrm{MED}[&W_1(\mathbf{n})\Diamond X'(n_1-1, n_2, n_3), W_2(\mathbf{n})\Diamond Z'(n_1, n_2-2, n_3), \\ &W_3(\mathbf{n})\Diamond X'(n_1, n_2, n_3-1), W_4(\mathbf{n})\Diamond Z'(n_1-1, n_2, n_3), \\ &W_5(\mathbf{n})\Diamond Z'(n_1-1, n_2-2, n_3)], \end{aligned} \tag{14.43}$$

where

$$\begin{aligned} Z'(n_1, n_2-2, n_3) &= X'(n_1, n_2-2, n_3) - X'(n_1, n_2-2, n_3-1) + \\ &\quad X'(n_1, n_2, n_3-1), \\ Z'(n_1-1, n_2, n_3) &= X'(n_1-1, n_2, n_3) - X'(n_1-1, n_2, n_3-1) + \\ &\quad X'(n_1, n_2, n_3-1), \\ Z'(n_1-1, n_2-2, n_3) &= X'(n_1, n_2-2, n_3) + X'(n_1-1, n_2, n_3) - \\ &\quad X'(n_1-1, n_2-2, n_3). \end{aligned} \tag{14.44}$$

A weight $W_i(\mathbf{n})$ stands for the ith weight at location $(\mathbf{n})$ in an image sequence. Thus, the values of the weights are changing. Adaptation of the weights is controlled by an algorithm originally designed for minimum mean absolute error stack filtering [76]. At each location new weights are calculated from the current weights and from reconstructed samples X' in the transmitter and the receiver, so no overhead information needs to be transmitted. In order to simplify our notation, we denote $X'(n_1-1, n_2, n_3), Z'(n_1, n_2-2, n_3), \ldots, Z'(n_1-1, n_2-2, n_3)$ in Equation 14.43 by $X'(n-1), X'(n-2), \ldots, X'(n-5)$, respectively. Weights of the predictor are calculated using the algorithm

$$\begin{aligned} W_i(\mathbf{n}') = Q\Bigg[W_i(\mathbf{n}) + 2\mu \Bigg((X_{max}(n) - X_{min}(n)) \times \left(1 - \sum_{j=1}^{5} W_j(\mathbf{n})\right) \\ -2|X'(\mathbf{n}) - X'(n-i)| + 2\sum_{j=1}^{5} W_j(\mathbf{n})|X'(n-i) - X'(n-j)| \Bigg) \Bigg], \end{aligned} \tag{14.45}$$

where μ is an adaptation step size, $Q[.]$ is a projection operation defined by

$$Q[X] = \begin{cases} X & \text{if } X \geq 0; \\ 0 & \text{otherwise,} \end{cases} \tag{14.46}$$

and

$$\begin{aligned} X_{max}(n) &= \max[X'(n-i)], \\ X_{min}(n) &= \min[X'(n-i)], \\ i &= 1, \ldots 5. \end{aligned} \tag{14.47}$$

In Equation 14.45, $(\mathbf{n}')$ means the next window location after position $(\mathbf{n})$ in an image sequence.

14.3 Experimental Results

In this section, the performance of the 3-D median filters and predictors is examined using computer simulations. First, performance of weighted median filters and idempotent weighted median filters is analyzed. Then, some median predictors are tested.

Figure 14.8: The seventh frame of the Costgirls sequence.

14.3.1 Image sequence filtering

In Section 14.2.2, a set of 3-D median filters was presented. Here the performance of the filters is analyzed in terms of noise reduction and detail preservation. A 19 frames long image sequence called Costgirls was used in the simulations. Only the luminance component was considered. The size of each frame in this sequence is 720×576 and each pixel has 8 bit accuracy. The seventh frame of the Costgirls sequence is shown in Figure 14.8. In the sequence, the canvas in the background is moving slowly, train and the hands of the girls are moving fast and floor and toys are stationary.

Fifteen reference filters were included in the comparison. These filters are: a temporal 3-point median filter (TMED3) [49], a spatial plus shaped 5-point median filter (SMED5), an averaging filter (AVE5), which operates in the window of the MED5 filter, the Bidirectional Multistage Median (BidiMM) filter ($N = 1$) [4, 5], the Planar filter [2, 3], center weighted median filters [5, 39], and adaptive center weighted median (ACWM) filters [39]. All CWM and ACWM filters operate in a $3 \times 3 \times 3$ window. In order to find the best CWM and ACWM filters, exhaustive tests were done. Filters CWM(27,7), CWM(27,11), CWM(27,13), CWM(27,15), CWM(27,19), CWM(27,21), ACWM(T=2), ACWM(T=3), ACWM(T=4), and ACWM(T=5) were tested. A CWM(27,X) filter is a WM filter, where all the weights are equal to 1 except the center weight

Table 14.1: Contaminated Gaussian noise attenuation capability of the filters.

Sequence: Costgirls, $(\nu_1, \nu_2) = (5, 30)$, $p = 0.1$		
Filter	Mean Absolute Error	Mean Square Error
TMED3	4.3	46.3
SMED5	4.2	46.7
AVE5	5.0	54.3
L-shaped	4.5	51.4
MED5	3.7	32.4
WM7	3.5	**25.4**
TMM	3.6	31.9
WM11	3.4	28.0
FMH7	3.6	30.2
FMH5	3.9	41.1
Planar	3.7	34.4
BidiMM	3.6	28.6
CWM(27,7)	3.8	45.0
CWM(27,11)	3.7	34.4
CWM(27,13)	3.8	32.7
CWM(27,15)	3.9	32.6
CWM(27,19)	4.2	36.1
CWM(27,21)	4.5	40.4
ACWM(T=2)	3.3	28.1
ACWM(T=3)	3.3	27.3
ACWM(T=4)	**3.2**	27.1
ACWM(T=5)	3.3	27.5

being equal to X. Variable T in ACWM filters is related to detail preservation capabilities. In addition, ACWM filters assume that the noise variance is known.

In the experiment, we used an input sequence corrupted by contaminated Gaussian noise of the form

$$f(x) = (1 - p)G(0, \nu_1) + pG(0, \nu_2), \tag{14.48}$$

where p determines the probability of the contamination and ν_1 and ν_2 are standard deviations of Gaussian distributions. An input sequence having parameters $p = 0.1$, $\nu_1 = 5$ and $\nu_2 = 30$ was filtered with different filters. Mean absolute error (MAE) and mean square error (MSE) values were calculated between the original and filtered image sequences. In Table 14.1, the average MAE and MSE values calculated over frames 2 to 18 are shown. The MSE values for each frame in the Costgirls sequence are shown in Figure 14.9 for the seven most promising filters. In this figure and in the following comparisons, the CWM(27,13) and the ACWM(T=4) filters are considered as representatives of their groups.

In general, large filter masks perform better than small. Properly designed spatiotemporal filters give better results than spatial or temporal filters. This can be seen by comparing the results of the MED5, the SMED5, and the TMED3 filters. CWM filters do not perform well. It is impossible to find a center weight so that the filter would attenuate noise effectively and preserve details.

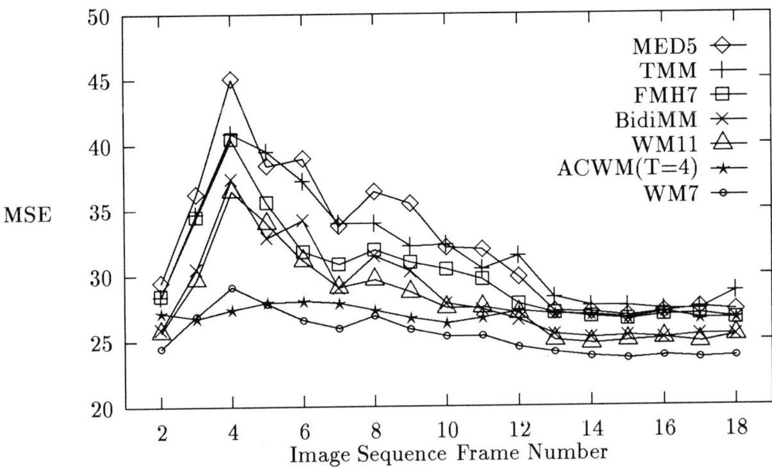

Figure 14.9: Contaminated Gaussian noise attenuation. Mean square error values of the Costgirls sequence for different filters.

ACWM filters, in which the center weight is changed according to local signal statistics, produce good results. They are, however, computationally more time consuming. Also the WM7 and the WM11 filters give very good results.

Part of frame 7 of the noisy Costgirls sequence is shown in Figure 14.10(a). Corresponding parts from sequences filtered with the TMED3, the L-shaped, the FMH7, the WM7, the BidiMM, the WM11, and the ACWM(T=4) filters are shown in Figures 14.10(b)–(d) and 14.11(a)–(d), respectively. The visual evaluation of the results supports the calculated error values with some exceptions. As can be seen from Figure 14.10(b), the TMED3 filter introduces severe distortion of moving areas. The blurring can be clearly seen by examining the thumb of the girl. Visual results of the L-shaped filter are much better than the results of the TMED3 filter. The noise attenuation capability of these 3-point median filters is low. The FMH7 filter is a combination of linear and median operations. As can be seen from Figure 14.10(d), this filter can effectively remove noise while retaining moving and stationary details. Both the WM7 and the BidiMM preserve details extremely well as can be seen from Figures 14.11(a) and 14.11(b). Figures 14.11(c) and 14.11(d) show the results of WM11 and ACWM(T=4) filters, respectively. The performance of these filters is about the same in stationary regions. In moving regions, however, the noise attenuation capability of the ACWM(T=4) filter deteriorates. This can be seen, e.g., by examining the hand of the girl. This is reasonable, since the adaptation algorithm pushes the ACWM(T=4) filter towards an identity filter when the sample variance inside the window is large. Large sample variance in moving regions of the Costgirls sequence is partly due to interlacing. Clearly, the design principles introduced for 3-D median filters give good results in filtering of noisy image

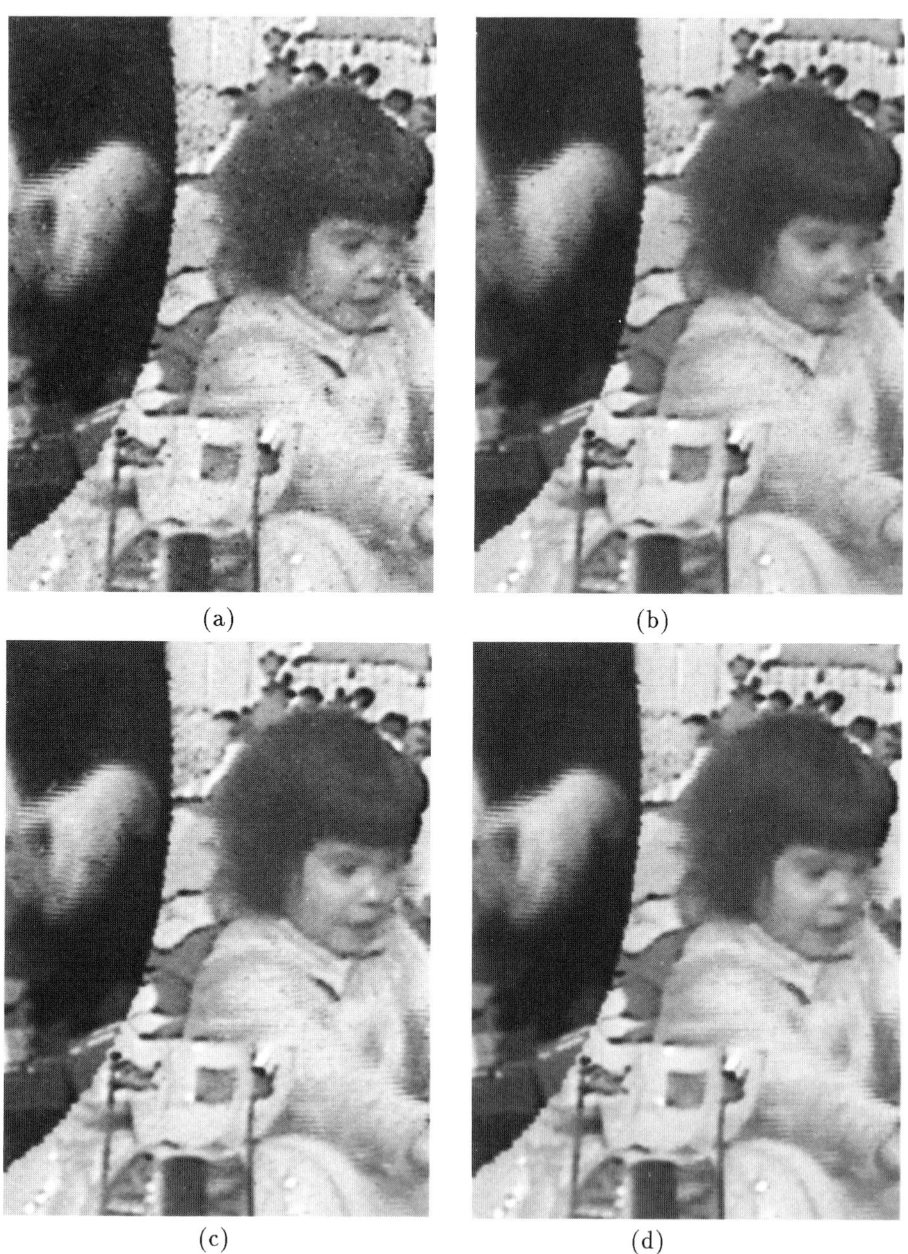

Figure 14.10: Contaminated Gaussian noise reduction. (a) Noisy image sequence, (b) the TMED3, (c) the L-shaped, (d) the FMH7 filter.

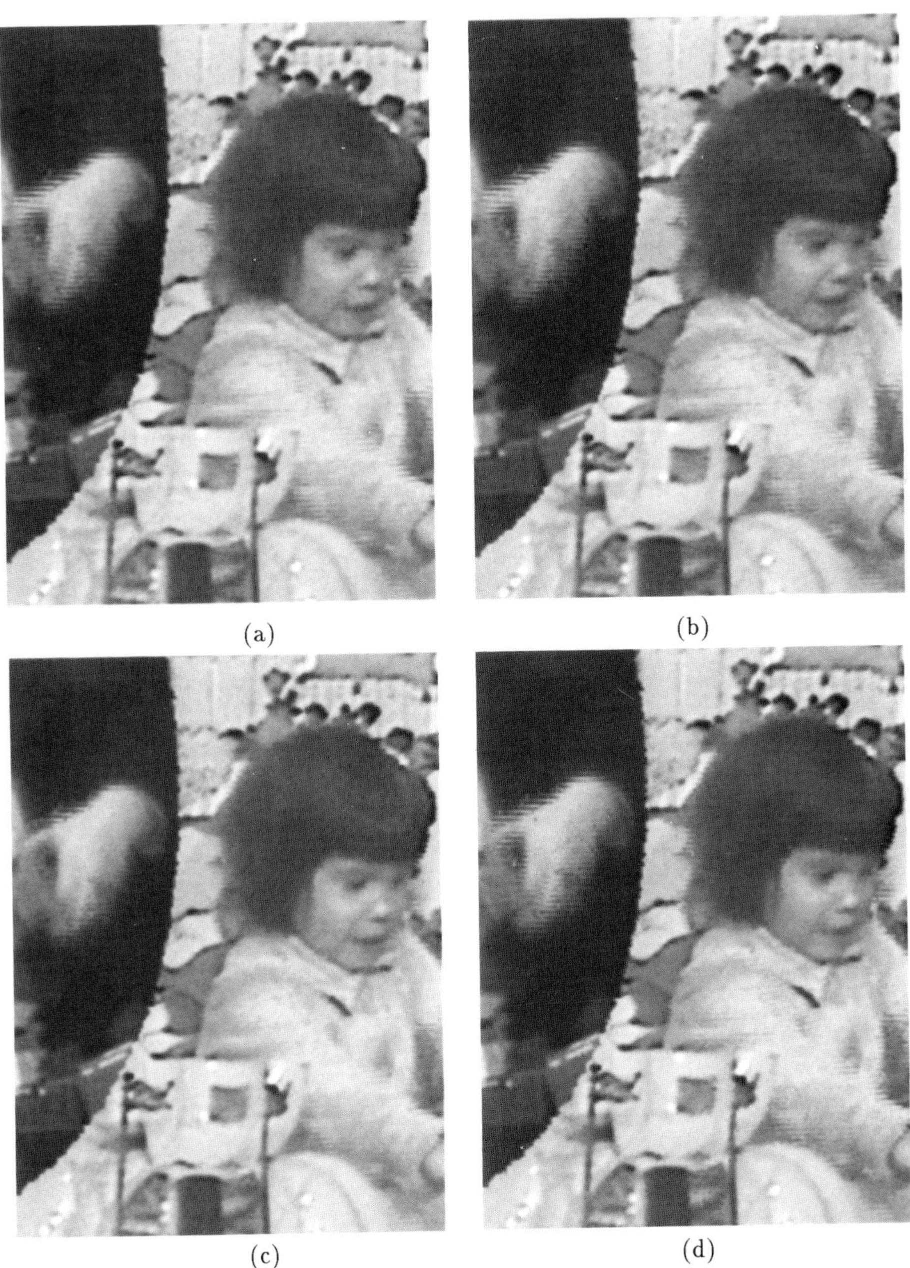

Figure 14.11: Contaminated Gaussian noise reduction continued. (a) The WM7, (b) the BidiMM, (c) the WM11, (d) the ACWM(T=4) filter.

sequences.

14.3.2 Preprocessing of image sequences

Next we study the detail preservation and impulse noise removal properties of idempotent weighted median filters. The WM7 filter is used as a reference filter. In the test, the Costgirls sequence was corrupted by salt-and-pepper noise where the probability of a corruption for a positive or a negative impulse of length one was 0.01. MAE and MSE values were calculated between the original and the filtered image sequences. Average MAE and MSE values over the frames 3 to 17 are presented in Table 14.2. As can be seen from these error values, the spatial and the spatiotemporal idempotent filters have very low MAE values. The WM7 filter has the lowest MSE value. The temporal idempotent filter gives the worst results.

Part of frame 7 of the noisy Costgirls sequence is shown in Figure 14.12(a). Corresponding parts from sequences filtered with the temporal and the spatiotemporal idempotent filters, and the WM7 filter are shown in Figures 14.12(b)–(d), respectively. The temporal idempotent filter gives worst visual results due to severe blurring of moving areas. Proper selection of the window shape is important also for median-based idempotent filters. The spatiotemporal idempotent filter preserves details very well. There are, however, some impulses left in the filtered sequence. If the current input sample is not unique inside the window, it is preserved by idempotent weighted median filters. Thus, an impulse is not removed if there is also another impulse having the same value inside the window. The WM7 filter removes impulses effectively and produces a visually pleasing image quality. Both the spatiotemporal idempotent filter and the WM7 filter perform well in noise free regions.

14.3.3 Predictive coding systems

The performance of median predictors was compared with some fixed and adaptive linear predictors.[4] The first reference predictor is the 3-D linear inter-intraframe predictor $P3$ [12], which is here called the 3DL predictor. The second reference predictor is the adaptive "steepest descent" 3-D linear inter-intraframe predictor (A3DL), where W = {1,2,3} [57]. Also the 3-D three-plane multistage median (3D3PMM) predictor [46, 69] was included in the comparison. For the sake of simplicity, a fixed four-bit Tapered quantizer [62] was used with all predictors. The input (decision) levels and output (representative) levels on the positive side of this symmetric quantizer are shown in Table 14.3.

Root mean square (RMS) values of unquantized prediction errors for different predictors are shown in Figure 14.13. Entropy values of quantized prediction errors are given in Figure 14.14. As can be seen from these figures, the 3DL predictor gives the worst results. It is noticeably inferior to other predictors. When comparing the performance of fixed median predictors, good performance of the

[4] The authors would like to thank Mr. Xudong Song for performing simulations of the predictors.

Table 14.2: Salt-and-Pepper noise attenuation capability of idempotent weighted median filters.

Sequence: Costgirls, prob. of corruption 0.01		
Filter	Mean Absolute Error	Mean Square Error
TEMP	1.0	21.1
SPAT	0.4	17.1
SPAT-TEMP	**0.3**	18.0
WM7	0.8	**7.5**

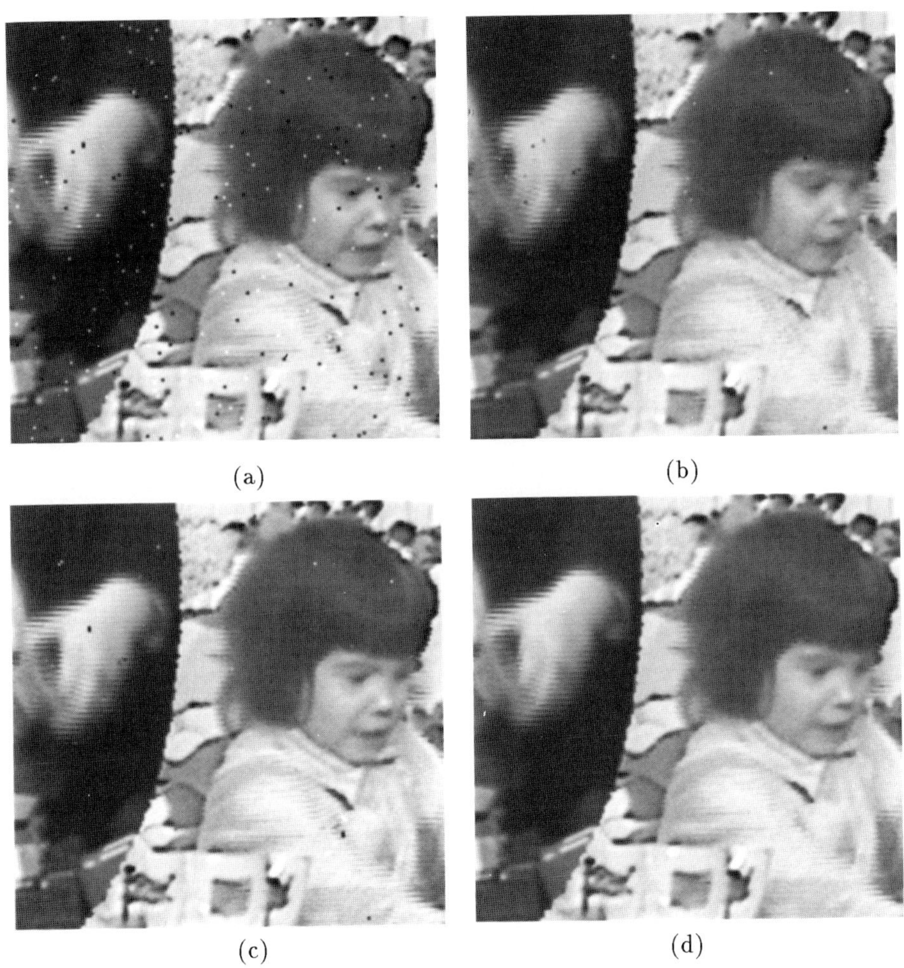

(a) (b) (c) (d)

Figure 14.12: Salt-and-pepper noise reduction. (a) Noisy image sequence, (b) the temporal idempotent filter, (c) the spatiotemporal idempotent filter, (d) the WM7 filter.

Table 14.3: Characteristics of the 4-Bit Tapered Quantizer on the positive side.

input level	255...96	96...48	48...24	24...12	12...6	6...3	3...1	1...-1
output level	128	64	32	16	8	4	2	0

3DM and the 3DMM predictors can be seen. They outperform the 3D3PMM predictor, which is not designed according to the principles presented in Section 14.2.1. As can be seen from the RMS and entropy values, the 3DM and 3DMM predictors have almost equal performance. Thus, for this sequence, use of a multistage median structure has not resulted in better performance. It is interesting to note that the median adaptive 3DM and 3DMM predictors outperform the A3DL predictor, which is an adaptive linear predictor. The A3DWM predictor gives the best results.

Due to the 4-bit Tapered quantizer, no visual impairments are seen in the coded image sequences. Coarser, carefully optimized quantizers [51, 58] should be used with the predictors in order to test the overall performance of the DPCM coder. Difference image sequences between the original and the coded image sequences were calculated for limited visual comparisons. Part of frame 7 of the Costgirls sequence is shown in Figure 14.15(a). Corresponding parts from the difference sequences are shown in Figures 14.15(b)–(f) for the 3DL, the 3D3PMM, the A3DL, the 3DM, and the A3DWM predictors, respectively. Before printing a difference image X, a transformation $20\times \mid X \mid +80$ enhancing the graylevel variations was done. Results from visual evaluation coincide with the RMS and entropy values.

14.4 Conclusions

Several three-dimensional median-related structures for image sequence enhancement and predictive coding were presented. The structures are based on weighted median operations and combinations of linear and median (FMH) operations. Idempotent weighted median structures were introduced for impulsive noise elimination from image sequences.

Guidelines for the design of good 3-D median structures were given for obtaining good performance in stationary and moving regions. Several 3-D median filters and predictors designed according to these design principles were presented. Good performance of these structures was verified using computer simulations.

Noise reduction and coding were treated separately in this chapter. In practical applications these operations are often connected and they both contribute to the overall performance of the system.

Recently, the theory of adaptive median filtering has been advanced and several promising algorithms have been proposed [70, 76, 77, 78]. Adaptive methods can be used as filter synthesis methods to find good fixed weights using a representative training sequence. The other possibility is to adapt the

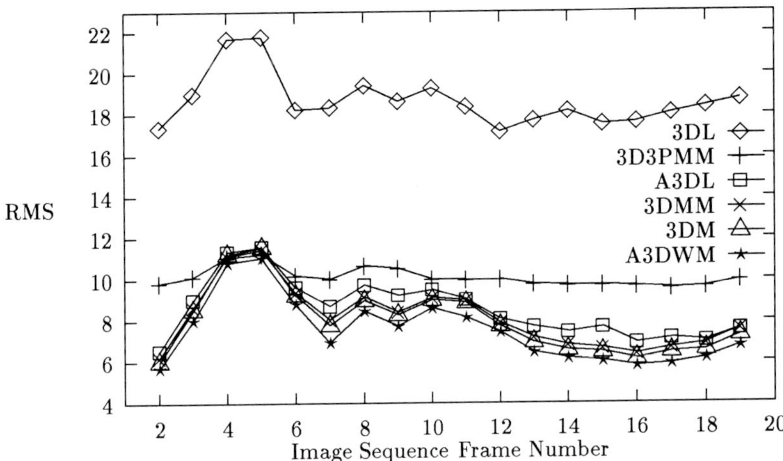

Figure 14.13: RMS error values of the Costgirls sequence for different predictors.

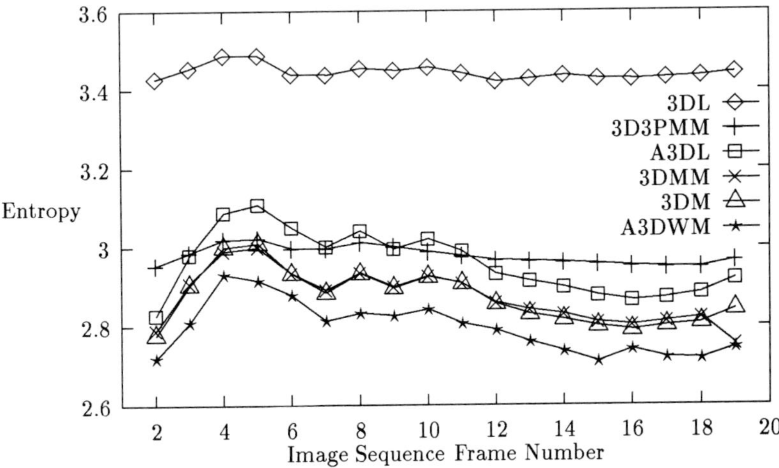

Figure 14.14: Entropy values of the Costgirls sequence for different predictors.

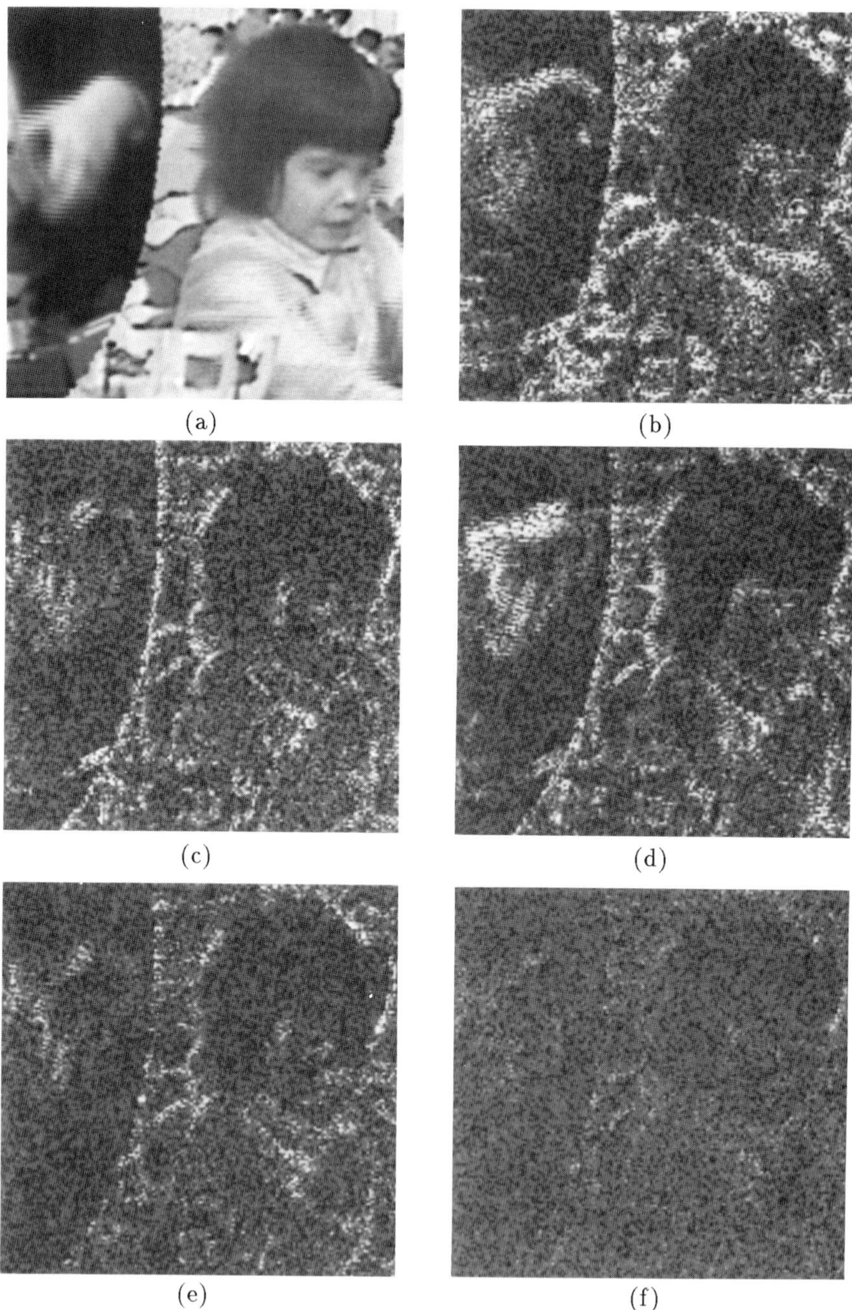

Figure 14.15: Visual evaluation of coded images. (a) The original Costgirls sequence. Difference images between the original and coded images. (b) The 3DL predictor, (c) the 3D3PMM predictor, (d) the A3DL predictor, (e) the 3DM, (f) the A3DWM predictor.

weights continuously. Application of adaptive algorithms to 3-D median-related structures is the subject of future work.

References

[1] B. Alp, J. Juhola, T. Jarske, and Y. Neuvo, "Multidimensional Reconstruction of Quincunx Coded Image Sequences," in *Proc. 1990 Picture Coding Symposium*, Cambridge, MA, USA, March 1990.

[2] B. Alp, P. Haavisto, T. Jarske, K. Öistämö, and Y. Neuvo, "Median Based Algorithms for Image Sequence Processing," in *Visual Communications and Image Processing '90*, pp. 122–134, Lausanne, Switzerland, Oct. 1990.

[3] B. Alp and Y. Neuvo, "3-Dimensional Median Filters for Image Sequence Processing," in *Proc. IEEE Int. Conf. Acoust., Speech, Signal Processing*, pp. 2917–2920, Toronto, Canada, May 1991.

[4] G. R. Arce and E. Malaret, "Motion-Preserving Ranked-Order Filters for Image Sequence Processing," in *Proc. IEEE Int. Symp. Circuits and Systems*, pp. 983–986, Portland, Oregon, May 1989.

[5] G. A. Arce, "Multistage Order Statistic Filters for Image Sequence Processing," *IEEE Trans. Signal Processing*, vol. 39, no. 5, pp. 1146–1163, May 1991.

[6] J. Astola, P. Heinonen, and Y. Neuvo, "On Root Structures of Median and Median-Type Filters," *IEEE Trans. Acoust., Speech, Signal Processing*, vol. ASSP-35, no. 8, pp. 1199–1201, Aug. 1987.

[7] J. Astola, P. Haavisto, P. Heinonen, and Y. Neuvo, "Median Type Filters for Color Signals," in *Proc. IEEE Int. Symp. Circuits and Systems*, pp. 1753–1756, Espoo, Finland, June 1988.

[8] J. Astola, P. Haavisto, and Y. Neuvo, "Vector Median Filters," *Proceedings of the IEEE*, vol. 78, no. 4, pp. 678–689, April 1990.

[9] E. Ataman, V. K. Aatre, and K. M. Wong, "Some Statistical Properties of Median Filters," *IEEE Trans. Acoust., Speech, Signal Processing*, vol. ASSP-29, no. 5, pp. 1073–1075, Oct. 1981.

[10] V. Barnett, "The Ordering of Multivariate Data," *J. R. Statist Soc. A*, 139, Part 3, pp. 318–355, 1976.

[11] J. M. Boyce, "Noise Reduction of Image Sequences Using Adaptive Motion Compensated Frame Averaging," in *Proc. IEEE Int. Conf. Acoust., Speech, Signal Processing*, pp. 461–464, San Francisco, CA, USA, March 1992.

[12] H. Buley and L. Stenger, "Inter/Intraframe Coding of Color TV Signals for Transmission at the Third Level of the Digital Hierarchy," *Proceedings of the IEEE*, vol. 73, no. 4, pp. 765–772, April 1985.

[13] E. J. Coyle, J.-H. Lin, and M. Gabbouj, "Optimal Stack Filtering and the Estimation and Structural Approaches to Image Processing," *IEEE Trans. Acoust., Speech, and Signal Processing*, vol. 37, no. 12, pp. 2037–2066, Dec. 1989.

[14] H. A. David, *Order Statistics*, Wiley, New York, 1970.

[15] T. J. Dennis, "Nonlinear Temporal Filter for Television Picture Noise Reduction," *IEE Proc., Part G*, vol. 127, pp. 52–56, Apr. 1980.

[16] E. Dubois and S. Sabri, "Noise Reduction in Image Sequences Using Motion-Compensated Temporal Filtering," *IEEE Trans. Communications*, vol. 32, no. 7, pp. 826–831, July 1984.

[17] E. Dubois, "Motion-Compensated Filtering of Time-Varying Images," *Multidimensional Systems and Signal Processing*, vol. 3, no. 2/3, pp. 211–240, May 1992.

[18] J. P. Fitch, E. J. Coyle, and N. C. Gallagher, Jr. "Median Filtering by Threshold Decomposition," *IEEE Trans. Acoust., Speech, and Signal Processing*, vol. ASSP-32, no. 6, pp. 1183–1188, Dec. 1984.

[19] J. P. Fitch, E. J. Coyle, and N. C. Gallagher, "Root Properties and Convergence Rates of Median Filters," *IEEE Trans. Acoust., Speech, and Signal Processing*, vol. ASSP-33, no. 1, pp. 230–240, Feb. 1985.

[20] M. Gabbouj and E. J. Coyle, "Minimum Mean Absolute Error Stack Filtering with Structural Constraints," *IEEE Trans. Acoust., Speech, Signal Processing*, vol. 38, no. 6, pp. 955–968, June 1990.

[21] M. Gabbouj, P. Haavisto, and Y. Neuvo, "Recent Advances in Median Filtering," in *Communication, Control, and Signal Processing*, E. Arikan (ed.), Elsevier Science Publishers B.V., vol. II, pp. 1080–1094, Ankara, Turkey, 1990.

[22] M. Gabbouj, E. J. Coyle, and N. C. Gallagher, Jr., "An Overview of Median and Stack Filtering," *Circuits, Systems, and Signal Processing*, vol. 11, no. 1, pp. 7–46, 1992.

[23] N. P. Galatsanos and R. T. Chin, "Digital Restoration of Multichannel Images," *IEEE Trans. Acoust., Speech, Signal Processing*, vol. ASSP-37, no. 3, pp. 415–421, March 1989.

[24] N. C. Gallagher, Jr. and G. L. Wise, "A Theoretical Analysis of the Properties of Median Filters," *IEEE Trans. Acoust., Speech, and Signal Processing*, vol. ASSP-29, no. 6, pp. 1136–1141, Dec. 1981.

[25] P. Haavisto, P. Heinonen, and Y. Neuvo, "Vector FIR-Median Hybrid Filters for Multispectral Signals," *Electronic Letters*, vol. 24, no. 1, pp. 7–8, Jan. 1988.

[26] P. Haavisto, M. Gabbouj, and Y. Neuvo, "Median Based Idempotent Filters," *Journal of Circuits, Systems, and Computers*, vol. 1, no. 2, pp. 125–148, 1991.

[27] M. Helsingius, O. Kalevo, P. Haavisto, and Y. Neuvo, "Producing a Very High Quality Still Image from HDMAC Transmission," in *Proc. International Broadcasting Convention IBC 92*, pp. 400–404, Amsterdam, The Netherlands, July 1992.

[28] R. C. Hardie and G. R. Arce, "Ranking in R^p and its Use in Multivariate Image Estimation," *IEEE Trans. Circuits and Systems for Video Technology*, vol. 1, no. 2, pp. 197–209, June 1991.

[29] P. Heinonen and Y. Neuvo, "Smoothed Median Filters with FIR Substructures," in *Proc. IEEE Int. Conf. Acoust., Speech, Signal Processing*, pp. 49–52, Tampa, FL, USA, Mar. 1985.

[30] P. Heinonen and Y. Neuvo, "FIR-Median Hybrid Filters," *IEEE Trans. Acoust., Speech, Signal Processing*, vol. ASSP-35, no. 6, pp. 832–838, June 1987.

[31] P. Heinonen and Y. Neuvo, "FIR-Median Hybrid Filters with Predictive Substructures," *IEEE Trans. Acoust., Speech, Signal Processing*, vol. ASSP-36, no. 6, pp. 892–899, June 1988.

[32] T. S. Huang and Y. P. Hsu, "Image Sequence Enhancement," in *Image Sequence Analysis*, T. S. Huang (ed.), Springer-Verlag, Berlin, 1981.

[33] P. J. Huber, *Robust Statistics*, Wiley, New York, 1981.

[34] T. Jarske and Y. Neuvo, "Adaptive DPCM with Median Type Predictors," *IEEE Trans. Consumer Electronics*, vol. 37, no. 3, pp. 348–352, Aug. 1991.

[35] T. Jarske, K. Saarinen, J. Juhola, and Y. Neuvo, "Quincunx Coding for Picture Memories," *131st SMPTE Technical Conference*, Los Angeles, USA, Oct. 1989.

[36] B. I. Justusson, "Median Filtering: Statistical Properties," *Topics in Applied Physics, Two-Dimensional Digital Signal Processing II*, T. S. Huang (ed.), pp. 161–196, Springer-Verlag, Berlin, Germany, 1981.

[37] D. S. Kalivas and A. A. Sawchuk, "Motion Compensated Enhancement of Noisy Image Sequences," in *Proc. Int. Conf. Acoust., Speech, Signal Processing*, pp. 2121–2124, Albuquerque, New Mexico, USA, 1990.

[38] S. A. Kassam and M. Aburdene, "Multivariate Median Filters and Their Extensions," in *Proc. IEEE Int. Symp. Circuits and Systems*, pp. 85–88, Singapore, June 1991.

[39] S.-J. Ko and Y. H. Lee, "Nonlinear Spatio-Temporal Noise Suppression Techniques with Applications in Image Sequence Processing," in *Proc. IEEE Int. Symp. Circuits and Systems*, pp. 662–665, Singapore, June 1991.

[40] S.-J. Ko and Y. H. Lee, "Center Weighted Median Filters and Their Applications to Image Enhancement," *IEEE Trans. Circuits and Systems*, vol. 38, no. 9, pp. 984–993, Sept. 1991.

[41] F. Kuhlmann and G. L. Wise, "On Second Moment Properties of Median Filtered Sequences of Independent Data," *IEEE Trans. Communications*, vol. 29, no. 9, pp. 1374–1379, Sept. 1981.

[42] Y. H. Lee, D. H. Kang, J. H. Choi, and K. D. Lee, "DPCM with Median Predictors," in *Proc. of Nonlinear Image Processing III*, pp. 199–209, San Jose, California, USA, Feb. 1992.

[43] E. L. Lehmann, *Theory of Point Estimation*, Wadsworth & Brooks/Cole Pacific Grove, California, 1991.

[44] M. P. McLoughlin and G. R. Arce, "Deterministic Properties of the Recursive Separable Median Filter," *IEEE Trans. Acoust., Speech, Signal Processing*, vol. ASSP-35, no. 1, pp. 98–106, Jan. 1987.

[45] R. Mickos, L. Öktem, T. G. Campbell, T. Sun, and Y. Neuvo, "3-D Median Based Prediction for Image Sequence Coding," in *Proc. IEEE Int. Symp. Circuits and Systems*, pp. 1656–1659, San Diego, CA, USA, May 1992.

[46] R. Mickos, X. Song, T. Sun, T. G. Campbell, and Y. Neuvo, "Median Structures for Image Sequence Prediction," in *Proc. Int. Conf. Consumer Electronics*, pp. 228–229, Rosemont, Illinois, USA, June 1992.

[47] H. G. Musmann, P. Pirsch, and H.-J. Grallert, "Advances in Picture Coding," *Proceedings of the IEEE*, vol. 73, no. 4, pp. 523–548, April 1985.

[48] S. Naimpally, L. Johnson, T. Darby, R. Meyer, L. Phillips, and J. Vantrease, "Integrated Digital IDTV Receiver with Features," *IEEE Trans. Consumer Electronics*, vol. 34, no. 3, pp. 410–419, Aug. 1988.

[49] S. S. H. Naqvi, N. C. Gallagher, and E. J. Coyle, "An Application of Median Filters to Digital Television," in *Proc. IEEE Int. Conf. Acoust., Speech, Signal Processing*, pp. 2451-2454, Tokyo, Japan, April 1986.

[50] A. N. Netravali and J. O. Limb, "Picture Coding: A Review," *Proceedings of the IEEE*, vol. 68, no. 3, pp. 366–406, March 1980.

[51] A. N. Netravali and B. G. Haskell, *Digital Pictures: Representation and Compression*, Plenum Press, New York, 1988.

[52] Y. Neuvo, "Interpolators for TV Scanning Rate Conversions," in *Proc. First World Electronic Media Symposium, ITU-COM 89*, pp. 203–207, Geneva, Switzerland, Oct. 1989.

[53] A. Nieminen, P. Heinonen, and Y. Neuvo, "A New Class of Detail Preserving Filters for Image Processing," *IEEE Trans. Pattern Analysis and Machine Intelligence*, vol. PAMI-9, no. 1, pp. 74–90, Jan. 1987.

[54] T. A. Nodes and N. C. Gallagher, Jr., "Two-Dimensional Root Structures and Convergence Properties of the Separable Median Filter," *IEEE Trans. Acoust., Speech, and Signal Processing*, vol. ASSP-31, no. 6, pp. 1350–1365, Dec. 1983.

[55] T. A. Nodes and N. C. Gallagher, Jr., "The Output Distribution of Median Type Filters," *IEEE Trans. Communications*, vol. COM-32, no. 5, pp. 532–541, May 1984.

[56] T. A. Nodes and N. C. Gallagher, Jr., "Median Filters: Some Modifications and Their Properties," *IEEE Trans. Acoust., Speech, Signal Processing*, vol. ASSP-30, no. 5, pp. 739–746, Oct. 1982.

[57] P. Pirsch, "Adaptive Intra-Interframe DPCM Coder," *The Bell System Technical Journal*, vol. 61, no. 5, pp. 747–764, May-June 1982.

[58] P. Pirsch, "Design of DPCM Quantizers for Video Signals Using Subjective Tests," *IEEE Trans. Communications*, vol. COM-29, no. 7, pp. 990–1000, July 1981.

[59] I. Pitas, "Marginal Order Statistics in Color Image Filtering," *Optical Engineering*, vol. 29, no. 5, pp. 495–503, May 1990.

[60] I. Pitas and A. N. Venetsapoulos, *Nonlinear Digital Filters*, Kluwer Academic Publishers, Boston, 1990.

[61] I. Pitas and P. Tsakalides, "Multivariate Ordering in Color Image Filtering," *IEEE Trans. Circuits and Systems for Video Technology*, vol. 1, no. 3, pp. 247–259, Sept. 1991.

[62] K. A. Prabhu, "A Predictor Switching Scheme for DPCM Coding of Video Signals," *IEEE Trans. Communications*, vol. COM-33, no. 4, pp. 373–379, April 1985.

[63] K. Saarinen and Y. Neuvo, "An Adaptive Weighted Median Filtering Based on the LMS-Algorithm," in *Communication, Control, and Signal Processing*, E. Arikan (ed.), Elsevier Science Publishers B.V, vol. II, pp. 1241–1248, Ankara, Turkey, July 1990.

[64] J. Salo, Y. Neuvo, and V. Hämeenaho, "Improving TV Picture Quality with Linear-Median Type Operations," *IEEE Trans. Consumer Electronics*, vol. 34, no. 3, pp. 373–379, Aug. 1988.

[65] W. F. Schreiber, "Psychophysics and the Improvement of Television Image Quality," *SMPTE Journal*, pp. 717–725, 1984.

[66] J. Serra and L. Vincent, "An Overview of Morphological Filtering," *Circuits, Systems, and Signal Processing*, vol. 11, no. 1, pp. 47–109, 1992.

[67] M. I. Sezan, M. K. Ozkan, and S. V. Fogel, "Temporally Adaptive Filtering of Noisy Image Sequences Using a Robust Motion Estimation Algorithm," in *Proc. IEEE Int. Conf. Acoust., Speech, Signal Processing*, pp. 2429–2432, Toronto, Canada, May 1991.

[68] X. Song, L. Yin, and Y. Neuvo, "Image Sequence Coding Using Adaptive Weighted Median Prediction," in *Signal Processing VI: Proc. EUSIPCO-92*, pp. 1307–1310, Brussels, Belgium, Aug. 1992.

[69] X. Song and Y. Neuvo, "A Three-Dimensional Weighted Prediction for TV," accepted to *IEEE Int. Conf. on Systems Engineering*, Kobe, Japan, Sept. 1992.

[70] T. Sun and Y. Neuvo, "A Simple Synthesis Method for Weighted Median Filters," in *Signal Processing VI: Proc. EUSIPCO-92*, pp. 1405–1408, Brussels, Belgium, Aug. 1992.

[71] S. G. Tyan, "Median Filtering: Deterministic Properties," *Topics in Applied Physics, Two-Dimensional Digital Signal Processing II*, T. S. Huang (ed.), pp. 197–217, Springer-Verlag, Berlin, Germany, 1981.

[72] T. Viero and Y. Neuvo, "Non-Moving Regions Preserving Median Filters for Image Sequence Processing," in *Proc. IEEE Int. Conf. Systems Engineering*, pp. 245–248, Dayton, OH, USA, Aug. 1991.

[73] T. Viero, K. Öistämö, and Y. Neuvo, "Median and Vector Median Filters for Image Sequence Filtering," in *Proc. European Conference on Circuit Theory and Design*, pp. 585–594, Copenhagen, Denmark, Sept. 1991.

[74] P. D. Wendt, E. J. Coyle, and N. C. Gallagher, Jr., "Some Convergence Properties of Median Filters," *IEEE Trans. Circuits and Systems*, vol. CAS-33, no. 3, pp. 276–286, Mar. 1986.

[75] P. D. Wendt, E. J. Coyle, and N. C. Gallagher, JR., "Stack Filters," *IEEE Trans. Acoust., Speech, Signal Processing*, vol. ASSP-34, no. 4, pp. 898–911, Aug. 1986.

[76] L. Yin, J. Astola, and Y. Neuvo, "Adaptive Weighted Median Filtering Under the Mean Absolute Error Criterion," in *Proc. IEEE Workshop on Visual Signal Processing and Communications*, pp. 184–187, Hsinchu, Taiwan, June 1991.

[77] L. Yin, J. Astola, and Y. Neuvo, "Optimal Weighted Order Statistic Filter Under the Mean Absolute Error Criterion," in *Proc. Int. Conf. Acoust., Speech, Signal Processing*, pp. 2529–2532, Toronto, Canada, 1991.

[78] L. Yin, J. Astola, and Y. Neuvo, "Adaptive Stack Filtering with Applications to Image Processing," to appear in *IEEE Trans. Signal Processing*, Jan. 1993.

[79] O. Yli-Harja, J. Astola, and Y. Neuvo, "Analysis of the Properties of Median and Weighted Median Filters Using Threshold Logic and Stack Filter Representation," *IEEE Trans. Signal Processing*, vol. 39, no. 2, pp. 395–410, Feb. 1991.

[80] K. Öistämö and Y. Neuvo, "Video Signal Processing Using Vector Median," in *Visual Communications and Image Processing '90*, pp. 1171–1183, Lausanne, Switzerland, October 1990.

[81] K. Öistämö and Y. Neuvo, "Reconstruction of Quincunx Coded Image Sequences Using Vector Median," in *Visual Communications and Image Processing '91*, pp. 735–742, Boston, MA, USA, November 1991.

15

Video Compression for Digital Advanced Television Systems

John G. Apostolopoulos and Jae S. Lim

Advanced Television Signal Processing Group
Research Laboratory of Electronics
Massachusetts Institute of Technology
Cambridge, MA USA

15.1 Introduction

Television began in 1941 with the adoption of a monochrome television standard by the National Television Systems Committee (NTSC) in the U.S. A fully compatible color television signal that fits within the same monochrome channel was adopted in 1953. Originally, television was perceived simply as a form of entertainment. Today, television in its many forms is also used for business, medicine, science, education, the military, and multi-media applications. These are a multitude of applications, with many more to come. With the various applications arise different sets of requirements. For example, accurately perceiving the beating of a human heart or the movements of a gymnast in the Olympics requires high temporal resolution, while examining museum pieces in an art class requires high spatial resolution. The advanced video systems that fulfill these requirements are called Advanced Television (ATV) Systems.

The evolution of ATV systems has been based primarily on a number of improvements over conventional television. These include higher spatial and temporal resolution as well as improved scanning formats. The video of conventional television is highly data intensive, and the improvements of ATV make this even more so. For economically feasible transmission and storage of ATV signals, some form of compression is usually required. Video compression removes the redundancy within the video signal so that only the perceptually important information is transmitted or stored. The actual form of compression to be employed is dependent upon the application. For example, for storing irreplaceable imagery from the space program, one would use a lossless algorithm. However, for transmission of a nightly talk show, a high-quality though lossy video compression algorithm may be acceptable. Many excellent books and papers have presented overviews of image and video compression [1, 2, 3, 4]. This chapter will focus on the application of video compression principles to the efficient delivery and storage of ATV video signals. We will present the important principles of video compression, and develop a framework for creating a high-quality, low-bit-rate digital ATV system. Since the term "ATV Systems" is rather broad, we will make our discussion more concrete by placing it, where appropriate, within the context of the design of a digital High Definition Television (HDTV) system. In particular, we hope to share some of the knowledge and experience we have gained through the design of the MIT/GI Channel-Compatible DigiCipher (CCDC) digital HDTV system [5]. The final section will tie together the compression principles and the system issues related to designing a successful ATV system through an overview of the CCDC HDTV system.

15.1.1 What is ATV?

An appreciation of the important features of ATV, and the amount of information it contains, can be obtained by a comparison with our conventional television system. The current color television system in the U.S. is an interlaced 525 scan line, 60 fields/sec or 30 frames/sec, 4:3 aspect ratio (width to height) system. The perceptual effects of interlace result in a spatial resolution of about 340 lines with about 420 resolvable elements per line. This is rather poor resolution which is especially evident on large screen displays. The interlaced scanning format can result in a number of degrading artifacts such as interline flicker and improper motion rendition. Also, the chrominance resolution is severely restricted, and the system is highly susceptible to transmission impairments such as multipath and interference.

An ATV system will process a multitude of different video formats. Whereas the current television system converts the various video sources to a single format for processing and display, an ATV system will process and display different source formats so as to exploit their individual characteristics. Many of the possible features of ATV are exemplified by HDTV. In the CCDC HDTV system, the baseline video format is a progressively scanned, 720×1280 square pixel, 60 frames/sec video signal, providing six times the spatial resolution and improved motion rendition over the current system. This increased spatio-temporal resolution significantly enhances the feeling of realism for the viewer. Also enhancing the realism is the 16:9 aspect ratio of the video signal which better emulates the greater horizontal than vertical human field of view. The improved chrominance reproduction and high-fidelity (CD-quality) digital audio add to the sharpness and the crispness of the entire experience. Furthermore, the displayed ATV video signal will not be plagued by transmission impairments such as multipath and random noise. Through the choice of progressive scanning and square pixels, the HDTV system is also easily interoperable with computer systems, facilitating the merging of ATV and computer-related tasks and services.

15.1.2 The technological challenge

All of these improvements lead to a great increase in the amount of information which must be either transmitted or stored; this produces a need for compression in many situations. Let us briefly examine the history of television and HDTV.

When television was first invented, spectrum was readily available and signal processing was very expensive. Therefore, the video signal was simply scanned and used to modulate a radio frequency carrier. Little signal processing or compression was used, resulting in a very inefficient use of the spectrum. The advent of HDTV, with its enormous amount of information to be delivered, coupled with a scarcity of available bandwidth, demanded a much more efficient method of delivering the HDTV signals to the home. In 1990 the Federal Communication Commission (FCC) ruled that, in order to apportion the limited RF spectrum among the services that may need it, without realigning the current services, any HDTV system used for terrestrial broadcast in the U.S. must use the same 6 MHz channel bandwidth as the current color television system. Comparing the raw data rate for the HDTV video signal described previously, about 1.3 Gb/s, with the transmission capacity of today's state of the art digital transmission systems, about 20 Mb/s within a terrestrial 6 MHz channel, it is evident that a large amount of compression is required. The transmission channel constraints result in a corresponding bit rate of approximately .35 bit/pixel, or a required compression ratio of about 70:1. This extremely high compression mandates the use of novel and sophisticated digital signal processing concepts to compress the video while retaining the video's original high resolution and excellent motion rendition. The proposed solution is the creation of a "digital" HDTV system.

An ATV system consists of two primary sections, a compression portion to compress the data that must be transmitted or stored (*source coding*), and a communication portion to efficiently deliver or store the compressed information (*channel coding*). The current television system in the U.S. uses analog processing for both portions. Digital ATV systems, including the proposed digital HDTV system, employ digital signal processing to compress the data and digital modulation for transmission or storage.

What are the benefits of all-digital processing? The combination of digital source coding and digital channel coding affords tremendous flexibility for both the representation and communication of the video signal. Applying sophisticated digital signal processing concepts for compression of the video signal enables the creation of an efficient and compact signal representation, so that the high-quality video can be transmitted within a much smaller bandwidth. Digital channel coding delivers or stores this digital information efficiently and robustly, even under severe channel conditions. The application of powerful Error Correction Coding (ECC) concepts ensures that, under predefined channel conditions, the compressed signal can be recovered with a arbitrarily small probability of error. The assumption that the encoder will know exactly what the decoder will receive facilitates the use of highly aggressive forms of video compression. The digital nature of the signal allows easy integration with the growing digital communication networks, including simple multiplexing of different signals and services. Also, computers can easily access and process the video signal.

System issues. There are a number of system issues that may directly affect the design of an ATV video compression algorithm. The most important issues include whether the system is to be employed for: (1) broadcast environment or point-to-point communication, (2) real-time or non-real-time application, and (3) fixed or variable bit rate representation. In this chapter, we will focus primarily on an ATV system to be employed in a broadcast environment, requiring real-time processing and coupling with a constant bit rate channel.

In a broadcast environment, as in our current television system, there will be few encoders and many decoders. The encoders may be expensive, but the decoders should be realizable at low cost. Therefore, most of the complexity should be localized at the encoder rather than at the decoder. Within a broadcast environment, consumers randomly turn on their receivers and change the channel, without the encoder having any knowledge of these changes. This necessitates fast receiver initialization and channel acquisition. The real-time application requires that the encoder/decoder time delay is significantly small so as not to hinder any "live" applications. Coupling a possibly variable bit rate video encoder to a constant bit rate channel requires a buffering mechanism. To achieve the necessary buffering without using large amounts of memory and while still attaining continuous high-quality video, a sophisticated buffer control algorithm is required.

A number of additional ATV features and issues will be examined in Section 15.6, following our discussion of the principles of video compression. These issues include interoperability, extensibility, scope of services and features, and bit stream integrity. Implementation considerations such as computation and memory requirements, delay, algorithm complexity, and amenability to parallel processing are also important, although to a lesser extent, as VLSI continues to increase in performance and drop in price.

Much research has been performed during the past few years toward the creation of international standards for compression of digital imagery and digital video. The primary proposed standards[1] are illustrated in Table 15.1. JPEG performs still-image compression at a number of bit rates depending on the particular application. CCITT's Recommendation H.261 (also referred to as "p×64") operates at p = 1 to 32 multiples of the baseline ISDN data rate. MPEG-1 achieves VHS quality video and audio within its bit rate, while MPEG-2 addresses the compression of higher resolution video. Finally, HDTV is aimed at even higher quality video at bit rates of approximately 20 Mb/s. As a result of the evolution of image and video compression, the various proposed standards have similar exterior structures and have converged to some general principles and techniques for compression, but the overall design of the video compression algorithm is driven by the high-level issues and bit rates of the specific applications.

[1] JPEG is the Joint Photographic Experts Group. MPEG is the Moving Pictures Expert Group. CCITT is the International Telegraph and Telephone Consultative Committee.

Standard	Application	Bit rate
JPEG	Continuous-tone still-image compression	Variable
H.261	Visual telephony and video teleconferencing	p$\times$64 kb/s
MPEG-1	Full motion video on digital storage media	1.5 Mb/s
MPEG-2	Higher resolution video than MPEG-1	$\geq$ 2 Mb/s
HDTV	Terrestrial broadcast of HDTV	20 Mb/s

Table 15.1: Proposed video and image compression standards.

15.1.3 Chapter Overview

Video compression has many applications. In this chapter, we will examine it in the context of designing a high-quality, low bit rate ATV system. Along the way, specific examples illustrating its application to terrestrial (over-the-air) broadcast of digital HDTV will be considered. We will discuss the fundamental source coding principles as applied to video and the system issues that arise for ATV applications. Section 2 presents an overview of video compression. The principles of video compression are examined, and a general framework for designing any digital compression algorithm is identified and discussed. This framework is composed of three distinct, though interrelated, operations: *Representation* of the signal, *Quantization*, and *Codeword Assignment*. Section 3 discusses efficient representations for the video signal, and investigates their applicability to processing along the temporal, spatial, and color space dimensions of the video. Section 4 examines the quantization of the important parameters of the representation. Codeword assignment of the quantized parameters is studied in Section 5, as well as the buffer control issues which may be important if entropy coding is used. Section 6 identifies important ATV features and issues, such as system interoperability and extensibility, scope of services and features, and bit stream integrity. The video compression principles and system issues are brought together in Section 7 where a proposed digital HDTV system is presented and its associated video compression algorithm is discussed. Section 8 provides some conclusions on the application of video compression to ATV.

15.2 Overview of Video Compression

One goal of source coding or compression is to reduce the *redundancy* existing in the source in order to transmit or store the information at a lower bit rate. The redundancy in a video signal is evident in consecutive frames of video, which are quite alike. Also within any single frame there are large regions (objects or background areas) that have similar characteristics. This is redundant information. It would be wasteful of channel bandwidth to transmit the same information repeatedly. Another significant attribute of video compression that is not applicable to all source coding schemes is the realization of what is *perceptually relevant* and what is not. Since the criterion for quality of the video signal is the human visual system, rather than any analytical metric such as mean square error, what the human visual system finds relevant or irrelevant is extremely important. For example, human spatial acuity is much less for randomly moving imagery than for still imagery. This tradeoff of spatial and temporal resolution can be used to great advantage in video compression systems. Therefore, the goal of source coding as applied to ATV video is to reduce the redundancy and irrelevancy, or equivalently, to transmit only the essential and relevant information.

Before becoming immersed in the details of video compression, let us examine the impact that classical information theory and rate-distortion theory may have upon our work. The birth of information theory began with Claude Shannon's celebrated papers in 1948. Shannon defined the information content of a source, *entropy*, and showed that with a bit rate equal to or greater than its entropy, and with a sufficiently long time delay, the source could be coded with zero error. This theory provides an elegant framework for the communication of digital data streams, but it lacks an adequate structure for the communication of high-quality video at efficient bit rates. This is for two primary reasons [6]. First, the theory is nonconstructive, giving bounds on rate-distortion performance without suggesting methods for achieving these bounds. Second, the classical theory was derived with a simplistic model of the source signal, and without a comprehensive (and to this day incomplete and not understood) psychophysical model of the video signal and its interaction with the human visual system (human perception). Yet, classical source coding theory is helpful in that it exemplifies the qualitative benefits of such delayed coding or block coding techniques as transform coding or vector quantization. If an appropriate perceptual model of video can be created, it can be applied to a rate-distortion formulation of the video, theoretically resulting in improved perceptual/bit rate efficiency.

Our goal for source coding is the perceptually transparent encoding of ATV or, if this is impossible to achieve, to minimize the perceived distortion. We begin by discussing how to initially represent the continuous, time-and-space-varying video signal. The different forms of redundancy and irrelevancy of a video signal are briefly examined, and a general framework that may be applied for their reduction will be discussed.

15.2.1 Representing the video signal

A video signal is a continuous function of time, space, and wavelength, whose amplitude must be discretized for digital processing and transmission. This intrinsically involves sampling and quantization of each of the video dimensions. Temporal sampling is performed when creating the individual frames or fields of video. Spatial sampling of these frames or fields results in image samples, often referred to as picture elements or *pixels*. The temporal and spatial sampling structures may be independent, or they may be coupled together as in interlaced scanning. To represent color, video is usually modeled as the additive combination of three primary colors: red, green, and blue. Each image sample is composed of three color components with a finite number of (often 8) bits of accuracy each.

Perhaps the most controversial issue today in terms of representing a video signal is the specific choice of the spatio-temporal sampling structure to be used. When television was invented, engineers were faced with the problem of achieving a high display rate to avoid flicker. Frame memories were unheard of in those days, and the options were limited to trading off the limited bandwidth between the spatial and temporal resolution. In an attempt to achieve the required temporal display frequency without sacrificing the spatial resolution, *interlaced scanning* was developed. In interlaced scanning, the video is split into even and odd fields, composed of even and odd scan lines respectively. The odd field is acquired, transmitted, and displayed first, and then the even field. One may argue that this results in twice the spatial resolution for the same temporal resolution, or twice the temporal resolution for the same spatial resolution. However, interlaced scanning also results in a number of degrading artifacts. These include improper motion rendition, interline flicker, line crawl, and vertical aliasing. Also, interlaced scanning complicates video processing and general interoperability with computers.

Today the source video rate may be totally decoupled from the display rate. There is no need to use interlaced scanning to acquire the original video or to display it. A simpler approach is *progressive scanning*. In progressive scanning, consecutive scan lines within each frame are read sequentially. An entire frame of video is sampled at one time, instead of splitting the video signal into even and odd fields through interlacing. By sampling an entire frame at the same time, progressive scanning eliminates the interlacing artifacts. Progressive scanning with square pixels lends itself easily to interoperability with computers, which use the same format. Similarly, progressive scanning and square pixels are extremely useful for format conversion, computer graphics, and general video processing, since they allow processing to be performed without any added complexity from the structure of the sample points. In the remainder of this chapter, we will assume a progressively scanned video sequence for processing. An excellent discussion of the sampling and reconstruction of video with different spatiotemporal sampling structures is contained in [7].

15.2.2 Redundancy and Irrelevancy

The goal of video compression is to reduce the redundancy and the irrelevancy inherent in the video signal. The sources of redundancy include:

- Temporal: Most frames are highly correlated with their neighbors.
- Spatial: Nearby pixels are correlated with each other.
- Color space: RGB components are correlated among themselves.

A high-performance video compression system identifies and exploits each of the redundancies within the video signal.

The redundancies are easy to pinpoint and exploit for compression. However, the question of what is relevant or what we can see is much harder to quantify. The human visual system (HVS) is a complex biological process that does not lend itself easily to analytical modeling. The various forms of relevancy can be decomposed into areas based on our temporal perception, spatial perception, and color perception. Also, the spatial and temporal masking phenomena of the HVS are important because they can be used to direct our attention away from a particular distortion within the video, thereby hiding or masking it. The important perceptual elements include sensitivity to motion, spatial frequency, color, and brightness. A discussion of the important features of human visual perception can be found in [1] and in Chapter 5. The goal of our work is to use some basic spatiotemporal models of the HVS to minimize the perceived distortion in the reconstructed video signal. We will view the irrelevancy as a form of *perceptual redundancy*, where an element is represented with more resolution than is perceptually required.

Before discussing different approaches toward reducing redundancy and irrelevancy within a video signal, it is worthwhile to briefly examine the tradeoffs between resolution of a video signal and its potential compression. Subsampling of a video signal is sometimes used as a first step in a compression algorithm. However, reducing the spatial resolution or frame rate by a factor of two, for example, will not necessarily halve the required bit rate to achieve a given reconstructed video quality. Because there is reduced correlation among image samples in space and time, less compression is possible. Similarly, doubling the resolution will not double the required bit rate; there will be more correlation between image samples and more compression will be possible.

15.2.3 Principles of Video Compression

Any digital compression system can be expressed as the combination of three distinct, though interrelated, operations: *Representation, Quantization,* and *Codeword Assignment.* These operations are depicted in a general digital compression system shown in Figure 15.1. In the first stage, the signal is expressed in a more efficient representation which facilitates the process of compression. The representation may contain more pieces of information to describe the signal than the signal itself, but most of the important information will be concentrated in only a small fraction of this description. In an efficient representation, only this small fraction of the data must be transmitted for an appropriate reconstruction of the signal. The primary approaches taken toward creating an efficient and compact representation may be classified broadly into the categories of *predictive processing, transform/subband filtering,* and *model-based processing.* These approaches may be intermixed and applied along the temporal, spatial, and color space dimensions of the video signal. The second operation, quantization, performs the discretization of the representation information in order to enable transmission over the digital channel. The information may be quantized one parameter at a time, *scalar quantization*, or a group or vector of parameters may be quantized jointly, *vector quantization.* The third operation takes the quantized parameters and assigns to each an appropriate codeword for transmission. *Fixed-length codeword assignment* may be performed, or more complex *entropy coding* may be applied to exploit the statistical redundancy of the quantized parameters and reduce the average bit rate.

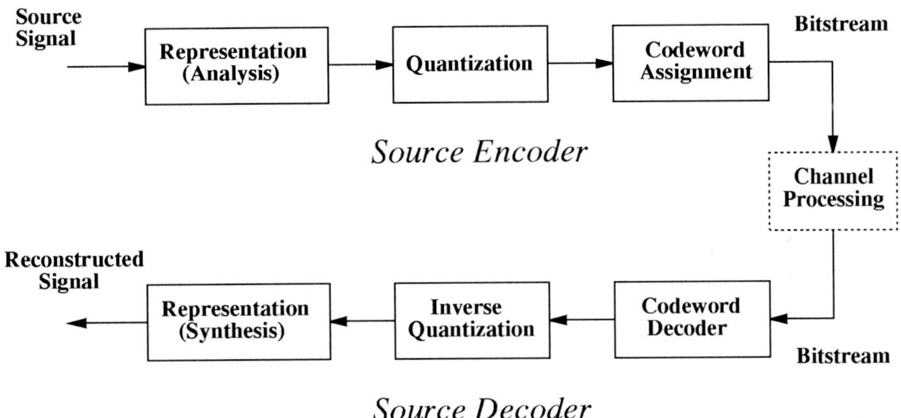

Figure 15.1: An overview of a digital compression system showing the three basic operations at both the encoder and the decoder. The operations at the decoder are the inverse of those at the encoder.

Each operation may be tuned to reduce both the statistical redundancy and the perceptual irrelevancy inherent in the video signal. The first and third operations may be performed in a lossless manner, with any loss of information being localized solely within the quantization operation. By isolating the potential loss of information in a single operation, a much simpler design process and fine tuning of the system are possible. The three operations are so closely related that highly optimizing one may reduce the potential gain from the others. Optimizing each of the three independently will not achieve the same performance as a joint optimization of all three. To achieve high performance, these operations must be designed in a manner to enhance the effectiveness of each while attenuating their weaknesses.

15.3 Representations

When a representation is developed for a video compression application, the underlying goal is to reduce the redundancy by compacting the maximum amount of perceptually important information possible into a small fraction of the parameters. This small fraction may be further processed for transmission while the remaining elements may be simply discarded. The chosen representation, as well as the amount of required compression, are dependent upon the specific application and its implementation constraints. The three categories of approaches taken toward reducing redundancy within a video signal are: predictive processing, transform/subband filtering schemes, and model-based processing. These approaches employ progressively greater complexity and sophistication to achieve increased compression of the relevant data. This section begins by discussing the various approaches for creating a compact representation, and will identify the important information that each attempts to represent efficiently. The possible application of each approach along temporal, spatial, and color space dimensions is investigated. Important issues that arise when employing each approach in a broadcast environment will be identified and briefly discussed.

15.3.1 General approaches

The simplest digital technique for representing a signal is pulse code modulation (PCM). PCM can reduce an individual sample's redundancy in the sense that it does not represent the sample with more fidelity than is necessary. However, since PCM does not exploit any inter-sample redundancy that may exist, it cannot achieve the same compression as the following approaches.

Predictive methods. Typical video is usually modeled as stationary in its characteristics over short segments along the spatial and temporal dimensions. Predictive methods attempt to exploit this predictability of the video signal. A prediction of the current value to be encoded is formed from the previously encoded values. The error in the prediction, or *residual*, is the new information which is then transmitted. Forming a prediction and then encoding the residual is the main idea behind all differential pulse code modulation (DPCM) systems. Their coding efficiency is primarily determined by the accuracy of their predictions. The space-and-time varying characteristics of video can be exploited through adaptive DPCM (ADPCM), where the process of forming a prediction is adaptive to the local video characteristics. The structure and important features of DPCM, and those of other approaches, are illustrated in Figure 15.2. Because of the recursive nature of predictive coding schemes, it is essential that the decoder accurately track the encoder. If they become unsynchronized, the prediction at the decoder will not match the prediction at the encoder and the whole process will fail.

Transform/subband filtering methods. Transform/subband filtering methods take a different approach to increase the coding efficiency. Transform schemes are based on the idea that an image can be linearly transformed into another domain where most of the energy (and information) is concentrated in a small fraction of the transform coefficients. Coding and transmission of these few energetic coefficients may then result in the reconstruction of a high-quality image with minimal distortion. In order to exploit the varying spatial characteristics of an image, it is typically partitioned into 8×8 or 16×16 blocks which are independently transformed and adaptively processed. Subband filtering schemes process an image by filtering it into separate frequency bands or subbands. A one-dimensional four-band analysis/synthesis subband filtering scheme is shown in Figure 15.2. The power of subband filtering resides in the fact that, once the signal is divided into many subbands, each subband can be adaptively encoded in order to exploit its specific characteristics.

Both transform and subband filtering schemes can be viewed as processes that decompose an image or video into its frequency or subband components. Both

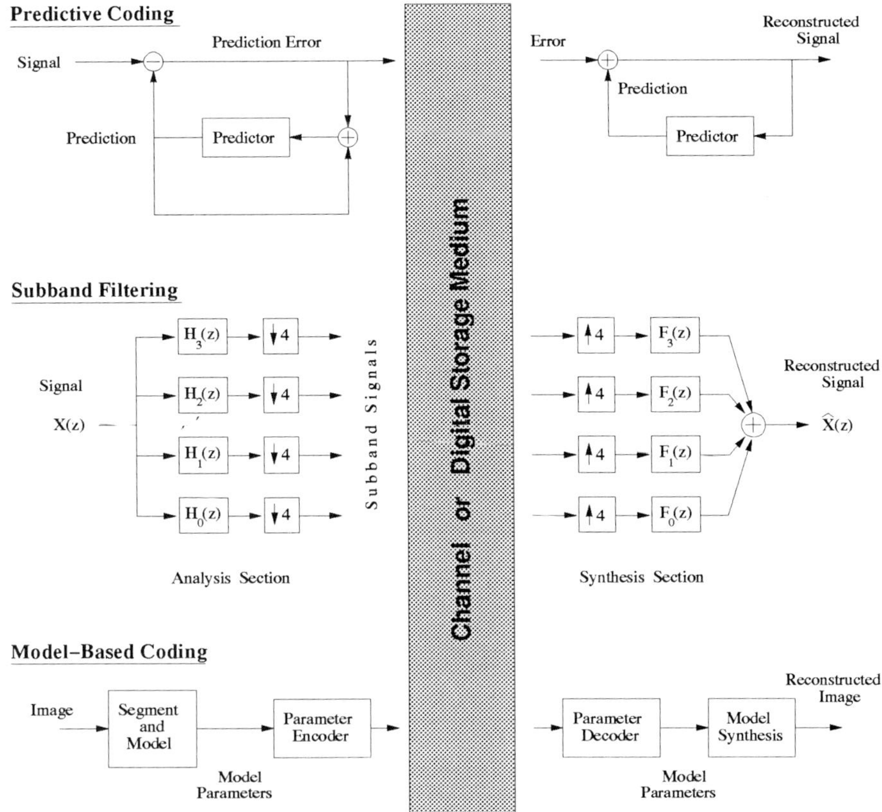

Figure 15.2: Sample schemes of the three general approaches for representing the signal at both the encoder and the decoder. The quantization and codeword assignment operations are not depicted in order to simplify the figures.

representations are inherently the same, but have been traditionally classified as separate entities due to the different methods by which each is computed. The application of transform/subband filtering schemes to imagery and video results in energy compaction because of the nonuniform spectra of those signals. The transform/subband coefficients can be adaptively encoded in order to exploit the specific perceptual and statistical characteristics of each coefficient.

Model-based methods. Model-based methods have the potential to achieve substantially greater compression than the previously discussed methods. At the core of these approaches are models of different features of video, such as contours, textures, 3-D objects in motion, human faces, etc. Based on these models, a compression algorithm can decompose a video into these features, parameterize the features, and use these few parameters to synthesize the features at the decoder. The efficiency of model-based coding is intrinsically dependent upon how accurately the model matches the video to be encoded. Model-based coding can be used alone or, for more robust performance, it can be applied in conjunction with other compression techniques. Currently, model-based coding has yet to reach the maturity and the quality performance required for an ATV system. Current research in model-based coding is described in Chapters 10 and 11, and therefore will not be discussed further here.

Creating a framework for the compression of high-quality ATV video is an art form, requiring delicate merging and perceptual adaptation of the powerful approaches described above. These approaches will now be discussed in the context of temporal, spatial, and color space processing required for a high-quality digital ATV system.

15.3.2 Temporal Processing

A video sequence is a series of still images shown in rapid succession to give the impression of continuous motion. Even though each of the frames is distinct, the high frame rate necessary to achieve proper motion rendition usually results in much temporal redundancy among the adjacent frames. Temporal processing attempts to exploit this redundancy.

Is temporal processing necessary? That is the first question that arises. Is spatial domain processing by itself sufficient? Processing a frame by itself (individually), without taking into account the temporal dimension of the video, is called *intraframe processing.* Processing a frame while exploiting the temporal dimension of the video is called *interframe processing.* Intraframe processing has a number of benefits, including (1) it does not require the extra complexity of temporal domain processing and (2) it obviates the need for extra frame stores at the encoder and the decoder. Intraframe processing requires only one or two frame stores (one for decoding and one for display). The tradeoff is, of course, that through intraframe encoding only the spatial domain redundancies can be exploited, excluding the temporal domain redundancies.

Temporal processing can result in large coding gains. Even simple temporal processing can dramatically reduce the required bit rate or the complexity of the spatial processing in order to achieve the same video quality. Simply stated, using temporal processing can result in a substantially higher performance than not using it. With today's rapidly declining memory and computation costs, temporal processing is an essential ingredient of a high-performance digital ATV system.

Predictive processing. Two simple methods to reduce redundancy along the temporal dimension include using the previously encoded frame as the prediction of the current frame, or encoding only those regions within the current frame that have changed from the previous frame. Both approaches reduce the temporal redundancy over the stationary regions within the video, but they do not perform well whenever there is any motion, which in some cases may be over most or all of the video frame. Much improved performance can be achieved by compensating for the inherent motion among neighboring frames of video. The processing of individual frames while compensating for the presence of motion is called *motion-compensated* (MC) processing. The process of estimating the motion is known as *motion estimation* (ME).

We will briefly discuss motion estimation as applied to ATV systems, and elaborate on two methods of motion compensation that are specifically applicable to ATV. The first method is motion-compensated prediction, or *MC-prediction*, where motion compensation is used to produce an accurate prediction of the current frame to be encoded. The second is motion-compensated interpolation, or *MC-interpolation*, where frames may be discarded at the transmitter and, through the use of motion compensation, these skipped frames may be accurately interpolated at the receiver.

Motion Estimation. In ME, the same imagery is assumed to appear in consecutive video frames, although possibly at different spatial locations. The motion may be global, or local within the frame. To optimize ME performance, an estimate of the motion is computed for each local region within a frame. The most common model for the local motion is simple translational motion. This model is highly restrictive, and cannot represent the large number of possible motions, such as rotations, scale changes, and other complex motions. Nevertheless, by assuming these motions only locally and by identifying and processing those regions where the model fails, excellent performance can be achieved.

One approach for performing ME is based on *block matching* methods. In block matching, the current frame is partitioned into rectangular regions or blocks, and a search is performed for the displacement which produces the "best match" among possible blocks in an adjacent frame. Hence the term "block matching". ME for consumer ATV systems is today nearly universally based on *block matching* methods. This is because block matching achieves high performance while also exhibiting a simple, periodic structure which allows straightforward VLSI implementation.

A number of important issues arise when designing a block matching scheme for an ATV system. The displacement or *motion vector* may be estimated by maximizing the similarity (e.g., normalized correlation) between blocks, or by minimizing the dissimilarity (e.g., mean square error (MSE) or mean absolute error (MAE)) between blocks. Of these different decision metrics, MAE is often chosen because it achieves the same performance as the others, without requiring any multiplications in the calculation. Choosing the size of the block is a tradeoff between the benefits of a higher resolution motion field (improved prediction/interpolation) and the amount of information required to describe it. Similarly, choosing the *search range* in the reference frame to search for a match trades off improved ability to track fast motion, such as sporting events, for a greater number of candidate matches that must be examined. Subpixel accuracy of the motion field may also be used to enhance performance. However, it requires spatial interpolation to determine the noninteger spaced image samples.

The motion vector for the best match may be found in a brute force yet straightforward manner by examining every possible candidate within the search area. This method is called *full search* or *exhaustive search* and it ensures the best match within the reference area. As an alternative to the large computational requirements of the exhaustive search method, adaptive methods may be applied which efficiently search for a minimum by evaluating a reduced number of possible displacements. Hierarchical or multigrid approaches [8] may also be employed to reduce the computational requirements of ME. In these approaches, a low-resolution version of the video is used to produce an initial coarse estimate of the motion. This estimate is subsequently refined using higher-resolution versions of the video.

Motion-Compensated Prediction. The temporal redundancy inherent in a video signal can be exploited with MC-prediction. Through a block matching ME algorithm, each block of the current frame can be predicted based upon a translation of a block from a reference frame. In causal or forward MC-prediction, the reference is a past encoded frame (typically, the immediate predecessor), but for noncausal and the subclass of bidirectional MC-prediction, the reference frames can be both preceding and following frames.

The process of causal or forward MC-prediction is illustrated in Figure 15.3. The frame to be encoded is partitioned into blocks and each block is predicted by displacing a block from the reference frame. Since the prediction is usually imperfect, the MC-prediction error or MC-residual is further processed using spatial redundancy reduction techniques before undergoing transmission or storage. The prediction and differential coding nature of causal MC-prediction allow it to be thought of as a "smart" form of DPCM along the temporal dimension. Typically, causal prediction algorithms of this form are highly effective for video compression.

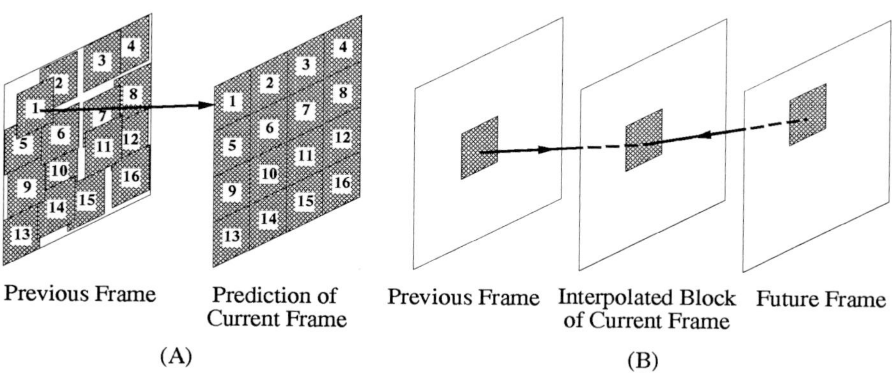

Figure 15.3: Causal MC-prediction is illustrated (a) where a previous frame is used to predict the current frame. MC-interpolation is used (b) to form a block in the current frame from a block in each of the previous and future frames.

MC-prediction is a nonlinear process which may inject high-frequency artifacts within the feedback loop. To prevent the buildup of high-frequency errors and quantization noise, a spatial lowpass filter is sometimes inserted within the feedback loop. This filter should be carefully designed to reduce the unwanted components while not unnecessarily reducing the sharpness of the reconstructed video. The filter may be adaptively turned on or off over each local region or have its characteristics altered in other ways to fit different situations. For example, when a local region undergoes complex motion, the LPF may be activated, while for simple panning and zooming for which the MC-prediction performs extremely well, the filter may not be used [9].

Motion-Compensated Interpolation. In MC-interpolation, frames of video can be periodically discarded at the encoder, and these discarded frames can be reconstructed at the receiver from the remaining encoded frames. Simple temporal interpolation schemes perform well when the video is stationary, but fail in the presence of motion. Performing the interpolation via replication of frames results in jerkiness in the areas where there is motion (aliasing artifacts because of the temporal undersampling). Employing linear interpolation along the temporal dimension is also inappropriate because the averaging of different frames results in blurred motion and may produce false contour artifacts from the motion of objects within the video.

MC-processing can account for the motion within the video and prevents these artifacts. By estimating the motion, each local region of the discarded frame can be interpolated using the corresponding local regions in the preceding and subsequent encoded frames. This is shown in Figure 15.3 where a block of a discarded frame is modeled as the average of translated blocks from previous and subsequent encoded frames. MC-interpolation can be thought of as a form of bidirectional MC-prediction, where a prediction is made given a reference on either side. Unlike MC-prediction, where the ME process attempts to find the "best predictor" for each block, in MC-interpolation the goal is a motion field that approximates the "true" motion. A truer motion field has been observed to give improved perceptual results, especially when the interpolation error is not coded [8, 10]. In general, the interpolation error may or may not be encoded. For a high-performance ATV system, however, steps must be taken to ensure that, if the interpolation fails, no degrading artifacts will result.

MC-interpolation can achieve a significant amount of compression, but it reduces the temporal correlation between the remaining frames, making them more difficult to code. Information regarding how to interpolate the discarded frames as well as the interpolation error must be transmitted. Also an extra frame memory is required for the bidirectional processing of MC-interpolation as compared to the forward-only processing of causal MC-prediction. Currently, it is not evident that reconstructing a high-quality discarded frame from two coded frames will yield higher performance and be more efficient (in terms of computation/memory/bit rate) than MC-predictive coding of the discarded original frame.

Before performing any form of MC-processing, it is essential to consider the characteristics of the video signal. For example, at a scene change or with the appearance of new imagery, causal MC-prediction may produce an error that can be more difficult to code than the original frame. In this case the MC-prediction should be suppressed and the original image itself should be coded. Similarly, with MC-interpolation, there are instances when one of the two references may be inappropriate to use and the current frame should then be extrapolated from the other. Each local region where the MC-processing fails should be identified, and the MC-processing should be applied in an adaptive framework over each frame in order to achieve a high-quality ATV system.

Temporal Transform/Subband filtering. Applying these approaches along the temporal dimension is not a viable compression technique for digital ATV. Nonuniform motion results in the signal energy being distributed throughout the 3-D frequency domain. This inefficiency hinders the performance of these approaches. Also, a temporal transform/subband filtering scheme requires storage of a large number of frames. On the other hand, MC-processing performs exceedingly well using only 2 frame stores, and is therefore the method of choice.

System issues. MC-processing, in particular MC-prediction, offers one of the best performance/complexity/cost tradeoffs for temporal processing of a video signal. For now, we will assume that a form of MC-processing is applied along the temporal dimension. This leads to a combined system/video processing issue which must be addressed. The decoder prediction loop *must accurately track* the encoder loop. These prediction loops are highly susceptible to noise. Digital transmission or storage facilitates the use of predictive coding schemes, since they ensure that if the transmitted signal is received above a threshold level, it will not be corrupted by noise. With typical analog transmission schemes, the received data would be a function of the channel, and this may introduce noise into the feedback loop. This noise may be augmented by additional noise at each iteration, resulting in a failure of the prediction.

The issue of the decoder loop accurately tracking the encoder loop arises when considering receiver initialization and channel acquisition (when the receiver is turned on or the channel is changed), and when uncorrectable channel errors occur. With the DPCM style MC-prediction, an initial frame must be available at the decoder to start the prediction loop. Therefore, a mechanism must be built into the system so that if the decoder loses synchronization for any reason, it can rapidly reacquire tracking. Two possible approaches are (1) periodic intra-encoding of an entire frame and (2) application of leakage within the prediction loop. The first approach produces a periodic reinitialization of the temporal prediction at both the encoder and the decoder at predetermined times. The decoders can then reacquire tracking at any of the predetermined instances. In the second approach, the prediction is multiplied by a factor less than 1 (the leakage factor), such that, even if the prediction is perfect, a portion of the current frame will always leak through the prediction and be encoded for transmission. This leakage of the current frame can be built up at the receiver in order to acquire an accurate prediction. Notice that both of these approaches attempt to solve the problem by encoding the original, as opposed to the error signal. Therefore, these approaches may also be used to remedy the situations where the MC-processing fails and the original frame itself should be coded. To account for local failures of the MC-processing, the chosen approach should be applied in a spatially adaptive manner for the processing of each frame.

15.3.3 Spatial Processing

Applying MC-processing, or any other form of temporal processing, reduces the temporal redundancy of the video signal, but spatial redundancy still exists within the MC-residual. This is especially true if no MC-processing is performed and the original frame itself is to be coded. For simplicity, and to remind us that we are not processing a typical image, the term *residual* will be used to represent the frame of data that is to be spatially processed, irrespective of whether or not MC-processing has been applied. This subsection will discuss methods to reduce the spatial redundancy of the residual.

Predictive processing. Predictive techniques may be used to predict each residual sample. Prediction may be based on previously encoded samples along the current and previous scanning lines. DPCM and ADPCM inherently operate at integer bit rates; they do not operate at fractional bit rates (under 1 bit/sample), unless some form of block coding of the error is utilized. This is because the error in predicting each residual sample must be transmitted, and each error requires a minimum of one bit to be represented. Predictive processing is not a viable approach to encode the residual. A more efficient representation for the residual is produced through transform/subband filtering.

Transform/Subband Filtering. The strong spatial correlation within an image or residual manifests itself as a concentration of energy within a small fraction of the transform/subband coefficients. By encoding and transmitting only these coefficients, high-quality reconstructed video can be achieved. An image is a large two-dimensional signal, so computing the 2-D transform or performing a 2-D convolution upon the image can be computationally very expensive. Therefore, the only transform/subband filtering schemes that are usually considered are 1-D schemes that may be applied separably along the rows and the columns of the image. The existence and application of fast computational algorithms is also extremely important. Even though transform and subband filtering schemes are inherently equivalent, we will adhere to convention and begin our discussion with what are referred to as transform domain schemes, and then progress to subband filtering schemes. The Discrete Fourier Transform (DFT) is usually the first transform to come to mind, however, it has some disadvantages including complex coefficients and an inherent inefficiency in its energy compaction [1].

Discrete Cosine Transform. The DCT is the real analog of the DFT. The DCT has real coefficients, a fast computational implementation, and it eliminates the artificial discontinuity inherent in computing the DFT, thereby yielding improved energy compaction. The DCT, under certain conditions, approaches the optimum performance of any linear transform. The Karhunen-Loève (KL) transform is the optimum of all linear transforms in terms of energy compaction and decorrelation of the transform coefficients. However, unlike the KL transform, the DCT has predetermined (signal-independent) basis functions, thereby achieving high performance without requiring further computations.

The near-optimal energy compaction and decorrelation properties of the DCT coupled with its fast computational implementation have resulted in its extensive study and application to image compression. The DCT of a 2-D image is usually computed by applying the 1-D DCT separably to the rows and the columns of the image. The chosen size of the DCT may be the entire frame, but much improved performance can be achieved by segmenting the frame into numerous smaller regions, each of which is independently and adaptively processed. Segmenting the image is one of the most important ingredients of a high-performance video encoder. By computing the DCT of the entire frame, the whole frame is treated equally. For typical video, however, the characteristics may vary considerably over the spatial extent of each frame and from frame to frame. To exploit the non-stationary nature of the video signal, each frame is typically partitioned into 8×8 or 16×16 blocks which are *independently transformed* and *adaptively processed* to exploit their individual characteristics. The application of the DCT in this manner is often referred to as the *Block DCT*. Partitioning a frame into small blocks before applying the transform affords other benefits, including reduced computational and memory requirements.

The Block DCT has been observed to achieve the highest performance of any block transform and is the most frequently used. Since adjacent blocks are processed independently, the coding distortion may manifest itself as discontinuities along the boundaries between adjacent blocks of the decoded image. These discontinuities form a structured *blocking artifact* that the human visual system easily detects and finds unpleasant. Sophisticated adaptive processing is required to ensure that this artifact does not occur. Other approaches to reduce blocking artifacts include Lapped Transforms [11], and subband filtering and multiscale schemes.

Subband filtering. A typical subband filtering scheme was illustrated in Figure 15.2. The encoder or analysis stage of the filterbank decomposes the input signal into separate frequency bands or subbands. Since each subband contains only 1/Mth the bandwidth of the original signal, each subband can be decimated by a factor of M and still satisfy Nyquist's condition. This produces a *critically sampled* representation, with the same number of coefficients as original image samples. These subband filtering schemes are referred to as *uniform band filterbanks* because each subband has identical bandwidth. The similarity between transform and subband filtering schemes can be easily seen here. A block transform groups an image first in terms of spatial location and then by frequency content, while subband filtering groups an image first in terms of frequency content and then by spatial location. The Block DCT, for example, can be computed either through a transform type operation or through a subband filtering operation. The subband filter impulse responses are typically larger than the decimation factor, while in a typical transform they would be equal. A block transform would typically exhibit blocking artifacts as the primary perceptual degradation, while subband filtering schemes would exhibit ripple artifacts as a result of their sharp filter cutoffs. Each transform/subband coefficient may be adaptively encoded to exploit its specific perceptual and statistical characteristics.

The conventional approach used to exploit the locally-stationary nature of video (used in JPEG, H.261, MPEG, and all the current digital HDTV proposals within the U.S.) is to partition each frame into blocks of pixels and adaptively encode each block based on its local characteristics. In this manner, an artificial block-like structure is imposed on the video. This allows simple, spatially adaptive processing, but only within the rigid, periodic structure. In view of the inherently nonstationary nature of a video signal, a more flexible analysis structure has the potential for better modeling of the video characteristics and achieving improved performance.

Multiscale, Multiresolution and Wavelet Transforms. Researchers of the human visual system have proposed that a nonuniform frequency decomposition may better match the characteristics of the HVS. In particular, they have theorized that humans may view the world at different scales or resolutions, thereby seeing detail at many scales. *Multiscale*, *Multiresolution*, and *Wavelet Transforms* are different names for conceptually similar representations. These schemes decompose a signal into basis functions that are of different lengths or scales. One of the earliest representations that utilized multiple-length basis functions was the Laplacian pyramid [12]. This scheme allowed for perfect reconstruction, but the basis functions were not orthogonal and the representation was overcomplete, with 4/3 as many coefficients as original image pixels. As a result, it was inefficient for coding. The newer schemes produce the same number of coefficients as original signal samples [13].

The multiscale or multiresolution schemes may be viewed as decompositions of the signal into subbands of equal bandwidth on a logarithmic scale. The Wavelet transform decomposes a signal into basis functions which are dilations and translations of a single prototype wavelet function. Instead of a constant spatial and frequency resolution for all basis functions, these schemes attempt to trade off one for the other. The basis functions with large regions of support enable precise frequency localization while those with small regions of support facilitate precise spatial localization. These schemes thereby achieve precise spatial localization required to efficiently represent the high-frequency features, such as edges or other transients, while also yielding precise frequency localization for the important low-frequency components within the video signal. Through these important attributes, the nonstationary nature of the video signal can be exploited without imposing any artificial and restrictive block-type structure on the video. The multiple-scale nature of these schemes may also enable a system to encode the perceptually important information first, and may facilitate simple scalability.

The Multiscale representation may be realized by recursively applying band-splitting quadrature mirror filters (QMF) to successive lowpass subbands. The

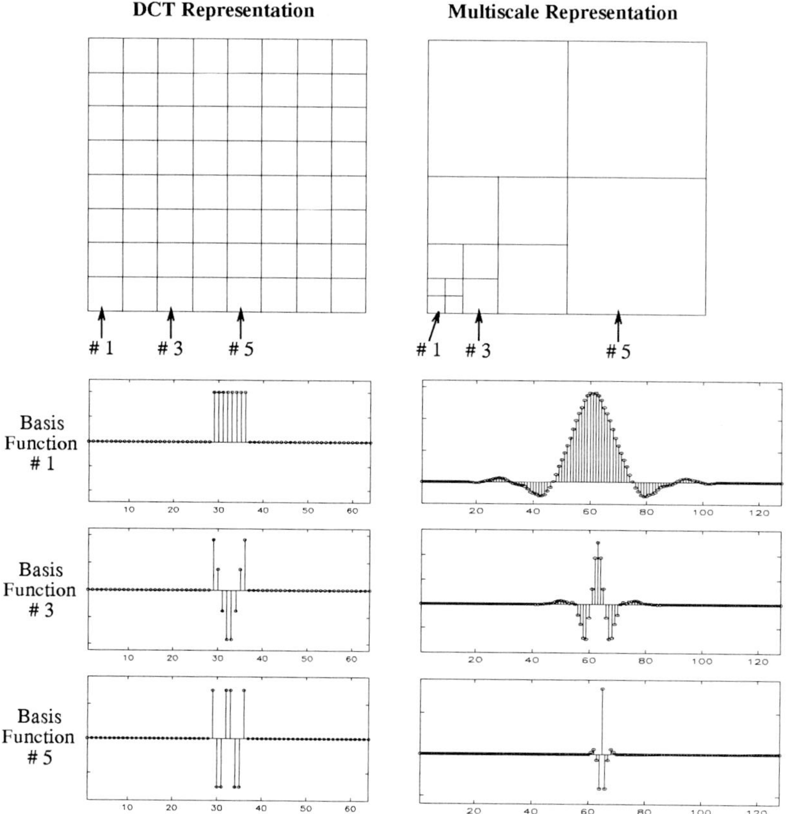

Figure 15.4: The Block DCT and Multiscale representations. The low-horizontal low-vertical spatial frequency components are located in the lower left corner of the representations. Three of the five 1-D Multiscale basis functions are displayed as are three of the eight 1-D Block DCT basis functions. For display purposes, the horizontal scale of the basis functions differ.

bandsplitting nature of the QMF filters produces an octave subband decomposition of the frequency domain. Correspondingly, the lengths of the basis functions are multiples of two of each other. To illustrate the differences between the 8-point Block DCT and the Multiscale scheme, Figure 15.4 shows the spatial frequency decomposition and some of the associated basis functions for each. Separable filtering is assumed and the 1-D basis functions are plotted.

When temporal processing fails. As discussed previously, there are many instances when the temporal processing may fail, either locally or globally. For a high-quality ATV system, it is very important that the system identify these instances and process them appropriately. Two possible approaches include spatially-adaptive inter/intra processing and spatially-adaptive leakage. Note that the latter is a more general version of the former. On the other hand, the former approach may be more convenient (less sophistication/computation) to perform. We will assume that the former is utilized. When examining the different spatial processing schemes in the context of spatially-adaptive inter/intra processing, a number of issues are evident. Since adjacent blocks are processed independently with the Block DCT, alternating inter/intra processing among adjacent blocks is easy to perform. However, for subband filtering schemes with impulse responses longer than the decimation factor, or Multiscale schemes, adjacent regions are processed together. These schemes require a more complex process for efficiently encoding adjacent regions (e.g., one region in an inter manner and an adjacent region in an intra manner). More sophisticated processing is required to implement spatially-adaptive inter/intra processing for any of these schemes as opposed to the Block DCT. A number of approaches for solving these problems can be found in [14].

15.3.4 Color Space Processing

Our discussion to this point has not taken into account the color nature of a video signal. We may view our discussion as a method to compress a monochrome video signal and simply apply the same processing steps to each of the three color components that comprise a color video signal. However, this would be very inefficient because the three color components, Red, Green, and Blue (RGB), are highly correlated with each other. More importantly, the human visual perception differs for the luminance (intensity) and chrominance characteristics of a video signal. To reduce the correlation among the RGB components and to enable the ATV system to exploit the differing perceptual sensitivity to the luminance and chrominance characteristics, a *color space conversion* is usually performed.

The goal is to convert the RGB color space to a domain where the differences in the HVS response can be exploited. Typically, this is accomplished through a linear transformation to the YIQ (NTSC) or YUV (SMPTE 240M colorimetry standard) color spaces. Y corresponds to the luminance (intensity or black and white picture) while I and Q or U and V correspond to the chrominance. The HVS has reduced perceptual sensitivity to the chrominance components and, with this representation, it can be easily exploited in the quantization operation. Similarly, the HVS has reduced spatial frequency response to the chrominance as compared to the luminance components. This characteristic can be exploited through a reduced sampling density for the chrominance components. For example, the chrominance may be decimated by a factor of 2 along both the horizontal and vertical dimensions, producing components that are one-quarter the spatial resolution of the luminance. However, for a high-performance ATV system, retaining the full chrominance resolution does not require much capacity and may be beneficial for some applications such as computer-generated graphics or encoding text containing saturated colors.

When performing ME on color video, the motion field may be computed from the luminance component only, and applied to both the luminance and chrominance components. This procedure eliminates the computationally expensive task of estimating the motion for each chrominance component by exploiting the significant correlation between movement among the different color planes. This method may fail in an isoluminance situation, when adjacent objects of similar luminance but differing chrominance move in different directions. Nevertheless, in general this algorithm performs extremely well.

By applying very simple processing, the differing human visual perception to, and the correlation among, the different color components can be exploited. A significant result is that a three-component color video signal can be coded with less than a 50 % increase in capacity over that required for a single-component (monochrome) video signal. Also, subjective tests have shown that more perceptually appealing video may be produced by coding a color video signal at a given rate than by coding a monochrome signal at the same rate.

15.4 Quantization

Through the processing discussed up to this point, an elegant representation in the form of the motion field, spatial frequency coefficients, and luminance/chrominance components has been created; however, no compression has been achieved. In fact, an expansion of data has resulted, since there are currently more pieces of information used to describe the video than before. However, most of the perceptually important information has been compressed into only a few of these "pieces of information", and this data can be selected and encoded for transmission. In this and the following section, we will examine methods to create an efficient digital bitstream representation for this information.

For ATV applications where a constant bit rate is required, such as broadcast HDTV, the goal of the video compression is not to minimize the bit rate for a given video quality, but to maximize the video quality at a given bit rate. Therefore, these applications require a wise distribution of the limited number of available bits. By exploiting the statistical and perceptual redundancy within the new representation, an appropriate *bit allocation* can yield significant improvements in performance. Quantization is performed to discretize the values, and through quantization and codeword assignment, the actual bit rate compression is achieved. The quantization process can be made the only lossy step in the compression algorithm. This is very important, as it simplifies the design process and facilitates fine tuning of the system. Quantization may be applied to elements individually (*scalar quantization*) or to a group or vector of elements simultaneously (*vector quantization*).

15.4.1 Scalar Quantization

In scalar quantization, each element may be quantized with a *uniform* (linear) or *nonuniform* (nonlinear) quantizer. The quantizer may also include a *dead zone* (enlarged interval around zero) to quantize or *core* to zero small, noise-like perturbations of the element value. The close relationship between quantization and codeword assignment suggests that separate optimization of each may not necessarily yield the optimum performance. On the other hand, joint optimization of quantization and codeword assignment is a highly nonlinear and complex process. However, experiments have shown that a linear quantizer with an appropriate stepsize individually chosen for each element to be quantized, followed by proper entropy coding, may yield close to optimum performance. This will be discussed in the context of quantizing the spatial frequency coefficients.

When quantizing transform/subband coefficients, the differing perceptual importance of the various coefficients can be exploited by "allocating the bits" to shape the quantization noise into the perceptually less important areas. This can be accomplished by varying the relative stepsizes of the quantizers for the different coefficients. The perceptually important coefficients may be quantized with a finer stepsize than the others. For example, low spatial frequency coefficients may be quantized finely, while the less important high-frequency coefficients may be quantized more coarsely. Similarly, luminance, which is the most visually important component, may be quantized more finely than chrominance. A simple method to achieve different stepsizes is to normalize or weight each coefficient based on its visual importance. All of the normalized coefficients may then be quantized in the same manner, such as rounding to the nearest integer (uniform quantization). Normalization or weighting effectively scales the quantizer from one coefficient to another.

In typical signal compression applications, only a few variables are usually quantized to zero. However, in video compression, most of the transform/subband coefficients are quantized to zero. There may be a few nonzero low-frequency coefficients and a sparse scattering of nonzero high-frequency coefficients, but the great majority of coefficients will have been quantized to zero. To exploit this phenomenon, the 2-D array of transform/subband coefficients may be reformatted and prioritized into a 1-D sequence through a zigzag, serpentine, or Peano-Hilbert scanning. This results in most of the important nonzero coefficients (in terms of energy and visual perception) being grouped together early in the sequence. They will be followed by long runs of coefficients that are quantized to zero. These zero-valued coefficients can be efficiently represented through runlength encoding. In runlength encoding, the number (run) of consecutive zero coefficients before a nonzero coefficient is encoded, followed by the nonzero coefficient value. The runlength and the coefficient value can be entropy coded, either separately or jointly. The scanning separates most of the zero and the nonzero coefficients into groups, thereby enhancing the efficiency of the runlength encoding process. Also, a special End Of Block (EOB) marker is used to signify when all of the remaining coefficients in the sequence are equal to zero. This approach is extremely efficient, yielding a significant degree of compression.

15.4.2 Vector Quantization

Vector quantization encodes a block of elements simultaneously and attempts to exploit the redundancy inherent to the block or vector of elements. VQ differs from typical block coding methods, such as transform/subband filtering, in that it involves quantization and is therefore inherently lossy. Unlike transform/subband filtering methods, which by themselves can only reduce the linear redundancies among the vector elements, VQ can exploit both the linear and the nonlinear dependencies which may exist. VQ may be applied to a block of pixels, a group of transform coefficients, or any other block of information. In VQ, for each input vector a "best match" is found among a number of possible matches in a codebook. An index to the appropriate entry in the codebook is transmitted to the decoder, where the matched vector is extracted from a duplicate codebook and used in the synthesis process, as illustrated in Figure 15.5. Compression is primarily achieved because there are a limited number of vectors within the codebook; hence, the associated index requires fewer bits to describe than the original vector. However, computational complexity and storage requirements increase exponentially with vector length; therefore, the application of VQ is usually restricted to short vector lengths and very high compression. An important feature of VQ that facilitates its use for broadcast ATV, such as HDTV, is that its computational complexity is localized at the encoder, while the decoder executes a simple codebook lookup operation.

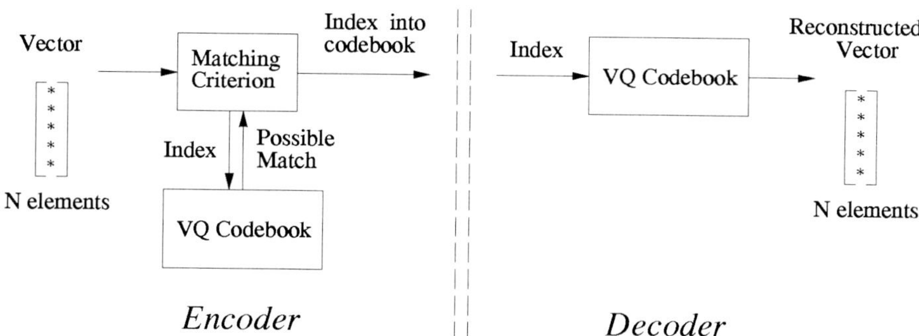

Figure 15.5: The process of vector quantization at both encoder and decoder.

As for any other statistical coding scheme, the performance of VQ depends upon the design and training of the codebook. VQ performs exceptionally well when the signal to be encoded is well-matched to the training data, but may perform badly when the match is poor. For example, applying VQ for encoding pixel values is difficult. Other schemes, such as transform/subband filtering, may be more robust with respect to the training than VQ for encoding a block of pixels. Also, transform/subband filtering schemes allow for simple adaptivity to local video characteristics, as well as simple, graceful variation of the bit rate, which is an important property for buffer control purposes. These attributes may also be achieved with VQ, but with much greater complexity and memory requirements. Therefore, current approaches to VQ have shown the most promise for encoding transform/subband coefficients or other information rather than for the direct encoding of pixel values. The application of VQ for video compression is described in detail in Chapter 9.

15.5 Codeword Assignment

Quantization creates an efficient discrete representation for the data to be transmitted. Codeword assignment takes the quantized values and produces a digital bitstream for transmission. The quantized values can be simply represented using *uniform* or *fixed length codewords*. Using this approach, every quantized value will be represented with the same number of bits. Greater efficiency, in terms of bit rate, can be achieved by employing *entropy coding*. Entropy coding attempts to exploit the statistical properties of the signal to be encoded. A signal, whether it is a pixel value or a transform coefficient, has a certain amount of information, or *entropy*, based on the probability of the different possible values or events occurring. For example, an event that occurs infrequently conveys much more new information than one that occurs often. By realizing that some events occur more frequently than others, the *average* bit rate may be reduced.

There are two important issues that arise when considering the application of entropy coding. First, entropy coding involves increased complexity and memory requirements over fixed length codes. Second, entropy coding coupled with the nonstationarity of the video signal results in a time-varying bit rate. (Other aspects of the source coding may also raise this issue.) A buffer control mechanism is necessary when the variable bit rate source coder is to be coupled with a constant bit rate channel. These issues must be weighted against the sizeable decrease in bit rate that may be obtained through entropy coding.

Entropy Coding. An entropy coder is used to reduce the statistical redundancy inherent in the elements encoded for transmission. The primary redundancy is the nonuniform probability distribution over the possible range of each element. The more the probability distribution deviates from a uniform distribution, the greater improvement can be achieved via entropy coding. Other sources of statistical redundancy which may exist include the statistical dependence among the encoded elements.

Entropy coding is comprised of *variable length coding* and *tree coding* schemes. *Huffman coding* and *arithmetic coding*, respectively, are the most common of these schemes. In Huffman coding, a codebook is generated which minimizes the entropy subject to the codeword constraints of integer lengths and unique decodability. Events which are more likely to occur will be assigned shorter length codewords while those which are less likely to occur will be assigned longer length codewords. Huffman coding results in a variable bit rate per event; more importantly, the *average* bit rate is reduced. The training or generation of the codebook is achieved by using a representative set of data to estimate the probability of each event. Optimal performance can be achieved by designing an individual codebook for each element to be encoded. However, this results in a large number of codebooks. Close to optimal performance is achieved by using a few codebooks where elements with similar statistics are grouped and encoded together. Similarly, the size of each codebook can be reduced by grouping together very unlikely events into a single entry within the codebook. When any event belonging to this group occurs, the codeword for this group is transmitted followed by an exact description of the event.

Huffman coding of individual elements has a number of limitations. Foremost among these, it is difficult to make Huffman coding *adaptive* to changing source statistics. A codebook may perform well for one signal, but may fail badly for another. In general, the better matched the codebook is to one signal, the less robust it may be to others. Also, Huffman coding designates codewords that are an integer number of bits in length. However, the entropy of an event may be less than one bit or a noninteger number of bits. Therefore, Huffman coding tends to be inefficient for these sources.

Arithmetic coding differs from Huffman coding in that, instead of one element being encoded at a time, a sequence of elements are encoded into a codeword sequence. Arithmetic coding can be made adaptive to changing source statistics more easily than Huffman coding. The arithmetic decoder varys its modeling statistics as it tracks the encoder. This can result in a significant reduction in the required bit rate over static Huffman coding. Also, because of its tree coding nature, arithmetic coding does not contain the integer bit length inefficiencies of Huffman coding. However, arithmetic coding is a more computationally complex process to perform. As technology progresses, arithmetic coding may achieve an important role in ATV compression, but currently it is limited to lower rate applications.

Whichever scheme is chosen, the key ingredient for high performance is adaptivity to the varying source characteristics. For example, the runlengths of zeros for the 8×8 Block DCT may vary from 0 to 64, while the range of the coefficient amplitudes may be much larger. Each of these parameters can benefit from separately designed codebooks. The luminance and chrominance components have different statistics; separate codebooks for each would result in higher performance. Similarly, the different transform/subband coefficients, low-frequency and high-frequency, have different statistics that may be exploited.

Some further examples of the application of entropy coding are in order. Huffman coding is assumed. The transform/subband coefficients should be grouped into separate sets, where each set contains coefficients with similar statistics. For some of the coefficients, such as the DC, there may exist significant correlation among adjacent spatial regions. Individual Huffman coding of each coefficient cannot exploit this dependence. To account for the redundancy among adjacent coefficients, differential Huffman coding can be applied. In a simple version of this, the previous DC coefficient can be used as a prediction of the current DC coefficient. The error in the prediction will then be Huffman coded. Similarly, differential Huffman coding can be used to exploit the spatial correlation of the motion field. Wise usage of differential encoding before Huffman coding can produce significant reductions in the required bit rate. The sign bit is also typically uniformly distributed. Huffman coding of the sign bit is therefore unnecessary.

Buffer Control. Whenever entropy coding is employed, the bit rate produced by the encoder is variable and is a function of the video statistics. If the application requires a constant bit rate output, a buffer is necessary to couple the two. The buffering must be carefully designed. Random spikes in the bit rate can overflow the buffer while dips in the bit rate can produce an underflow. What is needed is some form of buffer control that would allow efficient allocation of bits to encode the video while ensuring that no overflow or underflow occurs.

The buffer control typically involves a feedback mechanism to the compression algorithm whereby the amplitude resolution (quantization) and/or spatial, temporal and color resolution may be varied in accordance with the instantaneous bit rate requirements. The goal is to keep the average bit rate constant and equal to the available channel rate. If the bit rate increases significantly, the quantization can be made coarser to reduce it. If the bit rate decreases significantly, a finer quantization can be performed to increase it. When discussing the average bit rate, it may be considered over the entire frame (global buffer control) or over a local region (local buffer control). Global buffer control has the advantage of optimally allocating the bit rate over the entire frame, resulting in the highest performance and ensuring uniform video quality over the entire frame. With local buffer control, it is more difficult to achieve these results, but it trades off memory for computation and may yield a more cost effective solution. The buffer control mechanism is an essential part of any high-performance constant-output bit rate ATV system, and should be an integral part of the design of any such system.

15.6 Important ATV Features and Issues

A number of important features are essential for a successful ATV system. These include *Interoperability*, *Extensibility*, and *Scope of Services and Features*. Interoperability will be especially important in the future because there will be many different methods of information exchange. An ATV system may have to process video from a number of sources which have different formats, e.g., spatial resolution, frame rate, aspect ratio, color/monochrome, etc. The ATV system should be able to process each of these sources to achieve high-quality video. Various sources may include a header/descriptor within the video stream which identifies the video format and provides other information. If a header/descriptor is unavailable, the ATV system may need to decipher the format from the video signal. The ATV system should facilitate interaction with computers and communications over the growing digital networks. Simple scalability may allow a range of price/performance receivers.

Useful extensibility of an ATV system may come in many forms. These include the capabilities to process at higher bit rates such that no visible artifacts exist, or so high-quality studio processing can be performed, or so even higher resolution television may be transmitted in the future. Also, there must be provisions to add new features to the system or enhancements to the compression algorithms. This is partly facilitated in that an ATV standard will specify the core encoder and decoder algorithms and the bit stream syntax, but not how the information is generated at the encoder or used at the decoder. For example, improved methods for estimating the motion field or concealing channel errors may be developed. These schemes, as well as optional pre- and post-processing modules, may be added in a compatible manner.

The scope of services and features that an ATV system can facilitate will partly determine the overall usefulness of the system. An ATV system must be able to dynamically allocate its data capacity among the video and other services to be transmitted. For example, the audio may range from one channel to five-channel surround sound to multi-lingual stereo. Similarly, there should be provisions for an ancillary data channel, text, closed captioning, and encryption and addressing for pay-TV. Amenability to typical VCR functions, such as random access, fast forward/reverse search, slow motion, freeze frame, etc., is extremely important. The capability to splice or edit the video at the bitstream level may also be important. Sustained high quality even after multiple concatenated encode/decode operations facilitates other applications.

A crucial system issue concerns *bit stream integrity.* The compressed video bit stream is highly susceptible to channel errors, with greater susceptibility at higher compression. An effective and robust delivery system is required. Error correction coding as well as appropriate error concealment techniques at the decoder can guard against errors. Catastrophic errors may occur for any entropy coding scheme, since whenever an uncorrected channel error occurs, the decoding process becomes unsynchronized with the encoding process. This effect must be minimized.

15.7 Digital HDTV System Example

The fundamental source coding principles and system issues for digital ATV have been identified and discussed. This section illustrates how these principles may be integrated into a framework that will satisfy all of the system issues for creating a high-performance digital HDTV system. The Channel-Compatible DigiCipher (CCDC) digital HDTV system was developed by the Massachusetts Institute of Technology and General Instrument Corporation for possible adoption as the U.S. HDTV standard. The CCDC HDTV system is composed of many sophisticated subsystems functioning together; however, in this chapter, we will focus on the video compression subsystem. An in-depth technical discussion of the complete system may be found in [5].

System Overview. The CCDC HDTV system is composed of three primary modules: video coding, audio coding, and transmission. The baseline video signal is a progressively scanned, 720×1280 square pixel, 60 frames/sec, and 16:9 aspect ratio signal. The audio system delivers 4 or 6 channels of compact-disc quality audio, where the choice is left up to the broadcaster. Through the use of adaptive transform coding and psychoacoustic modeling, the audio is encoded at a rate of 128 kb/s per monophonic channel. The compressed video and audio data and auxiliary data services are multiplexed into a single bit stream for transmission. The system is designed to operate over a number of delivery media, where the primary form would be "over the air" terrestrial broadcast. Efficient and robust delivery of the compressed bit stream over the harsh simulcast terrestrial channel is performed by utilizing a number of powerful techniques. Concatenated block and trellis coding is used to reintroduce some redundancy to combat channel noise. Multiple interleaving is applied for robustness against burst noise and interference. A fast adaptive equalizer is used to correct for multipath and other channel distortions while enabling rapid acquisition. Quadrature amplitude modulation (QAM) is used to transmit the data over the channel. The system can operate at 32-QAM (preferred mode) or 16-QAM, depending on the desired system threshold and coverage considerations.

The 32-QAM and 16-QAM modes correspond to spectrum efficiencies of about 5 and 4 b/s/Hz, respectively. Using a symbol rate of 5.28 Ms/s within the 6 MHz channel produces total transmission rates of 26.4 and 21.1 Mb/s, respectively. After allocations for error correction coding, audio, and auxiliary services, the available data rates for the video are 18.88 and 13.60 Mb/s. In comparison with the raw data rate for the baseline video signal, 1.32 Gb/s, these rates correspond to compression ratios of about 70 and 97.

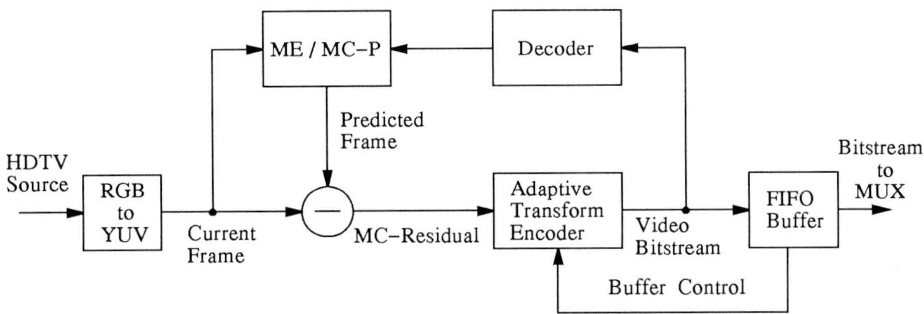

Figure 15.6: The CCDC HDTV video encoder.

Video Compression Overview. The CCDC video compression algorithm is designed to achieve very high quality at the given low bit rate. Compression of the HDTV video signal is achieved by exploiting the redundancy and the irrelevancy inherent in the video signal. An overview of the video encoder is shown in Figure 15.6. The input HDTV source signal is converted from an analog RGB format to a digital signal in the pre-processor (not shown in the figure). To reduce the color space redundancy, the RGB components are transformed to the YUV color space. To exploit the temporal redundancy, ME and MC-prediction are employed to form a prediction of the current frame from the previously encoded frame. The prediction error, or MC-residual, is spatially processed by an adaptive transform encoder. The MC-residual is partitioned into 8×8 blocks and the 2-D DCT is computed for each block. Adaptive, perceptually-tuned quantization is applied to the DCT coefficients. The quantized coefficients and other information are Huffman coded for increased efficiency. The encoder duplicates the decoder processing to ensure tracking between the two. A first-in-first-out (FIFO) buffer is used to couple the variable bit rate output of the video encoder to the constant bit rate channel. This is accomplished via a buffer control mechanism whereby the fullness of the FIFO regulates the coarseness/fineness of the coefficient quantization and thereby the video bit rate.

The Video Encoder. A detailed diagram of the video encoder is shown in Figure 15.7. The analog video signal undergoes anti-aliasing filtering and A/D conversion in the pre-processor. An RGB to YUV matrix conversion is performed (approximately the SMPTE 240M standard). The chrominance signals (U and V) are filtered and decimated by a factor of two along both the horizontal and the vertical dimensions, producing chrominance signals with one-fourth the sampling density of the luminance signal.

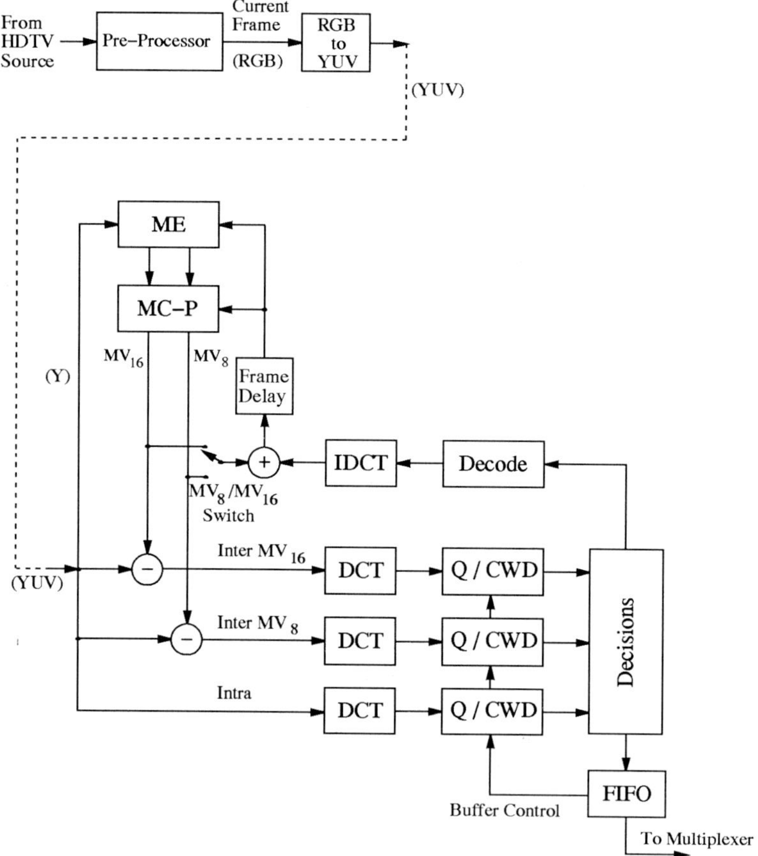

Figure 15.7: A detailed view of the CCDC HDTV video encoder.

A full search block matching scheme is employed to perform the motion estimation. The ME is performed on the luminance frames only, and the computed motion vectors are applied to both the luminance and chrominance components. The criterion for choosing a match is minimum absolute error (MAE) between blocks. The system is designed to exploit both the higher prediction efficiency provided by a dense motion field estimate and the lower amount of information to be transmitted by a sparser motion field estimate. Over each local region of the video, the resolution of the motion field is chosen based upon the local characteristics. The motion field resolution (blocksize) is chosen between 16×16 and 8×8 blocksize (MV_{16} and MV_8), where the criterion for selection will be discussed shortly. The search range for the motion vectors is +15/-16 pixels horizontally and +7/-8 pixels vertically. The larger horizontal search range corresponds to the increased likelihood of rapid motion along the horizontal direction. The search range will allow the encoder to track objects moving at up to 0.75 frame width and 0.67 frame height per second. This is especially useful for encoding video with large frame-to-frame movements, as in sporting events. The motion vectors are estimated with $\frac{1}{2}$ pixel accuracy. The subpixel accuracy is achieved by using bilinear interpolation to compute the noninteger sample points. To exploit the spatial correlation of the motion field, the motion vectors are differentially Huffman coded for transmission. The prediction used for each motion vector is the horizontally preceding motion vector.

MC-prediction displaces blocks from the previously encoded frame to create the prediction of the current frame. The size of the displaced blocks is dependent upon the local video characteristics. MC-prediction typically performs very well. Nevertheless, the CCDC system compares for each 8×8 pixel block the efficiency when using MC-prediction and when using intraframe encoding. The system selects for each block the approach that achieves the highest performance. A comparison is therefore made among three possibilities: (1) MC-prediction with 16×16 blocksize (MV_{16}), (2) MC-prediction with 8×8 blocksize (MV_8), and (3) purely intraframe encoding of the block. To determine which approach is the most efficient for each block, each block is processed independently with each of the three approaches. The criterion for efficiency is the minimum required number of bits to achieve the same reconstructed video quality. Through these processing steps, the CCDC system can exploit the advantages of both coarse and fine motion field estimates, as well as suppress the MC-processing and use purely intraframe processing if that leads to improved performance.

The capability of spatially adaptive inter/intra encoding of different blocks lends itself very easily to an elegant solution for a number of problems. When using a predictive coding scheme like MC-prediction, the decoder must be able to accurately track the encoder. This is especially true in a broadcast environment where consumers randomly change channels and uncorrected channel errors may occur. Two previously discussed approaches to achieve this are periodically intraframe encoding an entire frame and applying leakage within the prediction loop. Both of these approaches have drawbacks. Periodically intra-encoding an entire frame may result in artifacts in the reconstructed frame because of the higher bit rate required for intra-encoding and the limited distribution capabilities of the buffer. Leakage is designed for one-way processing (forward only) and hinders simple execution of VCR functions such as reverse playback.

The CCDC system uses spatially adaptive inter/intra processing to solve these problems in a very simple and elegant manner. For each frame, successive columns of the video are encoded in an intra manner. Every predetermined length of time the entire video frame is encoded using intraframe encoding, and the corresponding increase in bit rate is uniformly distributed over time. For example, with the prototype CCDC system, the baseline video signal has 1/20th of each frame intraframe encoded, producing a "refresh" of the video 3 times per second. The increase in required bit rate is therefore spread over 20 frames, as opposed to a single frame receiving the full hit in performance. Through this approach of partial refreshing of each frame, the video acquisition time is .33 sec, excluding synchronization of the communication systems, and the maximum propagation of uncorrected channel errors is also limited to .33 sec. Inter/intra decision-making therefore provides highly adaptive and efficient video compression, robustness against uncorrected channel errors, and also enables simple VCR functionality.

As previously discussed, there are many approaches for transform/subband filtering the residual. Of these, the Block DCT is currently the most mature and well understood. This, coupled with its high performance, simple spatially adaptive processing, and hardware availability, led to its choice for processing the residual. The MC-residual is partitioned into 8×8 pixel blocks, each of which is independently transformed using the 2-D DCT.

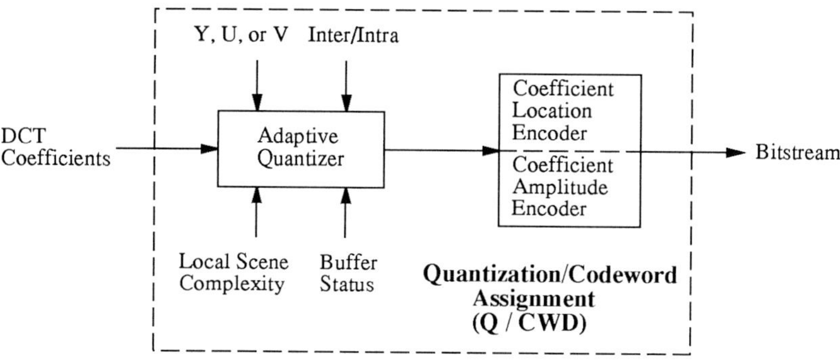

Figure 15.8: Quantization and codeword assignment for the DCT coefficients.

The key step toward reducing the bit rate is the quantization and entropy coding of the DCT coefficients. This process is illustrated in Figure 15.8. The quantization is performed in an adaptive manner for each local region in order to exploit the human visual perception. Each 8×8 block of DCT coefficients is weighted or normalized by a matrix of 8×8 weighting factors. The DCT coefficients are then quantized with a uniform quantizer. The matrix of weighting factors is chosen based upon the inter/intra and luminance/chrominance characteristics of each block, as well as measures of the local scene complexity and buffer fullness. The weighting factors effectively scale each individual DCT coefficient quantizer, making it more coarse (increasing the stepsize) or fine (decreasing the stepsize) in order to take advantage of the differing sensitivity of the human visual system. For example, each set of 8×8 weighting factors is designed to vary with frequency; more coarsely quantizing the high-frequency coefficients as compared to the low-frequency coefficients to exploit the HVS's reduced sensitivity to high-frequency quantization noise. Different weighting matrices are also used to exploit the differing luminance/chrominance and inter/intra HVS responses. A local scene complexity indicator determines the complexity within each 16×16 pixel area, and, based on the determined complexity, scales the DCT quantizers within the area to exploit the spatial masking that may exist. A feedback mechanism from the buffer also regulates the coarseness/fineness of the quantization to ensure high video quality while preventing buffer overflow or underflow.

The location of the nonzero quantized DCT coefficients is encoded through a Vector Coding (VC) approach. In this scheme, each 8×8 block of coefficients is divided into 4 regions each containing 16 coefficients. The division is such that the first region contains the low-frequency coefficients which are most likely to be nonzero, and the last region contains the high-frequency coefficients which are least likely to be nonzero. A 16 bit pattern identifies the location of the nonzero coefficients within each group. These patterns are Huffman coded for transmission. In addition, the codewords for the first three regions indicate if all of the coefficients in the following regions are zero. If this is the case, then the codewords for those regions are not transmitted. Following the VC codeword identifying the locations of the nonzero coefficients, the amplitudes of the nonzero coefficients are transmitted. The DC coefficient is differentially Huffman encoded and the AC coefficients are Huffman coded with a codebook chosen based upon the position of the coefficient. Improved performance for encoding the location and amplitude information is achieved through the use of different codebooks based upon the inter/intra and luminance/chrominance characteristics of each block.

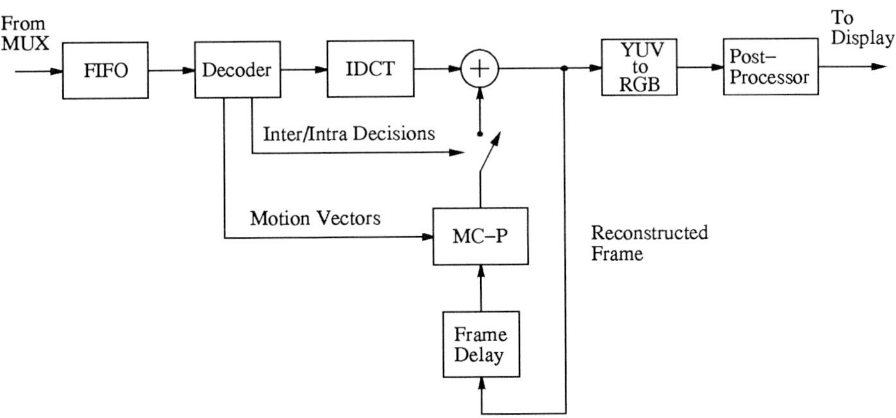

Figure 15.9: A detailed view of the CCDC HDTV video decoder.

The Video Decoder. The video decoding process at the receiver is the inverse of the encoding process. This is shown is Figure 15.9. Using the location information, the nonzero DCT coefficients are identified, reconstructed, and appropriately inverse weighted. An inverse Block DCT operation produces the residual signal, which is combined in a spatially adaptive manner with the previously reconstructed frame to reconstruct the current frame. Each step of the video processing algorithm has been designed to keep the decoder as simple and inexpensive as possible.

ATV Video Features and Issues. The CCDC system incorporates *Source Adaptive Processing* to identify a video signal and process it based on its source format. For example, the system can identify a 60 frames/sec input as a 3:2 pulldown representation for film (24 frames/sec). The system can then extract the appropriate frames and distribute the available bit rate among the frames to achieve the highest video quality. The choice of progressive scanning and square pixels avoids the artifacts of interlaced scanning while greatly simplifying interoperability with computers, format conversion, frame capture, and special effects. The inter/intra refreshing mechanism enables simple processing for VCR functions such as random access and fast forward/reverse search.

The CCDC system has a hierarchy of error correction, detection, and concealment mechanisms to combat any channel errors that may occur. The communication system provides the first level of protection via concatenated trellis/Reed-Solomon coding and multiple interleaving. Errors that penetrate this layer can be detected in the video encoder as incorrect Huffman bitstreams. In this way the error can be localized to 1/180th of the baseline video frame and appropriate error concealment measures can be taken. The CCDC system incorporates easy error detection mechanisms and lends itself toward simple error concealment measures which may be included as a receiver option. This allows price/complexity/performance categories in the receiver design. (Most receivers close to a transmitter would not need any concealment capabilities.) The CCDC system also provides quick recovery from channel errors as their duration is usually limited to a maximum of .33 sec by the inter/intra refreshing mechanism.

15.8 Concluding Remarks

Applying sophisticated video compression principles to a video signal enables the creation of an efficient and compact bitstream representation, whereby the high-quality video can be transmitted within a much smaller bandwidth. In this chapter we have examined the important video compression principles and system issues for high-quality ATV video, and have attempted to place them in a framework for creating a successful digital ATV system. The integration of the compression principles and system issues was illustrated within a sample digital HDTV system.

Acknowledgements: The authors would like to thank the entire Advanced Television and Signal Processing Group at MIT, and in particular Peter Monta and Julien Nicolas, for countless late-night brainstorming sessions on these matters. We would like to thank the VideoCipher Division of General Instrument, and in particular Woo Paik and Edward Krause, for many in-depth discussions on optimizing performance and practical issues for the Channel-Compatible DigiCipher HDTV system. We would also like to thank Deborah Gage for editing the manuscript. This research has been sponsored principally by the Center for Advanced Television Studies (CATS). Current Consortium members are Ampex, Capital Cities/ABC, Eastman Kodak, General Instrument, Motorola, PBS, and Tektronix. The views expressed are those of the authors and may not represent those of the sponsors.

References

[1] J. S. Lim, *Two-Dimensional Signal and Image Processing*. Englewood Cliffs, N.J.: Prentice Hall, Inc., 1990.

[2] A. Netravali and B. Haskell, *Digital Pictures, Representation and Compression*. New York: Plenum Press, 1988.

[3] M. Rabbani and P. Jones, *Digital Image Compression Techniques*. Bellingham, Washington: SPIE Optical Engineering Press, 1991.

[4] H. Musmann, P. Pirsch, and H. Grallert, "Advances in picture coding," *Proceedings of the IEEE*, vol. 73, pp. 523–548, April 1985.

[5] American Television Alliance, Massachusetts Institute of Technology and General Instrument Corp., *Channel-Compatible DigiCipher HDTV System*, April 1992.

[6] N. Jayant, "Signal compression: Technology targets and research directions," *IEEE Journal on Selected Areas in Communications*, vol. 10, pp. 796–818, June 1992.

[7] E. Dubois, "The sampling and reconstruction of time-varying imagery with application in video systems," *Proceedings of the IEEE*, vol. 73, pp. 502–522, April 1985.

[8] M. Bierling, "Displacement estimation by hierarchical blockmatching," *3rd SPIE Symposium on Visual Communications*, November 1988.

[9] K. Rao and P. Yip, *Discrete Cosine Transform: Algorithms, Advantages, Applications*. Boston: Academic Press, Inc., 1990.

[10] D. J. LeGall, "MPEG: A video compression standard for multimedia applications," *Communications of the ACM*, vol. 34, pp. 47–58, April 1991.

[11] H. Malvar and D. Staelin, "The LOT: Transform coding without blocking effects," *IEEE Transactions on Acoustics, Speech, and Signal Processing*, vol. 37, pp. 553–559, April 1989.

[12] P. Burt and E. Adelson, "The Laplacian pyramid as a compact image code," *IEEE Transactions on Communications*, vol. COM-31, pp. 532–540, April 1983.

[13] E. Adelson, E. Simoncelli, and R. Hingorani, "Orthogonal pyramid transforms for image coding," *SPIE Vol. 845 Visual Communications and Image Processing II*, 1987.

[14] P. Monta, *Signal Processing for High Definition Television*. PhD thesis, MIT, To be submitted 1992.

Index